Why is knowledge of science and mathematics important in engineering?

A career in any engineering field will require both basic and advanced mathematics and science. Without mathematics and science to determine principles, calculate dimensions and limits, explore variations, prove concepts, and so on, there would be no mobile telephones, televisions, stereo systems, video games, microwave ovens, computers, or virtually anything electronic. There would be no bridges, tunnels, roads, skyscrapers, automobiles, ships, planes, rockets or most things mechanical. There would be no metals beyond the common ones, such as iron and copper, no plastics, no synthetics. In fact, society would most certainly be less advanced without the use of mathematics and science throughout the centuries and into the future.

Electrical engineers require mathematics and science to design, develop, test, or supervise the manufacturing and installation of electrical equipment, components, or systems for commercial, industrial, military, or scientific use.

Mechanical engineers require mathematics and science to perform engineering duties in planning and designing tools, engines, machines, and other mechanically functioning equipment; they oversee installation, operation, maintenance, and repair of such equipment as centralised heat, gas, water, and steam systems.

Aerospace engineers require mathematics and science to perform a variety of engineering work in designing, constructing, and testing aircraft, missiles, and spacecraft; they conduct basic and applied research to evaluate adaptability of materials and equipment to aircraft design and manufacture and recommend improvements in testing equipment and techniques.

Nuclear engineers require mathematics and science to conduct research on nuclear engineering problems or apply principles and theory of nuclear science to problems concerned with release, control, and utilisation of nuclear energy and nuclear waste disposal.

Petroleum engineers require mathematics and science to devise methods to improve oil and gas well production and determine the need for new or modified tool designs; they oversee drilling and offer technical advice to achieve economical and satisfactory progress.

Industrial engineers require mathematics and science to design, develop, test, and evaluate integrated systems for managing industrial production processes, including human work factors, quality control, inventory control, logistics and material flow, cost analysis, and production coordination.

Environmental engineers require mathematics and science to design, plan, or perform engineering duties in the prevention, control, and remediation of environmental health hazards, using various engineering disciplines; their work may include waste treatment, site remediation, or pollution control technology.

Civil engineers require mathematics and science in all levels in civil engineering – structural engineering, hydraulics and geotechnical engineering are all fields that employ mathematical tools such as differential equations, tensor analysis, field theory, numerical methods and operations research.

Architects require knowledge of algebra, geometry, trigonometry and calculus. They use mathematics for several reasons, leaving aside the necessary use of mathematics in the engineering of buildings. Architects use geometry because it defines the spatial form of a building, and they use mathematics to design forms that are considered beautiful or harmonious. The front cover of this text shows a modern London architecture and the financial district, all of which at some stage required in its design a knowledge of mathematics.

Knowledge of mathematics and science is clearly needed by each of the disciplines listed above.

It is intended that this text – *Mathematics Pocket Book for Engineers and Scientists* – will provide a step by step, helpful reference, to essential mathematics topics needed by engineers and scientists.

Mathematics Pocket Book for Engineers and Scientists

Fifth Edition

John Bird

Routledge
Taylor & Francis Group

LONDON AND NEW YORK

Fifth edition published 2020
by Routledge
2 Park Square, Milton Park, Abingdon, Oxon, OX14 4RN

and by Routledge
52 Vanderbilt Avenue, New York, NY 10017

Routledge is an imprint of the Taylor & Francis Group, an informa business

First edition published as *Newnes Mathematics for Engineers Pocket Book* by Newnes 1983
Fourth edition published as *Engineering Mathematics Pocket Book* by Routledge 2008

British Library Cataloguing-in-Publication Data
A catalogue record for this book is available from the British Library

Library of Congress Cataloging-in-Publication Data
A catalog record has been requested for this book

ISBN: 978-0-367-26653-0 (hbk)
ISBN: 978-0-367-26652-3 (pbk)
ISBN: 978-0-429-29440-2 (ebk)

Typeset in Frutiger 45 Light by
Servis Fllmsetting Ltd, Stockport, Cheshire.

Printed and bound by CPI Group (UK) Ltd, Croydon, CR0 4YY

Mathematics Pocket Book for Engineers and Scientists

John Bird is the former Head of Applied Electronics in the Faculty of Technology at Highbury College, Portsmouth, UK. More recently, he has combined freelance lecturing at the University of Portsmouth, with examiner responsibilities for Advanced Mathematics with City and Guilds and examining for International Baccalaureate. He has over 45 years' experience of successfully teaching, lecturing, instructing, training, educating and planning trainee engineers study programmes. He is the author of 140 textbooks on engineering, science and mathematical subjects, with worldwide sales of over one million copies. He is a chartered engineer, a chartered mathematician, a chartered scientist and a Fellow of three professional institutions. He is currently lecturing at the Defence College of Marine Engineering in the Defence College of Technical Training at H.M.S. Sultan, Gosport, Hampshire, UK, one of the largest technical training establishments in Europe.

Contents

Preface

Mathematics Pocket Book for Engineers and Scientists 5th Edition is intended to provide students, technicians, scientists and engineers with a readily available reference to the essential engineering mathematics formulae, definitions, tables and general information needed during their work situation and/or studies – a handy book to have on the bookshelf to delve into as the need arises.

In this 5th edition, the text has been re-designed to make information easier to access. The importance of why each mathematical topic is needed in engineering and science is explained at the beginning of each section. Essential theory, formulae, definitions, laws and procedures are stated clearly at the beginning of each chapter, and then it is demonstrated how to use such information in practice.

The text is divided, for convenience of reference, into seventeen main sections embracing engineering conversions, constants and symbols, some algebra topics, some number topics, areas and volumes, geometry and trigonometry, graphs, complex numbers, vectors, matrices and determinants, Boolean algebra and logic circuits, differential and integral calculus and their applications, differential equations, Laplace transforms, z-transforms, Fourier series and statistics and probability. To aid understanding, over 675 application examples have been included, together with some 300 line diagrams.

The text assumes little previous knowledge and is suitable for a wide range of disciplines and/or courses of study. It will be particularly useful as a reference for those in industry involved in engineering and science and/or for students studying mathematics within Engineering and Science Degree courses, as well as for National and Higher National Technician Certificates and Diplomas, GCSE and A levels.

JOHN BIRD BSc(Hons), CEng, CSi, CMath, FIET, FIMA, FCollP
Royal Naval Defence College of Marine Engineering,
HMS Sultan,formerly University of Portsmouth
and Highbury College, Portsmouth

Section 1

Engineering conversions, constants and symbols

Why are engineering conversions, constants and symbols important?

In engineering there are many different quantities to get used to, and hence many units to become familiar with. For example, force is measured in Newtons, electric current is measured in amperes and pressure is measured in Pascals. Sometimes the units of these quantities are either very large or very small and hence prefixes are used. For example, 1000 Pascals may be written as 10^3 Pa which is written as 1 kPa in prefix form, the k being accepted as a symbol to represent 1000 or 10^3. Studying, or working, in an engineering and science discipline, you very quickly become familiar with the standard units of measurement, the prefixes used and engineering notation. An electronic calculator is extremely helpful with engineering notation.

Unit conversion is very important because the rest of the world other than three countries uses the metric system. So, converting units is important in science and engineering because it uses the metric system.

Without the ability to measure, it would be difficult for scientists to conduct experiments or form theories. Not only is measurement important in science and engineering, it is also essential in farming, construction, manufacturing, commerce, and numerous other occupations and activities.

Measurement provides a standard for everyday things and processes. Examples include weight, temperature, length, currency and time, and all play a very important role in our lives.

Chapter 1   General conversions and Greek alphabet

General conversions

Length (metric)
1 kilometre (km) = 1000 metres (m)
1 metre (m) = 100 centimetres (cm)
1 metre (m) = 1000 millimetres (mm)
$1\,cm = 10^{-2}\,m$
$1\,mm = 10^{-3}\,m$
1 micron (μ) = $10^{-6}\,m$
1 angstrom (A) = $10^{-10}\,m$

Length (imperial)
1 inch (in) = 2.540 cm or 1 cm = 0.3937 in
1 foot (ft) = 30.48 cm
1 mile (mi) = 1.609 km or 1 km = 0.6214 mi
1 cm = 0.3937 in
1 m = 39.37 in = 3.2808 ft = 1.0936 yd
1 km = 0.6214 mile
1 nautical mile = 1.15 mile

Area (metric)
$1\,m^2 = 10^6\,mm^2$
$1\,mm^2 = 10^{-6}\,m^2$
$1\,m^2 = 10^4\,cm^2$
$1\,cm^2 = 10^{-4}\,m^2$
1 hectare (ha) = $10^4\,m^2$

Area (imperial)
$1\,m^2 = 10.764\,ft^2 = 1.1960\,yd^2$
$1\,ft^2 = 929\,cm^2$
$1\,mile^2 = 640$ acres
1 acre = $43560\,ft^2 = 4840\,yd^2$
1 ha = 2.4711 acre = $11960\,yd^2 = 107639\,ft^2$

Volume
1 litre (l) = 1000 cm^3
1 litre = 1.057 quart (qt) = 1.7598 pint (pt) = 0.21997 gal
$1\,m^3 = 1000\,l$
1 British gallon = 4 qt = 4.545 l = 1.201 US gallon
1 US gallon = 3.785 l

Mass
1 kilogram (kg) = 1000 g = 2.2046 pounds (lb)
1 lb = 16 oz = 453.6 g
1 tonne (t) = 1000 kg = 0.9842 ton

Speed
1 km/h = 0.2778 m/s = 0.6214 m.p.h.
1 m.p.h. = 1.609 km/h = 0.4470 m/s
1 rad/s = 9.5493 rev/min
1 knot = 1 nautical mile per hour = 1.852 km/h = 1.15 m.p.h.
1 km/h = 0.540 knots
1 m.p.h. = 0.870 knots

Angular measure
1 rad = 57.296°

Greek alphabet

Letter Name	Upper Case	Lower Case
Alpha	A	α
Beta	B	β
Gamma	Γ	γ
Delta	Δ	δ
Epsilon	E	ϵ
Zeta	Z	ζ
Eta	H	η
Theta	θ	θ
Iota	I	ι
Kappa	K	κ
Lambda	Λ	λ
Mu	M	μ
Nu	N	ν
Xi	Ξ	ξ
Omicron	O	o
Pi	Π	π
Rho	P	ρ
Sigma	Σ	σ
Tau	T	τ
Upsilon	Y	υ
Phi	Φ	ϕ
Chi	X	χ
Psi	Ψ	ψ
Omega	Ω	ω

Chapter 2   Basic SI units, derived units and common prefixes

Basic SI units

Quantity	Unit
Length	metre, m
Mass	kilogram, kg
Time	second, s
Electric current	ampere, A
Thermodynamic temperature	kelvin, K
Luminous intensity	candela, cd
Amount of substance	mole, mol

SI supplementary units

Plane angle	radian, rad
Solid angle	steradian, sr

Derived units

Quantity	Unit
Electric capacitance	farad, F
Electric charge	coulomb, C
Electric conductance	siemens, S
Electric potential difference	volts, V
Electrical resistance	ohm, Ω
Energy	joule, J
Force	Newton, N
Frequency	hertz, Hz
Illuminance	lux, lx
Inductance	henry, H
Luminous flux	lumen, lm
Magnetic flux	weber, Wb
Magnetic flux density	tesla, T
Power	watt, W
Pressure	pascal, Pa

Some other derived units not having special names

Quantity	Unit
Acceleration	metre per second squared, m/s^2
Angular velocity	radian per second, rad/s
Area	square metre, m^2
Current density	ampere per metre squared, A/m^2
Density	kilogram per cubic metre, kg/m^3
Dynamic viscosity	pascal second, Pa s
Electric charge density	coulomb per cubic metre, C/m^3
Electric field strength	volt per metre, V/m
Energy density	joule per cubic metre, J/m^3
Heat capacity	joule per Kelvin, J/K

Quantity	Unit
Heat flux density	watt per square metre, W/m^3
Kinematic viscosity	square metre per second, m^2/s
Luminance	candela per square metre, cd/m^2
Magnetic field strength	ampere per metre, A/m
Moment of force	newton metre, Nm
Permeability	henry per metre, H/m
Permittivity	farad per metre, F/m
Specific volume	cubic metre per kilogram, m^3/kg
Surface tension	newton per metre, N/m
Thermal conductivity	watt per metre Kelvin, W/(mK)
Velocity	metre per second, m/s^2
Volume	cubic metre, m^3

Common prefixes

Prefix	Name	Meaning
Y	yotta	multiply by 10^{24}
Z	zeta	multiply by 10^{21}
E	exa	multiply by 10^{18}
P	peta	multiply by 10^{15}
T	tera	multiply by 10^{12}
G	giga	multiply by 10^{9}
M	mega	multiply by 10^{6}
k	kilo	multiply by 10^{3}
m	milli	multiply by 10^{-3}
μ	micro	multiply by 10^{-6}
n	nano	multiply by 10^{-9}
p	pico	multiply by 10^{-12}
f	femto	multiply by 10^{-15}
a	atto	multiply by 10^{-18}
z	zepto	multiply by 10^{-21}
y	yocto	multiply by 10^{-24}

Chapter 3   Some physical and mathematical constants

Below are listed some physical and mathematical constants, each stated correct to 4 decimal places, where appropriate.

Quantity	Symbol	Value
Speed of light in a vacuum	c	2.9979×10^8 m/s
Permeability of free space	μ_0	$4\pi \times 10^{-7}$ H/m
Permittivity of free space	ε_0	8.8542×10^{-12} F/m
Elementary charge	e	1.6022×10^{-19} C
Planck constant	h	6.6261×10^{-34} Js
Reduced Planck constant	$\hbar = \dfrac{h}{2\pi}$	1.0546×10^{-34} Js
Fine structure constant	$\alpha = \dfrac{e^2}{4\pi\varepsilon_0 \hbar c}$	7.2974×10^{-3}
Coulomb force constant	k_e	8.9875×10^9 Nm^2/C^2
Gravitational constant	G	6.6726×10^{-11} m^3/kg s^2
Atomic mass unit	u	1.6605×10^{-27} kg
Rest mass of electron	m_e	9.1094×10^{-31} kg
Rest mass of proton	m_p	1.6726×10^{-27} kg
Rest mass of neutron	m_n	1.6749×10^{-27} kg
Bohr radius	a_0	5.2918×10^{-11} m
Compton wavelength of electron	λ_C	2.4263×10^{-12} m
Avogadro constant	N_A	6.0221×10^{23}/mol
Boltzmann constant	k	1.3807×10^{-23} J/K
Stefan-Boltzmann constant	σ	5.6705×10^{-8} W/m^{2}K^4
Bohr constant	μ_B	9.2740×10^{-24} J/T
Nuclear magnetron	μ_N	5.0506×10^{-27} J/T
Triple point temperature	T_t	273.16 K
Molar gas constant	R	8.3145 J/K mol
Micron	μm	10^{-6} m
Characteristic impedance of vacuum	Z_o	$376.7303\,\Omega$

Astronomical constants		
Mass of earth	m_E	5.976×10^{24} kg
Radius of earth	R_E	6.378×10^6 m
Gravity of earth's surface	g	9.8067 m/s^2
Mass of sun	M_e	1.989×10^{30} kg
Radius of sun	R_e	6.9599×10^8 m
Solar effective temperature	T_e	5800 K
Luminosity of sun	L_e	3.826×10^{26} W
Astronomical unit	AU	1.496×10^{11} m
Parsec	pc	3.086×10^{16} m
Jansky	Jy	10^{-26} W/m^{2}HZ

Tropical year		$3.1557 \times 10^7 \, \text{s}$
Standard atmosphere	atm	$101325 \, \text{Pa}$

Mathematical constants		
Pi (Archimedes' constant)	π	3.1416
Exponential constant	e	2.7183
Apery's constant	$\zeta\,(3)$	1.2021
Catalan's constant	G	0.9160
Euler's constant	γ	0.5772
Feigenbaum's 1st constant	δ	4.6692
Feigenbaum's 2nd constant	α	2.5029
Gibb's constant	G	1.8519
Golden mean	ϕ	1.6180
Khintchine's constant	K	2.6855

Chapter 4   Recommended mathematical symbols

equal to	$=$
not equal to	$\neq$
identically equal to	$\equiv$
corresponds to	$\triangleq$
approximately equal to	$\approx$
approaches	$\rightarrow$
proportional to	$\propto$
infinity	∞
smaller than	$<$
larger than	$>$
smaller than or equal to	$\leq$
larger than or equal to	$\geq$
much smaller than	$\ll$
much larger than	$\gg$
plus	$+$
minus	$-$
plus or minus	$\pm$
minus or plus	$\mp$
a multiplied by b	ab or $a \times b$ or $a \cdot b$
a divided by b	$\dfrac{a}{b}$ or a/b or ab^{-1}
magnitude of a	$|a|$
a raised to power n	a^n
square root of a	$\sqrt{a}$ or $a^{\frac{1}{2}}$

n'th root of a	$\sqrt[n]{a}$ or $a^{\frac{1}{n}}$ or $a^{1/n}$
mean value of a	$\bar{a}$
factorial of a	$a!$
sum	Σ
function of x	$f(x)$
limit to which f(x) tends as x approaches a	$\lim_{x \to a} f(x)$
finite increment of x	Δx
variation of x	δx
differential coefficient of f(x) with respect to x	$\dfrac{df}{dx}$ or df/dy or $f'(x)$
differential coefficient of order n of f(x)	$\dfrac{d^n f}{dx^n}$ or $d^n f/dx^2$ or $f^n(x)$
partial differential coefficient of f(x, y, ...) w.r.t. x when y, ... are held constant	$\dfrac{\partial f(x,y,...)}{\partial x}$ or $\left(\dfrac{\partial f}{\partial x}\right)_y$ or f_x
total differential of f	df
indefinite integral of f(x) with respect to x	$\displaystyle\int f(x)\,dx$
definite integral of f(x) from $x = a$ to $x = b$	$\displaystyle\int_a^b f(x)\,dx$
logarithm to the base a of x	$\log_a x$
common logarithm of x	$\lg x$ or $\log_{10} x$
exponential of x	e^x or $\exp x$
natural logarithm of x	$\ln x$ or $\log_e x$
sine of x	$\sin x$
cosine of x	$\cos x$
tangent of x	$\tan x$
secant of x	$\sec x$
cosecant of x	$\operatorname{cosec} x$
cotangent of x	$\cot x$
inverse sine of x	$\sin^{-1} x$ or $\arcsin x$
inverse cosine of x	$\cos^{-1} x$ or $\arccos x$
inverse tangent of x	$\tan^{-1} x$ or $\arctan x$
inverse secant of x	$\sec^{-1} x$ or $\operatorname{arcsec} x$
inverse cosecant of x	$\operatorname{cosec}^{-1} x$ or $\operatorname{arccosec} x$
inverse cotangent of x	$\cot^{-1} x$ or $\operatorname{arccot} x$
hyperbolic sine of x	$\sinh x$
hyperbolic cosine of x	$\cosh x$
hyperbolic tangent of x	$\tanh x$
hyperbolic secant of x	$\operatorname{sech} x$
hyperbolic cosecant of x	$\operatorname{cosech} x$

hyperbolic cotangent of x	coth x
inverse hyperbolic sine of x	$\sinh^{-1}$ x or arsinh x
inverse hyperbolic cosine of x	$\cosh^{-1}$ x or arcosh x
inverse hyperbolic tangent of x	$\tanh^{-1}$ x or artanh x
inverse hyperbolic secant of x	sech^{-1} x or arsech x
inverse hyperbolic cosecant of x	cosech^{-1} x or arcosech x
inverse hyperbolic cotangent of x	$\coth^{-1}$ x or arcoth x
complex operator	i, j
modulus of z	$|z|$
argument of z	arg z
complex conjugate of z	z*
transpose of matrix A	A^{T}
determinant of matrix A	$|A|$
vector	**A** or $\vec{\textbf{A}}$
magnitude of vector **A**	$|\textbf{A}|$
scalar product of vectors **A** and **B**	**A • B**
vector product of vectors **A** and **B**	**A × B**

Chapter 5   Symbols for physical quantities

(a) Space and time

angle (plane angle)	$\alpha, \beta, \gamma, \theta, \phi$, etc.
solid angle	Ω, ω
length	l
breadth	b
height	h
thickness	d, δ
radius	r
diameter	d
distance along path	s, L
rectangular co-ordinates	x, y, z
cylindrical co-ordinates	r, ϕ, z
spherical co-ordinates	r, θ, ϕ
area	A
volume	V
time	t
angular speed, $\dfrac{d\theta}{dt}$	ω
angular acceleration, $\dfrac{d\omega}{dt}$	α
speed, $\dfrac{ds}{dt}$	u, v, w

acceleration, $\dfrac{du}{dt}$	a
acceleration of free fall	g
speed of light in a vacuum	c
Mach number	Ma

(b) Periodic and related phenomena

period	T
frequency	f
rotational frequency	n
circular frequency	ω
wavelength	λ
damping coefficient	δ
attenuation coefficient	α
phase coefficient	β
propagation coefficient	γ

(c) Mechanics

mass	m
density	ρ
relative density	d
specific volume	v
momentum	p
moment of inertia	I, J
second moment of area	I_a
second polar moment of area	I_p
force	F
bending moment	M
torque; moment of couple	T
pressure	p, P
normal stress	σ
shear stress	τ
linear strain	ε, e
shear strain	γ
volume strain	θ
Young's modulus	E
shear modulus	G
bulk modulus	K
Poisson ratio	μ, ν
compressibility	κ
section modulus	Z, W
coefficient of friction	μ
viscosity	η
fluidity	ϕ
kinematic viscosity	ν
diffusion coefficient	D
surface tension	γ, σ
angle of contact	θ
work	W

energy	E, W
potential energy	E_p, V, Φ
kinetic energy	E_k, T, K
power	P
gravitational constant	G
Reynold's number	Re

(d) Thermodynamics

thermodynamic temperature	T, Θ
common temperature	t, θ
linear expansivity	α, λ
cubic expansivity	α, γ
heat; quantity of heat	Q, q
work; quantity of work	W, w
heat flow rate	Φ, q
thermal conductivity	λ, k
heat capacity	C
specific heat capacity	c
entropy	S
internal energy	U, E
enthalpy	H
Helmholtz function	A, F
Planck function	Y
specific entropy	s
specific internal energy	u, e
specific enthalpy	h
specific Helmholz function	a, f

(e) Electricity and magnetism

Electric charge; quantity of electricity	Q
electric current	I
charge density	ρ
surface charge density	σ
electric field strength	E
electric potential	V, ϕ
electric potential difference	U, V
electromotive force	E
electric displacement	D
electric flux	ψ
capacitance	C
permittivity	ε
permittivity of a vacuum	ε_0
relative permittivity	ε_r
electric current density	J, j
magnetic field strength	H
magnetomotive force	F_m
magnetic flux	Φ
magnetic flux density	B
self inductance	L

mutual inductance	M
coupling coefficient	k
leakage coefficient	σ
permeability	μ
permeability of a vacuum	μ_0
relative permeability	μ_r
magnetic moment	m
resistance	R
resistivity	ρ
conductivity	γ, σ
reluctance	R_m, S
permeance	Λ
number of turns	N
number of phases	m
number of pairs of poles	p
loss angle	δ
phase displacement	ϕ
impedance	Z
reactance	X
resistance	R
quality factor	Q
admittance	Y
susceptance	B
conductance	G
power, active	P
power, reactive	Q
power, apparent	S

(f) Light and related electromagnetic radiations

radiant energy	Q, Q_e
radiant flux, radiant power	Φ, Φ_e, P
radiant intensity	I, I_e
radiance	L, L_e
radiant exitance	M, M_e
irradiance	E, E_e
emissivity	e
quantity of light	Q, Q_v
luminous flux	Φ, Φ_v
luminous intensity	I, I_v
luminance	L, L_v
luminous exitance	M, M_v
illuminance	E, E_v
light exposure	H
luminous efficacy	K
absorption factor, absorptance	α
reflexion factor, reflectance	ρ
transmission factor, transmittance	τ
linear extinction coefficient	μ
linear absorption coefficient	a
refractive index	n

refraction	R
angle of optical rotation	α

(g) Acoustics

speed of sound	c
speed of longitudinal waves	c_l
speed of transverse waves	c_t
group speed	c_g
sound energy flux	P
sound intensity	I, J
reflexion coefficient	ρ
acoustic absorption coefficient	α, α_a
transmission coefficient	τ
dissipation coefficient	δ
loudness level	L_N

(h) Physical chemistry

atomic weight	A_r
molecular weight	M_r
amount of substance	n
molar mass	M
molar volume	V_m
molar internal energy	U_m
molar enthalpy	H_m
molar heat capacity	C_m
molar entropy	S_m
molar Helmholtz function	A_m
molar Gibbs function	G_m
(molar) gas constant	R
compression factor	Z
mole fraction of substance B	x_B
mass fraction of substance B	w_B
volume fraction of substance B	ϕ_B
molality of solute B	m_B
amount of substance concentration of solute B	c_B
chemical potential of substance B	μ_B
absolute activity of substance B	λ_B
partial pressure of substance B in a gas mixture	p_B
fugacity of substance B in a gas mixture	f_B
relative activity of substance B	α_B
activity coefficient (mole fraction basis)	f_B
activity coefficient (molality basis)	γ_B
activity coefficient (concentration basis)	y_B
osmotic coefficient	ϕ, g
osmotic pressure	Π
surface concentration	Γ
electromotive force	E
Faraday constant	F
charge number of ion i	z_i
ionic strength	I
velocity of ion i	v_i

electric mobility of ion i	u_i
electrolytic conductivity	κ
molar conductance of electrolyte	Λ
transport number of ion i	t_i
molar conductance of ion i	λ_i
overpotential	η
exchange current density	j_0
electrokinetic potential	ζ
intensity of light	I
transmittance	T
absorbance	A
(linear) absorption coefficient	a
molar (linear) absorption coefficient	ε
angle of optical rotation	α
specific optical rotatory power	α_m
molar optical rotatory power	α_n
molar refraction	R_m
stoiciometric coefficient of molecules B	ν_B
extent of reaction	ξ
affinity of a reaction	$\mathbf{A}$
equilibrium constant	K
degree of dissociation	α
rate of reaction	ξ, J
rate constant of a reaction	k
activation energy of a reaction	E

(i) Molecular physics

Avogadro constant	L, N_A
number of molecules	N
number density of molecules	n
molecular mass	m
molecular velocity	$\mathbf{c, u}$
molecular position	$\mathbf{r}$
molecular momentum	$\mathbf{p}$
average velocity	$\langle \mathbf{c} \rangle, \langle \mathbf{u} \rangle, \mathbf{c_0, u_0}$
average speed	$\langle c \rangle, \langle u \rangle, \bar{c}, \bar{u}$
most probable speed	$\hat{c}, \hat{u}$
mean free path	l, λ
molecular attraction energy	ε
interaction energy between molecules i and j	ϕ_{ij}, V_{ij}
distribution function of speeds	$f(c)$
Boltzmann function	H
generalized co-ordinate	q
generalized momentum	p
volume in phase space	Ω
Boltzmann constant	k
partition function	Q, Z
grand partition function	Ξ
statistical weight	g
symmetrical number	σ, s
dipole moment of molecule	p, μ

quadrupole moment of molecule	Θ
polarizability of molecule	α
Planck constant	h
characteristic temperature	Θ
Debye temperature	Θ_D
Einstein temperature	Θ_E
rotational temperature	Θ_r
vibrational temperature	Θ_v
Stefan-Boltzmann constant	σ
first radiation constant	c_1
second radiation constant	c_2
rotational quantum number	J, K
vibrational quantum number	v

(j) Atomic and nuclear physics

nucleon number; mass number	A
atomic number; proton number	Z
neutron number	N
(rest) mass of atom	m_a
unified atomic mass constant	m_u
(rest) mass of electron	m_e
(rest) mass of proton	m_p
(rest) mass of neutron	m_n
elementary charge (of protons)	e
Planck constant	h
Planck constant divided by 2π	$\hbar$
Bohr radius	a_0
Rydberg constant	R_∞
magnetic moment of particle	μ
Bohr magneton	μ_B
Bohr magneton number, nuclear magneton	μ_N
nuclear gyromagnetic ratio	γ
g-factor	g
Larmor (angular) frequency	ω_L
nuclear angular precession frequency	ω_N
cyclotron angular frequency of electron	ω_c
nuclear quadrupole moment	Q
nuclear radius	R
orbital angular momentum quantum number	L, l_1
spin angular momentum quantum number	S, s_1
total angular momentum quantum number	J, j_1
nuclear spin quantum number	I, J
hyperfine structure quantum number	F
principal quantum number	n, n_1
magnetic quantum number	M, m_1
fine structure constant	α
electron radius	r_e
Compton wavelength	λ_C
mass excess	Δ
packing fraction	f

mean life	τ
level width	Γ
activity	A
specific activity	a
decay constant	λ
half-life	$T_{\frac{1}{2}}, t_{\frac{1}{2}}$
disintegration energy	Q
spin-lattice relaxation time	T_1
spin-spin relaxation time	T_2
indirect spin-spin coupling	J

(k) Nuclear reactions and ionising radiations

reaction energy	Q
cross-section	σ
macroscopic cross-section	Σ
impact parameter	b
scattering angle	θ, ϕ
internal conversion coefficient	α
linear attenuation coefficient	μ, μ_1
atomic attenuation coefficient	μ
mass attenuation coefficient	μ_m
linear stopping power	S, S_1
atomic stopping power	S_a
linear range	R, R_1
recombination coefficient	α

Section 2

Some algebra topics

Why is Algebra important?

Algebra is one of the most fundamental tools for engineers and scientists because it allows them to determine the value of something (such as length, material constant, temperature, mass, and so on) given values that they do know (possibly other length, material properties, mass). Although the types of problems that mechanical, chemical, civil, environmental, electrical engineers deal with vary, all engineers use algebra to solve problems. An example where algebra is frequently used is in simple electrical circuits, where the resistance is proportional to voltage. Using Ohm's Law, $V = I \times R$, an engineer or scientist simply multiplies the current in a circuit by the resistance to determine the voltage across the circuit. Engineers and scientists use algebra in many ways, and so frequently that they don't even stop to think about it. Algebra lays the foundation for the mathematics needed to become an engineer or scientist.

A basic form of mathematics, algebra is nevertheless among the most commonly used forms of mathematics in the workplace. Although relatively simple, algebra is a powerful problem-solving tool used in many fields of engineering and science. For example, in designing a rocket to go to the moon, engineers and scientists must use algebra to solve for flight trajectory, how long to burn each thruster and at what intensity, and at what angle to lift off. Engineers and scientists use mathematics all the time – and, in particular, algebra; becoming familiar with algebra will make all engineering mathematics studies so much easier.

Chapter 6   Introduction to algebra

Some rules of algebra

Algebra merely uses letters to represent numbers. If, say, a, b, c and d represent any four numbers then in algebra:

(i)   **a + a + a + a = 4a**

(ii) **5b means 5 × b**

(iii) **2a + 3b + a − 2b** = 2a + a + 3b − 2b = **3a + b**

  Only similar terms can be combined in algebra.

(iv) **4abcd = 4 × a × b × c × d**

(v) **(a)(c)(d) means a × c × d**

  Brackets are often used instead of multiplication signs.

(vi) **ab = ba**

(vii) **$b^2 = b \times b$**

(viii) **$a^3 = a \times a \times a$**

Application: Simplify $4x^2 - 2x - 3y + 5x + 7y$

Re-ordering $4x^2 - 2x - 3y + 5x + 7y$ gives $4x^2 + 5x - 2x + 7y - 3y = \mathbf{4x^2 + 3x + 4y}$

Application: Simplify $3xy - 7x + 5xy + 3x$

Re-ordering $3xy - 7x + 5xy + 3x$ gives $3xy + 5xy + 3x - 7x = \mathbf{8xy - 4x}$

Application: Simplify $ab \times b^2c \times a$

$ab \times b^2c \times a = a \times b \times b \times b \times c \times a = a \times a \times b \times b \times b \times c = a^2 \times b^3 \times c = \mathbf{a^2b^3c}$

Laws of indices

The laws of indices in algebraic terms are as follows:

(i) $a^m \times a^n = a^{m+n}$ For example, $a^3 \times a^4 = a^{3+4} = a^7$

(ii) $\dfrac{a^m}{a^n} = a^{m-n}$ For example, $\dfrac{c^5}{c^2} = c^{5-2} = c^3$

(iii) $(a^m)^n = a^{mn}$ For example, $(d^2)^3 = d^{2 \times 3} = d^6$

(iv) $a^{\frac{m}{n}} = \sqrt[n]{a^m}$ For example, $x^{\frac{4}{3}} = \sqrt[3]{x^4}$

(v) $a^{-n} = \dfrac{1}{a^n}$ For example, $3^{-2} = \dfrac{1}{3^2} = \dfrac{1}{9}$

(vi) $a^0 = 1$ For example, $17^0 = 1$

Application: Simplify $a^2 b^3 c \times ab^2 c^5$

$$a^2 b^3 c \times ab^2 c^5 = a^2 \times b^3 \times c \times a \times b^2 \times c^5 = a^2 \times b^3 \times c^1 \times a^1 \times b^2 \times c^5$$

Grouping together like terms gives: $a^2 \times a^1 \times b^3 \times b^2 \times c^1 \times c^5$

Using law (i) of indices gives: $a^{2+1} \times b^{3+2} \times c^{1+5} = a^3 \times b^5 \times c^6$

i.e. $\mathbf{a^2 b^3 c \times ab^2 c^5 = a^3 b^5 c^6}$

Application: Simplify $\dfrac{x^5 y^2 z}{x^2 y \, z^3}$

$$\dfrac{x^5 y^2 z}{x^2 y z^3} = \dfrac{x^5 \times y^2 \times z}{x^2 \times y \times z^3} = \dfrac{x^5}{x^2} \times \dfrac{y^2}{y^1} \times \dfrac{z}{z^3} = x^{5-2} \times y^{2-1} \times z^{1-3} \text{ by law (ii) of indices}$$
$$= x^3 \times y^1 \times z^{-2} \times \mathbf{x^3 yz^{-2}} \text{ or } \dfrac{x^3 \, y}{z^2} \text{ by law (v) of indices}$$

Application: Simplify $(p^3)^2 (q^2)^4$

Using law (iii) of indices gives: $(p^3)^2 (q^2)^4 = p^{3 \times 2} \times q^{2 \times 4} = \mathbf{p^6 q^8}$

Brackets

With algebra   (i) $2(a + b) = 2a + 2b$

(ii) $(a + b)(c + d) = a(c + d) + b(c + d) = ac + ad + bc + bd$

Application: Determine $2b(a - 5b)$

$2b(a - 5b) = 2b \times a + 2b \times -5b = 2ba - 10b^2 = \mathbf{2ab - 10b^2}$ (Note that 2ba is the same as 2ab)

Application: Determine $(3x + 4y)(x - y)$

$(3x + 4y)(x - y) = 3x(x - y) + 4y(x - y) = 3x^2 - 3xy + 4yx - 4y^2$

$$= 3x^2 - 3xy + 4xy - 4y^2 = \mathbf{3x^2 + xy - 4y^2}$$

Application: Simplify $3(2x - 3y) - (3x - 5y)$

$3(2x - 3y) - (3x - 5y) = 3 \times 2x - 3 \times 3y - 3x - - 5y = 6x - 9y - 3x + 5y$

$$= 6x - 3x + 5y - 9y = \mathbf{3x - 4y}$$

Application: Remove the brackets from the expression and simplify:

$$2[x^2 - 3x\,(y + x) + 4xy]$$

$2[x^2 - 3x\,(y + x) + 4xy] = 2[x^2 - 3xy - 3x^2 + 4xy]$ Whenever more than one type of brackets are involved, **always start with the inner brackets**

$$= 2[-2x^2 + xy] = -4x^2 + 2xy = \mathbf{2xy - 4x^2}$$

Factorisation

The factors of 8 are 1, 2, 4, and 8, because 8 may be divided by 1, 2, 4, and 8

The factors of 24 are 1, 2, 3, 4, 6, 8, 12 and 24 because 24 may be divided by 1, 2, 3, 4, 6, 8, 12 and 24

The common factors of 8 and 24 are 1, 2, 4 and 8 since 1, 2, 4 and 8 are factors of both 8 and 24.

The highest common factor (HCF) is the largest number that divides into two or more terms.

Hence, the HCF of 8 and 28 is 4

When two or more terms in an algebraic expression contain a common factor, then this factor can be shown outside of a bracket.

For example, $df + dg = d(f + g)$

which is just the reverse of $d(f + g) = df + dg$

This process is called factorisation

Application: Factorise $ab - 6ac$

'a' is common to both of the terms ab and $-6ac$. 'a' is therefore taken outside of the bracket. What goes inside the bracket?

(i) What multiplies 'a' to make ab?   Answer: b

(ii) What multiplies 'a' to make – 6ac?   Answer: – 6c

Hence, b – 6c appears in the bracket.

Thus, $ab - 6ac = \mathbf{a(b - 6c)}$

Application: Factorise $2x^2 + 14xy^3$

For the numbers 2 and 14, the highest common factor (HCF) is 2 (i.e. 2 is the largest number that divides into both 2 and 14)

For the x terms, x^2 and x, the HCF is x

Thus, the HCF of $2x^2$ and $14xy^3$ is 2x

2x is therefore taken outside of the bracket. What goes inside the bracket?

(i) What multiplies 2x to make $2x^2$? Answer: x

(ii) What multiplies 2x to make $14xy^3$? Answer: $7y^3$

Hence $x + 7y^3$ appears in the bracket

Thus, $2x^2 + 14xy^3 = \mathbf{2x(x+7y^3)}$

Laws of precedence

With the laws of precedence the order is

Brackets
Order   (or p**O**wer)
Division
Multiplication
Addition
Subtraction

The first letter of each word spells BODMAS

Application: Simplify $5x + 3x \times 4x - x$

$$5x + 3x \times 4x - x = 5x + 12x^2 - x \qquad \text{(M)}$$

$$= 5x - x + 12x^2$$

$$= \mathbf{4x + 12x^2} \qquad \text{(S)}$$

$$\text{or } \mathbf{4x(\ 1 + 3x)} \text{ by factorising}$$

Application: Simplify $(y + 4y) \times 2y - 5y$

$$(y + 4y) \times 2y - 5y = 5y \times 2y - 5y \qquad \text{(B)}$$

$$= \mathbf{10y^2 - 5y} \qquad \text{(M)}$$

$$\text{or } \mathbf{5y(2y - 1)} \text{ by factorising}$$

Application: Simplify $2y \div (8y + 3y - 5y)$

$$2y \div (8y + 3y - 5y) = 2y \div 6y \qquad \text{(B)}$$

$$= \frac{2y}{6y} \qquad \text{(D)}$$

$$= \frac{1}{3} \text{ by cancelling}$$

Chapter 7   Polynomial division

A **polynomial** is an expression of the form: $f(x) = a + bx + cx^2 + dx^3 + \ldots$ and **polynomial division** is sometimes required when resolving a fraction into partial fractions.

Application: Divide $2x^2 + x - 3$ by $x - 1$

$2x^2 + x - 3$ is called the **dividend** and $x - 1$ the **divisor**. The usual layout is shown below with the dividend and divisor both arranged in descending powers of the symbols.

$$
\begin{array}{r}
2x + 3 \\
x - 1 \overline{\smash{)}\ 2x^2 + x - 3} \\
\underline{2x^2 - 2x} \\
3x - 3 \\
\underline{3x - 3} \\
\cdot \ \cdot
\end{array}
$$

Dividing the first term of the dividend by the first term of the divisor, i.e. $2x^2/x$ gives $2x$, which is placed above the first term of the dividend as shown. The divisor is then multiplied by $2x$, i.e. $2x(x - 1) = 2x^2 - 2x$, which is placed under the dividend as shown. Subtracting gives $3x - 3$.

The process is then repeated, i.e. the first term of the divisor, x, is divided into $3x$, giving $+3$, which is placed above the dividend as shown. Then $3(x - 1) = 3x - 3$ which is placed under the $3x - 3$. The remainder, on subtraction, is zero, which completes the process.

Thus, $\mathbf{(2x^2 + x - 3) \div (x - 1) = (2x + 3)}$

The answer, $(2x + 3)$, is called the **quotient.**

Application: Divide $(x^2 + 3x - 2)$ by $(x - 2)$

$$
\begin{array}{r}
x + 5 \\
x - 2\overline{)\,x^2 + 3x - 2} \\
\underline{x^2 - 2x} \\
5x - 2 \\
\underline{5x - 10} \\
8
\end{array}
$$

Hence $\dfrac{x^2 + 3x - 2}{x - 2} = x + 5 + \dfrac{8}{x - 2}$

Application: Divide $(3x^3 + x^2 - 3x + 5)$ by $(x + 1)$

$$
\begin{array}{r}
(1)\quad\ (4)\ \ (7) \\
3x^2 - 2x + 5 \\
x + 1\overline{)\,3x^3 +\ \ x^2 + 3x + 5} \\
\underline{3x^3 + 3x^2} \\
-2x^2 + 3x + 5 \\
\underline{-2x^2 - 2x} \\
5x + 5 \\
\underline{5x + 5} \\
\cdot\ \overline{\quad}\ \cdot
\end{array}
$$

(1) x into $3x^3$ goes $3x^2$. Put $3x^2$ above $3x^3$
(2) $3x^2(x + 1) = 3x^3 + 3x^2$
(3) Subtract
(4) x into $-2x^2$ goes $-2x$. Put $-2x$ above the dividend
(5) $-2x(x+1) = -2x^2 - 2x$
(6) Subtract
(7) x into $5x$ goes 5. Put 5 above the dividend
(8) $5(x+1) = 5x+5$
(9) Substract

Thus $\dfrac{3x^3 + x^2 + 3x + 5}{x + 1} = \mathbf{3x^2 - 2x + 5}$

Chapter 8   The factor theorem

A factor of $(x - a)$ in an equation corresponds to a root of $x = a$

If $x = a$ is a root of the equation $f(x) = 0$, then $(x - a)$ is a factor of $f(x)$

Application: Factorise $x^3 - 7x - 6$ and use it to solve the cubic equation $x^3 - 7x - 6 = 0$

Let $f(x) = x^3 - 7x - 6$

If $x = 1$, then $f(1) = 1^3 - 7(1) - 6 = -12$

If $x = 2$, then $f(2) = 2^3 - 7(2) - 6 = -12$

If $x = 3$, then $f(3) = 3^3 - 7(3) - 6 = 0$

If $f(3) = 0$, then $(x - 3)$ is a factor – from the factor theorem.

We have a choice now. We can divide $x^3 - 7x - 6$ by $(x - 3)$ or we could continue our 'trial and error' by substituting further values for x in the given expression – and hope to arrive at $f(x) = 0$.

Let us do both ways. Firstly, dividing out gives:

$$
\begin{array}{r}
x^2 + 3x + 2 \\
x - 3 \overline{\smash{\big)}\ x^3 + 0 \quad -7x - 6} \\
\underline{x^3 - 3x^2} \\
3x^2 - 7x - 6 \\
\underline{3x^2 - 9x} \\
2x - 6 \\
\underline{2x - 6} \\
\cdot \quad \cdot
\end{array}
$$

Hence, $\dfrac{x^3 - 7x - 6}{x - 3} = x^2 + 3x + 2$

i.e. $x^3 - 7x - 6 = (x - 3)(x^2 + 3x + 2)$

$x^2 + 3x + 2$ factorises 'on sight' as $(x + 1)(x + 2)$

Therefore, **$x^3 - 7x - 6 = (x - 3)(x + 1)(x + 2)$**

A second method is to continue to substitute values of x into $f(x)$.

Our expression for $f(3)$ was $3^3 - 7(3) - 6$. We can see that if we continue with positive values of x the first term will predominate such that $f(x)$ will not be zero.

Therefore let us try some negative values for x.

$f(-1) = (-1)^3 - 7(-1) - 6 = 0$; hence $(x + 1)$ is a factor (as shown above).

Also $f(-2) = (-2)^3 - 7(-2) - 6 = 0$; hence $(x + 2)$ is a factor.

To solve $x^3 - 7x - 6 = 0$, we substitute the factors, i.e.

$$(x - 3)(x + 1)(x + 2) = 0$$

from which, **$x = 3$, $x = -1$ and $x = -2$**

Note that the values of x, i.e. 3, -1 and -2, are all factors of the constant term, i.e. the 6. This can give us a clue as to what values of x we should consider.

Chapter 9   The remainder theorem

If $(ax^2 + bx + c)$ is divided by $(x - p)$, the remainder will be $ap^2 + bp + c$

If $(ax^3 + bx^2 + cx + d)$ is divided by $(x - p)$, the remainder will be $ap^3 + bp^2 + cp + d$

Application: When $(3x^2 - 4x + 5)$ is divided by $(x - 2)$ find the remainder

$ap^2 + bp + c$, (where $a = 3$, $b = -4$, $c = 5$ and $p = 2$),

hence the remainder is $3(2)^2 + (-4)(2) + 5 = 12 - 8 + 5 = \mathbf{9}$

We can check this by dividing $(3x^2 - 4x + 5)$ by $(x - 2)$ by long division:

$$
\begin{array}{r}
3x + 2 \\
x - 2 \overline{\smash{)}\ 3x^2 - 4x + 5} \\
\underline{3x^2 - 6x} \\
2x + 5 \\
\underline{2x - 4} \\
9
\end{array}
$$

Application: When $(2x^2 + x - 3)$ is divided by $(x - 1)$, find the remainder

$ap^2 + bp + c$, (where $a = 2$, $b = 1$, $c = -3$ and $p = 1$), hence the **remainder is** $2(1)^2 + 1(1) - 3 = \mathbf{0}$,

which means that $(x - 1)$ is a factor of $(2x^2 + x - 3)$.

In this case, the other factor is $(2x + 3)$,
i.e. $(2x^2 + x - 3) = (x - 1)(2x + 3)$

Application: When $(3x^3 + 2x^2 - x + 4)$ is divided by $(x - 1)$, find the remainder

The remainder is $ap^3 + bp^2 + cp + d$ (where $a = 3$, $b = 2$, $c = -1$, $d = 4$ and $p = 1$), i.e. the remainder is:
$3(1)^3 + 2(1)^2 + (-1)(1) + 4 = 3 + 2 - 1 + 4 = \mathbf{8}$

Chapter 10   Continued fractions

A **continued fraction** is an expression obtained through an iterative process of representing a number as the sum of its integer part and the reciprocal of another number, then writing this other number as the sum of its integer part and another reciprocal, and so on. These approximations to fractions are used to obtain practical ratios for **gearwheels** or for a **dividing head** (used to give a required angular displacement).

Any fraction may be expressed in the form shown below for the fraction 26/55:

$$\frac{26}{55} = \frac{1}{\frac{55}{26}} = \frac{1}{2 + \frac{3}{26}} = \frac{1}{2 + \frac{1}{\frac{26}{3}}} = \frac{1}{2 + \frac{1}{8 + \frac{2}{3}}} = \frac{1}{2 + \frac{1}{8 + \frac{1}{\frac{3}{2}}}}$$

$$= \frac{1}{2 + \frac{1}{8 + \frac{1}{1 + \frac{1}{2}}}}$$

The latter factor can be expressed as:

$$\frac{1}{A + \cfrac{\alpha}{B + \cfrac{\beta}{C + \cfrac{\gamma}{D + \delta}}}}$$

Comparisons show that A, B, C and D are 2, 8, 1 and 2 respectively.

A fraction written in the general form is called a **continued fraction** and the integers A, B, C and D are called the **quotients** of the continued fraction. The quotients may be used to obtain closer and closer approximations, called **convergents**.

A tabular method may be used to determine the convergents of a fraction:

		1	2	3	4	5
a			2	8	1	2
b $\begin{cases} bp \\ bq \end{cases}$		$\frac{0}{1}$	$\frac{1}{2}$	$\frac{8}{17}$	$\frac{9}{19}$	$\frac{26}{55}$

The quotients 2, 8, 1 and 2 are written in cells a2, a3, a4 and a5 with cell a1 being left empty.

The fraction $\frac{0}{1}$ is always written in cell b1.

The reciprocal of the quotient in cell a2 is always written in cell b2, i.e. $\frac{1}{2}$ in this case.

The fraction in cell b3 is given by $\dfrac{(a3 \times b2p) + b1p}{(a3 \times b2q) + b1q}$, i.e. $\dfrac{(8 \times 1) + 0}{(8 \times 2) + 1} = \dfrac{8}{17}$

The fraction in cell b4 is given by $\dfrac{(a4 \times b3p) + b2p}{(a4 \times b3q) + b2q}$,

i.e. $\dfrac{(1 \times 8) + 1}{(1 \times 17) + 2} = \dfrac{9}{19}$, and so on.

Hence the convergents of $\dfrac{26}{55}$ are $\dfrac{1}{2}$, $\dfrac{8}{17}$, $\dfrac{9}{19}$ and $\dfrac{26}{55}$, each value approximating

closer and closer to $\dfrac{26}{55}$.

Chapter 11   Solving simple equations

Introduction

$3x - 4$ is an example of an **algebraic expression**.

$3x - 4 = 2$ is an example of an **algebraic equation** (i.e. it contains an '=' sign)

An equation is simply a statement that two expressions are equal.

Hence, $A = \pi r^2$ (where A is the area of a circle of radius r)

$\qquad F = \dfrac{9}{5} C + 32$ (which relates Fahrenheit and Celsius temperatures)

and

$\qquad y = 3x + 2$ (which is the equation of a straight line graph)

are all examples of **equations**.

Solving equations

To '**solve an equation**' means '**to find the value of the unknown**'.

Application: Solve the equation: $4x = 24$

Dividing each side of the equation by 4 gives: $\dfrac{4x}{4} = \dfrac{24}{4}$

i.e. $\qquad\qquad\qquad\qquad\qquad$ **x = 6** by cancelling

which is the solution to the equation $4x = 24$

Application: Solve the equation: $\dfrac{2x}{5} = 4$

Multiplying both sides by 5 gives: $5\left(\dfrac{2x}{5}\right) = 5(4)$

Cancelling and removing brackets gives: $2x = 20$

Dividing both sides of the equation by 2 gives: $\dfrac{2x}{2} = \dfrac{20}{2}$

Cancelling gives: $\mathbf{x = 10}$

Application: Solve the equation: $x + 3 = 8$

Subtracting 3 from both sides gives: $x + 3 - 3 = 8 - 3$

i.e. $x = 8 - 3$

i.e. $\mathbf{x = 5}$

Application: Solve the equation: $7x + 1 = 3x + 9$

In such equations the terms containing x are grouped on one side of the equation and the remaining terms grouped on the other side of the equation. Changing from one side of an equation to the other must be accompanied by a change of sign.

Since $7x + 1 = 3x + 9$

then $7x - 3x = 9 - 1$

i.e. $4x = 8$

Dividing both sides by 4 gives: $\dfrac{4x}{4} = \dfrac{8}{4}$

Cancelling gives: $\mathbf{x = 2}$

Application: Solve the equation: $3(x - 2) = 12$

Removing the bracket gives: $3x - 6 = 12$

Rearranging gives: $3x = 12 + 6$

i.e. $3x = 18$

Dividing both sides by 3 gives: $\mathbf{x = 6}$

Application: Solve the equation: $4(2y - 3) - 2(y - 4) = 3(y - 3) - 1$

Removing brackets gives: $8y - 12 - 2y + 8 = 3y - 9 - 1$

Rearranging gives: $8y - 2y - 3y = -9 - 1 + 12 - 8$

i.e. $3y = -6$

Dividing both sides by 3 gives: $\mathbf{y = \dfrac{-6}{3} = -2}$

Application: Solve the equation: $\dfrac{4}{t} = \dfrac{2}{5}$

The lowest common multiple (LCM) of the denominators, i.e. the lowest algebraic expression that both t and 5 will divide into, is 5t

Multiplying both sides by 5t gives: $\qquad 5t\left(\dfrac{4}{t}\right) = 5t\left(\dfrac{2}{5}\right)$

Cancelling gives: $\qquad\qquad\qquad\qquad 5(4) = t(2)$

i.e. $\qquad\qquad\qquad\qquad\qquad\qquad 20 = 2t \qquad\qquad\qquad (1)$

Dividing both sides by 2 gives: $\qquad\qquad \dfrac{20}{2} = \dfrac{2t}{2}$

Cancelling gives: $\qquad\qquad\qquad\qquad$ **10 = t or t = 10**

When there is just one fraction on each side of the equation as in this example, there is a quick way to arrive at equation (1) without needing to find the LCM of the denominators.

We can move from $\left(\dfrac{4}{t}\right) = \dfrac{2}{5}$ to: $4 \times 5 = 2 \times t$

by what is called '**cross-multiplication**'.

In general, if $\dfrac{a}{b} = \dfrac{c}{d}$ then: $ad = bc$

We can use cross-multiplication when there is one fraction only on each side of the equation.

Application: Solve the equation: $\dfrac{3}{x-2} = \dfrac{4}{3x+4}$

'Cross-multiplication' gives: $\qquad 3(3x+4) = 4(x-2)$

Removing brackets gives: $\qquad\quad 9x + 12 = 4x - 8$

Rearranging gives: $\qquad\qquad\quad 9x - 4x = -8 - 12$

i.e. $\qquad\qquad\qquad\qquad\qquad 5x = -20$

Dividing both sides by 5 gives: $\qquad \mathbf{x = \dfrac{-20}{5} = -4}$

Application: Solve the equation: $2\sqrt{d} = 6$

Whenever square roots are involved in an equation, the square root term needs to be isolated on its own before squaring both sides

'Cross-multiplying' gives: $\qquad \sqrt{d} = \dfrac{6}{2}$

Cancelling gives: $\qquad\qquad\quad \sqrt{d} = 3$

Squaring both sides gives:
$$\left(\sqrt{d}\right)^2 = (3)^2$$

i.e.
$$d = 9$$

Application: Solve the equation: $\dfrac{15}{4t^2} = \dfrac{2}{3}$

We need to rearrange the equation to get the t^2 term on its own.

'Cross-multiplying' gives: $15(3) = 2(4t^2)$

i.e. $45 = 8t^2$

Dividing both sides by 8 gives: $\dfrac{45}{8} = \dfrac{8t^2}{8}$

By cancelling: $5.625 = t^2$

or $t^2 = 5.625$

Taking the square root of both sides gives:
$$\sqrt{t^2} = \sqrt{5.625}$$

i.e. $t = \pm\,\mathbf{2.372}$, correct to 4 significant figures,

---

Practical problems involving simple equations

Application: Applying the principle of moments to a beam results in the following equation:
$$F \times 3 = (7.5 - F) \times 2$$

where F is the force in Newtons. Determine the value of F.

Removing brackets gives: $3F = 15 - 2F$

Rearranging gives: $3F + 2F = 15$

i.e. $5F = 15$

Dividing both sides by 5 gives: $\dfrac{5F}{5} = \dfrac{15}{5}$

from which,   **force, F = 3 N**

Application: $PV = mRT$ is the characteristic gas equation. Find the value of gas constant R when pressure, $P = 3 \times 10^6 Pa$, volume, $V = 0.90m^3$, mass, $m = 2.81kg$ and temperature, $T = 231\ K$.

Dividing both sides of PV = mRT by mT gives:

$$\frac{PV}{mT} = \frac{mRT}{mT}$$

Cancelling gives: $\frac{PV}{mT} = R$

Substituting values gives: $R = \dfrac{\left(3 \times 10^6\right)\left(0.90\right)}{\left(2.81\right)\left(231\right)}$

Using a calculator, **gas constant, R = 4160 J/(kg K)**, correct to 4 significant figures.

Application:A formula relating initial and final states of pressures, P_1 and P_2, volumes V_1 and V_2, and absolute temperatures, T_1 and T_2, of an ideal gas is: $\dfrac{P_1 V_1}{T_1} = \dfrac{P_2 V_2}{T_2}$

Find the value of P_2 given P_1 = 100 × 10³ Pa, V_1 = 1.0 m³, V_2 = 0.266 m³ , T_1 = 423 K and T_2 = 293 K

Since $\dfrac{P_1 V_1}{T_1} = \dfrac{P_2 V_2}{T_2}$ then $\dfrac{(100 \times 10^3)(1.0)}{423} = \dfrac{P_2(0.266)}{293}$

'Cross-multiplying' gives:

$$(100 \times 10^3)(1.0)(293) = P_2(0.266)(423)$$

$$P_2 = \frac{(100 \times 10^3)(1.0)(293)}{(0.266)(423)}$$

Hence, **P_2 = 260 × 10³ Pa or 2.6 × 10⁵ Pa**

Application: The stress f in a material of a thick cylinder can be obtained from: $\dfrac{D}{d} = \sqrt{\left(\dfrac{f+p}{f-p}\right)}$ Calculate the stress, given that D = 21.5, d = 10.75 and

p = 1800

Since $\dfrac{D}{d} = \sqrt{\left(\dfrac{f+p}{f-p}\right)}$ then $\dfrac{21.5}{10.75} = \sqrt{\left(\dfrac{f+1800}{f-1800}\right)}$

i.e. $2 = \sqrt{\left(\dfrac{f+1800}{f-1800}\right)}$

Squaring both sides gives: $4 = \dfrac{f+1800}{f-1800}$

'Cross-multiplying' gives: $4(f - 1800) = f + 1800$

$$4f - 7200 = f + 1800$$

$$4f - f = 1800 + 7200$$

$$3f = 9000$$

$$f = \frac{9000}{3} = 3000$$

Hence,   **stress, f = 3000**

Chapter 12   Transposing formulae

Introduction to transposing formulae

In the formula $I = \frac{V}{R}$, I is called the **subject of the formula.**

Similarly, in the formula $y = mx + c$, y is the subject of the formula.

When a symbol other than the subject is required to be the subject, then the formula needs to be rearranged to make a new subject.

This rearranging process is called **transposing the formula** or **transposition.**

There are no new rules for transposing formulae. The same rules as were used for simple equations in chapter 11 are used, i.e. **the balance of an equation must be maintained**.

Whatever is done to one side of an equation must be done to the other.

Application: If $a + b = z - x + y$, express x as the subject

A term can be moved from one side of an equation to the other side but with a change of sign.

Hence, rearranging gives: **x = z + y – a – b**

Application: Transpose $v = f \lambda$ to make λ the subject

$v = f \lambda$ relates velocity v, frequency f and wavelength λ

Rearranging gives: $f \lambda = v$

Dividing both sides by f gives: $\frac{f \lambda}{f} = \frac{v}{f}$

Cancelling gives: $\lambda = \frac{v}{f}$

Application: When a body falls freely through a height h, the velocity v is given by $v^2 = 2gh$. Express this formula with h as the subject.

Rearranging gives: $\qquad\qquad\qquad 2gh = v^2$

Dividing both sides by 2g gives: $\dfrac{2gh}{2g} = \dfrac{v^2}{2g}$

Cancelling gives: $\qquad\qquad\qquad\qquad h = \dfrac{v^2}{2g}$

Application: If $I = \dfrac{V}{R}$, rearrange to make V the subject

$I = \dfrac{V}{R}$ is Ohm's law, where I is the current, V is the voltage and R is the resistance.

Rearranging gives: $\qquad\qquad\qquad\qquad \dfrac{V}{R} = I$

Multiplying both sides by R gives: $\qquad R\left(\dfrac{V}{R}\right) = R(I)$

Cancelling gives: $\qquad\qquad\qquad\qquad\quad \mathbf{V = IR}$

Application: Transpose $y = mx + c$ to make m the subject

$y = mx + c$ is the equation of a straight-line graph, where y is the vertical axis variable, x is the horizontal axis variable, m is the gradient of the graph and c is the y-axis intercept.

Subtracting c from both sides gives: $\qquad y - c = mx$

or $\qquad\qquad\qquad\qquad\qquad\qquad mx = y - c$

Dividing both sides by x gives: $\qquad\qquad \mathbf{m = \dfrac{y - c}{x}}$

Application: Transpose the formula $v = u + \dfrac{Ft}{m}$, to make t the subject

$v = u + \dfrac{Ft}{m}$ relates final velocity v, initial velocity u, force F, mass m and time t. ($\dfrac{F}{m}$ is acceleration 'a')

Rearranging gives: $\qquad\qquad\qquad u + \dfrac{Ft}{m} = v$

and $\qquad\qquad\qquad\qquad\qquad \dfrac{Ft}{m} = v - u$

Multiplying each side by m gives: $\qquad m\left(\dfrac{Ft}{m}\right) = m(v - u)$

Cancelling gives: $\qquad\qquad\qquad\qquad Ft = m(v - u)$

Dividing both sides by F gives: $\qquad\qquad \dfrac{Ft}{F} = \dfrac{m(v - u)}{F}$

Cancelling gives: $t = \dfrac{m(v - u)}{F}$ or $t = \dfrac{m}{F}(v - u)$

This shows two ways of expressing the answer. There is often more than one way of expressing a transposed answer.

In this case, both equations for t are equivalent; neither one is more correct than the other.

> **Application:** In a right-angled triangle having sides x, y and hypotenuse z, Pythagoras' theorem states $z^2 = x^2 + y^2$. Transpose the formula to find y.

Rearranging gives: $x^2 + y^2 = z^2$

and $y^2 = z^2 - x^2$

Taking the square root of both sides gives: $y = \sqrt{z^2 - x^2}$

> **Application:** Given $t = 2\pi\sqrt{\dfrac{l}{g}}$, find g in terms of t, *l* and π

Whenever the prospective new subject is within a square root sign, it is best to isolate that term on the LHS and then to square both sides of the equation.

Rearranging gives: $2\pi\sqrt{\dfrac{l}{g}} = t$

Dividing both sides by 2π gives: $\sqrt{\dfrac{l}{g}} = \dfrac{t}{2\pi}$

Squaring both sides gives: $\dfrac{l}{g} = \left(\dfrac{t}{2\pi}\right)^2 = \dfrac{t^2}{4\pi^2}$

Cross-multiplying, (i.e. multiplying each term by $4\pi^2 g$), gives:

$$4\pi^2 l = gt^2$$

or $gt^2 = 4\pi^2 l$

Dividing both sides by t^2 gives: $\dfrac{gt^2}{t^2} = \dfrac{4\pi^2 l}{t^2}$

Cancelling gives: $g = \dfrac{4\pi^2 l}{t^2}$

> **Application:** The impedance Z of an a.c. circuit is given by: $Z = \sqrt{R^2 + X^2}$ where R is the resistance. Make the reactance, X , the subject.

Rearranging gives: $Z = \sqrt{R^2 + X^2}$

Squaring both sides gives: $\qquad\qquad R^2 + X^2 = Z^2$

Rearranging gives: $\qquad\qquad\qquad X^2 = Z^2 - R^2$

Taking the square root of both sides gives: $\mathbf{X = \sqrt{Z^2 - R^2}}$

Application: Make b the subject of the formula $a = \dfrac{x - y}{\sqrt{bd + be}}$

Rearranging gives: $\qquad\qquad\qquad \dfrac{x - y}{\sqrt{bd + be}} = a$

Multiplying both sides by $\sqrt{bd + be}$ gives: $\quad x - y = a\sqrt{bd + be}$

or $\qquad\qquad\qquad\qquad\qquad a\sqrt{bd + be} = x - y$

Dividing both sides by a gives: $\qquad \sqrt{bd + be} = \dfrac{x - y}{a}$

Squaring both sides gives: $\qquad\quad bd + be = \left(\dfrac{x - y}{a}\right)^2$

Factorising the LHS gives: $\qquad\quad b(d + e) = \left(\dfrac{x - y}{a}\right)^2$

Dividing both sides by (d + e) gives: $\quad b = \dfrac{\left(\dfrac{x - y}{a}\right)^2}{(d + e)}$ or $\mathbf{b = \dfrac{(x - y)^2}{a^2(d + e)}}$

Application: If $a = \dfrac{b}{1 + b}$ make b the subject of the formula.

Rearranging gives: $\qquad\qquad\qquad \dfrac{b}{1 + b} = a$

Multiplying both sides by (1 + b) gives: $\qquad b = a(1 + b)$

Removing the bracket gives: $\qquad\qquad b = a + ab$

Rearranging to obtain terms in b on the LHS gives: $\quad b - ab = a$

Factorising the LHS gives: $\qquad\qquad b(1 - a) = a$

Dividing both sides by (1 − a) gives: $\qquad \mathbf{b = \dfrac{a}{1 - a}}$

Application: Given that $\dfrac{D}{d} = \sqrt{\left(\dfrac{f+p}{f-p}\right)}$ express p in terms of D, d and f

Rearranging gives:

$$\sqrt{\left(\frac{f+p}{f-p}\right)} = \frac{D}{d}$$

Squaring both sides gives:

$$\left(\frac{f+p}{f-p}\right) = \frac{D^2}{d^2}$$

Cross-multiplying, i.e. multiplying each term by $d^2(f-p)$, gives:

$$d^2(f+p) = D^2(f-p)$$

Removing brackets gives: $d^2f + d^2p = D^2f - D^2p$

Rearranging, to obtain terms in p on the LHS gives:

$$d^2p + D^2p = D^2f - d^2f$$

Factorising gives: $p(d^2 + D^2) = f(D^2 - d^2)$

Dividing both sides by $(d^2 + D^2)$ gives:

$$\mathbf{p} = \frac{\mathbf{f(D^2 - d^2)}}{\mathbf{(d^2 + D^2)}}$$

Chapter 13   Solving simultaneous equations

Introduction

When an equation contains **two unknown quantities** it has an infinite number of solutions. When two equations are available connecting the same two unknown values then a unique solution is possible.

Equations which have to be solved together to find the unique values of the unknown quantities, which are true for each of the equations, are called **simultaneous equations**.

Two methods of solving simultaneous equations analytically are:

(a) by **substitution**, and (b) by **elimination**.

Further methods of solving simultaneous equations are explained in Chapters 83 to 86.

Solving simultaneous equations in two unknowns

Application: Solve the following equations for x and y, (a) by substitution, and (b) by elimination:

$$x + 2y = -1 \qquad (1)$$
$$4x - 3y = 18 \qquad (2)$$

(a) By substitution

From equation (1): $\qquad\qquad x = -1 - 2y$

Substituting this expression for x into equation (2) gives:

$$4(-1 - 2y) - 3y = 18$$

This is now a simple equation in y.

Removing the bracket gives:

$$-4 - 8y - 3y = 18$$
$$-11y = 18 + 4 = 22$$
$$y = \frac{22}{-11} = -2$$

Substituting y = −2 into equation (1) gives:

$$x + 2(-2) = -1$$
$$x - 4 = -1$$
$$x = -1 + 4 = 3$$

Thus, x = 3 and y = −2 is the solution to the simultaneous equations

(b) By elimination

$$x + 2y = -1 \qquad (1)$$
$$4x - 3y = 18 \qquad (2)$$

If equation (1) is multiplied throughout by 4 the coefficient of x will be the same as in equation (2), giving:

$$4x + 8y = -4 \qquad (3)$$

Subtracting equation (3) from equation (2) gives:

$$4x - 3y = 18 \qquad (2)$$
$$\underline{4x + 8y = -4} \qquad (3)$$
$$0 - 11y = 22$$

Hence, $\mathbf{y = \dfrac{22}{-11} = -2}$ (Note, in the above subtraction, 18 − − 4 = 18 + 4 = 22)

Application: Solve, by a substitution method, the simultaneous equations

$$3x - 2y = 12 \tag{1}$$

$$x + 3y = -7 \tag{2}$$

From equation (2), $x = -7 - 3y$

Substituting for x in equation (1) gives:

$$3(-7 - 3y) - 2y = 12$$

i.e. $-21 - 9y - 2y = 12$

$$-11y = 12 + 21 = 33$$

Hence, $y = \dfrac{33}{-11} = -3$

Substituting $y = -3$ in equation (2) gives:

$$x + 3(-3) = -7$$

i.e. $x - 9 = -7$

Hence $x = -7 + 9 = 2$

Thus, **x = 2, y = -3** is the solution of the simultaneous equations.

Application: Solve $3p = 2q$ (1)

$$4p + q + 11 = 0 \tag{2}$$

Rearranging gives:

$$3p - 2q = 0 \tag{3}$$

$$4p + q = -11 \tag{4}$$

Multiplying equation (4) by 2 gives:

$$8p + 2q = -22 \tag{5}$$

Adding equations (3) and (5) gives:

$$11p + 0 = -22$$

$$p = \dfrac{-22}{11} = -2$$

Substituting $p = -2$ into equation (1) gives:

$$3(-2) = 2q$$

$$-6 = 2q$$

$$q = \dfrac{-6}{2} = -3$$

Checking, by substituting $p = -2$ and $q = -3$ into equation (2) gives:

$$\text{LHS} = 4(-2) + (-3) + 11 = -8 - 3 + 11 = 0 = \text{RHS}$$

Hence, **the solution is p = − 2, q = − 3**

Practical problems involving simultaneous equations

There are several situations in engineering and science where the solution of simultaneous equations is required.

Application: The law connecting friction F and load L for an experiment is of the form $F = aL + b$, where a and b are constants. When $F = 5.6$ N, $L = 8.0$ N and when $F = 4.4$ N, $L = 2.0$ N. Find the values of a and b and the value of F when $L = 6.5$ N

Substituting $F = 5.6$, $L = 8.0$ into $F = aL + b$ gives:

$$5.6 = 8.0a + b \tag{1}$$

Substituting $F = 4.4$, $L = 2.0$ into $F = aL + b$ gives:

$$4.4 = 2.0a + b \tag{2}$$

Subtracting equation (2) from equation (1) gives:

$$1.2 = 6.0\,a$$

$$a = \frac{1.2}{6.0} = \frac{1}{5} \text{ or } \mathbf{0.2}$$

Substituting $a = \frac{1}{5}$ into equation (1) gives:

$$5.6 = 8.0\left(\frac{1}{5}\right) + b$$

$$5.6 = 1.6 + b$$

$$5.6 - 1.6 = b$$

i.e. $\qquad\qquad\qquad \mathbf{b = 4}$

Hence, $\mathbf{a = \dfrac{1}{5}}$ **and b = 4**

When L = 6.5, $\quad F = aL + b = \frac{1}{5}(6.5) + 4 = 1.3 + 4 \quad$ i.e. $\quad$ **F = 5.3 N**

Application: When Kirchhoff's laws are applied to the electrical circuit shown in Figure 13.1 the currents I_1 and I_2 are connected by the equations:

$$27 = 1.5I_1 + 8(I_1 - I_2) \tag{1}$$

$$-26 = 2I_2 - 8(I_1 - I_2) \tag{2}$$

Solve the equations to find the values of currents I_1 and I_2

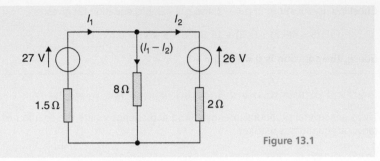

Figure 13.1

Removing the brackets from equation (1) gives:

$$27 = 1.5I_1 + 8I_1 - 8I_2$$

Rearranging gives: $9.5I_1 - 8I_2 = 27$ (3)

Removing the brackets from equation (2) gives:

$$-26 = 2I_2 - 8I_1 + 8I_2$$

Rearranging gives: $-8I_1 + 10I_2 = -26$ (4)

Multiplying equation (3) by 5 gives:

$$47.5\,I_1 - 40I_2 = 135$$ (5)

Multiplying equation (4) by 4 gives:

$$-32I_1 + 40I_2 = -104$$ (6)

Adding equations (5) and (6) gives:

$$15.5I_1 + 0 = 31$$

$$I_1 = \frac{31}{15.5} = 2$$

Substituting $I_1 = 2$ into equation (3) gives:

$$9.5(2) - 8I_1 = 27$$

$$19 - 8I_2 = 27$$

$$19 - 27 = 8I_2$$

$$-8 = 8I_2$$

and $$I_2 = -1$$

Hence, the solution is **$I_1 = 2$ and $I_2 = -1$**

Application: The resistance R Ω of a length of wire at t°C is given by:
$R = R_0(1 + \alpha t)$, where R_0 is the resistance at 0°C and α is the temperature coefficient of resistance in /°C. Find the values of α and R_0 if R = 30 Ω at 50°C, and R = 35 Ω at 100°C

Substituting $R = 30$, $t = 50$ into $R = R_0(1 + \alpha t)$ gives:

$$30 = R_0(1 + 50\alpha) \tag{1}$$

Substituting $R = 35$, $t = 100$ into $R = R_0(1 + \alpha t)$ gives:

$$35 = R_0(1 + 100\alpha) \tag{2}$$

Although these equations may be solved by the conventional substitution method, an easier way is to eliminate R_0 by division. Thus, dividing equation (1) by equation (2) gives:

$$\frac{30}{35} = \frac{R_0(1 + 50\alpha)}{R_0(1 + 50\alpha)} = \frac{1 + 50\alpha}{1 + 100\alpha}$$

'Cross-multiplying' gives:

$$30(1 + 100\alpha) = 35(1 + 50\alpha)$$

$$30 + 3000\alpha = 35 + 1750\alpha$$

$$3000\alpha - 1750\alpha = 35 - 30$$

$$1250\alpha = 5$$

i.e.

$$\alpha = \frac{5}{1250} = \frac{1}{250} \text{ or } \mathbf{0.004}$$

Substituting $\alpha = \frac{1}{250}$ into equation (1) gives:

$$30 = R_0 \left\{ 1 + (50)\left(\frac{1}{250}\right) \right\}$$

$$30 = R_0 (1.2)$$

$$\mathbf{R_0} = \frac{30}{1.2} = \mathbf{25}$$

Thus the solution is: $\boldsymbol{\alpha = 0.004/^\circ C}$ **and** $\mathbf{R_0 = 25\Omega}$

Solving simultaneous equations in three unknowns

Application: Solve the simultaneous equations:

$$x + y + z = 4 \tag{1}$$

$$2x - 3y + 4z = 33 \tag{2}$$

$$3x - 2y - 2z = 2 \tag{3}$$

There are a number of ways of solving these equations. One method is shown below.

The initial object is to produce two equations with two unknowns. For example, multiplying equation (1) by 4 and then subtracting this new equation from equation (2) will produce an equation with only x and y involved.

Multiplying equation (1) by 4 gives: $4x + 4y + 4z = 16$ (4)

Equation (2) – equation (4) gives: $-2x - 7y = 17$ (5)

Similarly, multiplying equation (3) by 2 and then adding this new equation to equation (2) will produce another equation with only x and y involved.

Multiplying equation (3) by 2 gives: $6x - 4y - 4z = 4$ (6)

Equation (2) + equation (6) gives: $8x - 7y = 37$ (7)

Rewriting equation (5) gives: $-2x - 7y = 17$ (5)

Now we can use the previous method for solving simultaneous equations in two unknowns.

Equation (7) – equation (5) gives: $10x = 20$

from which,   **$x = 2$**

(note that $8x - -2x = 8x + 2x = 10x$)

Substituting $x = 2$ into equation (5) gives: $-4 - 7y = 17$

from which, $-7y = 17 + 4 = 21$

and   **$y = -3$**

Substituting $x = 2$ and $y = -3$ into equation (1) gives:

$$2 - 3 + z = 4$$

from which,   **$z = 5$**

Hence, the solution of the simultaneous equations is: $x = 2$, $y = -3$ and $z = 5$

Chapter 14   Solving quadratic equations by factorising

Introduction

A **quadratic equation** is one in which the highest power of the unknown quantity is 2. For example, $x^2 - 3x + 1 = 0$ is a quadratic equation.

Factorisation

Multiplying out $(x + 1)(x - 3)$ gives $x^2 - 3x + x - 3$ i.e. $x^2 - 2x - 3$
The reverse process of moving from $x^2 - 2x - 3$ to $(x + 1)(x - 3)$ is called **factorising**.

For example, if $x^2 - 2x - 3 = 0$, then, by factorising:

$$(x + 1)(x - 3) = 0$$

Hence either $(x + 1) = 0$ i.e. $x = -1$

or $(x - 3) = 0$ i.e. $x = 3$

Hence, $x = -1$ and $x = 3$ are the **roots** of the quadratic equation $x^2 - 2x - 3 = 0$

The technique of factorising is often one of 'trial and error'.

Application: Solve the equation $x^2 + x - 6 = 0$ by factorisation.

The factors of x^2 are: x and x. These are placed in brackets: $(x\quad)(x\quad)$

The factors of -6 are: $+6$ and -1, or -6 and $+1$, or $+3$ and -2, or -3 and $+2$.

The only combination to give a middle term of $+x$ is $+3$ and -2

i.e. $x^2 + x - 6 = (x + 3)(x - 2)$

The quadratic equation, $x^2 + x - 6 = 0$ thus becomes $(x + 3)(x - 2) = 0$

Since the only way that this can be true is for either the first or the second, or both factors, to be zero, then either $(x + 3) = 0$ i.e. $x = -3$

or $(x - 2) = 0$ i.e. $x = 2$

Hence, the roots of $x^2 + x - 6 = 0$ are $x = -3$ and $x = 2$

Application: Solve the equation $x^2 + 2x - 8 = 0$ by factorisation

The factors of x^2 are: x and x. These are placed in brackets thus: $(x\quad)(x\quad)$

The factors of -8 are: $+8$ and -1, or -8 and $+1$, or $+4$ and -2, or -4 and $+2$.

The only combination to give a middle term of $+2x$ is $+4$ and -2,

i.e. $x^2 + 2x - 8 = (x + 4)(x - 2)$

(Note that the product of the two inner terms, $4x$, added to the product of the two outer terms, $-2x$, must equal the middle term, $+2x$ in this case.)

The quadratic equation $x^2 + 2x - 8 = 0$ thus becomes $(x + 4)(x - 2) = 0$

Since the only way that this can be true is for either the first or the second, or both factors to be zero, then either $(x + 4) = 0$ i.e. $x = -4$

or $(x - 2) = 0$ i.e. $x = 2$

Hence, the roots of $x^2 + 2x - 8 = 0$ are $x = -4$ and $x = 2$

Application: Determine the roots of $x^2 - 8x + 16 = 0$ by factorisation.

$$x^2 - 8x + 16 = (x - 4)(x - 4) \quad \text{i.e.} \quad (x - 4)^2 = 0$$

The left-hand side is known as a **perfect square**.

Hence, **x = 4** is the only root of the equation $x^2 - x + 16 = 0$

Application: Solve the equation: $x^2 - 5x = 0$

Factorising gives: $\qquad\qquad x(x - 5) = 0$

If $\qquad x(x - 5) = 0 \qquad$ then either $x = 0$ or $x - 5 = 0$

i.e. $\qquad\qquad\qquad\qquad$ **x = 0** or **x = 5**

These are the two roots of the given equation. Answers can always be checked by substitution into the original equation.

Application: Determine the roots of $4x^2 - 25 = 0$ by factorisation.

The left-hand side of $4x^2 - 25 = 0$ is the difference of two squares, $(2x)^2$ and $(5)^2$

By factorising, $4x^2 - 25 = (2x + 5)(2x - 5)$ i.e. $(2x + 5)(2x - 5) = 0$

Hence, either $(2x + 5) = 0$ i.e. $x = -\dfrac{5}{2} = -2.5$

or $\qquad (2x - 5) = 0$ i.e. $x = \dfrac{5}{2} = 2.5$

Application: Solve the equation $3x^2 - 11x - 4 = 0$ by factorisation.

The factors of $3x^2$ are: $3x$ and x. These are placed in brackets: $(3x \quad)(x \quad)$

The factors of -4 are: -4 and $+1$, or $+4$ and -1, or -2 and 2

Remembering that the product of the two inner terms added to the product of the two outer terms must equal $-11x$, the only combination to give this is $+1$ and -4

i.e. $\qquad\qquad\qquad 3x^2 - 11x - 4 = (3x + 1)(x - 4)$

The quadratic equation $3x^2 - 11x - 4 = 0$ thus becomes $(3x + 1)(x - 4) = 0$

Hence, either $\qquad (3x + 1) = 0 \qquad$ i.e. $x = -\dfrac{1}{3}$

or $\qquad\qquad (x - 4) = 0 \qquad$ i.e. $x = 4$

and both solutions may be checked in the original equation.

Application: Solve the quadratic equation $15x^2 + 2x - 8 = 0$ by factorising

The factors of $15x^2$ are: $15x$ and $x \quad$ or $\quad 5x$ and $3x$.

The factors of – 8 are: – 4 are +2, or 4 and – 2, or – 8 and +1, or 8 and – 1.

By trial and error the only combination that works is

$$15x^2 + 2x - 8 = (5x + 4)(3x - 2)$$

Hence $(5x + 4)(3x - 2) = 0$ from which

either $5x + 4 = 0$

or $3x - 2 = 0$

Hence, $x = -\dfrac{4}{5}$ or $x = \dfrac{2}{3}$

which may be checked in the original equation.

Chapter 15   Solving quadratic equations by completing the square

An expression such as x^2 or $(x + 2)^2$ or $(x - 3)^2$ is called a **perfect square**.

If $x^2 = 3$ then $x = \pm\sqrt{3}$

If $(x + 2)^2 = 5$ then $x + 2 = \pm\sqrt{5}$ and $x = -2 \pm \sqrt{5}$

If $(x - 3)^2 = 8$ then $x - 3 = \pm\sqrt{8}$ and $x = 3 \pm \sqrt{8}$

Hence, if a quadratic equation can be rearranged so that one side of the equation is a perfect square and the other side of the equation is a number, then the solution of the equation is readily obtained by taking the square root of each side as in the above examples. The process of rearranging one side of a quadratic equation into a perfect square before solving is called 'completing the square'.

$$(x + a)^2 = x^2 + 2ax + a^2$$

Thus in order to make the quadratic expression $x^2 + 2ax$ into a perfect square it is necessary to add (half the coefficient of $x)^2$ i.e. $\left(\dfrac{2a}{2}\right)^2$ or a^2

For example, $x^2 + 3x$ becomes a perfect square by adding $\left(\dfrac{3}{2}\right)^2$, i.e.

$$x^2 + 3x + \left(\dfrac{3}{2}\right)^2 = \left(x + \dfrac{3}{2}\right)^2$$

The method of 'completing the square' is demonstrated in the following Applications.

Application: Solve $2x^2 + 5x = 3$ by 'completing the square'

The procedure is as follows:

1. Rearrange the equation so that all terms are on the same side of the equals sign (and the coefficient of the x^2 term is positive).

   Hence, $2x^2 + 5x - 3 = 0$

2. Make the coefficient of the x^2 term unity. In this case this is achieved by dividing throughout by 2.

   Hence, $\dfrac{2x^2}{2} + \dfrac{5x}{2} - \dfrac{3}{2} = 0$

   i.e. $x^2 + \dfrac{5}{2}x - \dfrac{3}{2} = 0$

3. Rearrange the equations so that the x^2 and x terms are on one side of the equals sign and the constant is on the other side. Hence,

   $$x^2 + \frac{5}{2}x = \frac{3}{2}$$

4. Add to both sides of the equation (half the coefficient of x)^2. In this case the coefficient of x is $\dfrac{5}{2}$.

   Half the coefficient squared is therefore $\left(\dfrac{5}{4}\right)^2$

   Thus, $x^2 + \dfrac{5}{2}x + \left(\dfrac{5}{4}\right)^2 = \dfrac{3}{2} + \left(\dfrac{5}{4}\right)^2$

   The LHS is now a perfect square, i.e.

   $$\left(x + \frac{5}{4}\right)^2 = \frac{3}{2} + \left(\frac{5}{4}\right)^2$$

5. Evaluate the RHS. Thus

   $$\left(x + \frac{5}{4}\right)^2 = \frac{3}{2} + \frac{25}{16} = \frac{24 + 25}{16} = \frac{49}{16}$$

6. Take the square root of both sides of the equation (remembering that the square root of a number gives a ± answer). Thus

   $$\sqrt{\left(x + \frac{5}{4}\right)^2} = \sqrt{\left(\frac{49}{16}\right)}$$

   i.e. $x + \dfrac{5}{4} = \pm\dfrac{7}{4}$

7. Solve the simple equation, i.e.

   $$x = -\frac{5}{4} \pm \frac{7}{4}$$

   i.e. $x = -\dfrac{5}{4} + \dfrac{7}{4} = \dfrac{2}{4} = \dfrac{1}{2}$ or 0.5

and $$x = -\frac{5}{4} - \frac{7}{4} = -\frac{12}{4} = -3$$

Hence, **x = 0.5** or **x = – 3** i.e. **the roots of the equation $2x^2 + 5x = 3$ are 0.5 and – 3**

Application: Solve $2x^2 + 9x + 8 = 0$, correct to 3 significant figures, by 'completing the square'

Making the coefficient of x^2 unity gives: $x^2 + \dfrac{9}{2}x + 4 = 0$

and rearranging gives: $\qquad\qquad\qquad x^2 + \dfrac{9}{2}x = -4$

Adding to both sides (half the coefficient of x)^2 gives:

$$x^2 + \frac{9}{2}x + \left(\frac{9}{4}\right)^2 = \left(\frac{9}{4}\right)^2 - 4$$

The LHS is now a perfect square, thus

$$\left(x + \frac{9}{4}\right)^2 = \frac{81}{16} - 4 = \frac{81}{16} - \frac{64}{16} = \frac{17}{16}$$

Taking the square root of both sides gives:

$$x + \frac{9}{4} = \sqrt{\left(\frac{17}{16}\right)} = \pm 1.031$$

Hence, $\qquad\qquad\qquad\qquad x = -\dfrac{9}{4} \pm 1.031$

i.e. $\qquad\qquad$ **x = – 1.22** or **– 3.28**, correct to 3 significant figures

Chapter 16   Solution of quadratic equations by formula

If $ax^2 + bx + c = 0$ then $x = \dfrac{-b \pm \sqrt{b^2 - 4ac}}{2a}$

This is known as the quadratic formula.

Application: Solve $3x^2 - 11x - 4 = 0$ by using the quadratic formula

Comparing $3x^2 - 11x - 4 = 0$ with $ax^2 + bx + c = 0$ gives $a = 3$, $b = -11$ and $c = -4$

Hence, $x = \dfrac{-(-11) \pm \sqrt{(-11)^2 - 4(3)(-4)}}{2(3)} = \dfrac{+11 \pm \sqrt{121 + 48}}{6}$

$= \dfrac{11 \pm \sqrt{169}}{6} = \dfrac{11 \pm 13}{6}$

$= \dfrac{11 + 13}{6}$ or $\dfrac{11 - 13}{6}$

Hence, $x = \dfrac{24}{6} = 4$ or $\dfrac{-2}{6} = -\dfrac{1}{3}$

Application: Solve $4x^2 + 7x + 2 = 0$ giving the roots correct to 2 decimal places

Comparing $4x^2 + 7x + 2 = 0$ with $ax^2 + bx + c$ gives $a = 4$, $b = 7$ and $c = 2$

Hence, $x = \dfrac{-7 \pm \sqrt{7^2 - 4(4)(2)}}{2(4)} = \dfrac{-7 \pm \sqrt{17}}{8}$

$= \dfrac{-7 \pm 4.123}{8} = \dfrac{-7 + 4.123}{8}$ or $\dfrac{-7 - 4.123}{8}$

Hence, **$x = -0.36$** or **-1.39, correct to 2 decimal places.**

Application: The height s metres of a mass projected vertically upwards at time t seconds is $s = ut - \frac{1}{2}gt^2$. Determine how long the mass will take after being projected to reach a height of 16 m (a) on the ascent and (b) on the descent, when $u = 30$ m/s and $g = 9.81$ m/s^2

When height $s = 16$ m, $16 = 30t - \dfrac{1}{2}(9.81)t^2$

i.e. $4.905t^2 - 30t + 16 = 0$

Using the quadratic formula:

$$t = \dfrac{-(-30) \pm \sqrt{(-30)^2 - 4(4.905)(16)}}{2(4.905)}$$

$$= \dfrac{30 \pm \sqrt{586.1}}{9.81} = \dfrac{30 \pm 24.21}{9.81} = 5.53 \text{ or } 0.59$$

Hence the mass will reach a height of 16 m after 0.59 s on the ascent and after 5.53 s on the descent.

Application: A shed is 4.0 m long and 2.0 m wide. A concrete path of constant width is laid all the way around the shed and the area of the path is 9.50 m². Calculate its width, to the nearest centimetre

Figure 16.1 shows a plan view of the shed with its surrounding path of width t metres

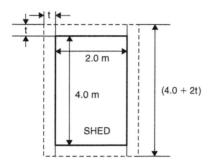

Figure 16.1

Area of path $= 2(2.0 \times t) + 2t(4.0 + 2t)$

i.e. $9.50 = 4.0t + 8.0t + 4t^2$

or $4t^2 + 12.0t - 9.50 = 0$

Hence $t = \dfrac{-(12.0) \pm \sqrt{(12.0)^2 - 4(4)(-9.50)}}{2(4)}$

$= \dfrac{-12.0 \pm \sqrt{296.0}}{8} = \dfrac{-12.0 \pm 17.20465}{8}$

Hence, $t = 0.6506$ m or -3.65058 m

Neglecting the negative result which is meaningless, the width of the path, **t = 0.651 m** or **65 cm**, correct to the nearest centimetre.

In this chapter and in the previous two, three methods of solving quadratic equations have been shown. However, **scientific notation electronic calculators** with an equation mode are also able to solve quadratic equations – and much more quickly!

Chapter 17   Logarithms

Introduction to logarithms

The theory of logarithms is important, for there are several scientific and engineering laws that involve the rules of logarithms.

Definition of a logarithm: If $y = a^x$ then $x = \log_a y$

Logarithms having a base of 10 are called common logarithms and $\log_{10}$ is usually abbreviated to lg. Logarithms having a base of e (where 'e' is a mathematical constant approximately equal to 2.7183) are called hyperbolic, Napierian or natural logarithms, and $\log_e$ is usually abbreviated to ln.

Application: Evaluate $\log_3 9$

Let $x = \log_3 9$ then $3^x = 9$ from the definition of a logarithm,

i.e. $\qquad\qquad 3^x = 3^2$ from which, $x = 2$

Hence, $\qquad\qquad$ **$\log_3 9 = 2$**

Application: Evaluate $\log_{16} 8$

Let $x = \log_{16} 8$ then $16^x = 8$ from the definition of a logarithm,

i.e. $\qquad\qquad (2^4)^x = 2^3$ i.e. $2^{4x} = 2^3$ from the laws of indices,

from which, $\qquad 4x = 3$ and $x = \dfrac{3}{4}$

Hence, $\qquad\qquad$ **$\log_{16} 8 = \dfrac{3}{4}$**

Application: Evaluate lg 0.001

Let $x = \lg 0.001 = \log_{10} 0.001$ then $10^x = 0.001$

i.e. $\qquad\qquad\qquad 10^x = 10^{-3}$ from which, $x = -3$

Hence, $\qquad\qquad$ **lg 0.001 = -3** (which may be checked by a calculator)

Application: Evaluate ln e

Let $x = \ln e = \log_e e$ then $e^x = e$

i.e. $\qquad\qquad\qquad e^x = e^1$ from which, $x = 1$

Hence, $\qquad\qquad$ **ln e = 1** (which may be checked by a calculator)

Application: Evaluate $\log_3 \dfrac{1}{81}$

Let $x = \log_3 \dfrac{1}{81}$ then $3^x = \dfrac{1}{81} = \dfrac{1}{3^4} = 3^{-4}$ from which, $x = -4$

Hence, $\qquad\qquad$ **$\log_3 \dfrac{1}{81} = -4$**

Application: Solve the equation: $\lg x = 4$

If $\lg x = 4$ then $\log_{10} x = 4$

and $\qquad\qquad x = 10^4$ i.e. **$x = 10{,}000$**

Application: Solve the equation: $\log_5 x = -2$

If $\log_5 x = -2$ then $x = 5^{-2} = \dfrac{1}{5^2} = \dfrac{1}{25}$

Laws of logarithms

There are three laws of logarithms, which apply to any base:

(i) $\qquad$ **$\log (A \times B) = \log A + \log B$**

(ii) $\qquad$ **$\log \left(\dfrac{A}{B}\right) = \log A - \log B$**

(iii) $\qquad$ **$\log A^n = n \log A$**

Application: Write $\log 3 + \log 8$ as the logarithm of a single number

$\log 3 + \log 8 = \log (3 \times 8)$ by the first law of logarithms

$\qquad\qquad = \textbf{log 24}$

Application: Write $\log 15 - \log 3$ as the logarithm of a single number

$\log 15 - \log 3 = \log\left(\dfrac{15}{3}\right)$ by the second law of logarithms

$\qquad\qquad = \textbf{log 5}$

Application: Write $2 \log 5$ as the logarithm of a single number

$2 \log 5 = \log 5^2$ by the third law of logarithms

$\qquad\qquad = \textbf{log 25}$

Application: Solve the equation: $\log(x - 1) + \log(x + 8) = 2 \log(x + 2)$

LHS $= \log(x - 1) + \log(x + 8) = \log(x - 1)(x + 8)$ from the first law of logarithms

$\qquad\qquad = \log(x^2 + 7x - 8)$

RHS $= 2 \log(x + 2) = \log(x + 2)^2$ from the third law of logarithms

$\qquad\qquad = \log(x^2 + 4x + 4)$

Hence, $\log(x^2 + 7x - 8) = \log(x^2 + 4x + 4)$

from which, $x^2 + 7x - 8 = x^2 + 4x + 4$

i.e. $7x - 8 = 4x + 4$

i.e. $3x = 12$

and $\mathbf{x = 4}$

Indicial equations

The laws of logarithms may be used to solve certain equations involving powers – called **indicial equations**.

Application: Solve $3^x = 27$

Logarithms to a base of 10 are taken of both sides,

i.e. $\log_{10} 3^x = \log_{10} 27$

and $x \log_{10} 3 = \log_{10} 27$ by the third law of logarithms

Rearranging gives: $x = \dfrac{\log_{10} 27}{\log_{10} 3} = \dfrac{1.43136...}{0.47712...} = \mathbf{3}$ which may be readily checked.

(Note, $\dfrac{\log 27}{\log 3}$ is **not** equal to $\log \dfrac{27}{3}$)

Application: Solve the equation: $2^x = 5$, correct to 4 significant figures.

Taking logarithms to base 10 of both sides of $2^x = 5$ gives:

$$\log_{10} 2^x = \log_{10} 5$$

i.e. $x \log_{10} 2 = \log_{10} 5$ by the third law of logarithms

Rearranging gives: $x = \dfrac{\log_{10} 5}{\log_{10} 2} = \dfrac{0.6989700..}{0.3010299..} = \mathbf{2.322}$, correct to 4 significant figures.

Application: Solve the equation: $x^{2.7} = 34.68$, correct to 4 significant figures.

Taking logarithms to base 10 of both sides gives:

$$\log_{10} x^{2.7} = \log_{10} 34.68$$

$$2.7 \log_{10} x = \log_{10} 34.68$$

Hence, $\log_{10} x = \dfrac{\log_{10} 34.68}{2.7} = 0.57040$

Thus, $x = $ antilog $0.57040 = 10^{0.57040} = \mathbf{3.719}$, correct to 4 significant figures.

Graphs of logarithmic functions

A graph of $y = \log_{10} x$ is shown in Figure 17.1 and a graph of $y = \log_e x$ is shown in Figure 17.2. Both are seen to be of similar shape; in fact, the same general shape occurs for a logarithm to any base.

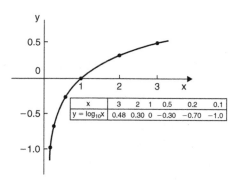

x	3	2	1	0.5	0.2	0.1
$y = \log_{10}x$	0.48	0.30	0	−0.30	−0.70	−1.0

Figure 17.1

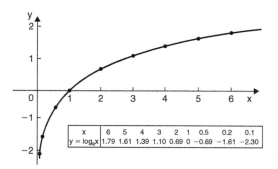

x	6	5	4	3	2	1	0.5	0.2	0.1
$y = \log_e x$	1.79	1.61	1.39	1.10	0.69	0	−0.69	−1.61	−2.30

Figure 17.2

In general, with a logarithm to any base a, it is noted that:
1. $\log_a 1 = 0$
2. $\log_a a = 1$
3. $\log_a 0 \rightarrow -\infty$

Chapter 18 Exponential functions

Introduction to exponential functions

An exponential function is one which contains e^x, e being a constant called the **exponent** and having an approximate value of 2.7183. The exponent arises from the **natural laws of growth and decay** and is used as a base for natural or Napierian logarithms.

Application: The instantaneous voltage v in a capacitive circuit is related to time t by the equation: $v = Ve^{-\frac{t}{CR}}$ where V, C and R are constants. Determine v, correct to 4 significant figures, when t = 50 ms, C = 10 μF, R = 47 kΩ and V = 300 volts.

$$v = Ve^{-\frac{t}{CR}} = 300e^{-\frac{50 \times 10^{-3}}{10 \times 10^{-6} \times 47 \times 10^{3}}}$$

Using a calculator, $v = 300e^{-0.1063829...} = 300(0.89908025...) = $ **269.7 volts**

The **power series for e^x** is:

$$e^x = 1 + x + \frac{x^2}{2!} + \frac{x^3}{3!} + \frac{x^4}{4!} + \cdots$$

(where 3! = 3 × 2×1 and is called 'factorial 3')
The series is valid for all values of x.

Graphs of exponential functions

Figure 18.1 shows graphs of $y = e^x$ and $y = e^{-x}$

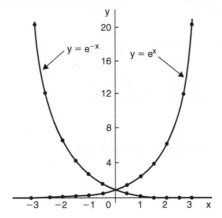

Figure 18.1

Application: The decay of voltage, v volts, across a capacitor at time t seconds is given by $v = 250e^{-t/3}$. Draw a graph showing the natural decay curve over the first 6 seconds. Determine (a) the voltage after 3.4 s, and (b) the time when the voltage is 150 volts

A table of values is drawn up as shown below.

t	0	1	2	3	4	5	6
$e^{-t/3}$	1.00	0.7165	0.5134	0.3679	0.2636	0.1889	0.1353
$v = 250e^{-t/3}$	250.0	179.1	128.4	91.97	65.90	47.22	33.83

The natural decay curve of $v = 250e^{-t/3}$ is shown in Figure 18.2.

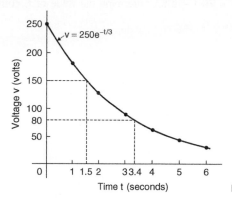

Figure 18.2

From the graph,

(a) when time t = 3.4 s, **voltage v = 80 volts**

(b) when voltage v = 150 volts, **time t = 1.5 seconds**

Chapter 19 Napierian logarithms

Introduction to Napierian logarithms

Logarithms having a base of 'e' are called **hyperbolic, Napierian** or **natural logarithms** and the Napierian logarithm of x is written as $\log_e x$, or more commonly as ln x. Logarithms were invented by **John Napier**, a Scotsman (1550–1617).

Check using your calculator, ln 1.812 = 0.59443, correct to 5 significant figures

$$\ln 1 = 0, \quad \ln e^3 = 3 \quad \text{and} \quad \ln e^1 = 1$$

From the last two examples we can conclude that: $\log_e e^x = x$

This is useful when **solving equations involving exponential functions**.

Application: Solve $e^{3x} = 8$

Taking Napierian logarithms of both sides, gives

$$\ln e^{3x} = \ln 8$$

ie $$3x = \ln 8$$

from which $$x = \frac{1}{3} \ln 8 = \mathbf{0.6931}, \text{ correct to 4 decimal places}$$

Application: Given $32 = 70(1 - e^{-\frac{t}{2}})$ determine the value of t, correct to 3 significant figures

Rearranging $32 = 70 (1 - e^{-\frac{t}{2}})$ gives: $\dfrac{32}{70} = 1 - e^{-\frac{t}{2}}$

and $e^{-\frac{t}{2}} = 1 - \dfrac{32}{70} = \dfrac{38}{70}$

Taking the reciprocal of both sides gives: $e^{\frac{t}{2}} = \dfrac{70}{38}$

Taking Napierian logarithms of both sides gives: $\ln e^{\frac{t}{2}} = \ln \left(\dfrac{70}{38}\right)$

i.e. $\dfrac{t}{2} = \ln \left(\dfrac{70}{38}\right)$

from which, $t = 2 \ln \left(\dfrac{70}{38}\right) = \textbf{1.22}$, correct to 3 significant figures

Application: The work done in an isothermal expansion of a gas from pressure p_1 to p_2 is given by:

$$w = w_0 \ln \left(\dfrac{p_1}{p_2}\right)$$

If the initial pressure $p_1 = 7.0\,\text{kPa}$, calculate the final pressure p_2 if $w = 3w_0$

If $w = 3w_0$ then $3\,w_0 = w_0 \ln \left(\dfrac{p_1}{p_2}\right)$

i.e. $3 = \ln \left(\dfrac{p_1}{p_2}\right)$

and $e^3 = \dfrac{p_1}{p_2} = \dfrac{7000}{p_2}$

from which,   **final pressure**, $p_2 = \dfrac{7000}{e^3} = 7000 e^{-3} = \textbf{348.5 Pa}$

Laws of growth and decay

The laws of exponential growth and decay are of the form $y = Ae^{-kx}$ and $y = A(1 - e^{-kx})$, where A and k are constants. When plotted, the form of each of these equations is as shown in **Figure 19.1**. The laws occur frequently in engineering and science and examples of quantities related by a natural law include

(i) Linear expansion $l = l_0\, e^{\alpha\theta}$

(ii) Change in electrical resistance with temperature $R_\theta = R_0\, e^{\alpha\theta}$

(iii) Tension in belts $T_1 = T_0\, e^{\mu\theta}$

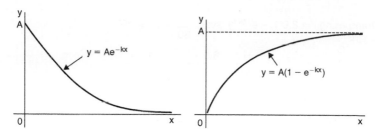

Figure 19.1

(iv) Newton's law of cooling $\qquad$ $\theta = \theta_0 \, e^{-kt}$

(v) Biological growth $\qquad$ $y = y_0 \, e^{kt}$

(vi) Discharge of a capacitor $\qquad$ $q = Q \, e^{-t/CR}$

(vii) Atmospheric pressure $\qquad$ $p = p_0 \, e^{-h/c}$

(viii) Radioactive decay $\qquad$ $N = N_0 \, e^{-\lambda t}$

(ix) Decay of current in an inductive circuit $\qquad$ $i = I \, e^{-Rt/L}$

(x) Growth of current in a capacitive circuit $\qquad$ $i = I(1 - e^{-t/CR})$

Application: In an experiment involving Newton's law of cooling, the temperature $\theta(°C)$ is given by $\theta = \theta_0 e^{-kt}$. Find the value of constant k when $\theta_0 = 56.6°C$, $\theta = 16.5°C$ and $t = 83.0$ seconds

Transposing $\theta = \theta_0 \, e^{-kt}$ gives $\dfrac{\theta}{\theta_0} = e^{-kt}$ from which,

$$\frac{\theta_0}{\theta} = \frac{1}{e^{-kt}} = e^{\,kt}$$

Taking Napierian logarithms of both sides gives: $\ln \dfrac{\theta_0}{\theta} = kt$

from which, $\mathbf{k} = \dfrac{1}{t} \ln \dfrac{\theta_0}{\theta} = \dfrac{1}{83.0} \ln\left(\dfrac{56.6}{16.5}\right) = \dfrac{1}{83.0}(1.2326486 \,..)$

$$= 1.485 \times 10^{-2}$$

Application: The current i amperes flowing in a capacitor at time t seconds is given by $i = 8.0(1 - e^{-t/CR})$, where the circuit resistance R is $25\,k\Omega$ and capacitance C is $16\,\mu F$. Determine (a) the current i after 0.5 seconds and (b) the time, to the nearest ms, for the current to reach 6.0 A

(a) Current $i = 8.0(1 - e^{-t/CR}) = 8.0\,[1 - e^{-0.5/(16\times10^{-6})(25\times10^{3})}]$

$$= 8.0(1 - e^{-1.25})$$

$$= 8.0(1 - 0.2865047\,..) = 8.0(0.7134952\,..)$$

$$= 5.71 \text{ amperes}$$

(b) Transposing $i = 8.0(1 - e^{-t/CR})$ gives: $\dfrac{i}{80} = 1 - e^{-t/CR}$

from which, $e^{-t/CR} = 1 - \dfrac{i}{8.0} = \dfrac{8.0 - i}{8.0}$

Taking the reciprocal of both sides gives: $e^{t/CR} = \dfrac{8.0}{8.0 - i}$

Taking Napierian logarithms of both sides gives:

$$\frac{t}{CR} = \ln\left(\frac{8.0}{8.0 - i}\right)$$

Hence $\quad t = CR \ln\left(\dfrac{8.0}{8.0 - i}\right)$

$$= (16 \times 10^{-6})(25 \times 10^3) \ln\left(\frac{8.0}{8.0 - 6.0}\right) \text{ when } i = 6.0 \text{ amperes,}$$

i.e. $\qquad t = 0.40 \ln\left(\dfrac{8.0}{2.0}\right) = 0.4 \ln 4.0$

$$= 0.4(1.3862943..)$$

$$= 0.5545 \text{ s}$$

$$= 555 \text{ ms, to the nearest millisecond.}$$

A graph of current against time is shown in Figure 19.2.

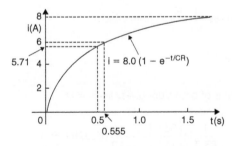

Figure 19.2

Chapter 20   Hyperbolic functions

$\sinh x = \dfrac{e^x - e^{-x}}{2}$

$\operatorname{cosech} x = \dfrac{1}{\sinh x} = \dfrac{2}{e^x - e^{-x}}$

$\cosh x = \dfrac{e^x + e^{-x}}{2}$

$\operatorname{sech} x = \dfrac{1}{\cosh x} = \dfrac{2}{e^x + e^{-x}}$

$\tanh x = \dfrac{\sinh x}{\cosh x} = \dfrac{e^x - e^{-x}}{e^x + e^{-x}}$

$\coth x = \dfrac{1}{\tanh x} = \dfrac{e^x + e^{-x}}{e^x - e^{-x}}$

$$\cosh x = 1 + \frac{x^2}{2!} + \frac{x^4}{4!} + \cdots \quad \text{(which is valid for all values of x)}$$

$$\sinh x = x + \frac{x^3}{3!} + \frac{x^5}{5!} + \cdots \quad \text{(which is valid for all values of x)}$$

Graphs of hyperbolic functions

A graph of **y = sinh x** is shown in Figure 20.1. Since the graph is symmetrical about the origin, sinh x is an **odd function**.

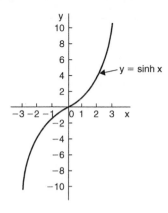

Figure 20.1

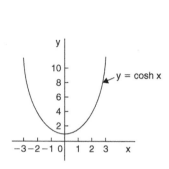

Figure 20.2

A graph of **y = cosh x** is shown in Figure 20.2. Since the graph is symmetrical about the y-axis, cosh x is an **even function**. The shape of y = cosh x is that of a heavy rope or chain hanging freely under gravity and is called a **catenary**. Examples include **transmission lines**, a **telegraph wire** or a **fisherman's line**, and are used in the **design of roofs and arches**. Graphs of y = tanh x, y = coth x, y = cosech x and y = sech x are shown in Figures 20.3 and 20.4.

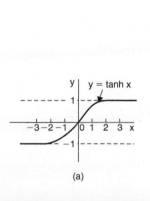

(a)

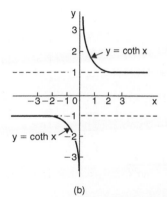

(b)

Figure 20.3

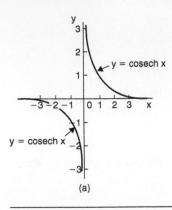

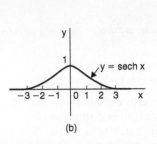

(a)   (b)   Figure 20.4

Hyperbolic identities

Trigonometric identity	Corresponding hyperbolic identity
$\cos^2 x + \sin^2 x = 1$	$ch^2 x - sh^2 x = 1$
$1 + \tan^2 x = \sec^2 x$	$1 - th^2 x = sech^2 x$
$\cot^2 x + 1 = \csc^2 x$	$coth^2 x - 1 = cosech^2 x$

Compound angle formulae	
$\sin(A \pm B) = $ $\sin A \cos B \pm \cos A \sin B$	$sh(A \pm B) = $ $sh\,A\,ch\,B \pm ch\,A\,sh\,B$
$\cos(A \pm B) = $ $\cos A \cos B \mp \sin A \sin B$	$ch(A \pm B) = $ $ch\,A\,ch\,B \pm sh\,A\,sh\,B$
$\tan(A \pm B) = \dfrac{\tan A \pm \tan B}{1 \mp \tan A \tan B}$	$\tan(A \pm B) = \dfrac{th\,A \pm th\,B}{1 \pm th\,A\,th\,B}$

Double angles	
$\sin 2x = 2 \sin x \cos x$	$sh\,2x = 2\,sh\,x\,ch\,x$
$\cos 2x = \cos^2 x - \sin^2 x$	$ch\,2x = ch^2 x + sh^2 x$
$= 2 \cos^2 x - 1$	$= 2\,ch^2 x - 1$
$= 1 - 2 \sin^2 x$	$= 1 + 2\,sh^2 x$
$\tan 2x = \dfrac{2 \tan x}{1 - \tan^2 x}$	$th\,2x = \dfrac{2\,th\,x}{1 + th^2 x}$

Solving equations involving hyperbolic functions

Equations of the form **a ch x + b sh x = c**, where a, b and c are constants may be solved either by:

(a) plotting graphs of $y = a \text{ ch } x + b \text{ sh } x$ and $y = c$ and noting the points of intersection, or more accurately,

(b) by adopting the following procedure:

1. Change sh x to $\left(\dfrac{e^x - e^{-x}}{2} \right)$ and ch x to $\left(\dfrac{e^x + e^{-x}}{2} \right)$

2. Rearrange the equation into the form $pe^x + qe^{-x} + r = 0$, where p, q and r are constants.

3. Multiply each term by e^x, which produces an equation of the form $p(e^x)^2 + re^x + q = 0$ (since $(e^{-x})(e^x) = e^0 = 1$)

4. Solve the quadratic equation $p(e^x)^2 + re^x + q = 0$ for e^x by factorising or by using the quadratic formula.

5. Given $e^x = $ a constant (obtained by solving the equation in 4), take Napierian logarithms of both sides to give x = ln(constant)

Application: Solve the equation sh x = 3, correct to 4 significant figures

Following the above procedure:

1. sh x $= \left(\dfrac{e^x - e^{-x}}{2} \right) = 3$

2. $e^x - e^{-x} = 6$, i.e. $e^x - e^{-x} - 6 = 0$

3. $(e^x)^2 - (e^{-x})(e^x) - 6e^x = 0$, i.e. $(e^x)^2 - 6e^x - 1 = 0$

4. $e^x = \dfrac{-(-6) \pm \sqrt{[(-6)^2 - 4(1)(-1)]}}{2(1)} = \dfrac{6 \pm \sqrt{40}}{2} = \dfrac{6 \pm 6.3246}{2}$

    Hence, $e^x = 6.1623$ or -0.1623

5. x = ln 6.1623 or x = ln(−0.1623) which has no solution since it is not possible in real terms to find the logarithm of a negative number.

    Hence x = ln 6.1623 = **1.818**, correct to 4 significant figures.

The above solution may be obtained much quicker with a calculator. Using a calculator:

(i) Press hyp   (ii) Choose 4, which is $\sinh^{-1}$ (iii) Type in 3   (iv) Close bracket

(v) Press = and the answer is 1.818448459

i.e. the solution of sh x = 3 is: **x = 1.818**, correct to 4 significant figures, as above.

Application: A chain hangs in the form given by $y = 40 \, \text{ch} \frac{x}{40}$. Determine, correct to 4 significant figures, (a) the value of y when x is 25 and (b) the value of x when y = 54.30

(a) $y = 40 \, \text{ch} \frac{x}{40}$ and when x = 25,

$$y = 40 \, \text{ch} \frac{25}{40} = 40 \, \text{ch} \, 0.625 = 40 \left(\frac{e^{0.625} + e^{-0.625}}{2} \right)$$

$$= 20(1.8682 + 0.5353) = 48.07$$

(b) When $y = 54.30$, $54.30 = 40 \, \text{ch} \frac{x}{40}$, from which

$$\text{ch} \frac{x}{40} = \frac{54.30}{40} = 1.3575$$

Following the above procedure:

1. $\dfrac{e^{x/40} + e^{-x/40}}{2} = 1.3575$

2. $e^{x/40} + e^{-x/40} = 2.715$ i.e. $e^{x/40} + e^{-x/40} - 2.715 = 0$

3. $(e^{x/40})^2 + 1 - 2.715 \, e^{x/40} = 0$ i.e. $(e^{x/40})^2 - 2.715 \, e^{x/40} + 1 = 0$

4. $e^{x/40} = \dfrac{-(-2.715) \pm \sqrt{[(-2.715)^2 - 4(1)(1)]}}{2(1)}$

$$= \frac{2.715 \pm \sqrt{(3.3712)}}{2} = \frac{2.715 \pm 1.8361}{2}$$

Hence $e^{x/40} = 2.2756$ or 0.43945

5. $\dfrac{x}{40} = \ln 2.2756$ or $\dfrac{x}{40} = \ln(0.43945)$

Hence, $\dfrac{x}{40} = 0.8222$ or $\dfrac{x}{40} = -0.8222$

Hence, x = 40(0.8222) or x = 40(−0.8222)

i.e. **x = ±32.89**, correct to 4 significant figures.

Chapter 21   Partial fractions

Provided that the numerator $f(x)$ is of less degree than the relevant denominator, the following identities are typical examples of the form of partial fraction used:

Linear factors
$$\frac{f(x)}{(x + a)(x - b)(x + c)} = \frac{A}{(x + a)} + \frac{B}{(x - b)} + \frac{C}{(x + c)}$$

Repeated linear factors
$$\frac{f(x)}{(x + a)^3} \equiv \frac{A}{(x + a)} + \frac{B}{(x + a)^2} + \frac{C}{(x + a)^3}$$

Quadratic factors
$$\frac{f(x)}{(ax^2 + bx + c)(x + d)} \equiv \frac{Ax + B}{(ax^2 + bx + c)} + \frac{C}{(x + d)}$$

Application: Resolve $\dfrac{11 - 3x}{x^2 + 2x - 3}$ into partial fractions

The denominator factorises as $(x - 1)(x + 3)$ and the numerator is of less degree than the denominator.

Thus $\dfrac{11 - 3x}{x^2 + 2x - 3}$ may be resolved into partial fractions.

Let $\dfrac{11 - 3x}{x^2 + 2x - 3} = \dfrac{11 - 3x}{(x - 1)(x + 3)} \equiv \dfrac{A}{(x - 1)} + \dfrac{B}{(x + 3)}$ where A and B are constants to be determined,

i.e. $\dfrac{11 - 3x}{(x - 1)(x + 3)} \equiv \dfrac{A(x + 3) + B(x - 1)}{(x - 1)(x + 3)}$ by algebraic addition

Since the denominators are the same on each side of the identity then the numerators are equal to each other.

Thus, $\qquad 11 - 3x = A(x + 3) + B(x - 1)$

To determine constants A and B, values of x are chosen to make the term in A or B equal to zero.

When $x = 1$, then $11 - 3(1) = A(1 + 3) + B(0)$

i.e. $\qquad\qquad 8 = 4A$
i.e. $\qquad\qquad A = 2$

When $x = -3$, then $11 - 3(-3) = A(0) + B(-3 - 1)$

i.e. $20 = -4B$

i.e. $B = -5$

Thus $\dfrac{11 - 3x}{x^2 + 2x - 3} = \dfrac{2}{(x - 1)} + \dfrac{-5}{(x + 3)} = \dfrac{2}{(x - 1)} - \dfrac{5}{(x + 3)}$

$\left[\text{Check:} \quad \dfrac{2}{(x - 1)} - \dfrac{5}{(x + 3)} = \dfrac{2(x + 3) - 5(x - 1)}{(x - 1)(x + 3)} = \dfrac{11 - 3x}{x^2 + 2x - 3} \right]$

Application: Express $\dfrac{x^3 - 2x^2 - 4x - 4}{x^2 + x - 2}$ in partial fractions

The numerator is of higher degree than the denominator. Thus dividing out gives:

$$
\begin{array}{r}
x - 3 \\
x^2 + x - 2 \overline{\smash{\big)}\ x^3 - 2x^2 - 4x - 4} \\
\underline{x^3 + x^2 - 2x} \\
-3x^2 - 2x - 4 \\
\underline{-3x^2 - 3x + 6} \\
x - 10
\end{array}
$$

Thus $\dfrac{x^3 - 2x^2 - 4x - 4}{x^2 + x - 2} \equiv x - 3 + \dfrac{x - 10}{x^2 + x - 2}$

$$\equiv x - 3 + \dfrac{x - 10}{(x - 2)(x - 1)}$$

Let $\dfrac{x - 10}{(x + 2)(x - 1)} \equiv \dfrac{A}{(x + 2)} + \dfrac{B}{(x - 1)} = \dfrac{A(x - 1) + B(x + 2)}{(x + 2)(x - 1)}$

Equating the numerators gives: $x - 10 = A(x - 1) + B(x + 2)$

Let $x = -2$, then $-12 = -3A$

i.e. $A = 4$

Let $x = 1$, then $-9 = 3B$

i.e. $B = -3$

Hence $\dfrac{x - 10}{(x + 2)(x - 1)} = \dfrac{4}{(x + 2)} - \dfrac{3}{(x - 1)}$

Thus $\dfrac{x^3 - 2x^2 - 4x - 4}{x^2 + x - 2} = x - 3 + \dfrac{4}{(x + 2)} - \dfrac{3}{(x - 1)}$

Application: Express $\dfrac{5x^2 - 2x - 19}{(x + 3)(x - 1)^2}$ as the sum of three partial fractions

The denominator is a combination of a linear factor and a repeated linear factor. Let

$$\frac{5x^2 - 2x - 19}{(x + 3)(x - 1)^2} = \frac{A}{(x + 3)} + \frac{B}{(x - 1)} + \frac{C}{(x - 1)^2}$$

$$= \frac{A(x - 1)^2 + B(x + 3)(x - 1) + C(x + 3)}{(x + 3)(x - 1)^2} \quad \text{by algebraic addition}$$

Equating the numerators gives:

$$5x^2 - 2x - 19 \equiv A(x - 1)^2 + B(x + 3)(x - 1) + C(x + 3) \tag{1}$$

Let $x = -3$, then $5(-3)^2 - 2(-3) - 19 = A(-4)^2 + B(0)(-4) + C(0)$

i.e. $32 = 16A$

i.e. $A = 2$

Let $x = 1$, then $5(1)^2 - 2(1) - 19 = A(0)^2 + B(4)(0) + C(4)$

i.e. $-16 = 4C$

i.e. $C = -4$

Without expanding the RHS of equation (1) it can be seen that equating the coefficients of x^2 gives:

$$5 = A + B, \text{ and since } A = 2, \mathbf{B = 3}$$

Hence $\dfrac{5x^2 - 2x - 19}{(x + 3)(x - 1)^2} \equiv \dfrac{2}{(x + 2)} + \dfrac{3}{(x - 1)} - \dfrac{4}{(x - 1)^2}$

Application: Resolve $\dfrac{3 + 6x + 4x^2 - 2x^3}{x^2(x^2 + 3)}$ into partial fractions

Terms such as x^2 may be treated as $(x + 0)^2$, i.e. they are repeated linear factors.

$(x^2 + 3)$ is a quadratic factor which does not factorise without containing surds and imaginary terms.

Let $\dfrac{3 + 6x + 4x^2 - 2x^3}{x^2(x^2 + 3)} = \dfrac{A}{x} + \dfrac{B}{x^2} + \dfrac{Cx + D}{(x^2 + 3)}$

$$= \frac{Ax(x^2 + 3) + B(x^2 + 3) + (Cx + D)x^2}{x^2(x^2 + 3)}$$

Equating the numerators gives:

$$3 + 6x + 4x^2 - 2x^3 = Ax(x^2 + 3) + B(x^2 + 3) + (Cx + D)x^2$$
$$= Ax^3 + 3Ax + Bx^2 + 3B + Cx^3 + Dx^2$$

Let $x = 0$, then $\qquad\qquad 3 = 3B$

i.e. $\qquad\qquad\qquad\qquad B = 1$

Equating the coefficients of x^3 terms gives: $\qquad -2 = A + C$ $\qquad\qquad$ (1)

Equating the coefficients of x^2 terms gives: $\qquad 4 = B + D$

Since $B = 1$, **D = 3**

Equating the coefficients of x terms gives: $\qquad 6 = 3A$

i.e. $\qquad\qquad\qquad\qquad A = 2$

From equation (1), since $A = 2$, **C = −4**

Hence $\dfrac{3 + 6x + 4x^2 - 2x^3}{x^2(x^2 + 3)} = \dfrac{2}{x} + \dfrac{1}{x^2} + \dfrac{-4x + 3}{x^2 + 3}$

$$= \dfrac{2}{x} + \dfrac{1}{x^2} + \dfrac{3 - 4x}{x^2 + 3}$$

Some number topics

Why are number topics important?

Number sequences are widely used in engineering and scientific applications including computer data structure and sorting algorithms, financial engineering, audio compression, and architectural engineering. Thanks to engineers, robots have migrated from factory shop floors – as industrial manipulators, to outer space – as interplanetary explorers, hospitals – as minimally invasive surgical assistants, homes – as vacuum cleaners and lawn mowers, and battlefields – as unmanned air, underwater, and ground vehicles. Arithmetic progressions are used in simulation engineering and in the reproductive cycle of bacteria. Some uses of A.P.s in daily life include uniform increase in the speed at regular intervals, completing patterns of objects, calculating simple interest, speed of an aircraft, increase or decrease in the costs of goods, sales and production, and so on. Geometric progressions are used in compound interest and the range of speeds on a drilling machine. In fact, G.P.s are used throughout mathematics, and they have many important applications in physics, engineering, biology, economics, computer science, queuing theory, and finance.

There are many, many different types of equations used in every branch of engineering and science. There are straight forwardmethods for solving simple, quadratic and simultaneous equations. However, there are many other types of equations than these three. Great progress has been made in the engineering and scientific disciplines regarding the use of iterative methods for linear systems. In engineering it is important that we can solve any equation; iterative methods help us do that.

There are infinite ways to represent a number. The four commonly associated with modern computers and digital electronics are decimal, binary, octal, and hexadecimal. All four number systems are equally capable of representing any number. Furthermore, a number can be perfectly converted between the various number systems without any loss of numeric value. At a first look, it seems like using any number system other than decimal is complicated and unnecessary. However, since the job of electrical and software engineers is to work with digital circuits, engineers and scientists require number systems that can best transfer information between the human world and the digital circuit world. Thus, the way in which a number is represented can make it easier for the engineer to perceive the meaning of the number as it applies to a digital circuit, i.e. the appropriate number system can actually make things less complicated.

Chapter 22   Simple number sequences

Simple sequences

A set of numbers which are connected by a definite law is called a series or a sequence of numbers.

Each of the numbers in the series is called a term of the series.

For example, 1, 3, 5, 7, .. is a series obtained by adding 2 to the previous term, and 2, 8, 32, 128, .. is a sequence obtained by multiplying the previous term by 4

Application: Determine the next two terms in the series: 3, 6, 9, 12, ...

We notice that the sequence 3, 6, 9, 12, ... progressively increases by 3, thus the next two terms will be **15 and 18**

Application: Find the next three terms in the series: 9, 5, 1, ...

We notice that each term in the series 9, 5, 1, ... progressively decreases by 4, thus the next two terms will be 1 – 4, i.e. **– 3** and – 3 – 4, i.e. **– 7**

Application: Determine the next two terms in the series: 2, 6, 18, 54, ...

We notice that the second term, 6, is three times the first term, the third term, 18, is three times the second term, and that the fourth term, 54, is three times the third term. Hence the fifth term will be 3 x 54 = **162**, and the sixth term will be 3 x 162 = **486**

The n'th term of a series

If a series is represented by a general expression, say, $2n + 1$, where n is an integer (i.e. a whole number), then by substituting $n = 1, 2, 3, ...$ the terms of the series can be determined; in this example, the first three terms will be:

$$2(1) + 1,   2(2) + 1,   2(3) + 1, ... ,   \text{i.e.   } 3, 5, 7, ...$$

What is the n'th term of the sequence 1, 3, 5, 7, ...? Firstly, we notice that the gap between each term is 2, hence the law relating the numbers is: '$2n +$ something'

The second term, $3 = 2n +$ something, hence when $n = 2$ (i.e. the second term of the series), then

3 = 4 + something, and the 'something' must be − 1

Thus, **the n'th term of 1, 3, 5, 7, … is 2n − 1**

Hence the fifth term is given by 2(5) − 1 = 9, and the twentieth term is 2(20) − 1 = 39, and so on.

Application: The n'th term of a sequence is given by 3n + 1. Write down the first four terms.

The first four terms of the series 3n + 1 will be:

$$3(1) + 1, \quad 3(2) + 1, \quad 3(3) + 1 \quad \text{and} \quad 3(4) + 1$$

i.e. **4, 7, 10 and 13**

Application: The n'th term of a series is given by 4n − 1. Write down the first four terms.

The first four terms of the series 4n − 1 will be:

$$4(1) − 1, \quad 4(2) − 1, \quad 4(3) − 1 \quad \text{and} \quad 4(4) − 1$$

i.e. **3, 7, 11 and 15**

Application: Find the n'th term of the series: 1, 4, 7, …

We notice that the gap between each of the given three terms is 3, hence the law relating the numbers is: '3n + something'

The second term, $\quad$ 4 = 3n + something,

so when n = 2, then $\quad$ 4 = 6 + something,

so the 'something' must be − 2 (from simple equations)

Thus, **the n'th term of the series 1, 4, 7, … is: 3n − 2**

Application: Find the n'th term of the sequence: 3, 9, 15, 21, …
$\qquad\qquad$ Hence determine the 15th term of the series.

We notice that the gap between each of the given four terms is 6, hence the law relating the numbers is: '6n + something'

The second term, $\quad$ 9 = 6n + something,

so when n = 2, then   9 = 12 + something,

so the 'something' must be − 3

Thus, **the n'th term of the series 3, 9, 15, 21, ... is: 6n − 3**

The 15th term of the series is given by 6n − 3 when n = 15

Hence, **the 15th term of the series 3, 9, 15, 21, ... is:** $6(15) − 3 = \mathbf{87}$

Application: Find the n'th term of the series: 1, 4, 9, 16, 25, ..

This is a special series and does not follow the pattern of the previous examples. Each of the terms in the given series are square numbers,

i.e. 1, 4, 9, 16, 25, ... $\equiv 1^2, 2^2, 3^2, 4^2, 5^2$, ..

Hence the n'th term is: n^2

Chapter 23   Arithmetic progressions

When a sequence has a constant difference between successive terms it is called an arithmetic progression (often abbreviated to AP).

If a = first term, d = common difference and n = number of terms, then the arithmetic progression is:

$$a, a + d, a + 2d, ...$$

The n'th term is:

$$a + (n − 1)d$$

The sum of n terms,

$$S_n = \frac{n}{2}[2a + (n − 1)d]$$

Application: Find the sum of the first 7 terms of the series 1, 4, 7, 10, 13, ...

The sum of the first 7 terms is given by

$$S_7 = \frac{7}{2}[2(1) + (7 − 1)3] \qquad \text{since } a = 1 \text{ and } d = 3$$

$$= \frac{7}{2}[2 + 18] = \frac{7}{2}[20] = 70$$

Application: Determine (a) the ninth, and (b) the sixteenth term of the series 2, 7, 12, 17, ...

2, 7, 12, 17, ... is an arithmetic progression with a common difference, d, of 5

(a) The n'th term of an AP is given by $a + (n - 1)d$
Since the first term $a = 2$, $d = 5$ and $n = 9$
then the 9th term is: $2 + (9 - 1)5 = 2 + (8)(5) = 2 + 40 = \mathbf{42}$

(b) The 16th term is: $2 + (16 - 1)5 = 2 + (15)(5) = 2 + 75 = \mathbf{77}$

Application: Find the sum of the first 12 terms of the series 5, 9, 13, 17, ...

5, 9, 13, 17, ... is an AP where $a = 5$ and $d = 4$

The sum of n terms of an AP, $S_n = \dfrac{n}{2}[2a + (n - 1)d]$

Hence the sum of the first 12 terms, $S_{12} = \dfrac{12}{2}[2(5) + (12 - 1)4]$

$$= 6[10 + 44] = 6(54)$$
$$= \mathbf{324}$$

Application: An oil company bores a hole 80 m deep. Estimate the cost of boring if the cost is £30 for drilling the first metre with an increase in cost of £2 per metre for each succeeding metre.

The series is: 30, 32, 34, ... to 80 terms,   i.e. $a = 30$, $d = 2$ and $n = 80$

Thus, **total cost**, $S_n = \dfrac{n}{2}\big[2a + (n-1)d\big] = \dfrac{80}{2}\big[2(30) + (80 - 1)(2)\big] = 40\big[60 + 158\big]$

$$= 40(218) = \mathbf{£8720}$$

Chapter 24   Geometric progressions

When a sequence has a constant ratio between successive terms it is called a geometric progression (often abbreviated to GP).

If a = first term, r = common ratio, and n = number of terms, then the geometric progression is:

$$a, ar, ar^2, ar^3, ...$$

The n'th term is: ar^{n-1}

The sum of n terms,

$$S_n = \frac{a(1 - r^n)}{(1 - r)} \qquad \text{which is valid when } r < 1$$

or

$$S_n = \frac{a(r^n - 1)}{(r - 1)} \quad \text{which is valid when } r > 1$$

If $-1 < r < 1$, $\qquad\qquad S_\infty = \frac{a}{(1 - r)}$

Application: Find the sum of the first 8 terms of the GP 1, 2, 4, 8, 16, ...

The sum of the first 8 terms is given by

$$S_8 = \frac{1(2^8 - 1)}{(2 - 1)} \quad \text{since } a = 1 \text{ and } r = 2$$

i.e. $\qquad\qquad S_8 = \frac{1(256 - 1)}{1} = 255$

Application: Determine the tenth term of the series 3, 6, 12, 24, ...

3, 6, 12, 24, ... is a geometric progression with a common ratio r of 2.

The n'th term of a GP is ar^{n-1}, where a is the first term.

Hence the 10th term is: $(3)(2)^{10-1} = (3)(2)^9 = 3(512) = \mathbf{1536}$

Application: A tool hire firm finds that their net return from hiring tools is decreasing by 10% per annum. Their net gain on a certain tool this year is £400. Find the possible total of all future profits from this tool (assuming the tool lasts forever)

The net gain forms a series: £400 + £400 × 0.9 + £400 × 0.9² + ·····,

which is a GP with a = 400 and r = 0.9

The sum to infinity,

$$S_\infty = \frac{a}{(1 - r)} = \frac{400}{(1 - 0.9)} = \mathbf{£4000} = \textbf{total future profits}$$

Application: A drilling machine is to have 6 speeds ranging from 50 rev/min to 750 rev/min. Determine their values, each correct to the nearest whole number, if the speeds form a geometric progression

Let the GP of n terms be given by a, ar, ar², ... ar^{n-1}

The first term a = 50 rev/min

The 6th term is given by ar^{6-1}, which is 750 rev/min,

i.e. $\qquad\qquad\qquad ar^5 = 750$

from which
$$r^5 = \frac{750}{a} = \frac{750}{50} = 15$$

Thus the common ratio, $r = \sqrt[5]{15} = 1.7188$

The first term is a = 50 rev/min
the second term is ar = (50)(1.7188) = 85.94,
the third term is ar^2 = (50)(1.7188)^2 = 147.71,
the fourth term is ar^3 = (50)(1.7188)^3 = 253.89,
the fifth term is ar^4 = (50)(1.7188)^4 = 436.39,
the sixth term is ar^5 = (50)(1.7188)^5 = 750.06

Hence, correct to the nearest whole number, the 6 speeds of the drilling machine are:

50, 86, 148, 254, 436 and 750 rev/min

Chapter 25   Inequalities

Introduction to inequalities

An inequality is any expression involving one of the symbols $<$, $>$, $\leq$ or $\geq$

p < q means p is less than q   p > q means p is greater than q

p ≤ q means p is less than or equal to q   p ≥ q means p is greater than or equal to q

Some simple rules

(i) When a quantity is **added or subtracted** to both sides of an inequality, the inequality still remains.

For example, if p < 3 then p + 2 < 3 + 2 (adding 2 to both sides)

and   p − 2 < 3 − 2 (subtracting 2 from both sides)

(ii) When **multiplying or dividing** both sides of an inequality by a **positive** quantity, say 5, the inequality **remains the same**. For example, if p > 4 then 5p > 20 and $\dfrac{p}{5} > \dfrac{4}{5}$

(iii) When **multiplying or dividing** both sides of an inequality by a **negative** quantity, say −3, the **inequality is reversed**. For example, if $p > 1$ then $-3p < -3$ and $\dfrac{p}{-3} < \dfrac{4}{-3}$ (Note $>$ has changed to $<$ in each example)

To **solve an inequality** means finding all the values of the variable for which the inequality is true.

Simple inequalities

Application: Solve the following inequalities: (a) $3 + x > 7$ (b) $3t < 6$
(c) $z - 2 \geq 5$ (d) $\dfrac{p}{3} \leq 2$

(a) Subtracting 3 from both sides of the inequality: $3 + x > 7$ gives: $3 + x - 3 > 7 - 3$ i.e. **$x > 4$**

Hence, all values of x greater than 4 satisfy the inequality.

(b) Dividing both sides of the inequality: $3t < 6$ by 3 gives: $\dfrac{3t}{3} < \dfrac{6}{3}$ i.e. **$t < 2$**

Hence, all values of t less than 2 satisfy the inequality.

(c) Adding 2 to both sides of the inequality: $z - 2 \geq 5$ gives: $z - 2 + 2 \geq 5 + 2$ i.e. **$z \geq 7$**

Hence, all values of z equal to or greater than 7 satisfy the inequality.

(d) Multiplying both sides of the inequality: $\dfrac{p}{3} \leq 2$ by 3 gives: $(3)\dfrac{p}{3} \leq (3)2$

i.e. **$p \leq 6$**

Hence, all values of p equal to or less than 6 satisfy the inequality.

Application: Solve the inequality: $4x + 1 > x + 5$

Subtracting 1 from both sides of the inequality: $4x + 1 > x + 5$ gives: $4x > x + 4$

Subtracting x from both sides of the inequality: $4x > x + 4$ gives: $3x > 4$

Dividing both sides of the inequality: $3x > 4$ by 3 gives: **$x > \dfrac{4}{3}$**

Hence all values of x greater than $\dfrac{4}{3}$ satisfy the inequality: $4x + 1 > x + 5$

Application: Solve the inequality: $3 - 4t \leq 8 + t$

Subtracting 3 from both sides of the inequality: $3 - 4t \leq 8 + t$ gives: $-4t \leq 5 + t$

Subtracting t from both sides of the inequality: $-4t \leq 5 + t$ gives: $-5t \leq 5$

Dividing both sides of the inequality: $-5t \leq 5$ by -5 gives:

$$t \geq -1 \text{ (remembering to reverse the inequality)}$$

Hence, all values of t equal to or greater than -1 satisfy the inequality.

Inequalities involving a modulus

The **modulus** of a number is the size of the number, regardless of sign. Vertical lines enclosing the number denote a modulus.

For example, $|4| = 4$ and $|-4| = 4$ (the modulus of a number is never negative)

The inequality: $|t| < 1$ means that all numbers whose actual size, regardless of sign, is less than 1, i.e. any value between -1 and $+1$.

Thus $|t| < 1$ **means** $-1 < t < 1$

Similarly, $|x| > 3$ means all numbers whose actual size, regardless of sign, is greater than 3, i.e. any value greater than 3 and any value less than -3.

Thus $|x| > 3$ **means** $x > 3$ **and** $x < -3$

Application: Solve the following inequality: $|3x + 1| < 4$

Since $|3x + 1| < 4$ then $-4 < 3x + 1 < 4$

Now $-4 < 3x + 1$ becomes $-5 < 3x$ i.e. $-\dfrac{5}{3} < x$ and $3x + 1 < 4$ becomes

$3x < 3$ i.e. $x < 1$

Hence, these two results together become $-\dfrac{5}{3} < x < 1$ and mean that the

inequality $|3x + 1| < 4$ is satisfied for any value of x greater than $-\dfrac{5}{3}$ but less than 1.

Application: Solve the inequality: $|1 + 2t| \leq 5$

Since $|1 + 2t| \leq 5$ then $-5 \leq 1 + 2t \leq 5$

Now $-5 \leq 1 + 2t$ becomes $-6 \leq 2t$ i.e. $-3 \leq t$ and $1 + 2t \leq 5$ becomes $2t \leq 4$ i.e. $t \leq 2$

Hence, these two results together become: $-3 \leq t \leq 2$

Inequalities involving quotients

If $\dfrac{p}{q} > 0$ then $\dfrac{p}{q}$ must be a **positive** value.

For $\dfrac{p}{q}$ to be positive, **either** p is positive **and** q is positive   **or**   p is negative **and** q is negative.

i.e. $\dfrac{+}{+} = +$ and $\dfrac{-}{-} = +$

If $\dfrac{p}{q} < 0$ then $\dfrac{p}{q}$ must be a **negative** value.

For $\dfrac{p}{q}$ to be negative, **either** p is positive **and** q is negative **or** p is negative **and** q is positive.

i.e. $\dfrac{+}{-} = -$ and $\dfrac{-}{+} = -$

Application: Solve the inequality: $\dfrac{t+1}{3t-6} > 0$

Since $\dfrac{t+1}{3t-6} > 0$ then $\dfrac{t+1}{3t-6}$ must be **positive**.

For $\dfrac{t+1}{3t-6}$ to be positive, **either** (i) $t + 1 > 0$ **and** $3t - 6 > 0$ **or**   (ii) $t + 1 < 0$ **and** $3t - 6 < 0$

(i) If $t + 1 > 0$ then $t > -1$ and if $3t - 6 > 0$ then $3t > 6$ and $t > 2$

Both of the inequalities $t > -1$ **and** $t > 2$ are only true when $t > 2$,

i.e. the fraction $\dfrac{t+1}{3t-6}$ is positive when **$t > 2$**

(ii) If $t + 1 < 0$ then $t < -1$ and if $3t - 6 < 0$ then $3t < 6$ and $t < 2$

Both of the inequalities $t < -1$ **and** $t < 2$ are only true when $t < -1$,

i.e. the fraction $\dfrac{t+1}{3t-6}$ is positive when **$t < -1$**

Summarising, $\dfrac{t+1}{3t-6} > 0$ when **$t > 2$ or $t < -1$**

Application: Solve the inequality: $\dfrac{2x+3}{x+2} \le 1$

Since $\dfrac{2x+3}{x+2} \leq 1$ then $\dfrac{2x+3}{x+2} - 1 \leq 0$

i.e. $\dfrac{2x+3}{x+2} - \dfrac{x+2}{x+2} \leq 0$ i.e. $\dfrac{2x+3-(x+2)}{x+2} \leq 0$ or $\dfrac{x+1}{x+2} \leq 0$

For $\dfrac{x+1}{x+2}$ to be negative or zero, **either** (i) $x+1 \leq 0$ **and** $x+2 > 0$

or (ii) $x+1 \geq 0$ **and** $x+2 < 0$

(i) If $x+1 \leq 0$ then $x \leq -1$ and if $x+2 > 0$ then $x > -2$

(Note that $>$ is used for the denominator, not $\geq$; a zero denominator gives a value for the fraction which is impossible to evaluate.)

Hence, the inequality $\dfrac{x+1}{x+2} \leq 0$ is true when x is greater than -2 and less than or equal to -1,

which may be written as **$-2 < x \leq -1$**

(ii) If $x+1 \geq 0$ then $x \geq -1$ and if $x+2 < 0$ then $x < -2$

It is not possible to satisfy both $x \geq -1$ and $x < -2$ thus no values of x satisfies (ii).

Summarising, $\dfrac{2x+3}{x+2} \leq 1$ when $-2 < x \leq -1$

Inequalities involving square functions

The following two general rules apply when inequalities involve square functions:

   (i) **if $x^2 > k$ then $x > \sqrt{k}$ or $x < -\sqrt{k}$**   (1)

   (ii) **if $x^2 < k$ then $-\sqrt{k} < x < \sqrt{k}$**   (2)

Application: Solve the inequality: $t^2 > 9$

Since $t^2 > 9$ then $t^2 - 9 > 0$ i.e. $(t+3)(t-3) > 0$ by factorising

For $(t+3)(t-3)$ to be positive, **either** (i) $(t+3) > 0$ **and** $(t-3) > 0$

or (ii) $(t+3) < 0$ **and** $(t-3) < 0$

(i) If $(t+3) > 0$ then $t > -3$ and if $(t-3) > 0$ then $t > 3$

 Both of these are true only when **$t > 3$**

(ii) If $(t+3) < 0$ then $t < -3$ and if $(t-3) < 0$ then $t < 3$

 Both of these are true only when **$t < -3$**

Summarising, $t^2 > 9$ when **t > 3 or t < − 3**

This demonstrates the general rule:

$$\text{if } x^2 > k \text{ then } x > \sqrt{k} \text{ or } x < -\sqrt{k} \qquad (1)$$

Application: Solve the inequality: $x^2 > 4$

From the general rule stated above in equation (1):
$$\text{if } x^2 > 4 \text{ then } x > \sqrt{4} \text{ or } x < -\sqrt{4}$$
i.e. the inequality: $x^2 > 4$ is satisfied when **x > 2 or x < − 2**

Application: Solve the inequality: $(2z + 1)^2 > 9$

From equation (1), if $(2z + 1)^2 > 9$ then $2z + 1 > \sqrt{9}$ or $2z + 1 < -\sqrt{9}$

i.e. $2z + 1 > 3$ or $2z + 1 < − 3$

i.e. $2z > 2$ or $2z < -4$

i.e. **z > 1 or z < − 2**

Application: Solve the inequality: $t^2 < 9$

Since $t^2 < 9$ then $t^2 - 9 < 0$ i.e. $(t + 3)(t - 3) < 0$ by factorising.

For $(t + 3)(t - 3)$ to be negative, **either** (i) $(t + 3) > 0$ **and** $(t - 3) < 0$

or (ii) $(t + 3) < 0$ **and** $(t - 3) > 0$

(i) If $(t + 3) > 0$ then $t > - 3$ and if $(t - 3) < 0$ then $t < 3$

Hence (i) is satisfied when $t > - 3$ and $t < 3$ which may be written as: **− 3 < t < 3**

(ii) If $(t + 3) < 0$ then $t < - 3$ and if $(t - 3) > 0$ then $t > 3$

It is not possible to satisfy both $t < - 3$ and $t > 3$, thus no values of t satisfies (ii).

Summarising, $t^2 < 9$ when **− 3 < t < 3** which means that all values of t between − 3 and +3 will satisfy the inequality.

This demonstrates the general rule:

$$\text{if } x^2 < k \text{ then } -\sqrt{k} < x < \sqrt{k} \qquad (2)$$

Quadratic inequalities

Inequalities involving quadratic expressions are solved using either **factorisation** or '**completing the square**'.

For example, $x^2 - 2x - 3$ is factorised as $(x + 1)(x - 3)$

and $6x^2 + 7x - 5$ is factorised as $(2x - 1)(3x + 5)$

If a quadratic expression does not factorise, then the technique of 'completing the square' is used. In general, the procedure for $x^2 + bx + c$ is:

$$x^2 + bx + c \equiv \left(x + \frac{b}{2}\right)^2 + c - \left(\frac{b}{2}\right)^2$$

For example, $x^2 + 4x - 7$ does not factorise; completing the square gives:

$$x^2 + 4x - 7 \equiv (x + 2)^2 - 7 - 2^2 \equiv (x + 2)^2 - 11$$

Similarly, $\qquad x^2 - 6x - 5 \equiv (x - 3)^2 - 5 - 3^2 \equiv (x - 3)^2 - 14$

Application: Solve the inequality. $x^2 + 2x - 3 > 0$

Since $x^2 + 2x - 3 > 0$ then $(x - 1)(x + 3) > 0$ by factorising.

For the product $(x - 1)(x + 3)$ to be positive, **either** (i) $(x - 1) > 0$ **and** $(x + 3) > 0$

$\qquad\qquad\qquad\qquad\qquad\qquad\qquad\qquad$ **or** (ii) $(x - 1) < 0$ **and** $(x + 3) < 0$

(i) Since $(x - 1) > 0$ then $x > 1$ and since $(x + 3) > 0$ then $x > -3$

$\qquad$ Both of these inequalities are satisfied only when **x > 1**

(ii) Since $(x - 1) < 0$ then $x < 1$ and since $(x + 3) < 0$ then $x < -3$

$\qquad$ Both of these inequalities are satisfied only when **x < -3**

Summarising, $x^2 + 2x - 3 > 0$ is satisfied when either **x > 1 or x < -3**

Application: Solve the inequality: $t^2 - 2t - 8 < 0$

Since $t^2 - 2t - 8 < 0$ then $(t - 4)(t + 2) < 0$ by factorising.

For the product $(t - 4)(t + 2)$ to be negative, **either** (i) $(t - 4) > 0$ **and** $(t + 2) < 0$

$\qquad\qquad\qquad\qquad\qquad\qquad\qquad\qquad$ **or** (ii) $(t - 4) < 0$ **and** $(t + 2) > 0$

(i) Since $(t - 4) > 0$ then $t > 4$ and since $(t + 2) < 0$ then $t < -2$

It is not possible to satisfy both $t > 4$ and $t < -2$, thus no values of t satisfies the inequality (i)

(ii) Since $(t - 4) < 0$ then $t < 4$ and since $(t + 2) > 0$ then $t > -2$

Hence, (ii) is satisfied when $-2 < t < 4$

Summarising, $t^2 - 2t - 8 < 0$ is satisfied when $-2 < t < 4$

Application: Solve the inequality: $x^2 + 6x + 3 < 0$

$x^2 + 6x + 3$ does not factorise; completing the square gives:

$$x^2 + 6x + 3 = (x + 3)^2 + 3 - 3^2 = (x + 3)^2 - 6$$

The inequality thus becomes: $(x + 3)^2 - 6 < 0$ or $(x + 3)^2 < 6$

From equation 2, $-\sqrt{6} < (x + 3) < \sqrt{6}$ from which, $(-\sqrt{6} - 3) < x < (\sqrt{6} - 3)$

Hence, $x^2 + 6x + 3 < 0$ is satisfied when **$-5.45 < x < -0.55$** correct to 2 decimal places.

Chapter 26   The binomial series

The **binomial series** or **binomial theorem** is a formula for raising a binomial expression to any power without lengthy multiplication.

$$(a + x)^n = a^n + na^{n-1} x + \frac{n(n - 1)}{2!} a^{n-2} x^2$$

$$+ \frac{n(n - 1)(n - 2)}{3!} a^{n-3} x^3 + \cdots + x^n$$

$$(1 + x)^n = 1 + nx + \frac{n(n - 1)}{2!} x^2 + \frac{n(n - 1)(n - 2)}{3!} x^3 + \cdots$$

which is valid for $-1 < x < 1$

The r'th term of the expansion $(a + x)^n$ is:

$$\frac{n(n - 1)(n - 2) \ldots \text{ to } (r - 1) \text{ terms}}{(r - 1)!} a^{n-(r-1)} x^{r-1}$$

Application: Using the binomial series, determine the expansion of $(2 + x)^7$

From above, when a = 2 and n = 7:

$$(2 + x)^7 = 2^7 + 7(2)^6 x + \frac{(7)(6)}{(2)(1)}(2)^5 x^2 + \frac{(7)(6)(5)}{(3)(2)(1)}(2)^4 x^3$$

$$+ \frac{(7)(6)(5)(4)}{(4)(3)(2)(1)}(2)^3 x^4 + \frac{(7)(6)(5)(4)(3)}{(5)(4)(3)(2)(1)}(2)^2 x^5$$

$$+ \frac{(7)(6)(5)(4)(3)(2)}{(6)(5)(4)(3)(2)(1)}(2) x^6 + \frac{(7)(6)(5)(4)(3)(2)(1)}{(7)(6)(5)(4)(3)(2)(1)} x^7$$

i.e. $(2 + x)^7 = 128 + 448x + 672x^2 + 560x^3 + 280x^4 + 84x^5 + 14x^6 + x^7$

Application: Determine the fifth term $(3 + x)^7$ without fully expanding

The r'th term of the expansion $(a + x)^n$ is given by:

$$\frac{n(n - 1)(n - 2)\ldots \text{ to } (r - 1) \text{ terms}}{(r - 1)!} a^{n-(r-1)} x^{r-1}$$

Substituting n = 7, a = 3 and r − 1 = 5 − 1 = 4 gives:

$$\frac{(7)(6)(5)(4)}{(4)(3)(2)(1)}(3)^{7-4} x^4$$

i.e. the fifth term of $(3 + x)^7 = 35(3)^3 x^4 = \mathbf{945x^4}$

Application: Expand $\dfrac{1}{(1 + 2x)^3}$ in ascending powers of x as far as the term in x^3, using the binomial series

Using the binomial expansion of $(1 + x)^n$, where n = −3 and x is replaced by 2x gives:

$$\frac{1}{(1 + 2x)^3} = (1 + 2x)^{-3}$$

$$= 1 + (-3)(2x) + \frac{(-3)(-4)}{2!}(2x)^2 + \frac{(-3)(-4)(-5)}{3!}(2x)^3 + \cdots$$

$$= 1 - 6x + 24x^2 - 80x^3 +$$

The expansion is valid provided $\left|2x\right| < 1$

i.e. $\left|x\right| < \dfrac{1}{2}$ or $-\dfrac{1}{2} < x < \dfrac{1}{2}$

Application: Using the binomial theorem, expand $\sqrt{4 + x}$ in ascending powers of x to four terms

$$\sqrt{4 + x} = \sqrt{\left[4\left(1 + \frac{x}{4}\right)\right]} = \sqrt{4}\sqrt{\left(1 + \frac{x}{4}\right)} = 2\left(1 + \frac{x}{4}\right)^{1/2}$$

Using the expansion of $(1 + x)^n$,

$$2\left(1 + \frac{x}{4}\right)^{1/2} = 2\left[1 + \left(\frac{1}{2}\right)\left(\frac{x}{4}\right) + \frac{(1/2)(-1/2)}{2!}\left(\frac{x}{4}\right)^2 + \frac{(1/2)(-1/2)(-3/2)}{3!}\left(\frac{x}{4}\right)^3 + \cdots\right]$$

$$= 2\left[1 + \frac{x}{8} - \frac{x^2}{128} + \frac{x^3}{1024} - \cdots\right]$$

$$= 2 + \frac{x}{4} - \frac{x^2}{64} + \frac{x^3}{512} - \cdots$$

This is valid when $\left|\dfrac{x}{4}\right| < 1,$ i.e. $|x| < 4$ or $-4 < x < 4$

Application: Simplify $\dfrac{\sqrt[3]{(1 - 3x)}\sqrt{(1 + x)}}{\left(1 + \dfrac{x}{2}\right)^3}$ given that powers of x above the first may be neglected

$$\frac{\sqrt[3]{(1 - 3x)}\sqrt{(1 + x)}}{\left(1 + \dfrac{x}{2}\right)^3} = (1 - 3x)^{\frac{1}{3}}(1 + x)^{\frac{1}{2}}\left(1 + \frac{x}{2}\right)^{-3}$$

$$\approx \left[1 + \left(\frac{1}{3}\right)(-3x)\right]\left[1 + \left(\frac{1}{2}\right)(x)\right]\left[1 + (-3)\left(\frac{x}{2}\right)\right]$$

when expanded by the binomial theorem as far as the x term only

$$= (1 - x)\left(1 + \frac{x}{2}\right)\left(1 - \frac{3x}{2}\right)$$

$$= \left(1 - x + \frac{x}{2} - \frac{3x}{2}\right) \quad \text{when powers of x higher than unity are neglected}$$

$$= (1 - 2x)$$

Application: The second moment of area of a rectangle through its centroid is given by $\dfrac{bl^3}{12}$. Determine the approximate change in the second moment of area if b is increased by 3.5% and l is reduced by 2.5%

New values of b and l are $(1 + 0.035)b$ and $(1 - 0.025)l$ respectively.

New second moment of area $= \dfrac{1}{12}[(1 + 0.035)b][(1 - 0.025)l]^3$

$$= \dfrac{bl^3}{12}(1 + 0.035)(1 - 0.025)^3$$

$$\approx \dfrac{bl^3}{12}(1 + 0.035)(1 - 0.075)$$

neglecting powers of small terms

$$\approx \dfrac{bl^3}{12}(1 + 0.035 - 0.075)$$

neglecting products of small terms

$$\approx \dfrac{bl^3}{12}(1 - 0.040) \text{ or } (0.96)\dfrac{bl^3}{12}$$

i.e. 96% of the original second moment of area

Hence the second moment of area is reduced by approximately 4%

Application: The resonant frequency of a vibrating shaft is given by: $f = \dfrac{1}{2\pi}\sqrt{\dfrac{k}{l}}$, where k is the stiffness and l is the inertia of the shaft. Using the binomial theorem, determine the approximate percentage error in determining the frequency using the measured values of k and l, when the measured value of k is 4% too large and the measured value of l is 2% too small

Let f, k and l be the true values of frequency, stiffness and inertia respectively. Since the measured value of stiffness, k_1, is 4% too large, then $k_1 = \dfrac{104}{100}k = (1 + 0.04)k$

The measured value of inertia, l_1, is 2% too small, hence $l_1 = \dfrac{98}{100}l = (1 - 0.02)l$

The measured value of frequency,

$$f_1 = \dfrac{1}{2\pi}\sqrt{\dfrac{k_1}{l_1}} = \dfrac{1}{2\pi}k_1^{1/2}\,l_1^{-1/2}$$

$$= \dfrac{1}{2\pi}[(1 + 0.04)k]^{1/2}\,[(1 - 0.02)l]^{-1/2}$$

$$= \frac{1}{2\pi}(1 + 0.04)^{1/2} \, k^{1/2} \, (1 - 0.02)^{-1/2} \, l^{-1/2}$$

$$= \frac{1}{2\pi} \, k^{1/2} \, l^{-1/2} \, (1 + 0.04)^{1/2} \, (1 - 0.02)^{-1/2}$$

i.e. $f_1 = f(1 + 0.04)^{1/2} (1 - 0.02)^{-1/2}$

$$\approx f[1 + (1/2)(0.04)][(1 + (-1/2)(-0.02)]$$

$$\approx f(1 + 0.02)(1 + 0.01)$$

Neglecting the products of small terms,

$$f_1 \approx (1 + 0.02 + 0.01)f \approx 1.03\,f$$

Thus the percentage error in f based on the measured values of k and l is approximately **3% too large**.

Chapter 27   Maclaurin's theorem

Maclaurin's theorem or **Maclaurin's series** may be stated as follows:

$$f(x) = f(0) + xf'(0) + \frac{x^2}{2!}f''(0) + \frac{x^3}{3!}f'''(0) + \cdots$$

Application: **Determine the first four terms of the power series for cos x**

The values of f(0), f'(0), f''(0), ... in the Maclaurin's series are obtained as follows:

$f(x) = \cos x$	$f(0) = \cos 0 = 1$
$f'(x) = -\sin x$	$f'(0) = -\sin 0 = 0$
$f''(x) = -\cos x$	$f''(0) = -\cos 0 = -1$
$f'''(x) = \sin x$	$f'''(0) = \sin 0 = 0$
$f^{iv}(x) = \cos x$	$f^{iv}(0) = \cos 0 = 1$
$f^{v}(x) = -\sin x$	$f^{v}(0) = -\sin 0 = 0$
$f^{vi}(x) = -\cos x$	$f^{vi}(0) = -\cos 0 = -1$

Substituting these values into the Maclaurin's series gives:

$$f(x) = \cos x = 1 + x(0) + \frac{x^2}{2!}(-1) + \frac{x^3}{3!}(0) + \frac{x^4}{4!}(1) + \frac{x^5}{5!}(0) + \frac{x^6}{6!}(-1) + \cdots$$

i.e. $\cos x = 1 - \dfrac{x^2}{2!} + \dfrac{x^4}{4!} - \dfrac{x^6}{6!} + \cdots$

Application: Determine the power series for cos 2θ

Replacing x with 2θ in the series obtained in the previous example gives:

$$\cos 2\theta = 1 - \frac{(2\theta)^2}{2!} + \frac{(2\theta)^4}{4!} - \frac{(2\theta)^6}{6!} + \cdots$$

$$= 1 - \frac{4\theta^2}{2} + \frac{16\theta^4}{24} - \frac{64\theta^6}{720} + \cdots$$

i.e. $\cos 2\theta = 1 - 2\theta^2 + \frac{2}{3}\theta^4 - \frac{4}{45}\theta^6 + \cdots$

Application: Expand ln (1 + x) to five terms

$f(x) = \ln(1 + x)$ $\qquad\qquad$ $f(0) = \ln(1 + 0) = 0$

$f'(x) = \dfrac{1}{(1 + x)}$ $\qquad\qquad$ $f'(0) = \dfrac{1}{1 + 0} = 1$

$f''(x) = \dfrac{-1}{(1 + x)^2}$ $\qquad\qquad$ $f''(0) = \dfrac{-1}{(1 + 0)^2} = -1$

$f'''(x) = \dfrac{2}{(1 + x)^3}$ $\qquad\qquad$ $f'''(0) = \dfrac{2}{(1 + 0)^3} = 2$

$f^{iv}(x) = \dfrac{-6}{(1 + x)^4}$ $\qquad\qquad$ $f^{iv}(0) = \dfrac{-6}{(1 + 0)^4} = -6$

$f^{v}(x) = \dfrac{24}{(1 + x)^5}$ $\qquad\qquad$ $f^{v}(0) = \dfrac{24}{(1 + 0)^5} = 24$

Substituting these values into the Maclaurin's series gives:

$$f(x) = \ln(1 + x) = 0 + x(1) + \frac{x^2}{2!}(-1) + \frac{x^3}{3!}(2) + \frac{x^4}{4!}(-6) + \frac{x^5}{5!}(24)$$

i.e. $\ln(1 + x) = x - \dfrac{x^2}{2} + \dfrac{x^3}{3} - \dfrac{x^4}{4} + \dfrac{x^5}{5} - \cdots$

Application: Find the expansion of $(2 + x)^4$ using Maclaurin's series

$$f(x) = (2 + x)^4 \qquad f(0) = 2^4 \qquad = 16$$
$$f'(x) = 4(2 + x)^3 \qquad f'(0) = 4(2)^3 \qquad = 32$$
$$f''(x) = 12(2 + x)^2 \qquad f''(0) = 12(2)^2 \qquad = 48$$
$$f'''(x) = 24(2 + x)^1 \qquad f'''(0) = 24(2) \qquad = 48$$
$$f^{iv}(x) = 24 \qquad\qquad f^{iv}(0) = 24$$

Substituting in Maclaurin's series gives:

$$(2 + x)^4 = f(0) + xf'(0) + \frac{x^2}{2!} f''(0) + \frac{x^3}{3!} f'''(0) + \frac{x^4}{4!} f^{iv}(0)$$

$$= 16 + (x)(32) + \frac{x^2}{2!} (48) + \frac{x^3}{3!} (48) + \frac{x^4}{4!} (24)$$

$$= 16 + 32x + 24x^2 + 8x^3 + x^4$$

Numerical integration using Maclaurin's series

Application: Evaluate $\displaystyle\int_{0.1}^{0.4} 2e^{\sin\theta}\, d\theta$, **correct to 3 significant figures**

A power series for $e^{\sin\theta}$ is firstly obtained using Maclaurin's series.

$$f(\theta) = e^{\sin\theta} \qquad\qquad f(0) = e^{\sin 0} = e^0 = 1$$
$$f'(\theta) = \cos\theta\, e^{\sin\theta} \qquad\qquad f'(0) = \cos 0\, e^{\sin 0} = (1)\, e^0 = 1$$

$$f''(\theta) = (\cos\theta)(\cos\theta\, e^{\sin\theta}) + (e^{\sin\theta})(-\sin\theta) \text{ by the product rule,}$$
$$= e^{\sin\theta}(\cos^2\theta - \sin\theta) \qquad f''(0) = e^0(\cos^2 0 - \sin 0) = 1$$
$$f'''(\theta) = (e^{\sin\theta})[(2\cos\theta\,(-\sin\theta) - \cos\theta] + (\cos^2\theta - \sin\theta)(\cos\theta\, e^{\sin\theta})$$
$$= e^{\sin\theta}\cos\theta[-2\sin\theta - 1 + \cos^2\theta - \sin\theta]$$

$$f'''(0) = e^0 \cos 0[(0 - 1 + 1 - 0)] = 0$$

Hence from the Maclaurin's series:

$$e^{\sin\theta} = f(0) + \theta\, f'(0) + \frac{\theta^2}{2!} f''(0) + \frac{\theta^3}{3!} f'''(0) + \cdots = 1 + \theta + \frac{\theta^2}{2} + 0$$

Thus $\displaystyle\int_{0.1}^{0.4} 2\, e^{\sin\theta}\, d\theta = \int_{0.1}^{0.4} 2\left(1 + \theta + \frac{\theta^2}{2}\right) d\theta$

$$= \int_{0.1}^{0.4} (2 + 2\theta + \theta^2)\, d\theta = \left[2\theta + \frac{2\theta^2}{2} + \frac{\theta^3}{3}\right]_{0.1}^{0.4}$$

$$= \left(0.8 + (0.4)^2 + \frac{(0.4)^3}{3} \right)$$

$$- \left(0.2 + (0.1)^2 + \frac{(0.1)^3}{3} \right)$$

$$= 0.981333 - 0.210333$$

$$= 0.771, \text{ correct to 33 significant figures}$$

Chapter 28   Limiting values – L'Hopital's rule

L'Hopital's rule states:

$$\lim_{\delta x \to a} \left\{ \frac{f(x)}{g(x)} \right\} = \lim_{\delta x \to a} \left\{ \frac{f'(x)}{g'(x)} \right\} \quad \text{provided } g'(a) \neq 0$$

Application: Determine $\lim_{\delta x \to 1} \left\{ \dfrac{x^2 + 3x - 4}{x^2 - 7x + 6} \right\}$

The first step is to substitute $x = 1$ into both numerator and denominator. In this case we obtain $\dfrac{0}{0}$.

It is only when we obtain such a result that we then use L'Hopital's rule.

Hence applying L'Hopital's rule,

$$\lim_{\delta x \to 1} \left\{ \frac{x^2 + 3x - 4}{x^2 - 7x + 6} \right\} = \lim_{\delta x \to 1} \left\{ \frac{2x + 3}{2x - 7} \right\} \quad \text{i.e. both numerator and denominator have been differentiated}$$

$$= \frac{5}{-5} = -1$$

Application: Determine $\lim_{\theta \to 0} \left\{ \dfrac{\sin \theta - \theta \cos \theta}{\theta^3} \right\}$

$$\lim_{\theta \to 0} \left\{ \frac{\sin \theta - \theta \cos \theta}{\theta^3} \right\} = \lim_{\theta \to 0} \left\{ \frac{\cos \theta - \left[(\theta)(-\sin \theta) + \cos \theta \right]}{3\theta^2} \right\}$$

$$= \lim_{\theta \to 0} \left\{ \frac{\theta \sin \theta}{3\theta^2} \right\} = \lim_{\theta \to 0} \left\{ \frac{\theta \cos \theta + \sin \theta}{6\theta} \right\}$$

$$= \lim_{\theta \to 0} \left\{ \frac{\theta(-\sin \theta) + \cos \theta(1) + \cos \theta}{6} \right\}$$

$$= \frac{1+1}{6} = \frac{2}{6} = \frac{1}{3}$$

Chapter 29   Solving equations by iterative methods (1) – the bisection method

In the **method of bisection** the mid-point of the interval, i.e. $x_3 = \dfrac{x_1 + x_2}{2}$, is taken, and from the sign of $f(x_3)$ it can be deduced whether a root lies in the half interval to the left or right of x_3. Whichever half interval is indicated, its mid-point is then taken and the procedure repeated. The method often requires many iterations and is therefore slow, but never fails to eventually produce the root. The procedure stops when two successive values of x are equal, to the required degree of accuracy.

Application: Using the bisection method, determine the positive root of the equation $x + 3 = e^x$, correct to 3 decimal places

Let $f(x) = x + 3 - e^x$ then, using functional notation:

$$f(0) = 0 + 3 - e^0 = +2$$

$$f(1) = 1 + 3 - e^1 = +1.2817..$$

$$f(2) = 2 + 3 - e^2 = -2.3890..$$

Since $f(1)$ is positive and $f(2)$ is negative, a root lies between $x = 1$ and $x = 2$. A sketch of $f(x) = x + 3 - e^x$, i.e. $x + 3 = e^x$ is shown in Figure 29.1.

Bisecting the interval between $x = 1$ and $x = 2$ gives $\dfrac{1+2}{2}$ i.e. 1.5

Hence **f(1.5)** $= 1.5 + 3 - e^{1.5} = $ **+0.01831..**

Since $f(1.5)$ is positive and $f(2)$ is negative, a root lies between $x = 1.5$ and $x = 2$.

Bisecting this interval gives $\dfrac{1.5+2}{2}$ i.e. 1.75

Hence **f(1.75)** $= 1.75 + 3 - e^{1.75} = $ **−1.00460..**

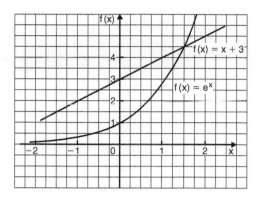

Figure 29.1

Since f(1.75) is negative and f(1.5) is positive, a root lies between x = 1.75 and x = 1.5

Bisecting this interval gives $\dfrac{1.75 + 1.5}{2}$ i.e. 1.625

Hence **f(1.625)** = $1.625 + 3 - e^{1.625}$ = **−0.45341..**

Since f(1.625) is negative and f(1.5) is positive, a root lies between x = 1.625 and x = 1.5

Bisecting this interval gives $\dfrac{1.625 + 1.5}{2}$ i.e. 1.5625

Hence **f(1.5625)** = $1.5625 + 3 - e^{1.5625}$ = **−0.20823..**

Since f(1.5625) is negative and f(1.5) is positive, a root lies between x = 1.5625 and x = 1.5

The iterations are continued and the results are presented in the table shown below.

The last two values of x_3 in the table are 1.504882813 and 1.505388282, i.e. both are equal to 1.505, correct to 3 decimal places. The process therefore stops.

Hence the root of x + 3 = e^x is x = 1.505, correct to 3 decimal places.

x_1	x_2	$x_3 = \dfrac{x_1 + x_2}{2}$	$f(x_3)$
		0	+2
		1	+1.2817..
		2	−2.3890..
1	2	1.5	+0.0183..
1.5	2	1.75	−1.0046..
1.5	1.75	1.625	−0.4534..
1.5	1.625	1.5625	−0.2082..

(Continued)

x_1	x_2	$x_3 = \dfrac{x_1 + x_2}{2}$	$f(x_3)$
1.5	1.5625	1.53125	−0.0927..
1.5	1.53125	1.515625	−0.0366..
1.5	1.515625	1.5078125	−0.0090..
1.5	1.5078125	1.50390625	+0.0046..
1.50390625	1.5078125	1.505859375	−0.0021..
1.50390625	1.505859375	**1.504882813**	+0.0012..
1.504882813	1.505859375	**1.505388282**	

Chapter 30   Solving equations by iterative methods (2) – an algebraic method of successive approximations

Procedure:

First approximation

(a) Using a graphical or functional notation method, determine an approximate value of the root required, say x_1

Second approximation

(b) Let the true value of the root be $(x_1 + \delta_1)$
(c) Determine x_2 the approximate value of $(x_1 + \delta_1)$ by determining the value of $f(x_1 + \delta_1) = 0$, but neglecting terms containing products of δ_1

Third approximation

(d) Let the true value of the root be $(x_2 + \delta_2)$
(e) Determine x_3, the approximate value of $(x_2 + \delta_2)$ by determining the value of $f(x_2 + \delta_2) = 0$, but neglecting terms containing products of δ_2
(f) The fourth and higher approximations are obtained in a similar way.

Using the techniques given in paragraphs (b) to (f), it is possible to continue getting values nearer and nearer to the required root. The procedure is repeated until the value of the required root does not change on two consecutive approximations, when expressed to the required degree of accuracy.

Application: Determine the value of the smallest positive root of the equation $3x^3 - 10x^2 + 4x + 7 = 0$, correct to 3 significant figures, using an algebraic method of successive approximations

The functional notation method is used to find the value of the first approximation.

$f(x) = 3x^3 - 10x^2 + 4x + 7$

$f(0) = 3(0)^3 - 10(0)^2 + 4(0) + 7 = 7$

$f(1) = 3(1)^3 - 10(1)^2 + 4(1) + 7 = 4$

$f(2) = 3(2)^3 - 10(2)^2 + 4(2) + 7 = -1$

Following the above procedure:

First approximation

(a) Let the first approximation be such that it divides the interval 1 to 2 in the ratio of 4 to -1, i.e. let x_1 be 1.8

Second approximation

(b) Let the true value of the root, x_2 , be $(x_1 + \delta_1)$

(c) Let $f(x_1 + \delta_1) = 0$, then since $x_1 = 1.8$,
$3(1.8 + \delta_1)^3 - 10(1.8 + \delta_1)^2 + 4(1.8 + \delta_1) + 7 = 0$

Neglecting terms containing products of δ_1 and using the binomial series gives:

$3[1.8^3 + 3(1.8)^2\delta_1] - 10[1.8^2 + (2)(1.8)\delta_1] + 4(1.8 + \delta_1) + 7 \approx 0$

$3(5.832 + 9.720\,\delta_1) - 32.4 - 36\,\delta_1 + 7.2 + 4\,\delta_1 + 7 \approx 0$

$17.496 + 29.16\,\delta_1 - 32.4 - 36\,\delta_1 + 7.2 + 4\,\delta_1 + 7 \approx 0$

$$\delta_1 \approx \frac{-17.496 + 32.4 - 7.2 - 7}{29.16 - 36 + 4} \approx -\frac{0.704}{2.84} \approx -0.2479$$

Thus, $x_2 \approx 1.8 - 0.2479 = 1.5521$

Third approximation

(d) Let the true value of the root, x_3, be $(x_2 + \delta_2)$

(e) Let $f(x_2 + \delta_2) = 0$, then since $x_2 = 1.5521$,

$3(1.5521 + \delta_2)^3 - 10(1.5521 + \delta_2)^2 + 4(1.5521 + \delta_2) + 7 = 0$

Neglecting terms containing products of δ_2 gives:

$11.217 + 21.681\,\delta_2 - 24.090 - 31.042\,\delta_2 + 6.2084 + 4\,\delta_2 + 7 \approx 0$

$$\delta_2 \approx \frac{-11.217 + 24.090 - 6.2084 - 7}{21.681 - 31.042 + 4} \approx \frac{-0.3354}{-5.361} \approx 0.06256$$

Thus $x_3 \approx 1.5521 + 0.06256 \approx 1.6147$

(f) Values of x_4 and x_5 are found in a similar way.

$f(x_3 + \delta_3) = 3(1.6147 + \delta_3)^3 - 10(1.6147 + \delta_3)^2 + 4(1.6147 + \delta_3) + 7 = 0$

giving $\delta_3 \approx 0.003175$ and $x_4 \approx 1.618$, i.e. 1.62 correct to 3 significant figures

$f(x_4 + \delta_4) = 3(1.618 + \delta_4)^3 - 10(1.618 + \delta_4)^2 + 4(1.618 + \delta_4) + 7 = 0$

giving $\delta_4 \approx 0.0000417$, and $x_5 \approx 1.62$, correct to 3 significant figures.

Since x_4 and x_5 are the same when expressed to the required degree of accuracy, then the required root is **1.62**, correct to 3 significant figures.

Chapter 31   Solving equations by iterative methods (3) – the Newton-Raphson method

The **Newton-Raphson formula**, often just referred to as **Newton's method**, may be stated as follows:

If r_1 is the approximate value of a real root of the equation $f(x) = 0$, then a closer approximation to the root, r_2, is given by:

$$r_2 = r_1 - \frac{f(r_1)}{f'(r_1)}$$

Application: Using Newton's method, find the positive root of

$(x + 4)^3 - e^{1.92x} + 5\cos\dfrac{x}{3} = 9,$ correct to 3 significant figures

The functional notational method is used to determine the approximate value of the root.

$f(x) = (x + 4)^3 - e^{1.92x} + 5\cos\dfrac{x}{3} - 9$

$f(0) = (0 + 4)^3 - e^0 + 5\cos 0 - 9 = 59$

$f(1) = 5^3 - e^{1.92} + 5\cos\dfrac{1}{3} - 9 \approx 114$

$f(2) = 6^3 - e^{3.84} + 5\cos\dfrac{2}{3} - 9 \approx 164$

$f(3) = 7^3 - e^{5.76} + 5\cos 1 - 9 \approx 19$

$f(4) = 8^3 - e^{7.68} + 5\cos\dfrac{4}{3} - 9 \approx -1660$

From these results, let a first approximation to the root be $r_1 = 3$

Newton's method states that a better approximation to the root,

$r_2 = r_1 - \dfrac{f(r_1)}{f'(r_1)}$

$f(r_1) = f(3) = 7^3 - e^{5.76} + 5\cos 1 - 9 = 19.35$

$$f'(x) = 3(x + 4)^2 - 1.92e^{1.92x} - \frac{5}{3}\sin\frac{x}{3}$$

$$f'(r_1) = f'(3) = 3(7)^2 - 1.92e^{5.76} - \frac{5}{3}\sin 1 = -463.7$$

Thus, $r_2 = 3 - \dfrac{19.35}{-463.7} = 3 + 0.042 = 3.042 = 3.04$,

correct to 3 significant figures.

Similarly,

$$r_3 = 3.042 - \frac{f(3.042)}{f'(3.042)} = 3.042 - \frac{(-1.146)}{(-513.1)} = 3.042 - 0.0022$$

$$= 3.0398 = 3.04, \text{ correct to 3 significant figures.}$$

Since r_2 and r_3 are the same when expressed to the required degree of accuracy, then the required root is **3.04**, correct to 3 significant figures.

Chapter 32   Computer numbering systems

Conversion of binary to decimal

Application: Change the binary number 1101.1 to its equivalent decimal form

$$1101.1 = 1 \times 2^3 + 1 \times 2^2 + 0 \times 2^1 + 1 \times 2^0 + 1 \times 2^{-1}$$

$$= 8 + 4 + 0 + 1 + \frac{1}{2}, \text{ that is 13.5}$$

i.e. **$1101.1_2 = 13.5_{10}$**, the suffixes 2 and 10 denoting binary and decimal systems of numbers respectively.

Application: Convert 101.0101_2 to a decimal number

$$101.0101_2 = 1 \times 2^2 + 0 \times 2^1 + 1 \times 2^0 + 0 \times 2^{-1}$$

$$+ 1 \times 2^{-2} + 0 \times 2^{-3} + 1 \times 2^{-4}$$

$$= 4 + 0 + 1 + 0 + 0.25 + 0 + 0.0625 = 5.3125_{10}$$

Conversion of decimal to binary

An integer decimal number can be converted to a corresponding binary number by repeatedly dividing by 2 and noting the remainder at each stage, as shown below.

Application: Change 39_{10} into binary

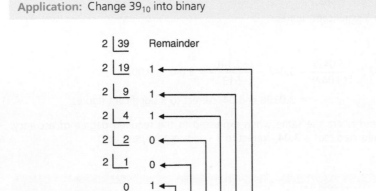

The result is obtained by writing the top digit of the remainder as the least significant bit, (a bit is a **b**inary dig**it** and the least significant bit is the one on the right). The bottom bit of the remainder is the most significant bit, i.e. the bit on the left.

Thus, $39_{10} = 100111_2$

Application: Change 0.625 in decimal into binary form

The fractional part of a denary number can be converted to a binary number by repeatedly multiplying by 2, as shown below for the fraction 0.625

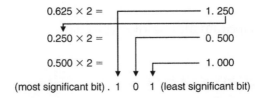

For fractions, the most significant bit of the result is the top bit obtained from the integer part of multiplication by 2. The least significant bit of the result is the bottom bit obtained from the integer part of multiplication by 2.

Thus $0.625_{10} = 0.101_2$

Binary addition

Binary addition of two/three bits is achieved according to the following rules:

sum	carry		sum	carry
0 + 0 = 0	0		0 + 0 + 0 = 0	0
0 + 1 = 1	0		0 + 0 + 1 = 1	0
1 + 0 = 1	0		0 + 1 + 0 = 1	0
1 + 1 = 0	1		0 + 1 + 1 = 0	1
			1 + 0 + 0 = 1	0
			1 + 0 + 1 = 0	1
			1 + 1 + 0 = 0	1
			1 + 1 + 1 = 1	1

Application: Perform the binary addition: 1001 + 10110

$$\begin{array}{r} 1001 \\ + \ 10110 \\ \hline \mathbf{11111} \end{array}$$

Application: Perform the binary addition: 11111 + 10101

$$\begin{array}{r} 11111 \\ + \ 10101 \\ \hline \end{array}$$

sum **110100**
carry 11111

Application: Perform the binary addition: 1101001 + 1110101

$$\begin{array}{r} 1101001 \\ + \ 1110101 \\ \hline \end{array}$$

sum **11011110**
carry 11   1

Application: Perform the binary addition: 1011101 + 1100001 + 110101

$$\begin{array}{r} 1011101 \\ 1100001 \\ + \quad 110101 \\ \hline \end{array}$$

sum **11110011**
carry 11111 1

Conversion of decimal to binary via octal

For decimal integers containing several digits, repeatedly dividing by 2 can be a lengthy process. In this case, it is usually easier to convert a decimal number to a binary number via the octal system of numbers. This system has a radix of 8, using the digits 0, 1, 2, 3, 4, 5, 6 and 7.

Application: Find the decimal number equivalent to the octal number 4317_8

$$4317_8 = 4 \times 8^3 + 3 \times 8^2 + 1 \times 8^1 + 7 \times 8^0$$
$$= 4 \times 512 + 3 \times 64 + 1 \times 8 + 7 \times 1 = 2255_{10}$$

Thus, $4317_8 = 2255_{10}$

Application: Convert 493_{10} into octal

An integer decimal number can be converted to a corresponding octal number by repeatedly dividing by 8 and noting the remainder at each stage.

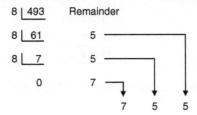

Thus, $493_{10} = 755_8$

Application: Convert 0.4375_{10} into octal

The fractional part of a decimal number can be converted to an octal number by repeatedly multiplying by 8, as shown below.

$$0.4375 \times 8 = \rule{1cm}{0.4pt} 3.5$$
$$0.5 \times 8 = \rule{1cm}{0.4pt} 4.0$$
$$.3 \quad 4$$

For fractions, the most significant bit is the top integer obtained by multiplication of the denary fraction by 8, thus

$$0.4375_{10} = 0.34_8$$

Conversion of octal to binary and decimal

The natural binary code for digits 0 to 7 is shown in Table 32.1, and an octal number can be converted to a binary number by writing down the three bits corresponding to the octal digit.

Table 32.1

Octal digit	Natural binary number
0	000
1	001
2	010
3	011
4	100
5	101
6	110
7	111

Application: Change 437_8 into binary

From Table 32.1, $\mathbf{437_8 = 100\ 011\ 111_2}$

Application: Change 26.35_8 into binary

From Table 32.1, $26.35_8 = 010\ 110.011\ 101_2$

The '0' on the extreme left does not signify anything, thus

$$\mathbf{26.35_8 = 10\ 110.011\ 101_2}$$

Application: Convert $11\ 110\ 011.100\ 01_2$ to a decimal number via octal

Grouping the binary number in three's from the binary point gives:

$$011\ 110\ 011.100\ 010_2$$

Using Table 32.1 to convert this binary number to an octal number gives:

363.42_8 and

$363.42_8 = 3 \times 8^2 + 6 \times 8^1 + 3 \times 8^0 + 4 \times 8^{-1} + 2 \times 8^{-2}$

$= 192 + 48 + 3 + 0.5 + 0.03125 = \mathbf{243.53125_{10}}$

Hence, $\mathbf{11\ 110\ 011.100\ 01_2 = 363.42_8} = 243.53125_{10}$

Hexadecimal numbers

A **hexadecimal numbering system** has a radix of 16 and uses the following 16 distinct digits:

0, 1, 2, 3, 4, 5, 6, 7, 8, 9, A, B, C, D, E and F

'A' corresponds to 10 in the decimal system, B to 11, C to 12, and so on.

Table 32.2 compares decimal, binary, octal and hexadecimal numbers.

Table 32.2

Decimal	Binary	Octal	Hexadecimal
0	0000	0	0
1	0001	1	1
2	0010	2	2
3	0011	3	3
4	0100	4	4
5	0101	5	5
6	0110	6	6
7	0111	7	7
8	1000	10	8
9	1001	11	9
10	1010	12	A
11	1011	13	B
12	1100	14	C
13	1101	15	D
14	1110	16	E
15	1111	17	F
16	10000	20	10
17	10001	21	11
18	10010	22	12
19	10011	23	13
20	10100	24	14
21	10101	25	15
22	10110	26	16
23	10111	27	17
24	11000	30	18
25	11001	31	19
26	11010	32	1A
27	11011	33	1B
28	11100	34	1C
29	11101	35	1D
30	11110	36	1E
31	11111	37	1F
32	100000	40	20

For example, $23_{10} = 10111_2 = 27_8 = 17_{16}$

Conversion from hexadecimal to decimal

Application: Change $1A_{16}$ into decimal form

$1A_{16} = 1 \times 16^1 + A \times 16^0 = 1 \times 16^1 + 10 \times 1 = 16 + 10 = 26$

i.e. $\quad$ $\mathbf{1A_{16} = 26_{10}}$

Application: Change $2E_{16}$ into decimal form

$\mathbf{2E_{16}} = 2 \times 16^1 + E \times 16^0 = 2 \times 16^1 + 14 \times 16^0 = 32 + 14 = \mathbf{46_{10}}$

Application: Change $1BF_{16}$ into decimal form

$$\mathbf{1BF_{16}} = 1 \times 16^2 + B \times 16^1 + F \times 16^0$$
$$= 1 \times 16^2 + 11 \times 16^1 + 15 \times 16^0$$
$$= 256 + 176 + 15 = \mathbf{447_{10}}$$

Application: Convert $1A4E_{16}$ into a decimal number

$$\mathbf{1A4E_{16}} = 1 \times 16^3 + A \times 16^2 + 4 \times 16^1 + E \times 16^0$$
$$= 1 \times 16^3 + 10 \times 16^2 + 4 \times 16^1 + 14 \times 16^0$$
$$= 1 \times 4096 + 10 \times 256 + 4 \times 16 + 14 \times 1$$
$$= 4096 + 2560 + 64 + 14 = 6734$$

Thus, $\mathbf{1A4E_{16} = 6734_{10}}$

Conversion from decimal to hexadecimal

This is achieved by repeatedly dividing by 16 and noting the remainder at each stage, as shown below.

Application: Change 26_{10} into hexadecimal

$$16\underline{|\,26} \quad \text{Remainder}$$
$$16\underline{|\,1} \quad\quad 10 \equiv A_{16}$$
$$0 \quad\quad\quad 1 \equiv 1_{16}$$

most significant bit $\rightarrow$ 1 A $\leftarrow$ least significant bit

Hence, $\mathbf{26_{10} = 1A_{16}}$

Application: Change 447_{10} into hexadecimal

$$
\begin{array}{rll}
16\,\underline{|\,447} & \text{Remainder} \\
16\,\underline{|\ \ 27} & 15 \equiv F_{16} \\
16\,\underline{|\ \ \ 1} & 11 \equiv B_{16} \\
\qquad\ \ 0 & 1 \equiv 1_{16} \\
& \qquad\quad 1\ \ B\ \ F
\end{array}
$$

Thus, $447_{10} = 1BF_{16}$

Conversion from binary to hexadecimal

The binary bits are arranged in groups of four, starting from right to left, and a hexadecimal symbol is assigned to each group.

Application: Convert the binary number 1110011110101001 into hexadecimal

The binary number 1110011110101001 is initially grouped in fours as:

$$1110 \quad 0111 \quad 1010 \quad 1001$$

and a hexadecimal symbol assigned to each group as:

$$E \qquad A \qquad 7 \qquad 9$$

from Table 32.2

Hence, $1110011110101001_2 = E7A9_{16}$

Conversion from hexadecimal to binary

Application: Convert $6CF3_{16}$ into binary form

$6CF3_{16} = 0110\ 1100\ 1111\ 0011$ from Table 32.2

i.e. $6CF3_{16} = 110110011110011_2$

Areas and volumes

Why are areas and volumes important?

To paint, wallpaper or panel a wall, you must know the total area of the wall, so you can buy the appropriate amount of finish. When designing a new building, or seeking planning permission, it is often necessary to specify the total floor area of the building. In construction, calculating the area of a gable end of a building is important when determining the amount of bricks and mortar to order. When using a bolt, the most important thing is that it is long enough for your application and it may also be necessary to calculate the shear area of the bolt connection. A race track is an oval shape, and it is sometimes necessary to find the perimeter of the inside of a race track. Arches are everywhere, from sculptures and monuments to pieces of architecture and strings on musical instruments; finding the height of an arch or its cross-sectional area is often required. Determining the cross-sectional areas of beam structures is vitally important in design engineering. There are thus many situations in engineering and science where determining area is important.

A circle is one of the fundamental shapes of geometry. Knowledge of calculations involving circles is needed with crank mechanisms, with determinations of latitude and longitude, with pendulums, and even in the design of paper clips. The floodlit area at a football ground, the area an automatic garden sprayer sprays and the angle of lap of a belt drive all rely on calculations involving the arc of a circle. The ability to handle calculations involving circles and its properties is clearly essential in several branches of engineering and science design.

There are many practical applications where volumes and surface areas of common solids are required. Examples include determining capacities of oil, water, petrol and fish tanks, ventilation shafts and cooling towers, determining volumes of blocks of metal, ball-bearings, boilers and buoys, and calculating the number of cubic metres of concrete needed for a path. Finding the surface areas of loudspeaker diaphragms and lampshades provide further practical examples. Understanding these calculations is essential for the many practical applications in engineering, construction, architecture and science.

Surveyors, farmers and landscapers often need to determine the area of irregularly shaped pieces of land to work with the land properly. There are many applications in business, economics and the sciences, including all aspects of engineering, where finding the areas of irregular shapes, the volumes of solids, and the lengths of irregular shaped curves are important applications. Typical earthworks include roads, railway beds, causeways, dams, canals, and berms. Other common earthworks are land grading to reconfigure the topography of a site, or to stabilise slopes. Engineers need to concern themselves with issues of geotechnical engineering (such as soil density and strength) and with quantity estimation to ensure that soil volumes in the cuts match those of the fills, while minimizing the distance of movement. Simpson's rule is a staple of scientific data analysis and engineering; it is widely used, for example, by naval architects to numerically determine hull offsets and cross-sectional areas to determine volumes and centroids of ships or lifeboats. There are therefore plenty of examples where irregular areas and volumes need to be determined by engineers and scientists.

Chapter 33 Areas of plane figures

(i) **Rectangle**   Area = l x b

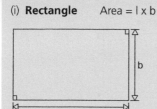

(ii) **Parallelogram**   Area = b x h

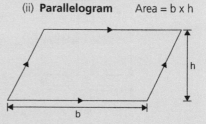

(iii) **Trapezium**   Area = $\frac{1}{2}$(a + b)h

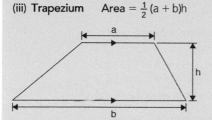

(iv) **Triangle**
Area = $\frac{1}{2}$ × b × h

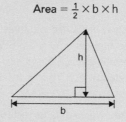

(v) **Ellipse**

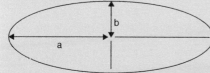

Area = πab
Perimeter $\approx \pi$ (a + b)

Figure 33.1

A **polygon** is a closed plane figure bounded by straight lines. A polygon, which has:

  (i) 3 sides is called a **triangle**
 (ii) 4 sides is called a **quadrilateral**
(iii) 5 sides is called a **pentagon**
(iv) 6 sides is called a **hexagon**
 (v) 7 sides is called a **heptagon**
(vi) 8 sides is called an **octagon**

Application: Find (a) the cross-sectional area of the girder shown in Figure 33.2(a), and (b) the area of the path shown in Figure 33.2(b)

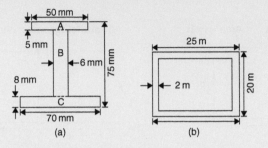

(a)   (b)

Figure 33.2

(a) The girder may be divided into three separate rectangles as shown.
Area of rectangle A = $50 \times 5 = 250\,mm^2$
Area of rectangle B = $(75 - 8 - 5) \times 6 = 62 \times 6 = 372\,mm^2$
Area of rectangle C = $70 \times 8 = 560\,mm^2$
Total area of girder = $250 + 372 + 560 = $ **1182 mm^2 or 11.82 cm^2**

(b) Area of path = area of large rectangle − area of small rectangle
$$= (25 \times 20) - (21 \times 16) = 500 - 336$$
$$= \mathbf{164\,m^2}$$

Application: Figure 33.3 shows the gable end of a building. Determine the area of brickwork in the gable end

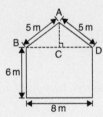

Figure 33.3

The shape is that of a rectangle and a triangle.
Area of rectangle = $6 \times 8 = 48\,m^2$

Area of triangle = $\dfrac{1}{2} \times$ base $\times$ height

CD = 4 m, AD = 5 m, hence AC = 3 m (since it is a 3, 4, 5 triangle)

Hence, area of triangle ABD = $\dfrac{1}{2} \times 8 \times 3 = 12\,m^2$

Total area of brickwork = $48 + 12 = $ **60 m^2**

Application: Calculate the area of a regular octagon, if each side is 5 cm and the width across the flats is 12 cm

An octagon is an 8-sided polygon. If radii are drawn from the centre of the polygon to the vertices then 8 equal triangles are produced (see Figure 33.4).

Area of one triangle $= \dfrac{1}{2} \times$ base $\times$ height

$$= \dfrac{1}{2} \times 5 \times \dfrac{12}{2} = 15 \text{ cm}^2$$

Area of octagon $= 8 \times 15 = 120 \text{ cm}^2$

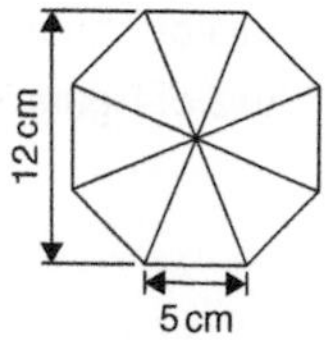

Figure 33.4

Application: Determine the area of a regular hexagon which has sides 8 cm long

A hexagon is a 6-sided polygon that may be divided into 6 equal triangles as shown in Figure 33.5. The angle subtended at the centre of each triangle is $360°/6 = 60°$.

The other two angles in the triangle add up to $120°$ and are equal to each other.

Hence each of the triangles is equilateral with each angle $60°$ and each side 8 cm.

Area of one triangle $= \dfrac{1}{2} \times$ base $\times$ height $= \dfrac{1}{2} \times 8 \times h$

h is calculated using Pythagoras' theorem:

$$8^2 = h^2 + 4^2$$

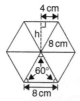

Figure 33.5

from which, $h = \sqrt{8^2 - 4^2} = 6.928 \text{ cm}$

Hence area of one triangle $= \dfrac{1}{2} \times 8 \times 6.928 = 27.71 \text{ cm}^2$

Area of hexagon $= 6 \times 27.71 = \mathbf{166.3 \text{ cm}^2}$

Areas of similar shapes

The areas of similar shapes are proportional to the squares of corresponding linear dimensions.

For example, Figure 33.6 shows two squares, one of which has sides three times as long as the other.

Area of Figure 33.6(a) $= (x)(x) = x^2$

Area of Figure 33.6(b) $= (3x)(3x) = 9x^2$

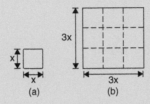

Figure 33.6

Hence Figure 33.6(b) has an area $(3)^2$, i.e. 9 times the area of Figure 33.6(a).

Application: A rectangular garage is shown on a building plan having dimensions 10 mm by 20 mm. If the plan is drawn to a scale of 1 to 250, determine the true area of the garage in square metres

Area of garage on the plan $= 10\,\text{mm} \times 20\,\text{mm} = 200\,\text{mm}^2$

Since the areas of similar shapes are proportional to the squares of corresponding dimensions then

$$\textbf{true area of garage} = 200 \times (250)^2 = 12.5 \times 10^6\,\text{mm}^2$$
$$= \frac{12.5 \times 10^6}{10^6}\,\text{m}^2 = \textbf{12.5}\,\textbf{m}^2$$

Chapter 34　Circles

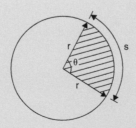

Figure 34.1

Area $= \pi r^2$ 　　　　　**Circumference** $= 2\pi r$

Radian measure: 2π radians $= 360°$

$$\text{Area of a sector} = \frac{\theta°}{360}(\pi r^2)$$

$$= \frac{\theta}{2\pi}(\pi r^2) = \frac{1}{2}r^2\theta \quad (\theta \text{ in radians})$$

$$\text{arc length, } s = \frac{\theta°}{360}(2\pi r)$$

$$= r\theta \quad (\theta \text{ in radians})$$

Application: Find the areas of the circles having (a) a radius of 5 cm, (b) a diameter of 15 mm, (c) a circumference of 70 mm

Area of a circle $= \pi r^2$ or $\dfrac{\pi d^2}{4}$

(a) Area $= \pi r^2 = \pi (5)^2 = 25\pi = \mathbf{78.54 \, cm^2}$

(b) Area $= \dfrac{\pi d^2}{4} = \dfrac{\pi (15)^2}{4} = \dfrac{225\pi}{4} = \mathbf{176.7 \, mm^2}$

(c) Circumference, $c = 2\pi r$, hence $r = \dfrac{c}{2\pi} = \dfrac{70}{2\pi} = \dfrac{35}{\pi}$ mm

Area of circle $= \pi r^2 = \pi\left(\dfrac{35}{\pi}\right)^2 = \dfrac{35^2}{\pi}$

$$= \mathbf{389.9 \, mm^2 \text{ or } 3.889 \, cm^2}$$

Application: A hollow shaft has an outside diameter of 5.45 cm and an inside diameter of 2.25 cm. Calculate the cross-sectional area of the shaft

The cross-sectional area of the shaft is shown by the shaded part in Figure 34.2 (often called an **annulus**).

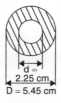

d = 2.25 cm
D = 5.45 cm   Figure 34.2

Area of shaded part = area of large circle − area of small circle

$$= \frac{\pi D^2}{4} - \frac{\pi d^2}{4} = \frac{\pi}{4}(D^2 - d^2)$$

$$= \frac{\pi}{4}(5.45^2 - 2.25^2) = \mathbf{19.35 \, cm^2}$$

Application: Convert (a) 125° (b) 69°47′ to radians

(a) Since 180° = π rad then 1° = π/180 rad, therefore

$$125° = 125\left(\frac{\pi}{180}\right)\text{rads} = \textbf{2.182 radians}$$

(b) $69°47′ = 69\dfrac{47°}{60} = 69.783°$

$$69.783° = 69.783\left(\frac{\pi}{180}\right) = \textbf{1.218 radians}$$

Application: Convert (a) 0.749 radians, (b) 3π/4 radians, to degrees and minutes

(a) Since π rad = 180° then 1 rad = 180°/π, therefore

$$0.749 = 0.749\left(\frac{180}{\pi}\right)^° = 42.915°$$

$0.915° = (0.915 \times 60)′ = 55′$, correct to the nearest minute,

hence **0.749 radians = 42°55′**

(b) Since $1 \text{ rad} = \left(\dfrac{180}{\pi}\right)^°$

then $\dfrac{3\pi}{4}\text{rad} = \dfrac{3\pi}{4}\left(\dfrac{180}{\pi}\right)^° = \dfrac{3}{4}(180)° = \textbf{135°}$

Application: Find the length of arc of a circle of radius 5.5 cm when the angle subtended at the centre is 1.20 radians

Length of arc, s = rθ, where θ is in radians, hence
$$s = (5.5)(1.20) = \textbf{6.60 cm}$$

Application: Determine the diameter and circumference of a circle if an arc of length 4.75 cm subtends an angle of 0.91 radians

Since s = rθ then $r = \dfrac{s}{\theta} = \dfrac{4.75}{0.91} = 5.22$ cm

Diameter = 2 × radius = 2 × 5.22 = **10.44 cm**

Circumference, c = πd = π(10.44) = **32.80 cm**

Application: Determine the angle, in degrees and minutes, subtended at the centre of a circle of diameter 42 mm by an arc of length 36 mm and the area of the minor sector formed

Since length of arc, $s = r\theta$ then $\theta = s/r$

Radius, $\quad r = \dfrac{\text{diameter}}{2} = \dfrac{42}{2} = 21\text{mm}$

hence $\quad \theta = \dfrac{s}{r} = \dfrac{36}{21} = 1.7143\,\text{radians}$

$1.7143\,\text{rad} = 1.7143 \times (180/\pi)° = 98.22° = \mathbf{98°13'}$

$\qquad\qquad = $ angle subtended at centre of circle

Area of sector $= \dfrac{1}{2}r^2\theta = \dfrac{1}{2}(21)^2(1.7143) = \mathbf{378\ mm^2}$

Application: A football stadium floodlight can spread its illumination over an angle of 45° to a distance of 55 m. Determine the maximum area that is floodlit

Floodlit area $= $ area of sector $= \dfrac{1}{2}r^2\theta = \dfrac{1}{2}(55)^2\left(45 \times \dfrac{\pi}{180}\right)$

$$= \mathbf{1188\ m^2}$$

Application: An automatic garden spray produces a spray to a distance of 1.8 m and revolves through an angle α which may be varied. If the desired spray catchment area is to be 2.5 m², determine the required angle α, correct to the nearest degree

Area of sector $= \dfrac{1}{2}r^2\theta$, hence $2.5 = \dfrac{1}{2}(1.8)^2\alpha$

from which, $\quad \alpha = \dfrac{2.5 \times 2}{1.8^2} = 1.5432\,\text{radians}$

$1.5432\,\text{rad} = \left(1.5432 \times \dfrac{180}{\pi}\right)° = 88.42°$

Hence **angle** $\pi = \mathbf{88°}$, correct to the nearest degree.

Application: The angle of a tapered groove is checked using a 20 mm diameter roller as shown in Figure 34.3. If the roller lies 2.12 mm below the top of the groove, determine the value of angle θ

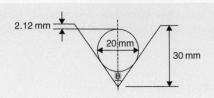

Figure 34.3

In Figure 34.4, triangle ABC is right-angled at C

Length BC = 10 mm (i.e. the radius of the circle),

and AB = 30 − 10 − 2.12 = 17.88 mm from Figure 34.4.

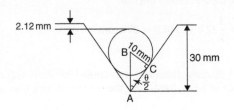

Figure 34.4

Hence,

$$\sin\frac{\theta}{2} = \frac{10}{17.88} \text{ and } \frac{\theta}{2} = \sin^{-1}\left(\frac{10}{17.88}\right) = 34° \text{ and } \textbf{angle } \boldsymbol{\theta} = \textbf{68°}$$

The equation of a circle

The equation of a circle, centre at the origin, radius r, is given by:

$$x^2 + y^2 = r^2$$

The equation of a circle, centre (a, b), radius r, is given by:

$$(x − a)^2 + (y − b)^2 = r^2$$

Figure 34.5 shows a circle $(x − 2)^2 + (y − 3)^2 = 4$

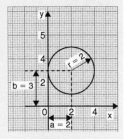

Figure 34.5

Application: Determine the radius and the co-ordinates of the centre of the circle given by the equation $x^2 + y^2 + 8x − 2y + 8 = 0$

$x^2 + y^2 + 8x - 2y + 8 = 0$ may be rearranged as:

$$(x + 4)^2 + (y - 1)^2 - 9 = 0$$

i.e. $\qquad\qquad\qquad (x + 4)^2 + (y - 1)^2 = 3^2$

which represents a circle, **centre (−4, 1)** and **radius 3** as shown in Figure 34.6.

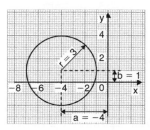

Figure 34.6

Chapter 35   Volumes and surface areas of regular solids

(i) Rectangular prism (or cubold)

Volume $= l \times b \times h$
Surface area $= 2(bh + hl + lb)$

(ii) Cylinder

Volume $= \pi r^2 h$
Surface area $= 2\pi rh + 2\pi r^2$

(ii) Pyramid

If area of base $= A$ and
perpendicular height $= h$ then:

Volume $= \frac{1}{3} \times A \times h$

Total surface area $=$ sum of areas of triangles forming sides $+$ area of base

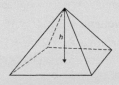

(iv) Cone

Volume $= \frac{1}{3}\pi r^2 h$
Curved surface area $= \pi rl$
Total surface area $= \pi rl + \pi r^2$

Figure 35.1

(v) Sphere

$$\text{Volume} = \frac{4}{3}\pi r^3$$
$$\text{Surface area} = 4\pi r^2$$

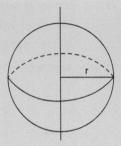

Figure 35.1 **(cont)**

Application: A water tank is the shape of a rectangular prism having length 2 m, breadth 75 cm and height 50 cm. Determine the capacity of the tank in (a) m^3 (b) cm^3 (c) litres

Volume of rectangular prism = l × b × h

(a) **Volume of tank** = 2 × 0.75 × 0.5 = **0.75 m³**

(b) $1\,m^3 = 10^6\,cm^3$, hence $0.75\,m^3 = 0.75 \times 10^6\,cm^3 =$ **750000 cm³**

(c) 1 litre = 1000 cm³, hence 750000 cm³ = $\dfrac{750000}{1000}$ litres

$$= 750 \text{ litres}$$

Application: Find the volume and total surface area of a cylinder of length 15 cm and diameter 8 cm

Volume of cylinder = $\pi r^2 h$

Since diameter = 8 cm, then radius r = 4 cm.

Hence, **volume** = $\pi \times 4^2 \times 15 =$ **754 cm³**

Total surface area = $2\pi rh + 2\pi r^2$ (i.e. including the two ends)

$$= (2 \times \pi \times 4 \times 15) + (2 \times \pi \times 4^2) = \textbf{477.5 cm}^2$$

Application: Determine the volume and the total surface area of the square pyramid shown in Figure 35.2 if its perpendicular height is 12 cm

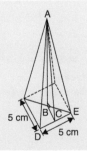

5 cm

5 cm

Figure 35.2

Volume of pyramid $= \dfrac{1}{3}$ (area of base) $\times$ perpendicular height

$$= \dfrac{1}{3}(5 \times 5) \times 12 = \mathbf{100\ cm^3}$$

The total surface area consists of a square base and 4 equal triangles.

Area of triangle ADE $= \dfrac{1}{2} \times$ base $\times$ perpendicular height

$$= \dfrac{1}{2} \times 5 \times AC$$

The length AC may be calculated using Pythagoras' theorem on triangle ABC, where

$AB = 12\ cm,\ BC = \dfrac{1}{2} \times 5 = 2.5\ cm,$

and $\quad AC = \sqrt{AB^2 + BC^2} = \sqrt{12^2 + 2.5^2} = 12.26\ cm$

Hence area of triangle ADE $= \dfrac{1}{2} \times 5 \times 12.26 = 30.65\ cm^2$

Total surface area of pyramid $= (5 \times 5) + 4(30.65) = \mathbf{147.6\ cm^2}$

Application: Determine the volume and total surface area of a cone of radius 5 cm and perpendicular height 12 cm

Figure 35.3

The cone is shown in Figure 35.3.

Volume of cone $= \dfrac{1}{3} \pi r^2 h = \dfrac{1}{3} \times \pi \times 5^2 \times 12 = \mathbf{314.2\ cm^3}$

Total surface area = curved surface area + area of base = $\pi rl + \pi r^2$

From Figure 35.3, slant height l may be calculated using Pythagoras' theorem:
$l = \sqrt{12^2 + 5^2} = 13\ cm$

Hence, **total surface area** $= (\pi \times 5 \times 13) + (\pi \times 5^2) = \mathbf{282.7\ cm^2}$

Application: A wooden section is shown in Figure 35.4. Find (a) its volume (in m³), and (b) its total surface area

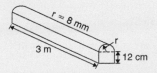

Figure 35.4

The section of wood is a prism whose end comprises a rectangle and a semicircle. Since the radius of the semicircle is 8 cm, the diameter is 16 cm. Hence the rectangle has dimensions 12 cm by 16 cm.

Area of end $= (12 \times 16) + \dfrac{1}{2}\pi 8^2 = 292.5\,cm^2$

Volume of wooden section $=$ area of end $\times$ perpendicular height

$$= 292.5 \times 300 = 87750\,cm^3$$

$$= \dfrac{87750\,m^3}{10^6} = \textbf{0.08775}\,\textbf{m}^3$$

The total surface area comprises the two ends (each of area 292.5 cm²), three rectangles and a curved surface (which is half a cylinder), hence

total surface area $= (2 \times 292.5) + 2(12 \times 300) + (16 \times 300)$
$$+ \dfrac{1}{2}(2\pi \times 8 \times 300)$$

$$= 585 + 7200 + 4800 + 2400\pi$$
$$= \textbf{20125}\,\textbf{cm}^2\ \text{or}\ \textbf{2.0125}\,\textbf{m}^2$$

Application: A rivet consists of a cylindrical head, of diameter 1 cm and depth 2 mm, and a shaft of diameter 2 mm and length 1.5 cm. Determine the volume of metal in 2000 such rivets

Radius of cylindrical head $= \dfrac{1}{2}\,cm = 0.5\,cm$
and height of cylindrical head 2 mm $= 0.2\,cm$

Hence, volume of cylindrical head $= \pi r^2 h = \pi\,(0.5)^2(0.2) = 0.1571\,cm^3$

Volume of cylindrical shaft $= \pi r^2 h = \pi\left(\dfrac{0.2}{2}\right)^2(1.5) = 0.0471\,cm^3$

Total volume of 1 rivet $= 0.1571 + 0.0471 = 0.2042\,cm^3$
Volume of metal in 2000 such rivets $= 2000 \times 0.2042 = \textbf{408.4}\,\textbf{cm}^3$

Application: A boiler consists of a cylindrical section of length 8 m and diameter 6 m, on one end of which is surmounted a hemispherical section of diameter 6 m, and on the other end a conical section of height 4 m and base diameter 6 m. Calculate the volume of the boiler and the total surface area

The boiler is shown in Figure 35.5.

Volume of hemisphere, P $= \dfrac{2}{3}\pi r^3 = \dfrac{2}{3} \times \pi \times 3^3 = 18\pi\,m^3$

Volume of cylinder, $Q = \pi r^2 h = \pi \times 3^2 \times 8 = 72\pi\, m^3$

Volume of cone, $R = \dfrac{1}{3}\pi r^2 h = \dfrac{1}{3} \times \pi \times 3^2 \times 4 = 12\pi\, m^3$

Total volume of boiler $= 18\pi + 72\pi + 12\pi = 102\pi = \mathbf{320.4\, m^3}$

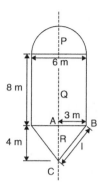

Figure 35.5

Surface area of hemisphere, $P = \dfrac{1}{2}(4\pi r^2) = 2 \times \pi \times 3^2 = 18\pi\, m^2$

Curved surface area of cylinder, $Q = 2\pi rh = 2 \times \pi \times 3 \times 8 = 48\pi\, m^2$

The slant height of the cone, I, is obtained by Pythagoras' theorem on triangle ABC,

i.e. $I = \sqrt{(4^2 + 3^2)} = 5\, m$

Curved surface area of cone, $R = \pi rl = \pi \times 3 \times 5 = 15\pi\, m^2$

Total surface area of boiler $= 18\pi + 48\pi + 15\pi = 81\pi = \mathbf{254.5\, m^2}$

Volumes of similar shapes

The volumes of similar bodies are proportional to the cubes of corresponding linear dimensions.

For example, Figure 35.6 shows two cubes, one of which has sides three times as long as those of the other.

Volume of Figure 35.6(a) $= (x)(x)(x) = x^3$

Volume of Figure 35.6(b) $= (3x)(3x)(3x)$
$$= 27x^3$$

Hence Figure 35.6(b) has a volume $(3)^3$, i.e. 27 times the volume of Figure 35.6(a).

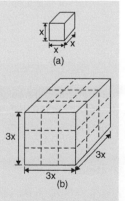

Figure 35.6

Application: A car has a mass of 1000 kg. A model of the car is made to a scale of 1 to 50. Determine the mass of the model if the car and its model are made of the same material

$$\frac{\text{Volume of model}}{\text{Volume of car}} = \left(\frac{1}{50}\right)^3$$ since the volume of similar bodies are proportional to

the cube of corresponding dimensions.

Mass = density × volume, and since both car and model are made of the same material then:

$$\frac{\text{Mass of model}}{\text{Mass of car}} = \left(\frac{1}{50}\right)^3$$

Hence, **mass of model** = (mass of car)$\left(\dfrac{1}{50}\right)^3 = \dfrac{1000}{50^3}$

$$= \textbf{0.008 kg or 8 g}$$

Chapter 36   Volumes and surface areas of frusta of pyramids and cones

For the **frustum of a cone** shown in Figure 36.1:

$$\text{Volume} = \frac{1}{3}\pi h(R^2 + Rr + r^2)$$

$$\textbf{Curved surface area} = \pi l(R + r)$$

$$\textbf{Total surface area} = \pi l(R + r) + \pi r^2 + \pi R^2$$

Figure 36.1

Application:

(a) Determine the volume of a frustum of a cone if the diameter of the ends are 6.0 cm and 4.0 cm and its perpendicular height is 3.6 cm.

(b) Find the total surface area of the frustum of the cone

(a) **Method 1**

A section through the vertex of a complete cone is shown in Figure 36.2.

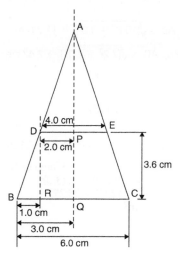

Figure 36.2

Using similar triangles $\dfrac{AP}{DP} = \dfrac{DR}{BR}$

Hence $\dfrac{AP}{2.0} = \dfrac{3.6}{1.0}$, from which AP $= \dfrac{(2.0)(3.6)}{1.0} = 7.2\,cm$

The height of the large cone $= 3.6 + 7.2 = 10.8\,cm$

Volume of frustum of cone = volume of large cone
$\qquad\qquad\qquad\qquad\qquad$ − volume of small cone cut off

$$= \frac{1}{3}\pi(3.0)^2(10.8) - \frac{1}{3}\pi(2.0)^2(7.2)$$
$$= 101.79 - 30.16 = \mathbf{71.6\,cm^3}$$

Method 2

From above, volume of the frustum of a cone $= \dfrac{1}{3}\pi h(R^2 + Rr + r^2)$,
$\qquad\qquad$ where R $= 3.0\,cm$, r $= 2.0\,cm$ and h $= 3.6\,cm$

Hence, **volume of frustum** $= \dfrac{1}{3}\pi(3.6)[(3.0)^2 + (3.0)(2.0) + (2.0)^2]$

$$= \frac{1}{3}\pi(3.6)(19.0) = \mathbf{71.6\,cm^3}$$

(b) Method 1

Curved surface area of frustum = curved surface area of large cone
$\qquad\qquad\qquad\qquad\qquad\qquad$ − curved surface area of small cone cut off

From Figure 36.2, using Pythagoras' theorem:

$AB^2 = AQ^2 + BQ^2$, from which, $AB = \sqrt{[10.8^2 + 3.0^2]} = 11.21\,\text{cm}$

and $AD^2 = AP^2 + DP^2$, from which, $AD = \sqrt{[7.2^2 + 2.0^2]} = 7.47\,\text{cm}$

Curved surface area of large cone $= \pi r l = \pi$ (BQ)(AB)

$$= \pi\,(3.0)(11.21) = 105.65\,\text{cm}^2$$

and curved surface area of small cone $= \pi$ (DP)(AD)

$$= \pi\,(2.0)(7.47) = 46.94\,\text{cm}^2$$

Hence, curved surface area of frustum $= 105.65 - 46.94 = 58.71\,\text{cm}^2$

Total surface area of frustum = curved surface area + area of two circular ends

$$= 58.71 + \pi(2.0)^2 + \pi(3.0)^2$$
$$= 58.71 + 12.57 + 28.27$$
$$= \mathbf{99.6\,cm^2}$$

Method 2

Total surface area of frustum $= \pi l(R + r) + \pi r^2 + \pi R^2$,

where $l = BD = 11.21 - 7.47 = 3.74\,\text{cm}$, $R = 3.0\,\text{cm}$ and $r = 2.0\,\text{cm}$.

Hence, **total surface area of frustum**

$$= \pi(3.74)(3.0 + 2.0) + \pi(2.0)^2 + \pi(3.0)^2 = \mathbf{99.6\,cm^2}$$

Application: A lampshade is in the shape of a frustum of a cone. The vertical height of the shade is 25.0 cm and the diameters of the ends are 20.0 cm and 10.0 cm, respectively. Determine the area of the material needed to form the lampshade, correct to 3 significant figures

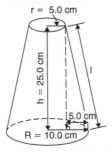

Figure 36.3

The curved surface area of a frustum of a cone $= \pi l(R + r)$

Since the diameters of the ends of the frustum are 20.0 cm and 10.0 cm, then from Figure 36.3, $r = 5.0\,\text{cm}$, $R = 10.0\,\text{cm}$ and $l = \sqrt{[25.0^2 + 5.0^2]} = 25.50\,\text{cm}$, from Pythagoras' theorem.

Hence curved surface area $= \pi(25.50)(10.0 + 5.0) = 1201.7\,\text{cm}^2$, i.e. **the area of material needed to form the lampshade is 1200 cm²**, correct to 3 significant figures.

Application: A cooling tower is in the form of a cylinder surmounted by a frustum of a cone as shown in Figure 36.4. Determine the volume of air space in the tower if 40% of the space is used for pipes and other structures

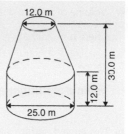

Figure 36.4

Volume of cylindrical portion $= \pi r^2 h = \pi \left(\dfrac{25.0}{2} \right)^2 (12.0) = 5890\,\text{m}^3$

Volume of frustum of cone $= \dfrac{1}{3}\pi h(R^2 + Rr + r^2)$

where h = 30.0 − 12.0 = 18.0 m, R = 25.0/2 = 12.5 m and r = 12.0/2 = 6.0 m.

Hence volume of frustum of cone

$$= \frac{1}{3}\pi(18.0)\,[(12.5)^2 + (12.5)(6.0) + (6.0)^2]$$
$$= 5038\,\text{m}^3$$

Total volume of cooling tower = 5890 + 5038 = 10928 m³

If 40% of space is occupied then
volume of air space = 0.6 × 10928 = **6557 m³**

Chapter 37   The frustum and zone of a sphere

With reference to the **zone of a sphere** shown in Figure 37.1:

Surface area of a zone of a sphere = 2πrh

Volume of frustum of sphere $= \dfrac{\pi h}{6}\,(h^2 + 3r_1^2 + 3r_2^2)$

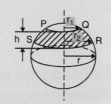

Figure 37.1

Application:

(a) Determine the volume of a frustum of a sphere of diameter 49.74 cm if the diameter of the ends of the frustum are 24.0 cm and 40.0 cm, and the height of the frustum is 7.00 cm.

(b) Determine the curved surface area of the frustum

(a) Volume of frustum of a sphere $= \dfrac{\pi h}{6}(h^2 + 3r_1^2 + 3r_2^2)$

where $h = 7.00$ cm, $r_1 = 24.0/2 = 12.0$ cm and
$r_2 = 40.0/2 = 20.0$ cm.

Hence,

volume of frustum $= \dfrac{\pi(7.00)}{6}[(7.00)^2 + 3(12.0)^2 + 3(20.0)^2]$

$= \mathbf{6161\,cm^3}$

(b) The curved surface area of the frustum = surface area of zone $= 2\pi rh$ (from above), where r = radius of sphere $= 49.74/2 = 24.87$ cm and $h = 7.00$ cm.

Hence, **surface area of zone** $= 2\pi(24.87)(7.00) = \mathbf{1094\,cm^2}$

Application: A frustum of a sphere, of diameter 12.0 cm, is formed by two parallel planes, one through the diameter and the other distance h from the diameter. The curved surface area of the frustum is required to be $\frac{1}{4}$ of the total surface area of the sphere. Determine (a) the volume and surface area of the sphere, (b) the thickness h of the frustum, (c) the volume of the frustum, and (d) the volume of the frustum expressed as a percentage of the sphere

---

(a) **Volume of sphere**, $V = \dfrac{4}{3}\pi r^3 = \dfrac{4}{3}\pi\left(\dfrac{12.0}{2}\right)^3 = \mathbf{904.8\ cm^3}$

   Surface area of sphere $= 4\pi r^2 = 4\pi\left(\dfrac{12.0}{2}\right)^2 = \mathbf{452.4\ cm^2}$

(b) Curved surface area of frustum $= \dfrac{1}{4} \times$ surface area of sphere

$= \dfrac{1}{4} \times 452.4 = 113.1 \text{cm}^2$

From above, $113.1 = 2\pi rh = 2\pi\left(\dfrac{12.0}{2}\right)h$

Hence, **thickness of frustum, h** $= \dfrac{113.1}{2\pi(6.0)} = \mathbf{3.0\ cm}$

(c) Volume of frustum, $V = \dfrac{\pi h}{6}(h^2 + 3r_1^2 + 3r_2^2)$

where $h = 3.0$ cm, $r_2 = 6.0$ cm and $r_1 = \sqrt{OQ^2 - OP^2}$, from Figure 37.2,

i.e. $r_1 = \sqrt{6.0^2 - 3.0^2} = 5.196$ cm

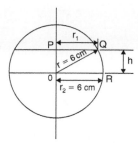

Figure 37.2

Hence,

volume of frustum $= \dfrac{\pi(3.0)}{6}[(3.0)^2 + 3(5.196)^2 + 3(6.0)^2]$

$\qquad\qquad = \dfrac{\pi}{2}[9.0 + 81 + 108.0] =$ **311.0 cm³**

(d) $\dfrac{\text{Volume of frustum}}{\text{Volume of sphere}} = \dfrac{311.0}{904.8} \times 100\% =$ **34.37%**

Application: A spherical storage tank is filled with liquid to a depth of 20 cm. Determine the number of litres of liquid in the container (1 litre = 1000 cm³), if the internal diameter of the vessel is 30 cm

The liquid is represented by the shaded area in the section shown in Figure 37.3.

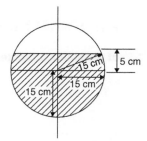

Figure 37.3

The volume of liquid comprises a hemisphere and a frustum of thickness 5 cm.

Hence volume of liquid $= \dfrac{2}{3}\pi r^3 + \dfrac{\pi h}{6}[h^2 + 3r_1^2 + 3r_2^2]$

where $r_2 = 30/2 = 15\,\text{cm}$ and $r_1 = \sqrt{15^2 - 5^2} = 14.14\,\text{cm}$

Volume of liquid $= \dfrac{2}{3}\pi(15)^3 + \dfrac{\pi(5)}{6}[5^2 + 3(14.14)^2 + 3(15)^2]$

$\qquad\qquad = 7069 + 3403 = 10470\,\text{cm}^3$

Since 1 litre = 1000 cm³,

the number of litres of liquid $= \dfrac{10470}{1000} =$ **10.47 litres**

Chapter 38   Areas and volumes of irregular figures and solids

Areas of irregular figures

Trapezoidal rule

To determine the areas PQRS in Figure 38.1:

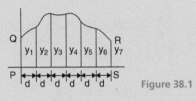

Figure 38.1

(i) Divide base PS into any number of equal intervals, each of width d (the greater the number of intervals, the greater the accuracy)

(ii) Accurately measure ordinates y_1, y_2, y_3, etc.

(iii) Area PQRS $= d \left[\dfrac{y_1 + y_7}{2} + y_2 + y_3 + y_4 + y_5 + y_6 \right]$

In general, the trapezoidal rule states:

$$\text{Area} = \text{(width of interval)} \left[\frac{1}{2} \text{(first + last ordinate)} + \text{sum of remaining ordinates} \right]$$

Mid-ordinate rule

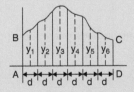

Figure 38.2

To determine the area ABCD of Figure 38.2:
 (i) Divide base AD into any number of equal intervals, each of width d (the greater the number of intervals, the greater the accuracy)
 (ii) Erect ordinates in the middle of each interval (shown by broken lines in Figure 38.2)
(iii) Accurately measure ordinates y_1, y_2, y_3, etc.
(iv) Area ABCD $= d(y_1 + y_2 + y_3 + y_4 + y_5 + y_6)$

In general, the mid-ordinate rule states:

$$\text{Area} = (\text{width of interval})(\text{sum of mid-ordinates})$$

Simpson's rule

To determine the area PQRS of Figure 38.1:

(i) Divide base PS into an **even** number of intervals, each of width d (the greater the number of intervals, the greater the accuracy)

(ii) Accurately measure ordinates y_1, y_2, y_3, etc.

(iii) Area PQRS $= \dfrac{d}{3}[(y_1 + y_7) + 4(y_2 + y_4 + y_6) + 2(y_3 + y_5)]$

In general, Simpson's rule states:

$$\text{Area} = \frac{1}{3}\begin{pmatrix}\text{width of} \\ \text{interval}\end{pmatrix}\left[\begin{pmatrix}\text{first + last} \\ \text{ordinate}\end{pmatrix} + 4\begin{pmatrix}\text{sum of even} \\ \text{ordinates}\end{pmatrix} + 2\begin{pmatrix}\text{sum of remaining} \\ \text{odd ordinates}\end{pmatrix}\right]$$

Application: A car starts from rest and its speed is measured every second for 6 s:

Time t(s)	0	1	2	3	4	5	6
Speed v (m/s)	0	2.5	5.5	8.75	12.5	17.5	24.0

Determine the distance travelled in 6 seconds (i.e. the area under the v/t graph), by (a) the trapezoidal rule, (b) the mid-ordinate rule, and (c) Simpson's rule

A graph of speed/time is shown in Figure 38.3.

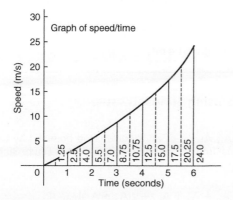

Figure 38.3

(a) Trapezoidal rule

The time base is divided into 6 strips each of width 1 s, and the length of the ordinates measured.

Thus

$$\text{area} = (1)\left[\left(\frac{0 + 24.0}{2}\right) + 2.5 + 5.5 + 8.75 + 12.5 + 17.5\right]$$

$$= 58.75 \text{ m}$$

(b) Mid-ordinate rule

The time base is divided into 6 strips each of width 1 second. Mid-ordinates are erected as shown in Figure 38.3 by the broken lines.

The length of each mid-ordinate is measured.

Thus

$$\text{area} = (1)[1.25 + 4.0 + 7.0 + 10.75 + 15.0 + 20.25] = 58.25 \text{ m}$$

(c) Simpson's rule

The time base is divided into 6 strips each of width 1 s, and the length of the ordinates measured.

Thus,

$$\text{area} = \frac{1}{3}(1)[(0 + 24.0) + 4(2.5 + 8.75 + 17.5) + 2(5.5 + 12.5)]$$

$$= 58.33 \text{ m}$$

Application: A river is 15 m wide. Soundings of the depth are made at equal intervals of 3 m across the river and are as shown below.

Depth (m)	0	2.2	3.3	4.5	4.2	2.4	0

Calculate the cross-sectional area of the flow of water at this point using Simpson's rule

From above,

$$\text{Area} = \frac{1}{3}(3)[(0 + 0) + 4(2.2 + 4.5 + 2.4) + 2(3.3 + 4.2)]$$

$$= (1)[0 + 36.4 + 15] = 51.4 \text{ m}^2$$

Volumes of irregular solids using Simpson's rule

If the cross-sectional areas A_1, A_2, A_3, ... of an irregular solid bounded by two parallel planes are known at equal intervals of width d (as shown in Figure 38.4), then by Simpson's rule:

$$\text{Volume, } V = \frac{d}{3}[(A_1 + A_7) + 4(A_2 + A_4 + A_6) + 2(A_3 + A_5)]$$

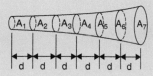

Figure 38.4

Application: A tree trunk is 12 m in length and has a varying cross-section. The cross-sectional areas at intervals of 2 m measured from one end are:

$$0.52, 0.55, 0.59, 0.63, 0.72, 0.84, 0.97 \, m^2$$

Estimate the volume of the tree trunk

A sketch of the tree trunk is similar to that shown in Figure 38.4 above, where $d = 2$ m, $A_1 = 0.52 \, m^2$, $A_2 = 0.55 \, m^2$, and so on. Using Simpson's rule for volumes gives:

$$\text{Volume} = \frac{2}{3}[(0.52 + 0.97) + 4(0.55 + 0.63 + 0.84) + 2(0.59 + 0.72)]$$

$$= \frac{2}{3}[1.49 + 8.08 + 2.62] = \mathbf{8.13 \, m^3}$$

Application: The areas of seven horizontal cross-sections of a water reservoir at intervals of 10 m are: 210, 250, 320, 350, 290, 230, 170 m^2. Calculate the capacity of the reservoir in litres

Using Simpson's rule for volumes gives:

$$\text{Volume} = \frac{10}{3}[(210 + 170) + 4(250 + 350 + 230) + 2(320 + 290)]$$

$$= \frac{10}{3}[380 + 3320 + 1220] = \mathbf{16400 \, m^3}$$

$16400 \, m^3 = 16400 \times 10^6 \, cm^3$

Since 1 litre = 1000 cm^3, capacity of reservoir $= \dfrac{16400 \times 10^6}{1000}$ litres

$$= 16400000$$

$$= \mathbf{1.64 \times 10^7 \, litres}$$

Prismoidal rule for finding volumes

With reference to Figure 38.5,

Volume, V $= \dfrac{x}{6}[A_1 + 4A_2 + A_3]$

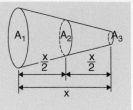

Figure 38.5

Application: A container is in the shape of a frustum of a cone. Its diameter at the bottom is 18 cm and at the top 30 cm. Determine the capacity of the container, correct to the nearest litre, by the prismoidal rule, if the depth is 24 cm.

The container is shown in Figure 38.6. At the mid-point, i.e. at a distance of 12 cm from one end, the radius r_2 is $(9 + 15)/2 = 12$ cm, since the sloping sides change uniformly.

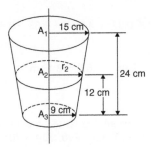

Figure 38.6

Volume of container by the prismoidal rule $= \dfrac{x}{6}[A_1 + 4A_2 + A_3]$

from above, where

$x = 24$ cm, $A_1 = \pi(15)^2$ cm^2, $A_2 = \pi(12)^2$ cm^2 and $A_3 = \pi(9)^2$ cm^2

Hence,

$$\textbf{volume of container} = \frac{24}{6}[\pi(15)^2 + 4\pi(12)^2 + \pi(9)^2]$$

$$= 4[706.86 + 1809.56 + 254.47]$$

$$= 11080 \text{ cm}^3 = \frac{11080}{1000} \text{ litres}$$

$$= \textbf{11 litres, correct to the nearest litre}$$

Application: The roof of a building is in the form of a frustum of a pyramid with a square base of side 5.0 m. The flat top is a square of side 1.0 m and all the sloping sides are pitched at the same angle. The vertical height of the flat top above the level of the eaves is 4.0 m. Calculate, using the prismoidal rule, the volume enclosed by the roof

Let area of top of frustum be $A_1 = (1.0)^2 = 1.0\,m^2$

Let area of bottom of frustum be $A_3 = (5.0)^2 = 25.0\,m^2$

Let area of section through the middle of the frustum parallel to A_1 and A_3 be A_2. The length of the side of the square forming A_2 is the average of the sides forming A_1 and A_3, i.e. (1.0 + 5.0)/2 = 3.0 m.

Hence $A_2 = (3.0)^2 = 9.0\,m^2$.

Using the prismoidal rule,

$$\text{volume of frustum} = \frac{x}{6}[A_1 + 4A_2 + A_3]$$

$$= \frac{4.0}{6}[1.0 + 4(9.0) + 25.0]$$

Hence, **volume enclosed by roof = 41.3 m³**

Chapter 39   The mean or average value of a waveform

The mean or average value, y, of the waveform shown in Figure 39.1 is given by:

$$y = \frac{\text{area under curve}}{\text{length of base, b}}$$

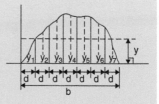

Figure 39.1

If the mid-ordinate rule is used to find the area under the curve, then:

$$y = \frac{\text{sum of mid-ordinates}}{\text{number of mid-ordinates}}$$

$$\left(= \frac{y_1 + y_2 + y_3 + y_4 + y_5 + y_6 + y_7}{7} \quad \text{for Figure 39.1} \right)$$

For a **sine wave**, the mean or average value:
1. over one complete cycle is zero (see Figure 39.2(a)),
2. over half a cycle is **0.637 × maximum value**, or **2/π × maximum value**,

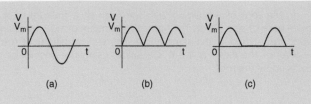

Figure 39.2

3. of a full-wave rectified waveform (see Figure 39.2(b)) is
   0.637 × maximum value
4. of a half-wave rectified waveform (see Fig. 39.2(c)) is
   0.318 × maximum value or **1/π × maximum value**

Application: Determine the average values over half a cycle of the periodic waveforms shown in Figure 39.3:

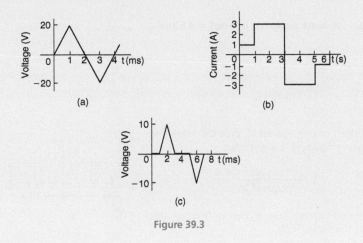

Figure 39.3

(a) Area under triangular waveform (a) for a half cycle is given by:

$$\text{Area} = \frac{1}{2}(\text{base})(\text{perpendicular height}) = \frac{1}{2}(2 \times 10^{-3})(20)$$
$$= 20 \times 10^{-3} \text{ Vs}$$

$$\text{Average value of waveform} = \frac{\text{area under curve}}{\text{length of base}}$$
$$= \frac{20 \times 10^{-3} \text{ Vs}}{2 \times 10^{-3} \text{ s}} = \textbf{10 V}$$

(b) Area under waveform (b) for a half cycle = $(1 \times 1) + (3 \times 2) = 7\,As$

Average value of waveform $= \dfrac{\text{area under curve}}{\text{length of base}} = \dfrac{7\,As}{3\,s}$

$$= 2.33\,A$$

(c) A half cycle of the voltage waveform (c) is completed in 4 ms.

Area under curve $= \dfrac{1}{2}\{(3-1)10^{-3}\}(10) = 10 \times 10^{-3}\,Vs$

Average value of waveform $= \dfrac{\text{area under curve}}{\text{length of base}}$

$$= \dfrac{10 \times 10^{-3}\,Vs}{4 \times 10^{-3}\,s} = 2.5\,V$$

Application: The power used in a manufacturing process during a 6 hour period is recorded at intervals of 1 hour as shown below.

Time (h)	0	1	2	3	4	5	6
Power (kW)	0	14	29	51	45	23	0

Determine (a) the area under the curve and (b) the average value of the power by plotting a graph of power against time and by using the mid-ordinate rule

The graph of power/time is shown in Figure 39.4.

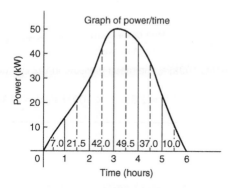

Figure 39.4

(a) The time base is divided into 6 equal intervals, each of width 1 hour.

Mid-ordinates are erected (shown by broken lines in Figure 39.4) and measured. The values are shown in Figure 39.4.

Area under curve = (width of interval)(sum of mid-ordinates)
$$= (1)[7.0 + 21.5 + 42.0 + 49.5 + 37.0 + 10.0]$$
$$= 167\,kWh \text{ (i.e. a measure of electrical energy)}$$

(b) Average value of waveform $= \dfrac{\text{area under curve}}{\text{length of base}} = \dfrac{167\,\text{kWh}}{6\,\text{h}}$

$= 27.83\,\text{kW}$

$\left(\text{Alternatively, average value} = \dfrac{\text{sum of mid-ordinates}}{\text{number of mid-ordinates}}\right)$

Application: An indicator diagram for a steam engine is shown in Figure 39.5. The base line has been divided into 6 equally spaced intervals and the lengths of the 7 ordinates measured with the results shown in centimetres.

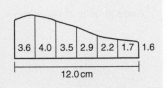

Determine (a) the area of the indicator diagram using Simpson's rule, and (b) the mean pressure in the cylinder given that 1 cm represents 100 kPa

Figure 39.5

(a) The width of each interval is $\dfrac{12.0}{6} = 2.0\,\text{cm}$. Using Simpson's rule,

area $= \dfrac{1}{3}(2.0)[(3.6 + 1.6) + 4(4.0 + 2.9 + 1.7) + 2(3.5 + 2.2)]$

$= \dfrac{2}{3}[5.2 + 34.4 + 11.4] = \textbf{34 cm}^2$

(b) Mean height of ordinates $= \dfrac{\text{area of diagram}}{\text{length of base}} = \dfrac{34}{12} = 2.83\,\text{cm}$

Since 1 cm represents 100 kPa, the mean pressure in the cylinder

$= 2.83\,\text{cm} \times 100\,\text{kPa/cm} = \textbf{283 kPa}$

Section 5

Geometry and trigonometry

Why is geometry and trigonometry important?

Knowledge of angles and triangles is very important in engineering and science. Trigonometry is needed in surveying and architecture, for building structures/ systems, designing bridges and solving scientific problems. Trigonometry is also used in electrical engineering; the functions that relate angles and side lengths in right angled triangles are useful in expressing how a.c. electric current varies with time. Engineers use triangles to determine how much force it will take to move along an incline, GPS satellite receivers use triangles to determine exactly where they are in relation to satellites orbiting hundreds of miles away. Whether you want to build a skateboard ramp, a stairway, or a bridge, you can't escape trigonometry. Further fields that use trigonometry include astronomy (especially for locating apparent positions of celestial objects, in which spherical trigonometry is essential) and hence navigation (on the oceans, in aircraft, and in space), music theory, acoustics, optics, analysis of financial markets, electronics, probability theory, statistics, biology, medical imaging (CAT scans and ultrasound), pharmacy, chemistry, number theory (and hence cryptology), seismology, meteorology, oceanography, many physical sciences, land surveying and geodesy (a branch of earth sciences), architecture, phonetics, economics, electrical engineering, mechanical engineering, civil engineering, computer graphics, cartography, crystallography and game development.

Trigonometric graphs are commonly used in all areas of science and engineering for modeling many different natural and mechanical phenomena such as waves, engines, acoustics, electronics, populations, UV intensity, growth of plants and animals, and so on. Periodic trigonometric graphs mean that the shape repeats itself exactly after a certain amount of time. Anything that has a regular cycle, like the tides, temperatures, rotation of the earth, and so on, can be modeled using a sine or cosine curve. The most common periodic signal waveform that is used in electrical and electronic engineering is the sinusoidal waveform. However, an alternating a.c. waveform may not always take the shape of a smooth shape based around the sine and cosine function; a.c. waveforms can also take the shape of square or triangular waves, i.e. complex waves. In engineering and science, it is therefore important to have some clear understanding of sine and cosine waveforms.

Applications where polar co-ordinates would be used include terrestrial navigation with sonar-like devices, and those in engineering and science involving energy radiation patterns. Applications where Cartesian co-ordinates would be used include any navigation on a grid and anything involving raster graphics (i.e. bitmap – a dot matrix data structure representing a generally rectangular grid of pixels). The ability to change from Cartesian to polar co-ordinates is vitally important when using complex numbers and their use in a.c. electrical circuit theory and with vector geometry.

In engineering, trigonometric identities occur often, examples being in the more advanced areas of calculus to generate derivatives and integrals, with tensors/vectors, and with differential equations and partial differential equations. One of the skills required for more advanced work in mathematics, especially in calculus, is the ability to use identities to write expressions in alternative forms. In software engineering, working, say, on the next 'Ratatouille' or 'Toy Story', trigonometric identities are needed for computer graphics; an RF engineer working on the next-generation mobile phone will also need trigonometric identities. In addition, identities are needed in electrical engineering when dealing with a.c. power, and wave addition/subtraction and the solutions of trigonometric equations often require knowledge of trigonometric identities. Compound angle (or sum and difference) formulae, and double angles are further commonly used identities. Compound angles are required for example in the analysis of acoustics (where a beat is an interference between two sounds of slightly different frequencies), and with phase detectors (which is a frequency mixer, analog multiplier or logic circuit that generates a voltage signal which represents the difference in phase between two signal inputs). Many rational functions of sine and cosine are difficult to integrate without compound angle formulae.

Chapter 40   Types and properties of angles

1. Any angle between 0° and 90° is called an **acute angle**.

2. An angle equal to 90° is called a **right angle**.

3. Any angle between 90° and 180° is called an **obtuse angle**.

4. Any angle greater than 180° and less than 360° is called a **reflex angle**.

5. An angle of 180° lies on a straight line.

6. If two angles add up to 90° they are called **complementary angles**.

7. If two angles add up to 180° they are called **supplementary angles**.

8. **Parallel lines** are straight lines that are in the same plane and never meet. (Such lines are denoted by arrows, as in Figure 40.1).

9. A straight line that crosses two parallel lines is called a **transversal** (see MN in Figure. 40.1).

10. With reference to Figure 40.1:
    (i) a = c, b = d, e = g and f = h. Such pairs of angles are called **vertically opposite angles**.
    (ii) a = e, b = f, c = g and d = h. Such pairs of angles are called **corresponding angles**.
    (iii) c = e and b = h. Such pairs of angles are called **alternate angles**.
    (iv) b + e = 180° and c + h = 180°. Such pairs of angles are called **interior angles**.

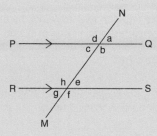

Figure 40.1

Chapter 41   Properties of triangles

1. The sum of the three angles of a triangle is equal to 180°.

2. An **acute-angled triangle** is one in which all the angles are acute, i.e. all the angles less than 90°.

3. A **right-angled triangle** is one that contains a right angle.

4. An **obtuse-angled triangle** is one that contains an obtuse angle, i.e. one angle which lies between 90° and 180°.

5. An **equilateral triangle** is one in which all the sides and all the angles are equal (i.e. each 60°).

6. An **isosceles triangle** is one in which two angles and two sides are equal.

7. A **scalene triangle** is one with unequal angles and therefore unequal sides.

8. With reference to Figure 41.1:
    (i) Angles A, B and C are called **interior angles** of the triangle.
    (ii) Angle θ is called an **exterior angle** of the triangle and is equal to the sum of the two opposite interior angles, i.e. θ = A + C
    (iii) a + b+c is called the **perimeter** of the triangle.

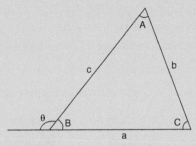

Figure 41.1

9. **Congruent triangles** – two triangles are congruent if:
    (i) the three sides of one are equal to the three sides of the other,
    (ii) they have two sides of the one equal to two sides of the other, and if the angles included by these sides are equal,
    (iii) two angles of the one are equal to two angles of the other and any side of the first is equal to the corresponding side of the other, or
    (iv) their hypotenuses are equal and if one other side of one is equal to the corresponding side of the other.

10. **Similar triangles**
    With reference to Figure 41.2, triangles ABC and PQR are similar and the corresponding sides are in proportion to each other,

i.e.
$$\frac{p}{a} = \frac{q}{b} = \frac{r}{c}$$

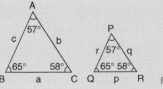

Figure 41.2

Application: A rectangular shed 2m wide and 3m high stands against a perpendicular building of height 5.5m. A ladder is used to gain access to the roof of the building. Determine the minimum distance between the bottom of the ladder and the shed

A side view is shown in Figure 41.3 where AF is the minimum length of ladder. Since BD and CF are parallel, $\angle ADB = \angle DFE$ (corresponding angles between parallel lines). Hence triangles BAD and EDF are similar since their angles are the same. $AB = AC - BC = AC - DE = 5.5 - 3 = 2.5\text{m}$

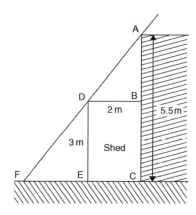

Figure 41.3

By proportion: $\dfrac{AB}{DE} = \dfrac{BD}{EF}$ i.e. $\dfrac{2.5}{3} = \dfrac{2}{EF}$

Hence $EF = 2\left(\dfrac{3}{2.5}\right) = \textbf{2.4m = minimum distance from bottom of ladder to the shed}$

Chapter 42 The theorem of Pythagoras

The **theorem of Pythagoras** states:

'In any right-angled triangle, the square on the hypotenuse is equal to the sum of the squares on the other two sides'

Hence $b^2 = a^2 + c^2$

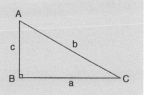

Figure 42.1

Application: Two aircraft leave an airfield at the same time. One travels due north at an average speed of 300 km/h and the other due west at an average speed of 220 km/h. Calculate their distance apart after 4 hours

After 4 hours, the first aircraft has travelled 4 × 300 = 1200 km, due north, and the second aircraft has travelled 4 × 220 = 880 km due west, as shown in Figure 42.2.

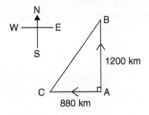

Figure 42.2

Distance apart after 4 hours = BC

From Pythagoras' theorem:

$$BC^2 = 1200^2 + 880^2$$
$$= 1440000 + 774400 \text{ and } BC = \sqrt{2214400}$$

Hence distance apart after 4 hours = 1488 km

Chapter 43   Trigonometric ratios of acute angles

With reference to the right-angled triangle shown in Figure 43.1:

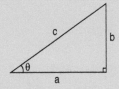

Figure 43.1

1. sine $\theta = \dfrac{\text{opposite side}}{\text{hypotenuse}}$ i.e. $\sin \theta = \dfrac{b}{c}$

2. cosine $\theta = \dfrac{\text{adjacent side}}{\text{hypotenuse}}$ i.e. $\cos \theta = \dfrac{a}{c}$

3. tangent $\theta = \dfrac{\text{opposite side}}{\text{adjacent side}}$ i.e. $\tan \theta = \dfrac{b}{a}$

4. secant $\theta = \dfrac{\text{hypotenuse}}{\text{adjacent side}}$ i.e. $\sec \theta = \dfrac{c}{a}$

5. cosecant $\theta = \dfrac{\text{hypotenuse}}{\text{opposite side}}$ i.e. cosec $\theta = \dfrac{c}{b}$

6. cotangent $\theta = \dfrac{\text{adjacent side}}{\text{opposite side}}$ i.e. cot $\theta = \dfrac{a}{b}$

From above,

7. $\dfrac{\sin \theta}{\cos \theta} = \dfrac{\dfrac{b}{c}}{\dfrac{a}{c}} = \dfrac{b}{a} = \tan \theta$ i.e. $\tan \theta = \dfrac{\sin \theta}{\cos \theta}$

8. $\dfrac{\cos \theta}{\sin \theta} = \dfrac{\dfrac{a}{c}}{\dfrac{b}{c}} = \dfrac{a}{b} = \cot \theta$ i.e. $\cot \theta = \dfrac{\cos \theta}{\sin \theta}$

9. $\sec \theta = \dfrac{1}{\cos \theta}$

10. $\mathbf{cosec}\ \theta = \dfrac{1}{\sin \theta}$ (Note 's' and 'c' go together)

11. $\cot \theta = \dfrac{1}{\tan \theta}$

Secants, cosecants and cotangents are called the **reciprocal ratios**.

Chapter 44   Evaluating trigonometric ratios

The easiest method of evaluating trigonometric functions of any angle is by using a **calculator**.

The following values, correct to 4 decimal places, may be checked:

sine 18° = 0.3090   cosine 56° = 0.5592   tangent 29° = 0.5543

sine 172° = 0.1392   cosine 115° = −0.4226   tangent 78° = −0.0349

sine 241.63° = −0.8799   cosine 331.78° = 0.8811
tangent 296.42° = −2.0127

Most calculators contain only sine, cosine and tangent functions. Thus to evaluate secants, cosecants and cotangents, reciprocals need to be used.

The following values, correct to 4 decimal places, may be checked:

$$\text{secant } 32° = \frac{1}{\cos 32°}$$
$$= 1.1792$$

$$\text{secant } 215.12° = \frac{1}{\cos 215.12°}$$
$$= -1.2226$$

$$\text{cosecant } 75° = \frac{1}{\sin 75°}$$
$$= 1.0353$$

$$\text{cosecant } 321.62° = \frac{1}{\sin 321.62°}$$
$$= -1.6106$$

$$\text{cotangent } 41° = \frac{1}{\tan 41°}$$
$$= 1.1504$$

$$\text{cotangent } 263.59° = \frac{1}{\tan 263.59°}$$
$$= 0.1123$$

Application: Determine the acute angles:

(a) $\sec^{-1} 2.3164$ (b) $\csc^{-1} 1.1784$ (c) $\cot^{-1} 2.1273$

(a) $\sec^{-1} 2.3164 = \cos^{-1}\left(\dfrac{1}{2.3164}\right) = \cos^{-1} 0.4317..$

$$= \mathbf{64.42°} \text{ or } \mathbf{64°25'} \text{ or } \mathbf{1.124\ rad}$$

(b) $\csc^{-1} 1.1784 = \sin^{-1}\left(\dfrac{1}{1.1784}\right) = \sin^{-1} 0.8486..$

$$= \mathbf{58.06°} \text{ or } \mathbf{58°4'} \text{ or } \mathbf{1.013\ rad}$$

(c) $\cot^{-1} 2.1273 = \tan^{-1}\left(\dfrac{1}{2.1273}\right) = \tan^{-1} 0.4700..$

$$= \mathbf{25.18°} \text{ or } \mathbf{25°11'} \text{ or } \mathbf{0.439\ rad}$$

Chapter 45   Fractional and surd forms of trigonometric ratios

In Figure 45.1, ABC is an equilateral triangle of side 2 units. AD bisects angle A and bisects the side BC. Using Pythagoras' theorem on triangle ABD gives:

$$AD = \sqrt{2^2 - 1^2} = \sqrt{3}$$

Hence,

Figure 45.1

$$\sin 30° = \frac{BD}{AB} = \frac{1}{2}, \cos 30° = \frac{AD}{AB} = \frac{\sqrt{3}}{2} \text{ and } \tan 30° = \frac{BD}{AD} = \frac{1}{\sqrt{3}}$$

$$\sin 60° = \frac{AD}{AB} = \frac{\sqrt{3}}{2}, \cos 60° = \frac{BD}{AB} = \frac{1}{2} \text{ and } \tan 60° = \frac{AD}{BD} = \sqrt{3}$$

In Figure 45.2, PQR is an isosceles triangle with PQ = QR = 1 unit. By

Pythagoras' theorem, $PR = \sqrt{1^2 + 1^2} = \sqrt{2}$

Hence, $\sin 45° = \dfrac{1}{\sqrt{2}}$, $\cos 45° = \dfrac{1}{\sqrt{2}}$ and

$\tan 45° = 1$

Figure 45.2

A quantity that is not exactly expressible as a rational number is called a **surd**. For example, $\sqrt{2}$ and $\sqrt{3}$ are called surds because they cannot be expressed as a fraction and the decimal part may be continued indefinitely.

From above, sin 30° = cos 60°, sin 45° = cos 45° and sin 60° = cos 30°.

In general, **sin θ = cos(90° − θ)** and **cos θ = sin(90° − θ)**

Chapter 46 Solution of right-angled triangles

To 'solve a right-angled triangle' means 'to find the unknown sides and angles'. This is achieved by using (i) the theorem of Pythagoras, and/or (ii) trigonometric ratios.

Application: Find the lengths of PQ and PR in triangle PQR shown in Figure 46.1

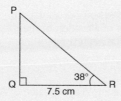

Figure 46.1

$$\tan 38° = \frac{PQ}{QR} = \frac{PQ}{7.5}, \text{ hence } PQ = 7.5 \tan 38°$$

$$= 7.5(0.7813) = \mathbf{5.860\,cm}$$

$$\cos 38° = \frac{QR}{PR} = \frac{7.5}{PR}, \text{ hence } PR = \frac{7.5}{\cos 38°} = \frac{7.5}{0.7880} = \mathbf{9.518\,cm}$$

[Check: Using Pythagoras' theorem $(7.5)^2 + (5.860)^2 = 90.59 = (9.518)^2$]

Angles of elevation and depression

If, in Figure 46.2, BC represents horizontal ground and AB a vertical flagpole, then the **angle of elevation** of the top of the flagpole, A, from the point C is

Figure 46.2

the angle that the imaginary straight line AC must be raised (or elevated) from the horizontal CB, i.e. angle θ.

If, in Figure 46.3, PQ represents a vertical cliff and R a ship at sea, then the **angle of depression** of the ship from point P is the angle through which the imaginary straight line PR must be lowered (or depressed) from the horizontal to the ship, i.e. angle φ.

Figure 46.3

(Note, ∠PRQ is also φ – alternate angles between parallel lines.)

Application: An electricity pylon stands on horizontal ground. At a point 80 m from the base of the pylon, the angle of elevation of the top of the pylon is 23°. Calculate the height of the pylon to the nearest metre.

Figure 46.4 shows the pylon AB and the angle of elevation of A from point C is 23°.

Now $\tan 23° = \dfrac{AB}{BC} = \dfrac{AB}{80}$

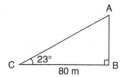

Figure 46.4

Hence, height of pylon, AB $= 80 \tan 23° = 80(0.4245) = 33.96$ m

$$= \textbf{34 m to the nearest metre}$$

Application: The angle of depression of a ship viewed at a particular instant from the top of a 75 m vertical cliff is 30°. The ship is sailing away from the cliff at constant speed and 1 minute later its angle of depression from the top of the cliff is 20°. Find (a) the initial distance of the ship from the base of the cliff, and (b) the speed of the ship in km/h and in knots

(a) Figure 46.5 shows the cliff AB, the initial position of the ship at C and the final position at D.
   Since the angle of depression is initially 30° then $\angle ACB = 30°$ (alternate angles between parallel lines).

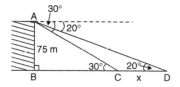

Figure 46.5

$\tan 30° = \dfrac{AB}{BC} = \dfrac{75}{BC}$ hence **the initial position of the ship from the base of**

cliff, BC $= \dfrac{75}{\tan 30°} = \dfrac{75}{0.5774} = 129.9$ m

(b) In triangle ABD, $\tan 20° = \dfrac{AB}{BD} = \dfrac{75}{BC + CD} = \dfrac{75}{129.9 + x}$

Hence $\qquad 129.9 + x = \dfrac{75}{\tan 20°} = \dfrac{75}{0.3640} = 206.0$ m

from which $\qquad x = 206.0 - 129.9 = 76.1$ m

Thus the ship sails 76.1 m in 1 minute, i.e. 60 s,

hence, **speed of ship** $= \dfrac{\text{distance}}{\text{time}} = \dfrac{76.1}{60}$ m/s

$$= \dfrac{76.1 \times 60 \times 60}{60 \times 1000}\text{ km/h}$$

$$= \textbf{4.57 km/h}$$

From chapter 1, page 2, 1 km/h $= 0.54$ knots

Hence, **speed of ship** $= 4.57 \times 0.54 = \textbf{2.47 knots}$

Chapter 47   Cartesian and polar co-ordinates

Changing from Cartesian into polar co-ordinates

In Figure 47.1, $r = \sqrt{x^2 + y^2}$

and $\theta = \tan^{-1}\dfrac{y}{x}$

The angle θ, which may be expressed in degrees or radians, must **always** be measured from the positive x-axis, i.e. measured from the line OQ in Figure. 47.1.

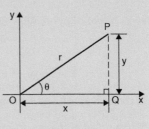

Figure 47.1

Application: Express in polar co-ordinates the position $(-4, 3)$

A diagram representing the point using the Cartesian co-ordinates $(-4, 3)$ is shown in Figure 47.2.

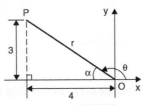

Figure 47.2

From Pythagoras' theorem, $r = \sqrt{4^2 + 3^2} = 5$

By trigonometric ratios, $\alpha = \tan^{-1}\dfrac{3}{4} = 36.87°$ or 0.644 rad

Hence $\theta = 180° - 36.87° = 143.13°$ or $\theta = \pi - 0.644 = 2.498$ rad

Hence the position of point P in polar co-ordinate form is (5, 143.13°) or (5, 2.498 rad)

Application: Express $(-5, -12)$ in polar co-ordinates

A sketch showing the position $(-5, -12)$ is shown in Figure 47.3.

$r = \sqrt{5^2 + 12^2} = 13$ and $\alpha = \tan^{-1}\dfrac{12}{5} = 67.38°$ or 1.176 rad

Hence $\theta = 180° + 67.38° = 247.38°$ or $\theta = \pi + 1.176 = 4.318$ rad

Thus $(-5, -12)$ in Cartesian co-ordinates corresponds to (13, 247.38°) or (13, 4.318 rad) in polar co-ordinates.

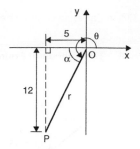

Figure 47.3

Changing from polar into Cartesian co-ordinates

From Figure 47.4,

$$x = r \cos \theta \quad \text{and} \quad y = r \sin \theta$$

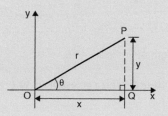

Figure 47.4

Application: Change (4, 32°) into Cartesian co-ordinates

A sketch showing the position (4, 32°) is shown in Figure 47.5.

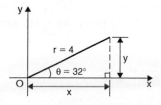

Figure 47.5

Now $x = r \cos \theta = 4 \cos 32° = 3.392$
and $y = r \sin \theta = 4 \sin 32° = 2.120$

Hence, (4, 32°) in polar co-ordinates corresponds to (3.392, 2.120) in Cartesian co-ordinates.

Application: Express (6, 137°) in Cartesian co-ordinates

A sketch showing the position (6, 137°) is shown in Figure 47.6.

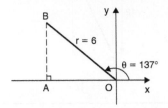

Figure 47.6

$$x = r \cos \theta = 6 \cos 137° = -4.388$$

which corresponds to length OA in Figure 47.6.

$$y = r \sin \theta = 6 \sin 137° = 4.092$$

which corresponds to length AB in Figure 47.6.

Thus, (6, 137°) in polar co-ordinates corresponds to (−4.388, 4.092) in Cartesian co-ordinates.

Chapter 48   Sine and cosine rules and areas of any triangle

Sine rule

With reference to triangle ABC of Figure. 48.1, the **sine rule** states:

$$\frac{a}{\sin A} = \frac{b}{\sin B} = \frac{c}{\sin C}$$

Figure 48.1

The rule may be used only when:

(i) 1 side and any 2 angles are initially given, or

(ii) 2 sides and an angle (not the included angle) are initially given.

Cosine rule

With reference to triangle ABC of Figure 48.1, the **cosine rule** states:

$$a^2 = b^2 + c^2 - 2bc \cos A$$
$$\text{or} \quad b^2 = a^2 + c^2 - 2ac \cos B$$
$$\text{or} \quad c^2 = a^2 + b^2 - 2ab \cos C$$

The rule may be used only when:

(i)   2 sides and the included angle are initially given, or

(ii) 3 sides are initially given.

Area of any triangle

The **area of any triangle** such as ABC of Figure. 48.1 is given by:

(i) $\dfrac{1}{2} \times$ base $\times$ perpendicular height, or

(ii) $\dfrac{1}{2}$ ab sin C   or $\dfrac{1}{2}$ ac sin B   or $\dfrac{1}{2}$ bc sin A,   or

(iii) $\sqrt{[s(s - a)(s - b)(s - c)]}$ where $s = \dfrac{a + b + c}{2}$

Application: A room 8.0 m wide has a span roof that slopes at 33° on one side and 40° on the other. Find the lengths of the roof slopes, correct to the nearest centimetre

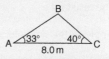

Figure 48.2

A section of the roof is shown in Figure 48.2.

Angle at ridge, B = 180° − 33° − 40° = 107°

From the sine rule: $\dfrac{8.0}{\sin 107°} = \dfrac{a}{\sin 33°}$

from which, $a = \dfrac{8.0 \sin 33°}{\sin 107°} = 4.556 \, \text{m}$

Also from the sine rule: $\dfrac{8.0}{\sin 107°} = \dfrac{c}{\sin 40°}$

from which, $c = \dfrac{8.0 \sin 40°}{\sin 107°} = 5.377 \, \text{m}$

Hence the roof slopes are 4.56 m and 5.38 m, correct to the nearest centimetre.

Application: Two voltage phasors are shown in Figure 48.3 where $V_1 = 40V$ and $V_2 = 100V$. Determine the value of their resultant (i.e. length OA) and the angle the resultant makes with V_1

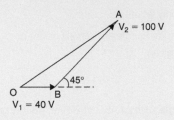

Figure 48.3

Angle OBA $= 180° - 45° = 135°$

Applying the cosine rule: $OA^2 = V_1{}^2 + V_2{}^2 - 2V_1V_2 \cos OBA$

$$= 40^2 + 100^2 - \{2(40)(100)\cos 135°\}$$
$$= 1600 + 10000 - \{-5657\}$$
$$= 1600 + 10000 + 5657 = 17257$$

The resultant $\qquad OA = \sqrt{17257} = 131.4\,V$

Applying the sine rule: $\dfrac{131.4}{\sin 135°} = \dfrac{100}{\sin AOB}$

from which, $\qquad \sin AOB = \dfrac{100 \sin 135°}{131.4} = 0.5381$

Hence, angle AOB $= \sin^{-1} 0.5381 = 32.55°$

Hence, **the resultant voltage is 131.4 volts at 32.55° to V_1**

Application: In Figure 48.4, PR represents the inclined jib of a crane and is 10.0 m long. PQ is 4.0 m long. Determine the inclination of the jib to the vertical and the length of tie QR

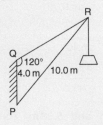

Figure 48.4

Applying the sine rule: $\dfrac{PR}{\sin 120°} = \dfrac{PQ}{\sin R}$

from which, $\sin R = \dfrac{PQ \sin 120°}{PR} = \dfrac{(4.0) \sin 120°}{10.0} = 0.3464$

Hence $\angle R = \sin^{-1} 0.3464 = 20.27°$

$\angle P = 180° - 120° - 20.27° = \mathbf{39.73°}$, **which is the inclination of the jib to the vertical.**

Applying the sine rule: $\dfrac{10.0}{\sin 120°} = \dfrac{QR}{\sin 39.73°}$

from which, **length of tie, QR** $= \dfrac{10.0 \sin 39.73°}{\sin 120°} = \mathbf{7.38\,m}$

Application: A crank mechanism of a petrol engine is shown in Figure 48.5. Arm OA is 10.0 cm long and rotates clockwise about O. The connecting rod AB is 30.0 cm long and end B is constrained to move horizontally. Determine the angle between the connecting rod AB and the horizontal, and the length of OB for the position shown in Figure 48.5

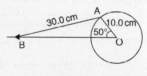

Figure 48.5

Applying the sine rule: $\qquad \dfrac{AB}{\sin 50°} = \dfrac{AO}{\sin B}$

from which, $\qquad\qquad \sin B = \dfrac{AO \sin 50°}{AB} = \dfrac{10.0 \sin 50°}{30.0}$

$$= 0.2553$$

Hence $B = \sin^{-1} 0.2553 = 14.79°$

Hence, the connecting rod AB makes an angle of 14.79° with the horizontal.

Angle OAB $= 180° - 50° - 14.79° = 115.21°$

Applying the sine rule: $\qquad \dfrac{30.0}{\sin 50°} = \dfrac{OB}{\sin 115.21°}$

from which, $\qquad\qquad \mathbf{OB} = \dfrac{30.0 \sin 115.21°}{\sin 50°} = \mathbf{35.43\,cm}$

Application: Determine in Figure 48.5 how far B moves when angle AOB changes from 50° to 120°

Figure 48.6 shows the initial and final positions of the crank mechanism.

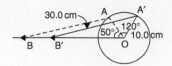

Figure 48.6

In triangle OA'B', applying the sine rule:

$$\frac{30.0}{\sin 120°} = \frac{10.0}{\sin A'B'O}$$

from which, $\sin A'B'O = \dfrac{10.0 \sin 120°}{30.0} = 0.2887$

Hence A'B'O = $\sin^{-1} 0.2887 = 16.78°$

Angle OA'B' = $180° - 120° - 16.78° = 43.22°$

Applying the sine rule: $\dfrac{30.0}{\sin 120°} = \dfrac{OB'}{\sin 43.22°}$

from which, $OB' = \dfrac{30.0 \sin 43.22°}{\sin 120°} = 23.72\,\text{cm}$

Since OB = 35.43 cm, from the previous example, and OB' = 23.72 cm then BB' = 35.43 − 23.72 = 11.71 cm.

Hence, B moves 11.71 cm when angle AOB changes from 50° to 120°

Application: The area of a field is in the form of a quadrilateral ABCD as shown in Figure 48.7. Determine its area

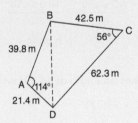

Figure 48.7

A diagonal drawn from B to D divides the quadrilateral into two triangles.

Area of quadrilateral ABCD

= area of triangle ABD + area of triangle BCD

$$= \frac{1}{2}(39.8)(21.4) \sin 114° + \frac{1}{2}(42.5)(62.3) \sin 56°$$

$$= 389.04 + 1097.5 = \mathbf{1487\,m^2}$$

Chapter 49　Graphs of trigonometric functions

Graphs of y = sin A, y = cos A and y = tan A are shown in Figure 49.1.

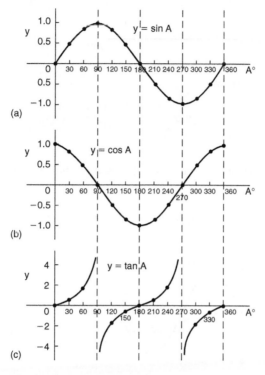

Figure 49.1

Chapter 50　Angles of any magnitude

Figure 50.1 summarises the trigonometric ratios for angles of any magnitude; the letters underlined spell the word CAST when starting in the fourth quadrant and moving in an anticlockwise direction.

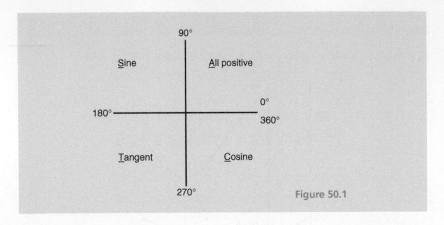

Figure 50.1

Application: Determine all the angles between 0° and 360° whose sine is −0.4638

The angles whose sine is −0.4638 occurs in the third and fourth quadrants since sine is negative in these quadrants – see Figure 50.2.

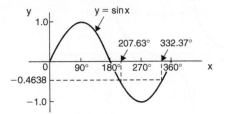

Figure 50.2

From Figure 50.3, $\theta = \sin^{-1} 0.4638 = 27.63°$. Measured from 0°, the two angles between 0° and 360° whose sine is −0.4638 are $180° + 27.63°$, i.e. **207.63°** and $360° − 27.63°$, i.e. **332.37°**

(Note that a calculator only gives one answer, i.e. −27.632588°)

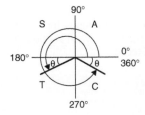

Figure 50.3

Application: Determine all the angles between 0° and 360° whose tangent is 1.7629

A tangent is positive in the first and third quadrants – see Figure 50.4.

From Figure 50.5, $\theta = \tan^{-1}1.7629 = 60.44°$

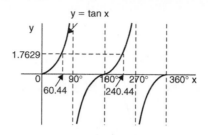

Figure 50.4

Measured from 0°, the two angles between 0° and 360° whose tangent is 1.7629 are **60.44°** and $180° + 60.44°$, i.e. **240.44°**

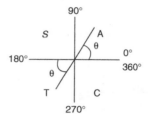

Figure 50.5

Application: Solve the equation $\cos^{-1}(-0.2348) = \alpha$ for angles of α between 0° and 360°

Cosine is positive in the first and fourth quadrants and thus negative in the second and third quadrants – from Figure 50.1 or from Figure 49.1(b).

In Figure 50.6, angle $\theta = \cos^{-1}(0.2348) = 76.42°$

Measured from 0°, the two angles whose cosine is -0.2348 are $\alpha = 180° - 76.42°$ i.e. **103.58°** and $\alpha = 180° + 76.42°$ i.e. **256.42°**

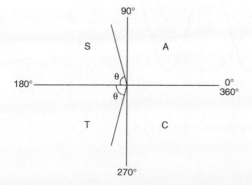

Figure 50.6

Chapter 51   Sine and cosine waveforms

Graphs of $y = \sin A$ and $y = \sin 2A$ are shown in Figure 51.1.

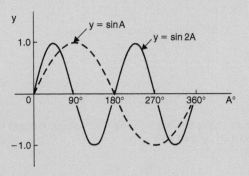

Figure 51.1

A graph of $y = \sin\dfrac{1}{2}A$ is shown in Figure 51.2.

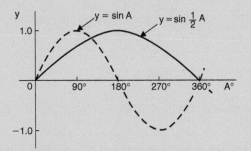

Figure 51.2

Graphs of $y = \cos A$ and $y = \cos 2A$ are shown in Figure 51.3.

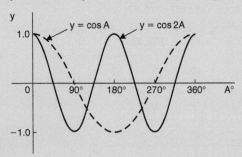

Figure 51.3

A graph of $y = \cos\dfrac{1}{2}A$ is shown in Figure 51.4.

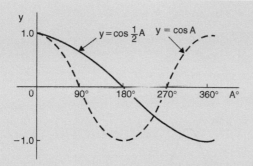

Figure 51.4

Period

If $y = \sin pA$ or $y = \cos pA$ (where p is a constant) then the period of the waveform is $360°/p$ (or $2\pi/p$ rad). Hence if $y = \sin 3A$ then the period is 360/3, i.e. 120°, and if $y = \cos 4A$ then the period is 360/4, i.e. 90°.

Amplitude is the name given to the maximum or peak value of a sine wave. If $y = 4\sin A$ the maximum value, and thus amplitude, is 4. Similarly, if $y = 5\cos 2A$, the amplitude is 5 and the period is $360°/2$, i.e. 180°.

Lagging and leading angles

The graph $y = \sin(A - 60°)$ **lags** $y = \sin A$ by 60° as shown in Figure 51.5.

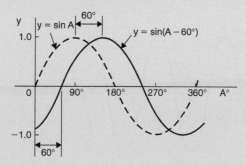

Figure 51.5

The graph of $y = \cos(A + 45°)$ **leads** $y = \cos A$ by 45° as shown in Figure 51.6.

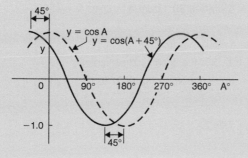

Figure 51.6

Application: Sketch y = 3 sin 2A from A = 0 to A = 360°

Amplitude = 3 and period = 360/2 = 180°
A sketch of y = 3 sin 2A is shown in Figure 51.7.

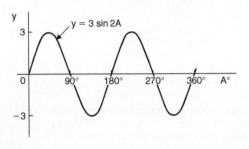

Figure 51.7

Application: Sketch y = 4 cos 3x from x = 0° to x = 360°

Amplitude = 4 and period = 360°/3 = 120°
A sketch of y = 4 cos 3x is shown in Figure 51.8.

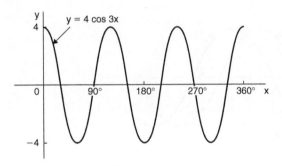

Figure 51.8

Application: Sketch y = 5 sin(A + 30°) from A = 0° to A = 360°

Amplitude = 5 and period = 360°/1 = 360°
5 sin(A + 30°) leads 5 sin A by 30° (i.e. starts 30° earlier)
A sketch of y = 5 sin (A + 30°) is shown in Figure 51.9.

Application: Sketch y = 7 sin(2A − π/3) in the range 0 ≤ A ≤ 360°

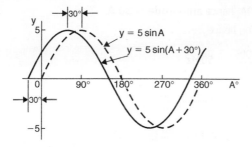

Figure 51.9

Amplitude = 7 and period = $2\pi/2 = \pi$ radians

In general, **y = sin(pt − α) lags y = sin pt by** α/p, hence $7\sin(2A − \pi/3)$ lags $7\sin 2A$ by $(\pi/3)/2$, i.e. $\pi/6$ rad or $30°$.

A sketch of $y = 7\sin(2A − \pi/3)$ is shown in Figure 51.10.

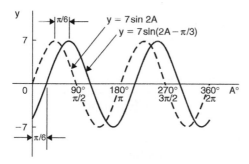

Figure 51.10

Sinusoidal form A sin(ωt ± α)

Given a general sinusoidal function **y = A sin(ωt ± α)**, then

(i) A = amplitude

(ii) ω = angular velocity = $2\pi f$ rad/s

(iii) $\dfrac{\omega}{2\pi}$ = frequency, f hertz

(iv) $\dfrac{1}{f}$ = periodic time T seconds

(v) α = angle of lead or lag (compared with y = A sin ωt), in radians.

Application: An alternating current is given by $i = 30\ \sin(100\pi t + 0.27)$ amperes. Find the amplitude, frequency, periodic time and phase angle (in degrees and minutes)

i = 30 sin (100πt + 0.27)A, hence **amplitude = 30 A**

Angular velocity ω = 100π, hence

Frequency, f $= \dfrac{\omega}{2\pi} = \dfrac{100\pi}{2\pi} = $ **50 Hz**

Periodic time, T $= \dfrac{1}{f} = \dfrac{1}{50} = $ **0.02s or 20 ms**

Phase angle, α = 0.27 rad $= \left(0.27 \times \dfrac{180}{\pi}\right)^{\circ}$

= **15.47° leading i = 30 sin(100πt)**

Application: An oscillating mechanism has a maximum displacement of 2.5 m and a frequency of 60 Hz. At time t = 0 the displacement is 90 cm. Express the displacement in the general form A sin(ωt ± α)

Amplitude = maximum displacement = 2.5 m

Angular velocity, ω = 2πf = 2π(60) = 120π rad/s

Hence, displacement = 2.5 sin(120πt + α) m

When t = 0, displacement = 90 cm = 0.90 m

Hence 0.90 = 2.5 sin (0 + α)   i.e. $\sin \alpha = \dfrac{0.90}{2.5} = 0.36$

Hence α = sin⁻¹0.36 = 21.10° = 0.368 rad

Thus, **displacement = 2.5 sin(120πt + 0.368) m**

Application: The instantaneous value of voltage in an a.c. circuit at any time t seconds is given by v = 340 sin(50πt − 0.541) volts. Determine

(a) the amplitude, frequency, periodic time and phase angle (in degrees),
(b) the value of the voltage when t = 0,
(c) the value of the voltage when t = 10 ms,
(d) the time when the voltage first reaches 200 V, and
(e) the time when the voltage is a maximum

(a) **Amplitude = 340 V**

Angular velocity, ω = 50π = 2πf

Frequency, f $= \dfrac{\omega}{2\pi} = \dfrac{50\pi}{2\pi} = $ **25 Hz**

Periodic time, T $= \dfrac{1}{f} = \dfrac{1}{25} = $ **0.04 s or 40 ms**

Phase angle = 0.541 rad $= \left(0.541 \times \dfrac{180}{\pi}\right)^{\circ}$

= **31° lagging** v = 340 sin(50πt)

(b) **When t = 0, v** $= 340\sin(0 - 0.541) = 340\sin(-31°)$
$$= \mathbf{-175.1\,V}$$

(c) **When t = 10 ms** then **v** $= 340\sin(50\pi \times 10 \times 10^{-3} - 0.541)$
$$= 340\sin(1.0298) = 340\sin 59°$$
$$= \mathbf{291.4\ volts}$$

(d) When $v = 200$ volts then $200 = 340\sin(50\pi t - 0.541)$
$$\frac{200}{340} = \sin(50\pi t - 0.541)$$

Hence $(50\pi t - 0.541) = \sin^{-1}\dfrac{200}{340} = 0.6288$ rad
$$50\pi t = 0.6288 + 0.541 = 1.1698$$

Hence when $v = 200\,V$, **time, t** $= \dfrac{1.1698}{50\pi} = \mathbf{7.447\ ms}$

(e) When the voltage is a maximum, $v = 340\,V$
Hence $\quad 340 = 340\sin(50\pi t - 0.541)$
$$1 = \sin(50\pi t - 0.541)$$
$$50\pi t - 0.541 = \sin^{-1}1 = 1.5708 \text{ rad}$$
$$50\pi t = 1.5708 + 0.541 = 2.1118$$

Hence, **time, t** $= \dfrac{2.1118}{50\pi} = \mathbf{13.44\ ms}$

A sketch of $v = 340\sin(50\pi t - 0.541)$ volts is shown in Figure 51.11.

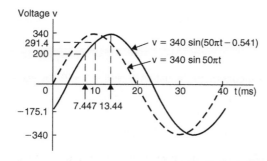

Figure 51.11

Chapter 52   Trigonometric identities and equations

$$\tan\theta = \frac{\sin\theta}{\cos\theta} \qquad \cot\theta = \frac{\cos\theta}{\sin\theta} \qquad \sec\theta = \frac{1}{\cos\theta}$$
$$\csc\theta = \frac{1}{\sin\theta} \qquad \cot\theta = \frac{1}{\tan\theta}$$

$\cos^2\theta + \sin^2\theta = 1$ $1 + \tan^2\theta = \sec^2\theta$ $\cot^2\theta + 1 = \text{cosec}^2\theta$

Equations of the type $a\sin^2 A + b\sin A + c = 0$

(i) **When $a = 0$**, $b\sin A + c = 0$, hence

$$\sin A = -\frac{c}{b} \quad \text{and} \quad \mathbf{A = sin^{-1}}\left(-\frac{c}{b}\right)$$

There are two values of A between 0° and 360° that satisfy such an equation, provided $-1 \le \dfrac{c}{b} \le 1$

(ii) **When $b = 0$**, $a\sin^2 A + c = 0$, hence

$$\sin^2 A = -\frac{c}{a}, \sin A = \sqrt{\left(-\frac{c}{a}\right)} \text{ and } \mathbf{A = sin^{-1}}\sqrt{\left(-\frac{c}{a}\right)}$$

If either a or c is a negative number, then the value within the square root sign is positive. Since when a square root is taken there is a positive and negative answer there are four values of A between 0° and 360° which satisfy such an equation, provided $-1 \le \dfrac{c}{a} \le 1$

(iii) **When a, b and c are all non-zero:**
$a\sin^2 A + b\sin A + c = 0$ is a quadratic equation in which the unknown is sin A. The solution of a quadratic equation is obtained either by factorising (if possible) or by using the quadratic formula:

$$\sin A = \frac{-b \pm \sqrt{(b^2 - 4ac)}}{2a}$$

(iv) Often the trigonometric identities $\cos^2 A + \sin^2 A = 1$, $1 + \tan^2 A = \sec^2 A$ and $\cot^2 A + 1 = \text{cosec}^2 A$ need to be used to reduce equations to one of the above forms.

Application: Solve the trigonometric equation $5\sin\theta + 3 = 0$ for values of θ from 0° to 360°

$5\sin\theta + 3 = 0$, from which $\sin\theta = -3/5 = -0.6000$

Hence, $\theta = \sin^{-1}(-0.6000)$. Sine is negative in the third and fourth quadrants (see Figure 52.1(a)). The acute angle $\sin^{-1}(0.6000) = 36.87°$ (shown as α in Figure 52.1(b)).

Hence $\theta = 180° + 36.87°$ i.e. **216.87°** or $\theta = 360° - 36.87°$ i.e. **323.13°**

Application: Solve $4\sec t = 5$ for values of t between 0° and 360°

$4\sec t = 5$, from which $\sec t = \dfrac{5}{4} = 1.2500$ and $t = \sec^{-1} 1.2500$

Secant $= \dfrac{1}{\cos\text{ine}}$ is positive in the first and fourth quadrants (see Figure 52.2).

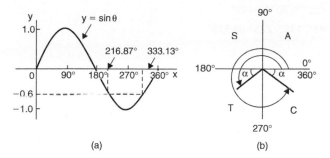

(a)   (b)   Figure 52.1

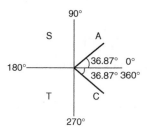

Figure 52.2

The acute angle, $\sec^{-1}1.2500 = \cos^{-1}\left(\dfrac{1}{1.2500}\right) = \cos^{-1}0.8 = 36.87°$.

Hence, **t = 36.87°** or $360° - 36.87° = $ **323.13°**

Application: Solve $2 - 4\cos^2 A = 0$ for values of A in the range $0° < A < 360°$

$2 - 4\cos^2 A = 0$, from which $\cos^2 A = \dfrac{2}{4} = 0.5000$

Hence $\cos A = \sqrt{0.5000} = \pm0.7071$ and $A = \cos^{-1}(\pm0.7071)$

Cosine is positive in quadrants one and four and negative in quadrants two and three. Thus in this case there are four solutions, one in each quadrant (see Figure 52.3).

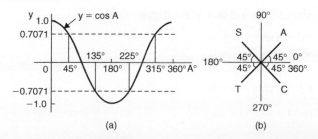

(a)   (b)   Figure 52.3

The acute angle $\cos^{-1}0.7071 = 45°$

Hence,   **A = 45°, 135°, 225° or 315°**

Application: Solve $\dfrac{1}{2}\cot^2 y = 1.3$ for $0° < y < 360°$

$\dfrac{1}{2}\cot^2 y = 1.3$, from which, $\cot^2 y = 2(1.3) = 2.6$

Hence $\cot y = \sqrt{2.6} = \pm 1.6125$, and $y = \cot^{-1}(\pm 1.6125)$. There are four solutions, one in each quadrant.

The acute angle $\cot^{-1}1.6125 = \tan^{-1}\left(\dfrac{1}{1.6125}\right) = 31.81°$

Hence,   **y = 31.81°, 148.19°, 211.81° or 328.19°**

Application: Solve the equation $8\sin^2\theta + 2\sin\theta - 1 = 0$, for all values of θ between 0° and 360°

Factorising $8\sin^2\theta + 2\sin\theta - 1 = 0$ gives $(4\sin\theta - 1)(2\sin\theta + 1) = 0$

Hence $4\sin\theta - 1 = 0$, from which, $\sin\theta = \dfrac{1}{4} = 0.2500$

or $2\sin\theta + 1 = 0$, from which, $\sin\theta = -\dfrac{1}{2} = -0.5000$

(Instead of factorising, the quadratic formula can, of course, be used).

$\theta = \sin^{-1}0.250 = 14.48°$ or $165.52°$, since sine is positive in the first and second quadrants, or

$\theta = \sin^{-1}(-0.5000) = 210°$ or $330°$, since sine is negative in the third and fourth quadrants.

Hence, **θ = 14.48°, 165.52°, 210° or 330°**

Application: Solve $5\cos^2 t + 3\sin t - 3 = 0$ for values of t from 0° to 360°

Since $\cos^2 t + \sin^2 t = 1$, then $\cos^2 t = 1 - \sin^2 t$

Substituting for $\cos^2 t$ in $5\cos^2 t + 3\sin t - 3 = 0$ gives

$$5(1 - \sin^2 t) + 3\sin t - 3 = 0$$
$$5 - 5\sin^2 t + 3\sin t - 3 = 0$$
$$-5\sin^2 t + 3\sin t + 2 = 0$$
$$5\sin^2 t - 3\sin t - 2 = 0$$

Factorising gives $(5\sin t + 2)(\sin t - 1) = 0$

Hence, $5\sin t + 2 = 0$, from which, $\sin t = -\dfrac{2}{5} = -0.4000$ or $\sin t - 1 = 0$, from which, $\sin t = 1$.

$t = \sin^{-1}(-0.4000) = 203.58°$ or $336.42°$, since sine is negative in the third and fourth quadrants, or $t = \sin^{-1}1 = 90°$

Hence, **$t = 90°, 203.58°$ or $336.42°$** as shown in Figure 52.4.

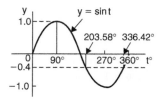

Figure 52.4

Application: Solve $18\sec^2 A - 3\tan A = 21$ for values of A between $0°$ and $360°$

$1 + \tan^2 A = \sec^2 A$. Substituting for $\sec^2 A$ in $18\sec^2 A - 3\tan A = 21$ gives

$$18(1 + \tan^2 A) - 3\tan A = 21$$

i.e. $$18 + 18\tan^2 A - 3\tan A - 21 = 0$$

$$18\tan^2 A - 3\tan A - 3 = 0$$

Factorising gives: $(6\tan A - 3)(3\tan A + 1) = 0$

Hence, $6\tan A - 3 = 0$, from which, $\tan A = \dfrac{3}{6} = 0.5000$

or $3\tan A + 1 = 0$, from which, $\tan A = -\dfrac{1}{3} = -0.3333$

Thus, $A = \tan^{-1}(0.5000) = 26.57°$ or $206.57°$, since tangent is positive in the first

and third quadrants, or $A = \tan^{-1}(-0.3333) = 161.57°$ or $341.57°$, since tangent is negative in the second and fourth quadrants.

Hence, **$A = 26.57°, 161.57°, 206.57°$ or $341.57°$**

Chapter 53   The relationship between trigonometric and hyperbolic functions

$$\cos \theta = \frac{1}{2}(e^{j\theta} + e^{-j\theta})$$

$$\sin \theta = \frac{1}{2j}(e^{j\theta} - e^{-j\theta})$$

$$\cos j\theta = \cosh \theta \tag{1}$$

$$\sin j\theta = j \sinh \theta \tag{2}$$

$$\cosh j\theta = \cos \theta$$

$$\sinh j\theta = j \sin \theta$$

$$\tan j\theta = j \tanh \theta$$

$$\tanh j\theta = j \tan \theta$$

Application: Verify that $\cos^2 j\theta + \sin^2 j\theta = 1$

From equation (1), $\cos j\theta = \cosh \theta$, and from equation (2), $\sin j\theta = j \sinh \theta$
Thus, $\cos^2 j\theta + \sin^2 j\theta = \cosh^2\theta + j^2 \sinh^2\theta$, and since $j^2 = -1$ (from chapter 64),
$$\cos^2 j\theta + \sin^2 j\theta = \cosh^2\theta - \sinh^2\theta$$
But, $\cosh^2\theta - \sinh^2\theta = 1$, from page 60,
hence $\qquad\qquad\qquad\qquad$ $\mathbf{\cos^2 j\theta + \sin^2 j\theta = 1}$

Application: Determine the corresponding hyperbolic identity by writing jA for θ in $\cot^2\theta + 1 = \operatorname{cosec}^2\theta$

Substituting jA for θ gives:

$\cot^2 jA + 1 = \operatorname{cosec}^2 jA,$ $\quad$ i.e. $\quad \dfrac{\cos^2 jA}{\sin^2 jA} + 1 = \dfrac{1}{\sin^2 jA}$

But from equation (1), $\cos jA = \cosh A$
and from equation (2), $\sin jA = j \sinh A$

Hence $\dfrac{\cosh^2 A}{j^2 \sinh^2 A} + 1 = \dfrac{1}{j^2 \sinh^2 A}$

and since $j^2 = -1$, $\qquad -\dfrac{\cosh^2 A}{\sinh^2 A} + 1 = -\dfrac{1}{\sinh^2 A}$

Multiplying throughout by -1, gives:

$\dfrac{\cosh^2 A}{\sinh^2 A} - 1 = \dfrac{1}{\sinh^2 A}$ $\quad$ i.e. $\quad \coth^2 A - 1 = \operatorname{cosech}^2 A$

Application: Develop the hyperbolic identity corresponding to
$\sin 3\theta = 3 \sin \theta - 4 \sin^3 \theta$ by writing jA for θ

Substituting jA for θ gives: $\sin 3jA = 3 \sin jA - 4 \sin^3 jA$
and since from equation (2), $\sin jA = j \sinh A,$

$$j \sinh 3A = 3j \sinh A - 4j^3 \sinh^3 A$$

Dividing throughout by j gives:

$$\sinh 3A = 3 \sinh A - j^2 4 \sinh^3 A$$

But $j^2 = -1$, hence $\sinh 3A = 3 \sinh A + 4 \sinh^3 A$

Chapter 54   Compound angles

Compound angle addition and subtraction formulae

sin (A + B) = sin A cos B + cos A sin B

sin (A − B) = sin A cos B − cos A sin B

cos (A + B) = cos A cos B − sin A sin B

cos (A − B) = cos A cos B + sin A sin B

$$\textbf{tan (A + B)} = \frac{\tan A + \tan B}{1 - \tan A \tan B}$$

$$\textbf{tan (A − B)} = \frac{\tan A - \tan B}{1 + \tan A \tan B}$$

If $R\sin(\omega t + \alpha) = a \sin \omega t + b \cos \omega$ then:

$a = R \cos \alpha, b = R \sin \alpha, R = \sqrt{a^2 + b^2}$ and $\alpha = \tan^{-1} b/a$

Application: Solve the equation $4 \sin(x - 20°) = 5 \cos x$ for values of x between 0° and 90°

$4 \sin(x - 20°) = 4[\sin x \cos 20° - \cos x \sin 20°],$

from the formula for $\sin(A - B)$

$\qquad\qquad = 4[\sin x (0.9397) - \cos x (0.3420)]$

$\qquad\qquad = 3.7588 \sin x - 1.3680 \cos x$

Since $\quad 4 \sin(x - 20°) = 5 \cos x$

then $\quad 3.7588 \sin x - 1.3680 \cos x = 5 \cos x$

Rearranging gives: $3.7588 \sin x = 5 \cos x + 1.3680 \cos x$

$\qquad\qquad\qquad\qquad = 6.3680 \cos x$

and $\qquad \dfrac{\sin x}{\cos x} = \dfrac{6.3680}{3.7588} = 1.6942$

i.e. $\qquad \tan x = 1.6942$, and $x = \tan^{-1} 1.6942 = \mathbf{59.45°}$

[Check: LHS $= 4 \sin(59.45° - 20°) = 4 \sin 39.45° = 2.54$

$\qquad$ RHS $= 5 \cos x = 5 \cos 59.45° = 2.54$]

Application: Find an expression for $3\sin \omega t + 4\cos \omega t$ in the form $R\sin(\omega t + \alpha)$ and sketch graphs of $3\sin \omega t$, $4\cos \omega t$ and $R\sin(\omega t + \alpha)$ on the same axes

Let $3\sin \omega t + 4\cos \omega t = R\sin(\omega t + \alpha)$

then $3\sin \omega t + 4\cos \omega t = R[\sin \omega t \cos \alpha + \cos \omega t \sin \alpha]$

$$= (R\cos \alpha)\sin \omega t + (R\sin \alpha)\cos \omega t$$

Equating coefficients of $\sin \omega t$ gives:

$$3 = R\cos \alpha, \text{ from which, } \cos \alpha = \frac{3}{R}$$

Equating coefficients of $\cos \omega t$ gives:

$$4 = R\sin \alpha, \text{ from which, } \sin \alpha = \frac{4}{R}$$

There is only one quadrant where both $\sin \alpha$ **and** $\cos \alpha$ are positive, and this is the first, as shown in Figure 54.1. From Figure 54.1, by Pythagoras' theorem:

$$R = \sqrt{3^2 + 4^2} = 5$$

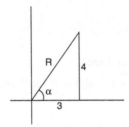

Figure 54.1

From trigonometric ratios: $\alpha = \tan^{-1}\dfrac{4}{3} = 53.13°$ or 0.927 radians

Hence $3\sin \omega t + 4\cos \omega t = 5\sin(\omega t + 0.927)$

A sketch of $3\sin \omega t$, $4\cos \omega t$ and $5\sin(\omega t + 0.927)$ is shown in Figure 54.2.

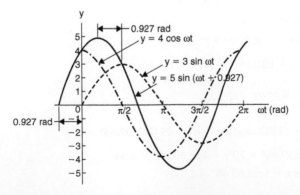

Figure 54.2

Application: Express 4.6 sin ωt − 7.3 cos ωt in the form R sin(ωt + α)

Let   4.6 sin ωt − 7.3 cos ωt = R sin(ωt + α)

then   4.6 sin ωt − 7.3 cos ωt = R[sin ωt cos α + cos ωt sin α]

$$= (R\cos α)\sin ωt + (R\sin α)\cos ωt$$

Equating coefficients of sin ωt gives:

$$4.6 = R\cos α, \text{ from which, } \cos α = \frac{4.6}{R}$$

Equating coefficients of cos ωt gives:

$$-7.3 = R\sin α, \text{ from which } \sin α = \frac{-7.3}{R}$$

There is only one quadrant where cosine is positive **and** sine is negative, i.e. the fourth quadrant, as shown in Figure 54.3. By Pythagoras' theorem:

$$R = \sqrt{4.6^2 + (-7.3)^2} = 8.628$$

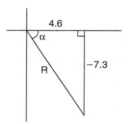

4.6

α

R

−7.3

Figure 54.3

By trigonometric ratios:

$$α = \tan^{-1}\left(\frac{-7.3}{4.6}\right) = -1.008 \text{ radians}$$

Hence, 4.6 sin ωt − 7.3 cos ωt = 8.628 sin(ωt − 1.008)

Application: Express −2.7 sin ωt − 4.1 cos ωt in the form R sin(ωt + α)

Let   −2.7 sin ωt − 4.1 cos ωt = R sin(ωt + α)

$$= R\{\sin ωt \cos α + \cos ωt \sin α]$$

$$= (R\cos α)\sin ωt + (R\sin α)\cos ωt$$

Equating coefficients gives:

$$-2.7 = R\cos α, \text{ from which, } \cos α = \frac{-2.7}{R}$$

and $$-4.1 = R\sin α, \text{ from which, } \sin α = \frac{-4.1}{R}$$

There is only one quadrant in which both cosine **and** sine are negative, i.e. the third quadrant, as shown in Figure 54.4.

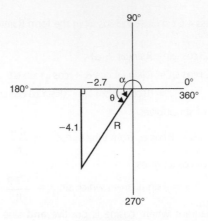

Figure 54.4

From Figure 54.4, $R = \sqrt{(-2.7)^2 + (-4.1)^2} = 4.909$

and $\theta = \tan^{-1}\dfrac{4.1}{2.7} = 56.63°$

Hence $\alpha = 180° + 56.63° = 236.63°$ or   4.130 radians

Thus, $-2.7\sin\omega t - 4.1\cos\omega t = 4.909\sin(\omega t + 4.130)$

An angle of 236.63° is the same as $-123.37°$ or -2.153 radians

Hence, $-2.7\sin\omega t - 4.1\cos\omega t$ may also be expressed as **$4.909\sin(\omega t - 2.153)$**, which is preferred since it is the **principal value** (i.e. $-\pi \leq \alpha \leq \pi$).

Double angles

> **$\sin 2A = 2\sin A \cos A$**
>
> **$\cos 2A = \cos^2 A - \sin^2 A = 1 - 2\sin^2 A = 2\cos^2 A - 1$**
>
> **$\tan 2A = \dfrac{2\tan A}{1 - \tan^2 A}$**

Application: $I_3 \sin 3\theta$ is the third harmonic of a waveform. Express the third harmonic in terms of the first harmonic $\sin\theta$, when $I_3 = 1$

When $I_3 = 1$, $I_3\sin 3\theta = \sin 3\theta = \sin(2\theta + \theta)$

$= \sin 2\theta\cos\theta + \cos 2\theta\sin\theta$,

from the sin(A + B) formula

$= (2\sin\theta\cos\theta)\cos\theta + (1 - 2\sin^2\theta)\sin\theta$,

from the double angle expansions

$$= 2\sin\theta\cos^2\theta + \sin\theta - 2\sin^3\theta$$

$$= 2\sin\theta(1 - \sin^2\theta) + \sin\theta - 2\sin^3\theta,$$

$$\text{(since } \cos^2\theta = 1 - \sin^2\theta\text{)}$$

$$= 2\sin\theta - 2\sin^3\theta + \sin\theta - 2\sin^3\theta$$

i.e. $\quad\quad$ **$\sin 3\theta = 3\sin\theta - 4\sin^3\theta$**

Changing products of sines and cosines into sums or differences

$$\sin A \cos B = \frac{1}{2}[\sin(A + B) + \sin(A - B)] \quad\quad\quad (1)$$

$$\cos A \sin B = \frac{1}{2}[\sin(A + B) - \sin(A - B)] \quad\quad\quad (2)$$

$$\cos A \cos B = \frac{1}{2}[\cos(A + B) + \cos(A - B)] \quad\quad\quad (3)$$

$$\sin A \sin B = -\frac{1}{2}[\cos(A + B) - \cos(A - B)] \quad\quad\quad (4)$$

Application: Express $\sin 4x \cos 3x$ as a sum or difference of sines and cosines

From equation (1), $\sin 4x \cos 3x = \dfrac{1}{2}[\sin(4x + 3x) + \sin(4x - 3x)]$

$$= \frac{1}{2}\textbf{(sin 7x + sin x)}$$

Application: Express $2\cos 5\theta \sin 2\theta$ as a sum or difference of sines or cosines

From equation (2),

$$2\cos 5\theta \sin 2\theta = 2\left\{\frac{1}{2}[\sin(5\theta + 2\theta) - \sin(5\theta - 2\theta)]\right\}$$

$$= \textbf{sin 7}\boldsymbol\theta - \textbf{sin 3}\boldsymbol\theta$$

Application: In an alternating current circuit, voltage $v = 5\sin\omega t$ and current $i = 10\sin(\omega t - \pi/6)$. Find an expression for the instantaneous power p at time t given that $p = vi$, expressing the answer as a sum or difference of sines and cosines

$p = vi = (5\sin \omega t)[10\sin (\omega t - \pi/6)] = 50\sin \omega t \sin(\omega t - \pi/6)$

From equation (4),

$$50\sin \omega t \sin(\omega t - \pi/6) = (50) \left\{ -\frac{1}{2}\{\cos(\omega t + \omega t - \pi/6) - \cos[\omega t - (\omega t - \pi/6)]\} \right\}$$

$$= -25\{\cos(2\omega t - \pi/6) - \cos \pi/6\}$$

i.e. instantaneous power, p = 25[cos π/6 − cos(2ωt − π/6)]

Changing sums or differences of sines and cosines into products

$$\sin X + \sin Y = 2\sin\left(\frac{X+Y}{2}\right)\cos\left(\frac{X-Y}{2}\right) \qquad (5)$$

$$\sin X - \sin Y = 2\cos\left(\frac{X+Y}{2}\right)\sin\left(\frac{X-Y}{2}\right) \qquad (6)$$

$$\cos X + \cos Y = 2\cos\left(\frac{X+Y}{2}\right)\cos\left(\frac{X-Y}{2}\right) \qquad (7)$$

$$\cos X - \cos Y = -2\sin\left(\frac{X+Y}{2}\right)\sin\left(\frac{X-Y}{2}\right) \qquad (8)$$

Application: Express $\sin 5\theta + \sin 3\theta$ as a product

From equation (5),

$$\sin 5\theta + \sin 3\theta = 2\sin\left(\frac{5\theta + 3\theta}{2}\right)\cos\left(\frac{5\theta - 3\theta}{2}\right) = \textbf{2 sin 4θ cos θ}$$

Application: Express $\sin 7x - \sin x$ as a product

From equation (7),

$$\sin 7x - \sin x = 2\cos\left(\frac{7x + x}{2}\right)\sin\left(\frac{7x - x}{2}\right) = \textbf{2 cos 4x sin 3x}$$

Application: Express $\cos 2t - \cos 5t$ as a product

From equation (8),

$$\cos 2t - \cos 5t = -2\sin\left(\frac{2t + 5t}{2}\right)\sin\left(\frac{2t - 5t}{2}\right)$$

$$= -2\sin\frac{7}{2}t\sin\left(-\frac{3}{2}t\right) = \mathbf{2\sin\frac{7}{2}t\sin\frac{3}{2}t}$$

$$\left[\text{since } \sin\left(-\frac{3}{2}t\right) = -\sin\frac{3}{2}t\right]$$

Section 6

Graphs

Why are graphs important?

Graphs have a wide range of applications in engineering and in physical sciences because of their inherent simplicity. A graph can be used to represent almost any physical situation involving discrete objects and the relationship among them. If two quantities are directly proportional and one is plotted against the other, a straight line is produced. Examples include an applied force on the end of a spring plotted against spring extension, the speed of a flywheel plotted against time, and strain in a wire plotted against stress (Hooke's law). In engineering and science, the straight-line graph is the most basic graph to draw and evaluate.

Graphs are important tools for analysing and displaying data between two experimental quantities. Many times, situations occur in which the relationship between the variables is not linear. By manipulation, a straight-line graph may be plotted to produce a law relating the two variables. Sometimes this involves using the laws of logarithms. The relationship between the resistance of wire and its diameter is not a linear one. Similarly, the periodic time of oscillations of a pendulum does not have a linear relationship with its length, and the head of pressure and the flow velocity are not linearly related.

Lighting engineers study photometrics, which is the process for measuring the intensity of light. Such intensity varies according to how close or how far you are from LEDs and other bulbs. Engineers use polar curves to plot the level of intensity at any given point, i.e. a polar curve tells an engineer how LEDs spread light across a room or surface. This distribution may be narrow, wide, indirect or direct. By using polar curves, an engineer can arrange the right number and type of LEDs for a lighting project. For example, a designer may want specific lighting effects in a room. Creating polar curves before installation gives the designer and lighting engineer a chance to check light distribution and intensity. This saves time and money and ensures success.

Being able to solve equations graphically provides another method to aid understanding and interpretation of equations. Engineers, including architects, surveyors and a variety of engineers in fields such as biomedical, chemical, electrical, mechanical and nuclear, all use equations which need solving by one means or another.

Graphs and diagrams provide a simple and powerful approach to a variety of problems that are typical to computer science in general, and software engineering in particular; graphical transformations have many applications in software engineering problems. Periodic functions are used throughout engineering and science to describe oscillations, waves and other phenomena that exhibit periodicity. Engineers and scientists use many basic mathematical functions to represent, say, the input/output of systems – linear, quadratic, exponential, sinusoidal, and so on, and knowledge of these are needed to determine how these are used to generate some of the more unusual input/output signals such as the square wave, saw-tooth wave and fully-rectified sine wave. Understanding of continuous and discontinuous functions, odd and even functions, and inverse functions are helpful in this – it's all part of the 'language of engineering and science'.

Chapter 55   The straight-line graph

The equation of a straight line graph is: **y = mx + c**
where m is the gradient and c the y-axis intercept.

With reference to Figure 55.1, gradient, $m = \dfrac{y_2 - y_1}{x_2 - x_1}$

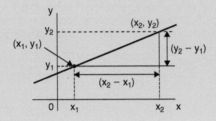

Figure 55.1

Application: Determine the gradient of the straight-line graph passing through the co-ordinates $(-2, 5)$ and $(3, 4)$

A straight line graph passing through co-ordinates (x_1, y_1) and (x_2, y_2) has a gradient given by:

$$m = \frac{y_2 - y_1}{x_2 - x_1} \quad \text{(see Figure 55.1)}$$

A straight line passes through $(-2, 5)$ and $(3, 4)$, from which, $x_1 = -2$, $y_1 = 5$, $x_2 = 3$ and $y_2 = 4$, hence

$$\text{gradient, } m = \frac{y_2 - y_1}{x_2 - x_1} = \frac{4 - 5}{3 - (-2)} = -\frac{1}{5} \text{ or } \mathbf{-0.2}$$

Application: The temperature in degrees Celsius and the corresponding values in degrees Fahrenheit are shown in the table below.

°C	10	20	40	60	80	100
°F	50	68	104	140	176	212

Plot a graph of degrees Celsius (horizontally) against degrees Fahrenheit (vertically). From the graph find (a) the temperature in degrees Fahrenheit at 55°C, (b) the temperature in degrees Celsius at 167°F, (c) the Fahrenheit temperature at 0°C, and (d) the Celsius temperature at 230°F

Axes with suitable scales are shown in Figure 55.2. The co-ordinates (10, 50), (20, 68), (40, 104), and so on are plotted as shown. When the co-ordinates are joined, a straight line is produced. Since a straight line results there is a linear relationship between degrees Celsius and degrees Fahrenheit.

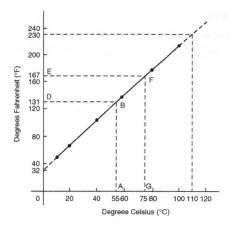

Figure 55.2

(a) To find the Fahrenheit temperature at 55°C, a vertical line AB is constructed from the horizontal axis to meet the straight line at B. The point where the horizontal line BD meets the vertical axis indicates the equivalent Fahrenheit temperature. **Hence 55°C is equivalent to 131°F.** This process of finding an equivalent value in between the given information in the above table is called **interpolation**.

(b) To find the Celsius temperature at 167°F, a horizontal line EF is constructed as shown in Figure 55.2. The point where the vertical line FG cuts the horizontal axis indicates the equivalent Celsius temperature. **Hence 167°F is equivalent to 75°C.**

(c) If the graph is assumed to be linear even outside of the given data, then the graph may be extended at both ends (shown by broken lines in Figure 55.2). From Figure 55.2, it is seen that **0°C corresponds to 32°F.**

(d) **230°F is seen to correspond to 110°C.**

The process of finding equivalent values outside of the given range is called **extrapolation**.

Application: Experimental tests to determine the breaking stress σ of rolled copper at various temperatures t gave the following results.

Stress σ N/cm²	8.46	8.04	7.78	7.37	7.08	6.63
Temperature t°C	70	200	280	410	500	640

Show that the values obey the law $\sigma = at + b$, where a and b are constants and determine approximate values for a and b. Use the law to determine the stress at 250°C and the temperature when the stress is 7.54 N/cm²

The co-ordinates (70, 8.46), (200, 8.04), and so on, are plotted as shown in Figure 55.3. Since the graph is a straight line then the values obey the law $\sigma = at + b$, and the gradient of the straight line, is:

$$a = \frac{AB}{BC} = \frac{8.36 - 6.76}{100 - 600} = \frac{1.60}{-500} = -0.0032$$

Vertical axis intercept, **b = 8.68**

Hence the law of the graph is: **σ = −0.0032t + 8.68**

When the temperature is 250°C, stress σ is given by

$$\sigma = -0.0032(250) + 8.68 = \textbf{7.88 N/cm}^2$$

Rearranging $\sigma = -0.0032t + 8.68$ gives:

$$0.0032t = 8.68 - \sigma, \quad \text{i.e.} \quad t = \frac{8.68 - \sigma}{0.0032}$$

Hence, when the stress, $\sigma = 7.54 \text{ N/cm}^2$,

$$\text{temperature } t = \frac{8.68 - 7.54}{0.0032} = 356.3°\text{C}$$

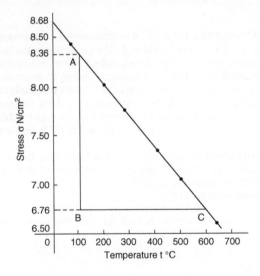

Figure 55.3

Chapter 56   Determination of law

Some examples of the reduction of equations to linear form include:

1. $y = ax^2 + b$ compares with $Y = mX + c$, where $m = a$, $c = b$ and $X = x^2$.
   Hence y is plotted vertically against x^2 horizontally to produce a straight line graph of gradient 'a' and y-axis intercept 'b'

2. $y + \dfrac{a}{x} + b$

   y is plotted vertically against $\dfrac{1}{x}$ horizontally to produce a straight line graph of gradient 'a' and y-axis intercept 'b'

3. $y = ax^2 + bx$

Dividing both sides by x gives $\dfrac{y}{x} = ax + b$

Comparing with $Y = mX + c$ shows that $\dfrac{y}{x}$ is plotted vertically against x horizontally to produce a straight line graph of gradient 'a' and $\dfrac{y}{x}$ axis intercept 'b'.

Determination of law involving logarithms

4. If $y = ax^n$ then $\lg y = \lg(ax^n) = \lg a + \lg x^n$

   i.e. $\lg y = n\lg x + \lg a$

5. If $y = ab^x$ then $\lg y = \lg(ab^x) = \lg a + \lg b^x = \lg a + x \lg b$

   i.e. $\lg y = (\lg b)x + \lg a$

6. If $y = ae^{bx}$ then $\ln y = \ln(ae^{bx}) = \ln a + \ln(e^{bx}) = \ln a + bx$

   i.e. $\ln y = bx + \ln a$

Application: Values of load L newtons and distance d metres obtained experimentally are shown in the following table.

Load, L (N)	32.3	29.6	27.0	23.2	18.3	12.8	10.0	6.4
distance, d (m)	0.75	0.37	0.24	0.17	0.12	0.09	0.08	0.07

Verify that the load and distance are related by a law of the form $L = \dfrac{a}{d} + b$ and determine approximate values of a and b. Hence calculate the load when the distance is 0.20 m and the distance when the load is 20 N

Comparing $L = \dfrac{a}{d} + b$ i.e. $L = a\left(\dfrac{1}{d}\right) + b$ with $Y = mX + c$ shows that L is to be plotted vertically against $\dfrac{1}{d}$ horizontally. Another table of values is drawn up as shown below.

L	32.3	29.6	27.0	23.2	18.3	12.8	10.0	6.4
d	0.75	0.37	0.24	0.17	0.12	0.09	0.08	0.07
$\dfrac{1}{d}$	1.33	2.70	4.17	5.88	8.33	11.11	12.50	14.29

A graph of L against $\dfrac{1}{d}$ is shown in Figure 56.1. A straight line can be drawn through the points, which verifies that load and distance are related by a law of the form $L = \dfrac{a}{d} + b$

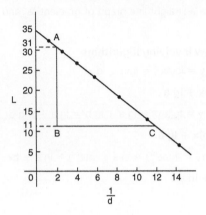

Figure 56.1

Gradient of straight line, $a = \dfrac{AB}{BC} = \dfrac{31-11}{2-12} = \dfrac{20}{-10} = -2$

L-axis intercept, **b = 35**

Hence, the law of the graph is: $L = -\dfrac{2}{d} + 35$

When the distance d is 0.20 m, load $L = \dfrac{-2}{0.20} + 35 = 25.0\,N$

Rearranging $L = -\dfrac{2}{d} + 35$ gives $\dfrac{2}{d} = 35 - L$ and $d = \dfrac{2}{35 - L}$

Hence, when the load L is 20 N,

$$\text{distance } d = \dfrac{2}{35-20} = \dfrac{2}{15} = 0.13\,m$$

Application: The current flowing in, and the power dissipated by a resistor are measured experimentally for various values and the results are as shown below.

Current, I amperes	2.2	3.6	4.1	5.6	6.8
Power, P watts	116	311	403	753	1110

Show that the law relating current and power is of the form $P = RI^{n}$, where R and n are constants, and determine the law

Taking logarithms to a base of 10 of both sides of $P = RI^n$ gives:

$\lg P = \lg(RI^n) = \lg R + \lg I^n = \lg R + n\lg I$ by the laws of logarithms

i.e. $\lg P = n\lg I + \lg R$, which is of the form $Y = mX + c$,

showing that $\lg P$ is to be plotted vertically against $\lg I$ horizontally.

A table of values for $\lg I$ and $\lg P$ is drawn up as shown below.

I	2.2	3.6	4.1	5.6	6.8
$\lg I$	0.342	0.556	0.613	0.748	0.833
P	116	311	403	753	1110
$\lg P$	2.064	2.493	2.605	2.877	3.045

A graph of $\lg P$ against $\lg I$ is shown in Figure 56.2 and since a straight line results the law $P = RI^n$ is verified.

$$\text{Gradient of straight line, } n = \frac{AB}{BC} = \frac{2.98 - 2.18}{0.80 - 0.40} = \frac{0.80}{0.40} = 2$$

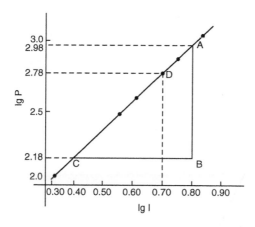

Figure 56.2

It is not possible to determine the vertical axis intercept on sight since the horizontal axis scale does not start at zero. Selecting any point from the graph, say point D, where $\lg I = 0.70$ and $\lg P = 2.78$, and substituting values into

$$\lg P = n\lg I + \lg R$$

gives: $2.78 = (2)(0.70) + \lg R$

from which $\lg R = 2.78 - 1.40 = 1.38$

Hence $R = \text{antilog } 1.38 \ (=10^{1.38}) = \mathbf{24.0}$

Hence the law of the graph is: $P = 24.0 \, I^2$

Application: The current i mA flowing in a capacitor which is being discharged varies with time t ms as shown below.

i mA	203	61.14	22.49	6.13	2.49	0.615
t ms	100	160	210	275	320	390

Show that these results are related by a law of the form $i = Ie^{t/T}$, where I and T are constants. Determine the approximate values of I and T.

Taking Napierian logarithms of both sides of $i = Ie^{t/T}$ gives

$$\ln i = \ln(Ie^{t/T}) = \ln I + \ln e^{t/T}$$

i.e.

$$\ln i = \ln I + \frac{t}{T} \qquad \text{(since } \ln e^x = x\text{)}$$

or

$$\ln i = \left(\frac{1}{T}\right)t + \ln I$$

which compares with $y = mx + c$, showing that $\ln i$ is plotted vertically against t horizontally. Another table of values is drawn up as shown below.

t	100	160	210	275	320	390
i	203	61.14	22.49	6.13	2.49	0.615
ln i	5.31	4.11	3.11	1.81	0.91	−0.49

A graph of $\ln i$ against t is shown in Figure 56.3 and since a straight line results the law $i = Ie^{t/T}$ is verified.

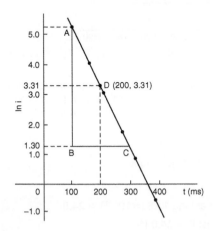

Figure 56.3

Gradient of straight line,

$$\frac{1}{T} = \frac{AB}{BC} = \frac{5.30 - 1.30}{100 - 300} = \frac{4.0}{-200} = -0.02$$

Hence, $T = \dfrac{1}{-0.02} = -50$

Selecting any point on the graph, say point D, where t = 200 and ln i = 3.31,

and substituting into $\ln i = \left(\dfrac{1}{T}\right) t + \ln I$

gives: $3.31 = -\dfrac{1}{50}(200) + \ln I$

from which, $\ln I = 3.31 + 4.0 = 7.31$

and   I = antilog 7.31 ($= e^{7.31}$) = 1495 or **1500** correct to 3 significant figures

Hence the law of the graph is i = 1500$e^{-t/50}$

Chapter 57   Graphs with logarithmic scales

Application: Experimental values of two related quantities x and y are shown below:

x	0.41	0.63	0.92	1.36	2.17	3.95
y	0.45	1.21	2.89	7.10	20.79	82.46

The law relating x and y is believed to be y = ax^b, where a and b are constants.

Verify that this law is true and determine the approximate values of a and b

If y = ax^b then lg y = b lg x + lg a, from page 175, which is of the form Y = mX + c, showing that to produce a straight line graph lg y is plotted vertically against lg x horizontally. x and y may be plotted directly on to log-log graph paper as shown in Figure 57.1. The values of y range from 0.45 to 82.46 and 3 cycles are needed (i.e. 0.1 to 1, 1 to 10 and 10 to 100). The values of x range from 0.41 to 3.95 and 2 cycles are needed (i.e. 0.1 to 1 and 1 to 10). Hence 'log 3 cycle × 2 cycle' is used as shown in Figure 57.1 where the axes are marked and the points plotted. Since the points lie on a straight line the law y = ax^b is verified.

To evaluate constants a and b:

Method 1. Any two points on the straight line, say points A and C, are selected, and AB and BC are measured (say in centimetres). Then, gradient,

$$b = \frac{AB}{BC} = \frac{11.5\,\text{units}}{5\,\text{units}} = 2.3$$

Since lg y = b lg x + lg a, when x = 1, lg x = 0 and lg y = lg a.

The straight line crosses the ordinate x = 1.0 at y = 3.5.

Hence, lg a = lg 3.5, i.e. **a = 3.5**

Method 2. Any two points on the straight line, say points A and C, are selected. A has co-ordinates (2, 17.25) and C has co-ordinates (0.5, 0.7).

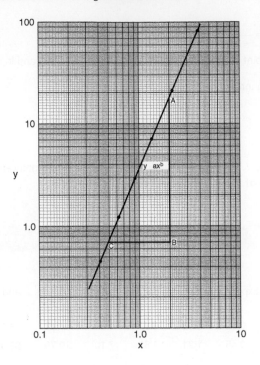

Figure 57.1

Since $y = ax^b$ then $17.25 = a(2)^b$ (1)

and $\quad\quad\quad\quad\quad\quad\quad 0.7 = a(0.5)^b$ (2)

i.e. two simultaneous equations are produced and may be solved for a and b.

Dividing equation (1) by equation (2) to eliminate a gives:

$$\frac{17.25}{0.7} = \frac{(2)^b}{(0.5)^b} = \left(\frac{2}{0.5}\right)^b$$

i.e. $\quad\quad\quad\quad\quad 24.643 = (4)^b$

Taking logarithms of both sides gives $\lg 24.643 = b\lg 4$,

i.e. $\quad b = \dfrac{\lg 24.643}{\lg 4} = 2.3$, correct to 2 significant figures.

Substituting $b = 2.3$ in equation (1) gives: $17.25 = a(2)^{2.3}$, i.e.

$$a = \frac{17.25}{(2)^{2.3}} = \frac{17.25}{4.925} = 3.5 \text{ correct to 2 significant figures.}$$

Hence the law of the graph is: $y = 3.5x^{2.3}$

Application: The pressure p and volume v of a gas are believed to be related by a law of the form $p = cv^n$, where c and n are constants. Experimental values of p and corresponding values of v obtained in a laboratory are:

p (Pascals)	2.28×10^5	8.04×10^5	2.03×10^6	5.05×10^6	1.82×10^7
v (m³)	3.2×10^{-2}	1.3×10^{-2}	6.7×10^{-3}	3.5×10^{-3}	1.4×10^{-3}

Verify that the law is true and determine approximate values of c and n

Since $p = cv^n$, then $\lg p = n \lg v + \lg c$, which is of the form $Y = mX + c$, showing that to produce a straight line graph, $\lg p$ is plotted vertically against $\lg v$ horizontally. The co-ordinates are plotted on 'log 3 cycle × 2 cycle' graph paper as shown in Figure 57.2. With the data expressed in standard form, the axes are marked in standard form also. Since a straight line results the law $p = cv^n$ is verified.

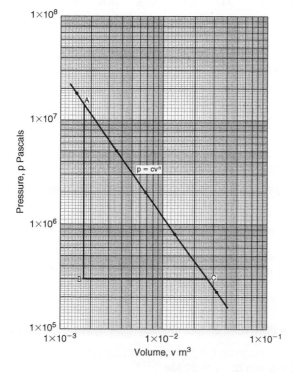

Figure 57.2

The straight line has a negative gradient and the value of the gradient is given

by: $\dfrac{AB}{BC} = \dfrac{14 \text{ units}}{10 \text{ units}} = 1.4.$ Hence $n = -1.4$

Selecting any point on the straight line, say point C, having co-ordinates $(2.63 \times 10^{-2}, 3 \times 10^{5})$, and substituting these values in $p = cv^n$ gives: $3 \times 10^{5} = c(2.63 \times 10^{-2})^{-1.4}$

Hence,

$$c = \frac{3 \times 10^{5}}{(2.63 \times 10^{-2})^{-1.4}} = \frac{3 \times 10^{5}}{(0.0263)^{-1.4}}$$

$$= \frac{3 \times 10^{5}}{1.63 \times 10^{2}} = 1840, \text{correct to 3 significant figures.}$$

Hence the law of the graph is: $p = 1840v^{-1.4}$ or $pv^{1.4} = 1840$

Application: The voltage, v volts, across an inductor is believed to be related to time, t ms, by the law $v = Ve^{t/T}$, where V and T are constants. Experimental results obtained are:

v volts	883	347	90	55.5	18.6	5.2
t ms	10.4	21.6	37.8	43.6	56.7	72.0

Show that the law relating voltage and time is as stated and determine the approximate values of V and T. Find also the value of voltage after 25 ms and the time when the voltage is 30.0 V

Since $v = Ve^{t/T}$ then $\ln v = \frac{1}{T}t + \ln V$, which is of the form $Y = mX + c$

Using 'log 3 cycle $\times$ linear' graph paper, the points are plotted as shown in Figure 57.3. Since the points are joined by a straight line the law $v = Ve^{t/T}$ is verified.

Gradient of straight line, $\frac{1}{T} = \frac{AB}{BC} = \frac{\ln 100 - \ln 10}{36.5 - 64.2} = \frac{2.3026}{-27.7}$

Hence $T = \frac{-27.7}{2.3026} = -12.0$, correct to 3 significant figures.

Since the straight line does not cross the vertical axis at $t = 0$ in Figure 57.3, the value of V is determined by selecting any point, say A, having co-ordinates (36.5, 100) and substituting these values into $v = Ve^{t/T}$.

Thus $100 = Ve^{36.5/-12.0}$

i.e. $V = \frac{100}{e^{-36.5/12.0}} = 2090$ volts, correct to 3 significant figures.

Hence the law of the graph is: $v = 2090e^{-t/12.0}$

When time $t = 25$ ms, voltage $v = 2090e^{-25/12.0} = 260$ V

When the voltage is 30.0 volts, $30.0 = 2090e^{-t/12.0}$

hence $e^{-t/12.0} = \frac{30.0}{2090}$ and $e^{t/12.0} = \frac{2090}{30.0} = 69.67$

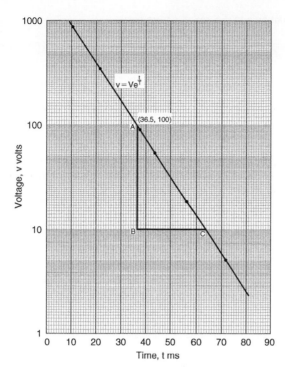

Figure 57.3

Taking Napierian logarithms gives: $\dfrac{t}{12.0} = \ln 69.67 = 4.2438$

from which, **time, t**= (12.0)(4.2438) = **50.9 ms**

Chapter 58   Graphical solution of simultaneous equations

Linear simultaneous equations in two unknowns may be solved graphically by:

1. plotting the two straight lines on the same axes, and
2. noting their point of intersection.

The co-ordinates of the point of intersection give the required solution.

Application: Solve graphically the simultaneous equations:

$$2x - y = 4$$
$$x + y = 5$$

Rearranging each equation into y = mx + c form gives:

$$y = 2x - 4 \tag{1}$$

$$y = -x + 5 \tag{2}$$

Only three co-ordinates need be calculated for each graph since both are straight lines.

x	0	1	2
y = 2x − 4	−4	−2	0

x	0	1	2
y = −x + 5	5	4	3

Each of the graphs is plotted as shown in Figure 58.1. The point of intersection is at (3, 2) and since this is the only point which lies simultaneously on both lines then **x = 3, y = 2** is the solution of the simultaneous equations.

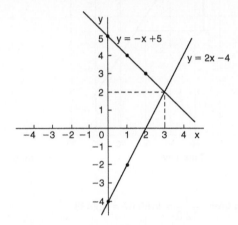

Figure 58.1

Chapter 59   Quadratic graphs

(i) y = ax²

Graphs of $y = x^2$, $y = 3x^2$ and $y = \frac{1}{2}x^2$ are shown in Figure 59.1.

(a)

(b)

(c)   Figure 59.1

All have minimum values at the origin (0, 0).

Graphs of $y = -x^2$, $y = -3x^2$ and $y = -\dfrac{1}{2}x^2$ are shown in Figure 59.2.

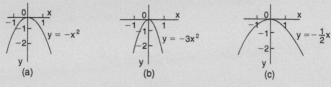

Figure 59.2

All have maximum values at the origin (0, 0).
When $y = ax^2$,

(a) curves are symmetrical about the y-axis,
(b) the magnitude of 'a' affects the gradient of the curve, and
(c) the sign of 'a' determines whether it has a maximum or minimum value

(ii) **$y = ax^2 + c$**
Graphs of $y = x^2 + 3$, $y = x^2 - 2$, $y = -x^2 + 2$ and $y = -2x^2 - 1$ are shown in Figure 59.3.

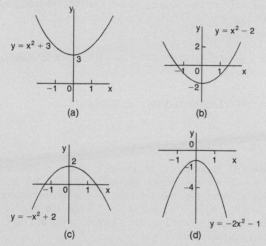

Figure 59.3

When $y = ax^2 + c$:

(a) curves are symmetrical about the y-axis,
(b) the magnitude of 'a' affects the gradient of the curve, and
(c) the constant 'c' is the y-axis intercept

(iii) **$y = ax^2 + bx + c$**

Whenever 'b' has a value other than zero the curve is displaced to the right or left of the y-axis. When b/a is positive, the curve is displaced b/2a to the left of the y-axis, as shown in Figure 59.4(a). When b/a is negative the curve is displaced b/2a to the right of the y-axis, as shown in Figure 59.4(b).

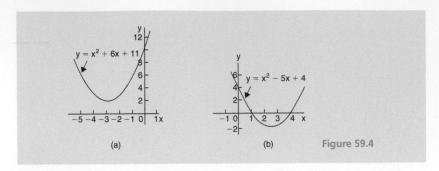

(a)   (b)   Figure 59.4

Graphical solutions of quadratic equations

Quadratic equations of the form $ax^2 + bx + c = 0$ may be solved graphically by:

(i) plotting the graph $y = ax^2 + bx + c$, and
(ii) noting the points of intersection on the x-axis (i.e. where $y = 0$).

The number of solutions, or roots of a quadratic equation, depends on how many times the curve cuts the x-axis and there can be no real roots (as in Figure 59.4(a)) or one root (as in Figures 59.1 and 59.2) or two roots (as in Figure 59.4(b)).

Application: Solve the quadratic equation $4x^2 + 4x - 15 = 0$ graphically given that the solutions lie in the range $x = -3$ to $x = 2$

Let $y = 4x^2 + 4x - 15$. A table of values is drawn up as shown below.

x	-3	-2	-1	0	1	2
$y = 4x^2 + 4x - 15$	9	-7	-15	-15	-7	9

A graph of $y = 4x^2 + 4x - 15$ is shown in Figure 59.5. The only points where $y = 4x^2 + 4x - 15$ and $y = 0$, are the points marked A and B. This occurs at **x = −2.5 and x = 1.5** and these are the solutions of the quadratic equation $4x^2 + 4x - 15 = 0$. (By substituting $x = -2.5$ and $x = 1.5$ into the original equation the solutions may be checked). The curve has a turning point at $(-0.5, -16)$ and the nature of the point is a **minimum**.

An alternative graphical method of solving $4x^2 + 4x - 15 = 0$ is to rearrange the equation as $4x^2 = -4x + 15$ and then plot two separate graphs – in this case $y = 4x^2$ and $y = -4x + 15$. Their points of intersection give the roots of equation $4x^2 = -4x + 15$, i.e. $4x^2 + 4x - 15 = 0$. This is shown in Figure 59.6, where the roots are $x = -2.5$ and $x = 1.5$ as before.

Application: Plot a graph of $y = 2x^2$ and hence solve the equations

(a) $2x^2 - 8 = 0$ and   (b) $2x^2 - x - 3 = 0$

A graph of $y = 2x^2$ is shown in Figure 59.7.

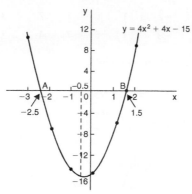

Figure 59.5

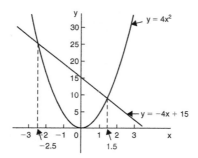

Figure 59.6

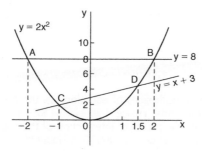

Figure 59.7

(a) Rearranging $2x^2 - 8 = 0$ gives $2x^2 = 8$ and the solution of this equation is obtained from the points of intersection of $y = 2x^2$ and $y = 8$, i.e. at co-ordinates $(-2, 8)$ and $(2, 8)$, shown as A and B, respectively, in Figure 59.7.

Hence the solutions of $2x^2 - 8 = 0$ are **x = -2 and x = +2**

(b) Rearranging $2x^2 - x - 3 = 0$ gives $2x^2 = x + 3$ and the solution of this equation is obtained from the points of intersection of $y = 2x^2$ and $y = x + 3$, i.e. at C and D in Figure 59.7. Hence the solutions of $2x^2 - x - 3 = 0$ are **x = -1 and x = 1.5**

Application: Plot the graph of $y = -2x^2 + 3x + 6$ for values of x from $x = -2$ to $x = 4$ and to use the graph to find the roots of the following equations

(a) $-2x^2 + 3x + 6 = 0$ (b) $-2x^2 + 3x + 2 = 0$
(c) $-2x^2 + 3x + 9 = 0$ (d) $-2x^2 + x + 5 = 0$

A table of values is drawn up as shown below.

x	-2	-1	0	1	2	3	4
$y = -2x^2 + 3x + 6$	-8	1	6	7	4	-3	-14

A graph of $-2x^2 + 3x + 6$ is shown in Figure 59.8.

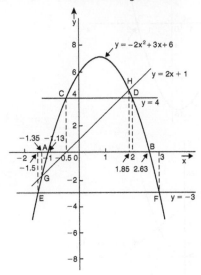

Figure 59.8

(a) The parabola $y = -2x^2 + 3x + 6$ and the straight line $y = 0$ intersect at A and B, where **$x = -1.13$ and $x = 2.63$** and these are the roots of the equation $-2x^2 + 3x + 6 = 0$

(b) Comparing

$$y = -2x^2 + 3x + 6 \qquad (1)$$

with

$$0 = -2x^2 + 3x + 2 \qquad (2)$$

shows that if 4 is added to both sides of equation (2), the right-hand side of both equations will be the same. Hence $4 = -2x^2 + 3x + 6$. The solution of this equation is found from the points of intersection of the line $y = 4$ and the parabola $y = -2x^2 + 3x + 6$, i.e. points C and D in Figure 59.8.

Hence the roots of $-2x^2 + 3x + 2 = 0$ are **$x = -0.5$ and $x = 2$**

(c) $-2x^2 + 3x + 9 = 0$ may be rearranged as $-2x^2 + 3x + 6 = -3$, and the solution of this equation is obtained from the points of intersection of the line $y = -3$ and the parabola $y = -2x^2 + 3x + 6$, i.e. at points E and F in Figure 59.8. Hence the roots of $-2x^2 + 3x + 9 = 0$ are **$x = -1.5$ and $x = 3$**

(d) Comparing

$$y = -2x^2 + 3x + 6 \qquad (3)$$

with

$$0 = -2x^2 + x + 5 \qquad (4)$$

shows that if $2x + 1$ is added to both sides of equation (4) the right-hand side of both equations will be the same. Hence equation (4) may be written as $2x + 1 = -2x^2 + 3x + 6$. The solution of this equation is found from the points of intersection of the line $y = 2x + 1$ and the parabola $y = -2x^2 + 3x + 6$, i.e. points G and H in Figure 59.8. Hence the roots of $-2x^2 + x + 5 = 0$ are **$x = -1.35$ and $x = 1.85$**

Chapter 60   Graphical solution of cubic equations

A **cubic equation** of the form $ax^3 + bx^2 + cx + d = 0$ may be solved graphically by:

(i) plotting the graph $y = ax^3 + bx^2 + cx + d$,

and   (ii) noting the points of intersection on the x-axis (i.e. where $y = 0$).

The number of solutions, or roots of a cubic equation depends on how many times the curve cuts the x-axis and there can be one, two or three possible roots, as shown in Figure 60.1.

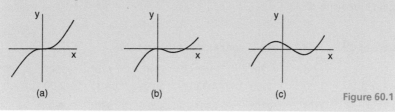

(a)   (b)   (c)   Figure 60.1

Application: Solve graphically the cubic equation $4x^3 - 8x^2 - 15x + 9 = 0$ given that the roots lie between $x = -2$ and $x = 3$. Find also the co-ordinates of the turning points on the curve.

Let $y = 4x^3 - 8x^2 - 15x + 9$. A table of values is drawn up as shown below.

x	−2	−1	0	1	2	3
y	−25	12	9	−10	−21	0

A graph of $y = 4x^3 - 8x^2 - 15x + 9$ is shown in Figure 60.2.

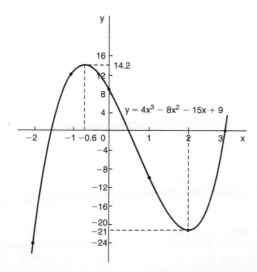

Figure 60.2

The graph crosses the x-axis (where y = 0) at **x = −1.5, x = 0.5 and x = 3** and these are the solutions to the cubic equation $4x^3 - 8x^2 - 15x + 9 = 0$.

The turning points occur at **(−0.6, 14.2)**, which is a **maximum**, and **(2, −21)**, which is a **minimum**.

Chapter 61   Polar curves

Application: Plot the polar graph of $r = 5\sin\theta$ between $\theta = 0°$ and $\theta = 360°$ using increments of 30°

A table of values at 30° intervals is produced as shown below.

θ	0	30°	60°	90°	120°	150°	180°
$r = 5\sin\theta$	0	2.50	4.33	5.00	4.33	2.50	0

θ	210°	240°	270°	300°	330°	360°
$r = 5\sin\theta$	−2.50	−4.33	−5.00	−4.33	−2.50	0

The graph is plotted as shown in Figure 61.1.

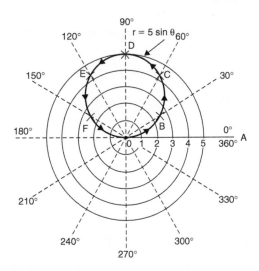

Figure 61.1

Initially the zero line OA is constructed and then the broken lines in Figure 61.1 at 30° intervals are produced. The maximum value of r is 5.00 hence OA is scaled and circles drawn as shown with the largest at a radius of 5 units. The polar co-ordinates (0, 0°), (2.50, 30°), (4.33, 60°), (5.00, 90°).... are plotted and shown as points O, B, C, D, ... in Figure 61.1. When polar co-ordinate (0, 180°) is plotted and the points joined with a smooth curve a complete circle is seen to have been produced. When plotting the next point, (−2.50, 210°), since r is negative it is plotted in the opposite direction to 210°, i.e. 2.50 units long on the 30° axis. Hence the point (−2.50, 210°) is equivalent to the point (2.50, 30°). Similarly, (−4.33, 240°) is the same point as (4.33, 60°).

When all the co-ordinates are plotted the graph r = 5 sin θ appears as a single circle; it is, in fact, two circles, one on top of the other.

In general, a polar curve **r = a sin θ** is as shown in Figure 61.2.

In a similar manner to that explained above, it may be shown that the polar curve **r = a cos θ** is as sketched in Figure 61.3.

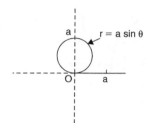

Figure 61.2

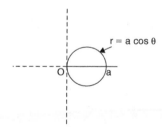

Figure 61.3

Application: Plot the polar graph of r = 4 sin²θ between θ = 0 and θ = 2π radians using intervals of $\dfrac{\pi}{6}$

A table of values is produced as shown below.

θ	0	$\dfrac{\pi}{6}$	$\dfrac{\pi}{3}$	$\dfrac{\pi}{2}$	$\dfrac{2\pi}{3}$	$\dfrac{5\pi}{6}$	π	$\dfrac{7\pi}{6}$	$\dfrac{4\pi}{3}$	$\dfrac{3\pi}{2}$	$\dfrac{5\pi}{3}$	$\dfrac{11\pi}{6}$	2π
r = 4 sin²θ	0	1	3	4	3	1	0	1	3	4	3	1	0

The zero line OA is firstly constructed and then the broken lines at intervals of $\dfrac{\pi}{6}$ rad (or 30°) are produced. The maximum value of r is 4 hence OA is scaled and circles produced as shown with the largest at a radius of 4 units.

The polar co-ordinates $(0, 0)$, $(1, \frac{\pi}{6})$, $(3, \frac{\pi}{3})$, ... $(0, \pi)$ are plotted and shown as points 0, B, C, D, E, F, 0, respectively. Then $(1, \frac{7\pi}{6})$, $(3, \frac{4\pi}{3})$,... $(0, 0)$ are plotted as shown by points G, H, I, J, K, 0 respectively. Thus two distinct loops are produced as shown in Figure 61.4.

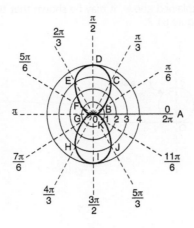

Figure 61.4

In general, a polar curve $r = a\sin^2\theta$ is as shown in Figure 61.5.

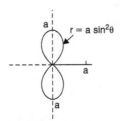

Figure 61.5

In a similar manner it may be shown that the polar curve $r = a\cos^2\theta$ is as sketched in Figure 61.6.

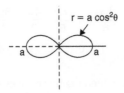

Figure 61.6

Application: Plot the polar graph of $r = 3\sin 2\theta$ between $\theta = 0°$ and $\theta = 360°$, using 15° intervals

As in previous applications a table of values may be produced.

The polar graph $r = 3\sin 2\theta$ is plotted as shown in Figure 61.7 and is seen to contain four similar shaped loops displaced at 90° from each other.

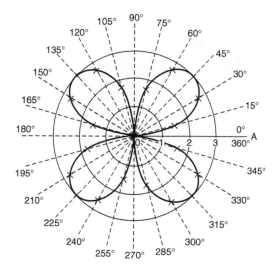

Figure 61.7

In general, a polar curve $r = a\sin 2\theta$ is as shown in Figure 61.8.

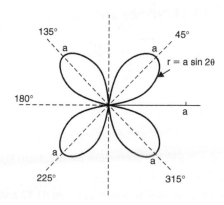

Figure 61.8

In a similar manner it may be shown that polar curves of $r = a\cos 2\theta$, $r = a\sin 3\theta$ and $r = a\cos 3\theta$ are as sketched in Figure 61.9.

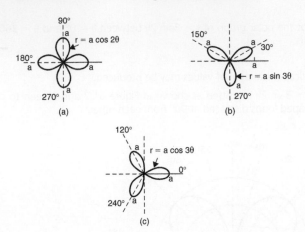

(a)   (b)

(c)

Figure 61.9

Application: Sketch the polar curve $r = 2\theta$ between $\theta = 0$ and $\theta = \dfrac{5\pi}{2}$ rad at intervals of $\dfrac{\pi}{6}$

A table of values may be produced and the polar graph of $r = 2\theta$ is shown in Figure 61.10 and is seen to be an ever-increasing spiral.

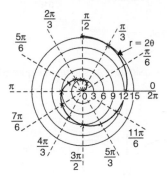

Figure 61.10

Application: Plot the polar curve $r = 5(1 + \cos \theta)$ from $\theta = 0°$ to $\theta = 360°$, using 30° intervals

A table of values may be produced and the polar curve $r = 5(1 + \cos \theta)$ is shown in Figure 61.11.

In general, a polar curve $r = a(1 + \cos \theta)$ is as shown in Figure 61.12 and the shape is called a **cardioid**.

In a similar manner it may be shown that the polar curve $r = a + b \cos \theta$ varies in shape according to the relative values of a and b. When $a = b$ the polar curve shown in Figure 61.12 results.

When $a < b$ the general shape shown in Figure 61.13(a) results and when $a > b$ the general shape shown in Figure 61.13(b) results.

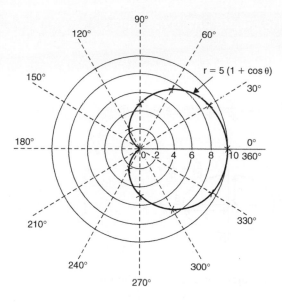

Figure 61.11

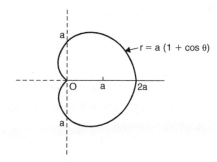

Figure 61.12

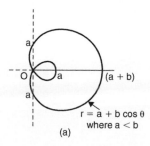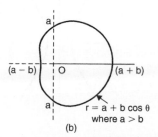

Figure 61.13

Chapter 62   The ellipse and hyperbola

Ellipse

The equation of an ellipse is $\dfrac{x^2}{a^2} + \dfrac{y^2}{b^2} = 1$ and the general shape is as shown in Figure 62.1.

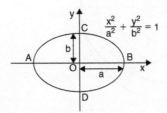

Figure 62.1

The length AB is called the **major axis** and CD the **minor axis**.

In the above equation, 'a' is the semi-major axis and 'b' is the semi-minor axis.

(Note that if b = a, the equation becomes $\dfrac{x^2}{a^2} + \dfrac{y^2}{a^2} = 1$, i.e. $x^2 + y^2 = a^2$, which is a circle of radius a.)

Hyperbola

The equation of a hyperbola is $\dfrac{x^2}{a^2} - \dfrac{y^2}{b^2} = 1$ and the general shape is shown in Figure 62.2. The curve is seen to be symmetrical about both the x- and y-axes. The distance AB in Figure 62.2 is given by 2a.

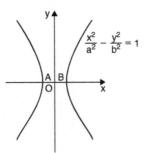

Figure 62.2

Rectangular hyperbola

The equation of a rectangular hyperbola is $xy = c$ or $y = \dfrac{c}{x}$ and the general shape is shown in Figure 62.3.

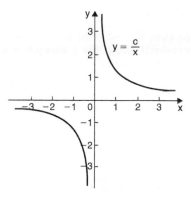

Figure 62.3

Chapter 63 Graphical functions

Periodic functions

A function f(x) is said to be **periodic** if f(x + T) = f(x) for all values of x, where T is some positive number. T is the interval between two successive repetitions and is called the **period** of the function f(x). For example, y = sin x is periodic in x with period 2π since sin x = sin(x + 2π) = sin(x + 4π), and so on. Similarly, y = cos x is a periodic function with period 2π since cos x = cos (x + 2π) = cos (x + 4π), and so on. In general, if y = sin ωt or y = cos ωt then the period of the waveform is $2\pi/\omega$. The function shown in Figure 63.1 is also periodic of period 2π and is defined by:

$$f(x) = \begin{cases} -1, \text{ when } -\pi \le x \le 0 \\ 1, \text{ when } \quad 0 \le x \le \pi \end{cases}$$

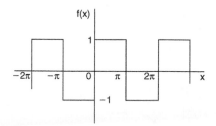

Figure 63.1

Continuous and discontinuous functions

If a graph of a function has no sudden jumps or breaks it is called a **continuous function**, examples being the graphs of sine and cosine functions. However, other graphs make finite jumps at a point or points in the interval. The square wave shown in Figure 63.1 has **finite discontinuities** as x = π, 2π, 3π, and so on, and is therefore a discontinuous function. y = tan x is another example of a discontinuous function.

Even and odd functions

A function $y = f(x)$ is said to be **even** if $f(-x) = f(x)$ for all values of x. Graphs of even functions are always **symmetrical about the y-axis** (i.e. is a mirror image). Two examples of even functions are $y = x^2$ and $y = \cos x$ as shown in Figure 63.2.

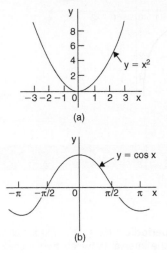

(a)

(b)   Figure 63.2

A function $y = f(x)$ is said to be **odd** if $f(-x) = -f(x)$ for all values of x. Graphs of odd functions are always **symmetrical about the origin**. Two examples of odd functions are $y = x^3$ and $y = \sin x$ as shown in Figure 63.3.

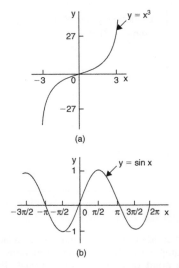

(a)

(b)   Figure 63.3

Inverse functions

> Given a function $y = f(x)$, its inverse may be obtained by inter-changing the roles of x and y and then transposing for y. The inverse function is denoted by $y = f^{-1}(x)$.

Application: Find the inverse of $y = 2x + 1$

(i) Transposing for x, i.e. $x = \dfrac{y-1}{2} = \dfrac{y}{2} - \dfrac{1}{2}$

and (ii) interchanging x and y, gives the inverse as $y = \dfrac{x}{2} - \dfrac{1}{2}$

Thus if $f(x) = 2x + 1$, then $f^{-1}(x) = \dfrac{x}{2} - \dfrac{1}{2}$

A graph of $f(x) = 2x + 1$ and its inverse $f^{-1}(x) = \dfrac{x}{2} - \dfrac{1}{2}$ is shown in Figure 63.4 and $f^{-1}(x)$ is seen to be a reflection of $f(x)$ in the line $y = x$.

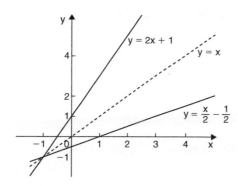

Figure 63.4

Application: Find the inverse of $y = x^2$

(i) Transposing for x, i.e. $x = \pm\sqrt{y}$

and (ii) interchanging x and y, gives the inverse $y = \pm\sqrt{x}$

Hence the inverse has two values for every value of x. Thus $f(x) = x^2$ does not have a single inverse. In such a case the domain of the original function may be restricted to $y = x^2$ for $x > 0$. Thus the inverse is then $f^{-1}(x) = +\sqrt{x}$

A graph of $f(x) = x^2$ and its inverse $f^{-1}(x) = \sqrt{x}$ for $x > 0$ is shown in Figure 63.5 and, again, $f^{-1}(x)$ is seen to be a reflection of $f(x)$ in the line $y = x$.

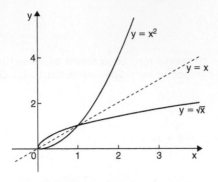

Figure 63.5

Inverse trigonometric functions

If $y = \sin x$, then x is the angle whose sine is y. Inverse trigonometric functions are denoted either by prefixing the function with 'arc' or more commonly $^{-1}$. Hence, transposing $y = \sin x$ for x gives $x = \sin^{-1} y$. Interchanging x and y gives the inverse $y = \sin^{-1} x$. Similarly, $y = \cos^{-1} x$, $y = \tan^{-1} x$, $y = \sec^{-1} x$, $y = \csc^{-1} x$ and $y = \cot^{-1} x$ are all inverse trigonometric functions. The angle is always expressed in radians.

Inverse trigonometric functions are periodic so it is necessary to specify the smallest or principal value of the angle. For $\sin^{-1} x$, $\tan^{-1} x$, $\csc^{-1} x$ and $\cot^{-1} x$, the principal value is in the range $-\dfrac{\pi}{2} < y < \dfrac{\pi}{2}$. For $\cos^{-1} x$ and $\sec^{-1} x$ the principal value is in the range $0 < y < \pi$.

Graphs of the six inverse trigonometric functions are shown in Figure 108.1, page 296.

Application: Determine the principal values of

(a) arcsin 0.5   (b) arctan(−1)   (c) $\arccos\left(-\dfrac{\sqrt{3}}{2}\right)$ (d) $\operatorname{arccosec}(\sqrt{2})$

Using a calculator,

(a) arcsin $0.5 \equiv \sin^{-1} 0.5 = 30° = \dfrac{\pi}{6}$ **rad or 0.5236 rad**

(b) arctan $(-1) \equiv \tan^{-1}(-1) = -45° = -\dfrac{\pi}{4}$ **rad or −0.7854 rad**

(c) $\arccos\left(-\dfrac{\sqrt{3}}{2}\right) \equiv \cos^{-1}\left(-\dfrac{\sqrt{3}}{2}\right) = 150° = \dfrac{5\pi}{6}$ **rad or 2.6180 rad**

(d) $\operatorname{arccosec}\left(\sqrt{2}\right) = \arcsin\left(\dfrac{1}{\sqrt{2}}\right) \equiv \sin^{-1}\left(\dfrac{1}{\sqrt{2}}\right) = 45° = \dfrac{\pi}{4}$ **rad or 0.7854 rad**

Asymptotes

If a table of values for the function $y = \dfrac{x+2}{x+1}$ is drawn up for various values of x and then y plotted against x, the graph would be as shown in Figure 63.6. The straight lines AB, i.e. $x = -1$, and CD, i.e. $y = 1$, are known as **asymptotes**.

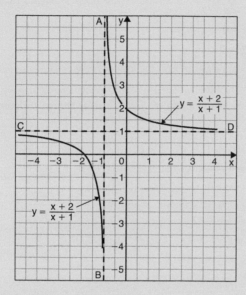

Figure 63.6

An asymptote to a curve is defined as a straight line to which the curve approaches as the distance from the origin increases. Alternatively, an asymptote can be considered as a tangent to the curve at infinity.

Asymptotes parallel to the x- and y-axes

For a curve $y = f(x)$:

(i) the asymptotes parallel to the x-axis are found by equating the coefficient of the highest power of x to zero

(ii) the asymptotes parallel to the y-axis are found by equating the coefficient of the highest power of y to zero

Other asymptotes

To determine asymptotes other than those parallel to x- and y-axes a simple procedure is:

(i) substitute $y = mx + c$ in the given equation

(ii) simplify the expression

(iii) equate the coefficients of the two highest powers of x to zero and determine the values of m and c. $y = mx + c$ gives the asymptote.

Application: Show that asymptotes occur at $y = 1$ and $x = -1$ for the curve $y = \dfrac{x + 2}{x + 1}$

Rearranging $y = \dfrac{x + 2}{x + 1}$ gives:

$$y(x + 1) = x + 2$$

i.e. $yx + y - x - 2 = 0$ (1)

and $x(y - 1) + y - 2 = 0$

The coefficient of the highest power of x (in this case x^1) is $(y - 1)$.

Equating to zero gives: $y - 1 = 0$

from which, **y = 1**, which is an asymptote of $y = \dfrac{x + 2}{x + 1}$ as shown in Figure 63.6.

Returning to equation (1): $yx + y - x - 2 = 0$

from which, $y(x + 1) - x - 2 = 0$

The coefficient of the highest power of y (in this case y^1) is $(x + 1)$.

Equating to zero gives: $x + 1 = 0$

from which, **x = −1**, which is another asymptote of $y = \dfrac{x + 2}{x + 1}$ as shown in Figure 63.6.

Application: Determine the asymptotes parallel to the x- and y-axes for the function $x^2y^2 = 9(x^2 + y^2)$

Asymptotes parallel to the x-axis:
Rearranging $x^2y^2 = 9(x^2 + y^2)$ gives $x^2y^2 - 9x^2 - 9y^2 = 0$
Hence $x^2(y^2 - 9) - 9y^2 = 0$
Equating the coefficient of the highest power of x to zero gives:
$y^2 - 9 = 0$ from which, $y^2 = 9$ and   **y = ±3**
Asymptotes parallel to the y-axis:
Since $x^2y^2 - 9x^2 - 9y^2 = 0$ then $y^2(x^2 - 9) - 9y^2 = 0$
Equating the coefficient of the highest power of y to zero gives:
$x^2 - 9 = 0$ from which, $x^2 = 9$ and **x = ±3**
Hence, asymptotes occur at y = ±3 and x = ±3

Application: Determine the asymptotes for the function:

$$y(x + 1) = (x - 3)(x + 2)$$

(i) Substituting $y = mx + c$ into $y(x + 1) = (x - 3)(x + 2)$
    gives $(mx + c)(x + 1) = (x - 3)(x + 2)$

(ii) Simplifying gives $mx^2 + mx + cx + c = x^2 - x - 6$
    and $(m - 1)x^2 + (m + c + 1)x + c + 6 = 0$

(iii) Equating the coefficient of the highest power of x to zero
gives $m - 1 = 0$ from which, **m = 1**
Equating the coefficient of the next highest power of x to zero
gives $m + c + 1 = 0$
and since $m = 1, 1 + c + 1 = 0$ from which, **c = −2**
Hence $y = mx + c = 1x - 2$
i.e.   **y = x − 2 is an asymptote**

To determine any asymptotes parallel to the x-axis:

Rearranging $y(x + 1) = (x - 3)(x + 2)$
gives $yx + y = x^2 - x - 6$

The coefficient of the highest power of x (i.e. x^2) is 1. Equating this to zero gives $1 = 0$, which is not an equation of a line. Hence there is no asymptote parallel to the x-axis

To determine any asymptotes parallel to the y-axis:

Since $y(x + 1) = (x - 3)(x + 2)$ the coefficient of the highest power of y is $x + 1$. Equating this to zero gives $x + 1 = 0$, from which, $x = -1$. Hence, **x = −1 is an asymptote**.

When **x = 0**, $y(1) = (-3)(2)$,   i.e. **y = −6**

When **y = 0**, $0 = (x - 3)(x + 2)$,   i.e.   **x = 3** and **x = −2**

A sketch of the function $y(x + 1) = (x - 3)(x + 2)$ is shown in Figure 63.7.

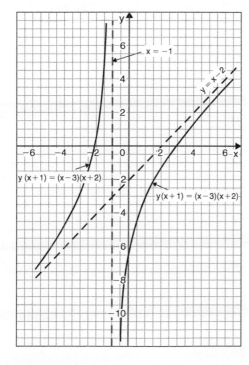

Figure 63.7

Section 7

Complex numbers

Why are complex numbers important?

Complex numbers are used in many scientific fields, including engineering, electromagnetism, quantum physics, and applied mathematics, such as chaos theory. Any physical motion which is periodic, such as an oscillating beam, string, wire, pendulum, electronic signal, or electromagnetic wave can be represented by a complex number function. This can make calculations with the various components simpler than with real numbers and sines and cosines. In control theory, systems are often transformed from the time domain to the frequency domain using the Laplace transform. In fluid dynamics, complex functions are used to describe potential flow in two dimensions. In electrical engineering, the Fourier transform is used to analyse varying voltages and currents. Complex numbers are used in signal analysis and other fields for a convenient description for periodically varying signals. This use is also extended into digital signal processing and digital image processing, which utilize digital versions of Fourier analysis (and wavelet analysis) to transmit, compress, restore, and otherwise process digital audio signals, still images, and video signals. Knowledge of complex numbers is clearly essential for further studies in so many engineering disciplines.

De Moivre's theorem has several uses, including finding powers and roots of complex numbers, solving polynomial equations, calculating trigonometric identities, and for evaluating the sums of trigonometric series. The theorem is also used to calculate exponential and logarithmic functions of complex numbers. De Moivre's theorem has applications in electrical engineering and physics.

Chapter 64   General complex number formulae

$z = a + jb = r(\cos \theta + j \sin \theta) = r\angle\theta = r\,e^{j\theta}$ where $j^2 = -1$

Modulus, $r = |z| = \sqrt{(a^2 + b^2)}$

Argument, $\theta = \arg z = \tan^{-1}\dfrac{b}{a}$

Addition: $(a + jb) + (c + jd) = (a + c) + j(b + d)$

Subtraction: $(a + jb) - (c + jd) = (a - c) + j(b - d)$

Complex equations: If $(m + jn) = (p + jq)$ then $m = p$ and $n = q$

Multiplication: $z_1 z_2 = r_1 r_2 \angle(\theta_1 + \theta_2)$

Division: $\dfrac{z_1}{z_2} = \dfrac{r_1}{r_2}\angle(\theta_1 - \theta_2)$

De Moivre's theorem: $[r\angle\theta]^n = r^n\angle n\theta = r^n(\cos n\theta + j\sin n\theta)$

Chapter 65 Cartesian form of a complex number

$(-1 + j2)$ and $(3 - j4)$ are examples of **Cartesian** (or **rectangular**) **complex numbers**. They are each of the form $a + jb$, 'a' being termed the **real part** and jb the **imaginary part**.

Application: Solve the quadratic equation $2x^2 + 3x + 5 = 0$

Using the quadratic formula,

$$x = \frac{-3 \pm \sqrt{[(3)^2 - 4(2)(5)]}}{2(2)} = \frac{-3 \pm \sqrt{-31}}{4}$$

$$= \frac{-3 \pm \sqrt{(-1)}\sqrt{31}}{4} = \frac{-3 \pm j\sqrt{31}}{4}$$

Hence, $x = -\dfrac{3}{4} + j\dfrac{\sqrt{31}}{4}$ or $\mathbf{-0.750 \pm j1.392}$, correct to 3 decimal places.

Application: Determine $(2 + j3) + (3 - j4) - (-5 + j)$

$$(2 + j3) + (3 - j4) - (-5 + j) = 2 + j3 + 3 - j4 + 5 - j$$
$$= (2 + 3 + 5) + j(3 - 4 - 1) = \mathbf{10 - j2}$$

Application: Determine $(3 + j2)(4 - j5)$

$$(3 + j2)(4 - j5) = 12 - j15 + j8 - j^2 10$$
$$= (12 - -10) + j(-15 + 8) \quad \text{where } j^2 = -1$$
$$= \mathbf{22 - j7}$$

Application: Solve the complex equation:
$$(1 + j2)(-2 - j3) = a + jb$$

$$(1 + j2)(-2 - j3) = a + jb$$
i.e. $\qquad -2 - j3 - j4 - j^2 6 = a + jb$
Hence, $\qquad\qquad 4 - j7 = a + jb$
Equating real and imaginary terms gives: $\quad \mathbf{a = 4}$ and $\mathbf{b = -7}$

Application: Solve the equation $(x - j2y) + (y - j3x) = 2 + j3$

Since $(x - j2y) + (y - j3x) = 2 + j3$
then $(x + y) + j(-2y - 3x) = 2 + j3$

Equating real and imaginary parts gives: $\qquad x + y = 2$ $\qquad\qquad$ (1)
and $\qquad\qquad\qquad\qquad\qquad -3x - 2y = 3$ $\qquad\qquad$ (2)
Multiplying equation (1) by 2 gives: $\qquad 2x + 2y = 4$ $\qquad\qquad$ (3)
Adding equations (2) and (3) gives: $\qquad -x = 7$ $\quad$ i.e. $\quad \mathbf{x = -7}$
From equation (1), $\mathbf{y = 9}$, which may be checked in equation (2).

Application: Determine $(3 + j4)(3 - j4)$

$(3 + j4)(3 - j4) = 9 - j12 + j12 - j^2 16 = 9 + 16 = \mathbf{25}$
$[(3 - j4)$ is called the **complex conjugate** of $(3 + j4)$; whenever a complex number is multiplied by its conjugate, a real number results. In general, $(a + jb)(a - jb)$ may be evaluated 'on sight' as $a^2 + b^2]$

Application: Determine $\dfrac{2 - j5}{3 + j4}$

$$\frac{2 - j5}{3 + j4} = \frac{2 - j5}{3 + j4} \times \frac{(3 - j4)}{(3 - j4)} = \frac{6 \quad j8 - j15 + j^2 20}{3^2 + 4^2}$$

$$= \frac{-14 - j23}{25}$$

$$= \frac{-14}{25} - j\frac{23}{25} \quad \text{or} \quad -0.56 - j0.92$$

Application: If $Z_1 = 1 - j3$ and $Z_2 = -2 + j5$ determine $\dfrac{Z_1 Z_2}{Z_1 + Z_2}$ in $(a + jb)$ form

$$\frac{Z_1 Z_2}{Z_1 + Z_2} = \frac{(1 - j3)(-2 + j5)}{(1 - j3) + (-2 + j5)} = \frac{-2 + j5 + j6 - j^2 15}{1 - j3 - 2 + j5}$$

$$= \frac{-2 + j5 + j6 + 15}{-1 + j2} = \frac{13 + j11}{-1 + j2}$$

$$= \frac{13 + j11}{-1 + j2} \times \frac{-1 - j2}{-1 - j2}$$

$$= \frac{-13 - j26 - j11 - j^2 22}{1^2 + 2^2}$$

$$= \frac{9 - j37}{5}$$

$$= \frac{9}{5} - j\frac{37}{5} \quad \text{or} \quad 1.8 - j7.4$$

Application: Show the following complex numbers on an Argand diagram $(3 + j2)$, $(-2 + j4)$, $(-3 - j5)$ and $(1 - j3)$

In Figure 65.1, the point A represents the complex number $(3 + j2)$ and is obtained by plotting the co-ordinates $(3, j2)$ as in graphical work. The Argand points B, C and D represent the complex numbers $(-2 + j4)$, $(-3 - j5)$ and $(1 - j3)$ respectively.

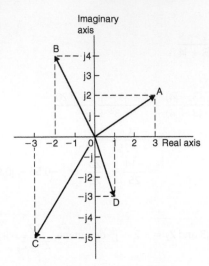

Figure 65.1

Chapter 66   Polar form of a complex number

A number written in the form $Z = r\angle\theta$ is known as the **polar form** of a complex number.

Application: Express   (a) $3 + j4$ and (b) $-3 + j4$ in polar form

(a) $3 + j4$ is shown in Figure 66.1 and lies in the first quadrant.

Modulus, $r = \sqrt{3^2 + 4^2} = 5$

and   argument $\theta = \tan^{-1}\dfrac{4}{3} = 53.13°$

Hence, $3 + j4 = 5\angle53.13°$

(b) $-3 + j4$ is shown in Figure 66.1 and lies in the second quadrant.

Modulus, $r = 5$ and angle $\alpha = 53.13°$, from part (a).

Argument $= 180° - 53.13° = 126.87°$ (i.e. the argument must be measured from the positive real axis)

Hence $-3 + j4 = 5\angle126.87°$

Similarly it may be shown that $(-3 - j4) = 5\angle233.13°$ **or** $5\angle-126.87°$, (by convention the **principal value** is normally used, i.e. the numerically least value, such that $-\pi < \theta < \pi$), and $(3 - j4) = 5\angle-53.13°$

Application: Change $7\angle-145°$ into $a + jb$ form:

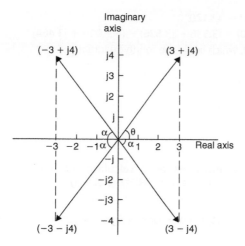

Figure 66.1

$7\angle-145°$ is shown in Figure 66.2 and lies in the third quadrant.

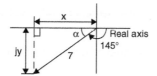

Figure 66.2

$$7\angle-145° = 7\cos(-145°) + j7\sin(-145°) = -5.734 - j4.015$$

Application: Determine $3\angle16° \times 5\angle-44° \times 2\angle80°$ in polar form

$$3\angle16° \times 5\angle-44° \times 2\angle80° = (3 \times 5 \times 2) \angle[16° + (-44°) + 80°]°$$
$$= 30\angle52°$$

Application: Determine $\dfrac{16\angle75°}{2\angle15°}$ in polar form

$$\frac{16\angle75°}{2\angle15°} = \frac{16}{2}\angle(75° - 15°) = 8\angle60°$$

Application: Evaluate, in polar form, $2\angle30° + 5\angle-45° - 4\angle120°$

$$2\angle30° = 2(\cos 30° + j\sin 30°) = 2\cos 30° + j2\sin 30°$$
$$= 1.732 + j1.000$$
$$5\angle-45° = 5(\cos(-45°) + j\sin(-45°)) = 5\cos(-45°) + j5\sin(-45°)$$
$$= 3.536 - j3.536$$
$$4\angle120° = 4(\cos 120° + j\sin 120°) = 4\cos 120° + j4\sin 120°$$
$$= -2.000 + j3.464$$

Hence, $2\angle 30° + 5\angle -45° - 4\angle 120°$

$$= (1.732 + j1.000) + (3.536 - j3.536) - (-2.000 + j3.464)$$
$$= 7.268 - j6.000, \text{ which lies in the fourth quadrant}$$

$$= \sqrt{7.268^2 + 6.000^2} \angle \tan^{-1}\left(\frac{-6.000}{7.268}\right)$$
$$= \mathbf{9.425\angle -39.54°}$$

Using a calculator

Using the **'Pol' and 'Rec' functions on a calculator** enables changing from Cartesian to polar and vice-versa to be achieved more quickly.

Application: Change $(3 + j4)$ to polar form using a calculator

1. Press shift + and Pol (appears 2. Type in 3,4) A comma is achieved by shift

3. Press = and r = 5 and θ = 53.13° appears. Hence, **(3 + j4) = 5∠53.13°**

Application: Change $13\angle 67.38°$ to Cartesian form using a calculator

1. Press shift − and Rec (appears 2. Type in 13,67.38)
3. Press = and X = 5 and Y = 12 appears.   Hence, **13∠67.38° = (5 + j12)**

Chapter 67 Applications of complex numbers

There are several applications of complex numbers in science and engineering, in particular in electrical alternating current theory and in mechanical vector analysis.

Application: Determine the value of current I and its phase relative to the 240 V supply for the parallel circuit shown in Figure 67.1

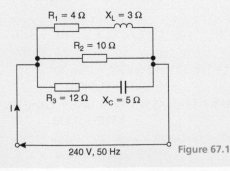

Figure 67.1

Current, $I = \dfrac{V}{Z}$. Impedance Z for the three-branch parallel circuit is given by:

$$\dfrac{1}{Z} = \dfrac{1}{Z_1} + \dfrac{1}{Z_2} + \dfrac{1}{Z_3},$$

where $Z_1 = 4 + j3\,\Omega$, $Z_2 = 10\,\Omega$ and $Z_3 = 12 - j5\,\Omega$

Admittance, $Y_1 = \dfrac{1}{Z_1} = \dfrac{1}{4 + j3} = \dfrac{1}{4 + j3} \times \dfrac{4 - j3}{4 - j3} = \dfrac{4 - j3}{4^2 + 3^2}$

$\qquad\qquad = 0.160 - j0.120$ siemens

Admittance, $Y_2 = \dfrac{1}{Z_2} = \dfrac{1}{10} = 0.10$ siemens

Admittance, $Y_3 = \dfrac{1}{Z_3} = \dfrac{1}{12 - j5} = \dfrac{1}{12 - j5} \times \dfrac{12 + j5}{12 + j5} = \dfrac{12 + j5}{12^2 + 5^2}$

$\qquad\qquad = 0.0710 + j0.0296$ siemens

Total admittance,

$$Y = Y_1 + Y_2 + Y_3$$
$$- (0.160 - j0.120) + (0.10) + (0.0710 + j0.0296)$$
$$= 0.331 - j0.0904 = 0.343\angle{-15.28°}\ \text{siemens}$$

Current, $I = \dfrac{V}{Z} = VY = (240\angle 0°)(0.343\angle{-15.28°})$

$\qquad\qquad\qquad = \mathbf{82.32\angle{-15.28°}\,A}$

Application: Determine the magnitude and direction of the resultant of the three coplanar forces shown in Figure 67.2

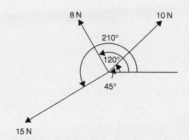

Figure 67.2

Force A, $f_A = 10\angle 45°$ N, force B, $f_B = 8\angle 120°$ N and force C, $f_C = 15\angle 210°$ N

The resultant force

$= f_A + f_B + f_C$

$= 10\angle 45° + 8\angle 120° + 15\angle 210°$

$= 10(\cos 45° + j \sin 45°) + 8(\cos 120° + j \sin 120°)$

$\qquad\qquad\qquad\qquad\qquad + 15(\cos 210° + j \sin 210°)$

$= (7.071 + j7.071) + (-4.00 + j6.928) + (-12.99 - j7.50)$

$= -9.919 + j6.499$

Magnitude of resultant force $= \sqrt{(-9.919)^2 + 6.499^2} = 11.86\,\text{N}$

Direction of resultant force $= \tan^{-1}\left(\dfrac{6.499}{-9.919}\right) = 146.77°$

(since $-9.919 + j6.499$ lies in the second quadrant).

Chapter 68 De Moivre's theorem

De Moivre's theorem states: $[r\angle\theta]^n = r^n\angle n\theta$

The theorem is used to determine powers and roots of complex numbers.

In general, **when finding the nth root of a complex number, there are n solutions.** For example, there are three solutions to a cube root, five solutions to a fifth root, and so on. In the solutions to the roots of a complex number, the modulus, r, is always the same, but the arguments, θ, are different. Arguments are symmetrically spaced on an Argand diagram and are $360°/n$ apart, where n is the number of the roots required. Thus if one of the solutions to the cube root of a complex number is, say, $5\angle20°$, the other two roots are symmetrically spaced $360°/3$, i.e. $120°$ from this root, and the three roots are $5\angle20°$, $5\angle140°$ and $5\angle260°$.

Application: Determine $[3\angle20°]^4$

$[3\angle20°]^4 = 3^4\angle(4 \times 20°) = \mathbf{81\angle80°}$ by de Moivre's theorem.

Application: Determine $(-2 + j3)^6$ in polar form

$(-2 + j3) = \sqrt{(-2)^2 + 3^2}\angle\tan^{-1}\left(\dfrac{3}{-2}\right)$

$\qquad\qquad = \sqrt{13}\angle123.69°$ since $-2 + j3$ lies in the second quadrant

$(-2 + j3)^6 = [\sqrt{13}\angle123.69°]^6$

$\qquad\qquad = (\sqrt{13})^6\angle(6 \times 123.69°)$ by De Moivre's theorem

$\qquad\qquad = 2197\angle742.14°$

$\qquad\qquad = 2197\angle382.14°$ (since $742.14° \equiv 742.14° - 360° = 382.14°$)

$\qquad\qquad = \mathbf{2197\angle22.14°}$ (since $382.14° \equiv 382.14° - 360° = 22.14°$)

Application: Determine the two square roots of the complex number $(5 + j12)$ in polar and Cartesian forms

$(5 + j12) = \sqrt{5^2 + 12^2} \angle\tan^{-1}\left(\dfrac{12}{5}\right) = 13\angle 67.38°$

When determining square roots two solutions result. To obtain the second solution one way is to express $13\angle 67.38°$ also as $13\angle(67.38° + 360°)$, i.e. $13\angle 427.38°$. When the angle is divided by 2 an angle less than 360° is obtained.

Hence $\sqrt{5^2 + 12^2} = \sqrt{13\angle 67.38°}$ and $\sqrt{13\angle 427.38°}$

$$= [13\angle 67.38°]^{1/2} \text{ and } [13\angle 427.38°]^{1/2}$$

$$= 13^{1/2} \angle\left(\frac{1}{2} \times 67.38°\right) \text{and } 13^{1/2} \angle\left(\frac{1}{2} \times 427.38°\right)$$

$$= \sqrt{13} \angle 33.69° \text{ and } \sqrt{13} \angle 213.69°$$

$$= 3.61\angle 33.69° \text{ and } 3.61\angle 213.69°$$

Thus, in polar form, the two roots are $3.61\angle 33.69°$ and $3.61\angle -146.31°$

$\sqrt{13} \angle 33.69° = \sqrt{13} \;(\cos 33.69° + j\sin 33.69°\,) = 3.0 + j2.0$

$\sqrt{13} \angle 213.69° = \sqrt{13} \;(\cos 213.69° + j\sin 213.69°) = -3.0 - j2.0$

Thus, in Cartesian form, the two roots are $\pm(3.0 + j2.0)$

From the Argand diagram shown in Figure 68.1 the two roots are seen to be 180° apart, which is always true when finding square roots of complex numbers.

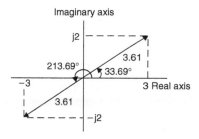

Figure 68.1

Application: Express the roots of $(-14 + j3)^{-2/5}$ in polar form

$$(-14 + j3) = \sqrt{205} \angle 167.905°$$

$$(-14 + j3)^{-2/5} = \sqrt{205}^{\,-2/5} \angle\left[\left(-\frac{2}{5}\right) \times 167.905°\right]$$

$$= 0.3449\angle -67.16°$$

There are five roots to this complex number, $\left(x^{-2/5} = \dfrac{1}{x^{2/5}} = \dfrac{1}{\sqrt[5]{x^2}}\right)$

The roots are symmetrically displaced from one another $\dfrac{360°}{5}$, i.e. 72° apart around an Argand diagram.

Thus, **the required roots are $0.3449\angle -67.16°$, $0.3449\angle 4.84°$, $0.3449\angle 76.84°$, $0.3449\angle 148.84°$ and $0.3449\angle 220.84°$**

Chapter 69 Exponential form of a complex number

There are three ways of expressing a complex number:

1. $z = (a + jb)$, called **Cartesian** or **rectangular form**,
2. $z = r(\cos\theta + j\sin\theta)$ or $r\angle\theta$, called **polar form**, and
3. $z = r\,e^{j\theta}$ called **exponential form**.

The **exponential form** is obtained from the polar form. For example, $4\angle 30°$ becomes $4e^{j\pi/6}$ in exponential form. (Note that in $r\,e^{j\theta}$, θ must be in radians).

Application: Express $(3 - j4)$ in polar and exponential forms

$$(3 - j4) = \mathbf{5\angle-53.13°} = \mathbf{5\angle-0.927} \text{ in polar form}$$
$$= \mathbf{5e^{-j0.927}} \text{ in exponential form}$$

Application: Express $7.2e^{j1.5}$ in rectangular form

$$7.2e^{j1.5} = 7.2\angle 1.5\,\text{rad} \ (=7.2\angle 85.944°) \text{ in polar form}$$
$$= 7.2\cos 1.5 + j7.2\sin 1.5$$
$$= \mathbf{(0.509 + j7.182)} \text{ in rectangular form}$$

Application: If $z = 2e^{1+j\pi/3}$ express z in Cartesian form

$$z = 2e^{1+j\pi/3} = (2e^{1})(e^{j\pi/3}) \text{ by the laws of indices}$$
$$= (2e^{1})\angle\frac{\pi}{3} \ (\text{or } 2e\angle 60°) \text{ in polar form}$$
$$= 2e\left(\cos\frac{\pi}{3} + j\sin\frac{\pi}{3}\right) = \mathbf{(2.718 + j4.708)} \text{ in Cartesian form}$$

Application: If $z = 4e^{j1.3}$ find $\ln z$ in polar form

If $z = 4e^{j1.3}$ then $\ln z = \ln(4e^{j1.3})$
$$= \mathbf{\ln 4 + j1.3} \ (\text{or } \mathbf{1.386 + j1.300}) \text{ in Cartesian form (by the}$$
$$\text{laws of logarithms)}$$
$$= \mathbf{1.90\angle 43.17°} \text{ or } \mathbf{1.90\angle 0.753} \text{ in polar form.}$$

Application: Determine $\ln(3 + j4)$

$$\ln(3 + j4) = \ln[5\angle0.927] = \ln[5e^{j0.927}] = \ln 5 + \ln(e^{j0.927})$$
$$= \ln 5 + j0.927 = 1.609 + j0.927$$
$$= \mathbf{1.857\angle29.95°} \quad \text{or} \quad \mathbf{1.857\angle0.523}$$

Section 8

Vectors

Why are vectors important?

Vectors are an important part of the language of science, mathematics, and engineering. They are used to discuss multivariable calculus, electrical circuits with oscillating currents, stress and strain in structures and materials, flows of atmospheres and fluids, together with many other applications. Resolving a vector into components is a precursor to computing things with or about a vector quantity. Because position, velocity, acceleration, force, momentum, and angular momentum are all vector quantities, resolving vectors into components is a most important skill required in any engineering and scientific studies.

In electrical engineering, a phasor is a rotating vector representing a quantity such as an alternating current or voltage that varies sinusoidally. Sometimes it is necessary when studying sinusoidal quantities to add together two alternating waveforms, for example in an ac series circuit where the quantities are not in-phase with each other. Electrical engineers, electronics engineers, electronic engineering technicians and aircraft engineers all use phasor diagrams to visualize complex constants and variables. So, given oscillations to add and subtract, the required rotating vectors are constructed, called a phasor diagram, and graphically the resulting sum and/or difference oscillation are added or calculated. Phasors may be used to analyse the behaviour of electrical and mechanical systems that have reached a kind of equilibrium called sinusoidal steady state. Hence, discovering different methods of combining sinusoidal waveforms is of some importance in certain areas of engineering.

Common applications of the scalar product in engineering and physics are to test whether two vectors are perpendicular or to find the angle between two vectors when they are expressed in Cartesian form or to find the component of one vector in the direction of another. In mechanical engineering the scalar and vector products are used with forces, displacement, moments, velocities and the determination of work done. In electrical engineering, scalar and vector product calculations are important in electromagnetic calculations; most electromagnetic equations use vector calculus which is based on the scalar and vector product. Such applications include electric motors and generators and power transformers. Knowledge of scalar and vector products thus have important engineering and scientific applications.

Chapter 70   Scalars and vectors

Some physical quantities are entirely defined by a numerical value and are called **scalar quantities** or **scalars**. Examples of scalars include time, mass, temperature, energy and volume. Other physical quantities are defined by both a numerical value **and** a direction in space and these are called **vector quantities** or **vectors**. Examples of vectors include force, velocity, moment and displacement.

Various ways of showing vector quantities include:

1. **bold print**.
2. two capital letters with an arrow above them to denote the sense of direction, e.g. $\overrightarrow{AB}$, where A is the starting point and B the end point of the vector,
3. a line over the top of letters, e.g. $\overline{AB}$ or $\bar{a}$
4. letters with an arrow above, e.g. $\vec{a}$, $\vec{A}$
5. underlined letters, e.g. $\underline{a}$
6. $xi + jy$, where i and j are axes at right-angles to each other; for example, $3i + 4j$ means 3 units in the i direction and 4 units in the j direction, as shown in Figure 70.1

Figure 70.1

7. a column matrix $\begin{pmatrix} a \\ b \end{pmatrix}$; for example, the vector **OA** of Figure 70.1 could be represented by $\begin{pmatrix} 3 \\ 4 \end{pmatrix}$

Thus, in Figure 70.1, $OA \equiv \overline{OA} \equiv \overrightarrow{OA} \equiv 3i + 4j \equiv \begin{pmatrix} 3 \\ 4 \end{pmatrix}$

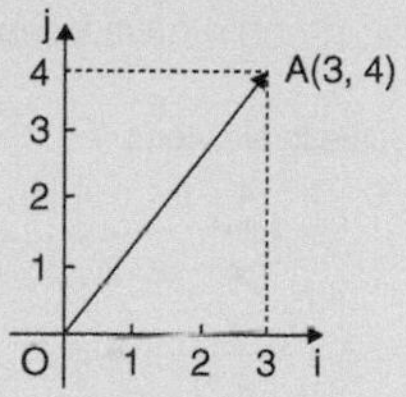

The one adopted in this text is to denote vector quantities in **bold print**.

Chapter 71   Vector addition

The resultant of adding two vectors together, say **V₁** at an angle θ_1 and **V₂** at angle $(-\theta_2)$, as shown in Figure 71.1(a), can be obtained by drawing **oa** to represent **V₁** and then drawing **ar** to represent **V₂**. The resultant of **V₁ + V₂** is given by **or**. This is shown in Figure 71.1(b), the vector equation being **oa + ar = or**. This is called the **'nose-to-tail' method** of vector addition.

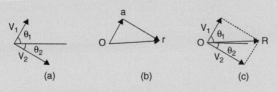

(a)   (b)   (c)

Figure 71.1

Alternatively, by drawing lines parallel to V_1 and V_2 from the noses of V_2 and V_1, respectively, and letting the point of intersection of these parallel lines be R, gives **OR** as the magnitude and direction of the resultant of adding V_1 and V_2, as shown in Figure 71.1(c). This is called the **'parallelogram' method** of vector addition.

Application: A force of 4 N is inclined at an angle of 45° to a second force of 7 N, both forces acting at a point. Find the magnitude of the resultant of these two forces and the direction of the resultant with respect to the 7 N force by both the 'triangle' and the 'parallelogram' methods

The forces are shown in Figure 71.2(a). Although the 7 N force is shown as a horizontal line, it could have been drawn in any direction.

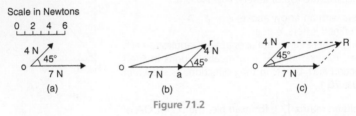

Figure 71.2

Using the **'nose-to-tail' method**, a line 7 units long is drawn horizontally to give vector **oa** in Figure 71.2(b). To the nose of this vector **ar** is drawn 4 units long at an angle of 45° to **oa**. The resultant of vector addition is **or** and by measurement is **10.2 units long and at an angle of 16° to the 7 N force**. Figure 71.2(c) uses the **'parallelogram' method** in which lines are drawn parallel to the 7 N and 4 N forces from the noses of the 4 N and 7 N forces, respectively. These intersect at R. Vector **OR** give the magnitude and direction of the resultant of vector addition and, as obtained by the 'nose-to-tail' method, is **10.2 units long at an angle of 16° to the 7 N force**.

Application: Use a graphical method to determine the magnitude and direction of the resultant of the three velocities shown in Figure 71.3

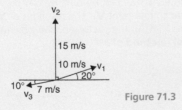

Figure 71.3

It is easier to use the 'nose-to-tail' method when more than two vectors are being added. The order in which the vectors are added is immaterial. In this case the order

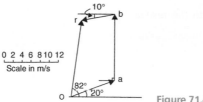

Figure 71.4

taken is v_1, then v_2, then v_3 but just the same result would have been obtained if the order had been, say, v_1, v_3 and finally v_2.

v_1 is drawn 10 units long at an angle of 20° to the horizontal, shown by **oa** in Figure 71.4. v_2 is added to v_1 by drawing a line 15 units long vertically upwards from a, shown as **ab**. Finally, v_3 is added to $v_1 + v_2$ by drawing a line 7 units long at an angle at 190° from b, shown as **br**. The resultant of vector addition is **or** and by measurement is 17.5 units long at an angle of 82° to the horizontal.

Thus, **$v_1 + v_2 + v_3$ = 17.5 m/s at 82° to the horizontal.**

Chapter 72   Resolution of vectors

A vector can be resolved into horizontal and vertical components. For the vector shown as **F** in Figure 72.1, the horizontal component is $F\cos\theta$ and the vertical component is $F\sin\theta$.

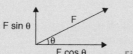

Figure 72.1

Application: Calculate the resultant of the two forces shown in Figure 71.2(a)

Horizontal component of force,
$$H = 7\cos 0° + 4\cos 45° = 7 + 2.828 = \mathbf{9.828\,N}$$

Vertical component of force,
$$V = 7\sin 0° + 4\sin 45° = 0 + 2.828 = \mathbf{2.828\,N}$$

The magnitude of the resultant of vector addition
$$= \sqrt{H^2 + V^2} = \sqrt{9.828^2 + 2.828^2} = \sqrt{104.59} = 10.23\,N$$

The direction of the resultant of vector addition $= \tan^{-1}\left(\dfrac{V}{H}\right)$

$$= \tan^{-1}\left(\dfrac{2.828}{9.828}\right)$$

$$= 16.05°$$

Thus, the resultant of the two forces is a single vector of 10.23 N at 16.05° to the 7 N vector

Application: Calculate the resultant velocity of the three velocities shown in Figure 71.3

Horizontal component of the velocity,
$$H = 10 \cos 20° + 15 \cos 90° + 7 \cos 190°$$
$$= 9.397 + 0 + (-6.894) = \mathbf{2.503\, m/s}$$

Vertical component of the velocity,
$$V = 10 \sin 20° + 15 \sin 90° + 7 \sin 190°$$
$$= 3.420 + 15 + (-1.216) = \mathbf{17.205\, m/s}$$

Magnitude of the resultant of vector addition
$$= \sqrt{H^2 + V^2} = \sqrt{2.503^2 + 17.205^2} = \sqrt{302.28} = 17.39\, m/s$$

Direction of the resultant of vector addition
$$= \tan^{-1}\left(\frac{V}{H}\right) = \tan^{-1}\left(\frac{17.205}{2.503}\right) = \tan^{-1} 6.8738 = 81.72°$$

Thus, the resultant of the three velocities is a single vector of 17.39 m/s at 81.72° to the horizontal.

Chapter 73   Vector subtraction

In Figure 73.1, a force vector **F** is represented by **oa**. The vector **(−oa)** can be obtained by drawing a vector from o in the opposite sense to **oa** but having the same magnitude, shown as **ob** in Figure 73.1, i.e. **ob = (−oa)**

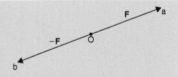

Figure 73.1

For two vectors acting at a point, as shown in Figure 73.2(a), the resultant of vector addition is **os = oa + ob**. Figure 73.2(b) shows vectors **ob + (−oa)**, that is, **ob − oa** and the vector equation is **ob − oa = od**. Comparing **od** in Figure 73.2(b) with the broken line ab in Figure 73.2(a) shows that the second diagonal of the 'parallelogram' method of vector addition gives the magnitude and direction of vector subtraction of **oa** from **ob**.

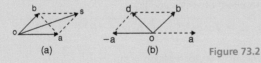

(a)   (b)   Figure 73.2

Application: Accelerations of $a_1 = 1.5\, m/s^2$ at 90° and $a_2 = 2.6\, m/s^2$ at 145° act at a point. Find $\mathbf{a_1 + a_2}$ and $\mathbf{a_1 - a_2}$ by (i) drawing a scale vector diagram and (ii) by calculation

(i) The scale vector diagram is shown in Figure 73.3. By measurement,

$$a_1 + a_2 = 3.7\,m/s^2 \text{ at } 126°$$

and $\qquad a_1 - a_2 = 2.1\,m/s^2 \text{ at } 0°$

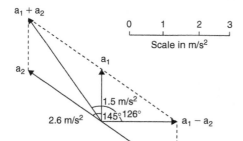

Figure 73.3

(ii) Resolving horizontally and vertically gives:

Horizontal component of $a_1 + a_2$,

$$H = 1.5\cos 90° + 2.6\cos 145° = -2.13$$

Vertical component of $a_1 + a_2$,

$$V = 1.5\sin 90° + 2.6\sin 145° = 2.99$$

Magnitude of $a_1 + a_2 = \sqrt{(-2.13)^2 + 2.99^2} = 3.67\,m/s^2$

Direction of $a_1 + a_2 = \tan^{-1}\left(\dfrac{2.99}{-2.13}\right)$ and must lie in the second quadrant since H is negative and V is positive.

$\tan^{-1}\left(\dfrac{2.99}{-2.13}\right) = -54.53°$, and for this to be in the second quadrant, the true angle is 180° displaced, i.e. 180° − 54.53° or 125.47°.

Thus **$a_1 + a_2 = 3.67\,m/s^2$ at 125.47°**

Horizontal component of $a_1 - a_2$, that is, $a_1 + (-a_2)$
$$= 1.5\cos 90° + 2.6\cos(145° - 180°)$$
$$= 2.6\cos(-35°) = 2.13$$

Vertical component of $a_1 - a_2$, that is,

$$a_1 + (-a_2) = 1.5\sin 90° + 2.6\sin(-35°) = 0$$

Magnitude of $a_1 - a_2 = \sqrt{2.13^2 + 0^2} = 2.13\,m/s^2$

Direction of $a_1 - a_2 = \tan^{-1}\left(\dfrac{0}{2.13}\right) = 0°$

Thus, **$a_1 - a_2 = 2.13\,m/s^2$ at 0°**

Application: Calculate the resultant of $v_1 - v_2 + v_3$ when $v_1 = 22$ units at 140°, $v_2 = 40$ units at 190° and $v_3 = 15$ units at 290°

The vectors are shown in Figure 73.4.

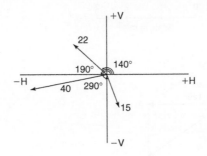

Figure 73.4

The horizontal component of

$$v_1 - v_2 + v_3 = (22\cos 140°) - (40\cos 190°) + (15\cos 290°)$$
$$= (-16.85) - (-39.39) + (5.13) = \mathbf{27.67\ units}$$

The vertical component of

$$v_1 - v_2 + v_3 = (22\sin 140°) - (40\sin 190°) + (15\sin 290°)$$
$$= (14.14) - (-6.95) + (-14.10) = \mathbf{6.99\ units}$$

The magnitude of the resultant, R, is given by:

$$|R| = \sqrt{27.67^2 + 6.99^2} = 28.54\ units$$

The direction of the resultant, **R**, is given by:

$$\arg R = \tan^{-1}\left(\frac{6.99}{27.67}\right) = 14.18°$$

Thus, $v_1 - v_2 + v_3 = \mathbf{28.54\ units\ at\ 14.18°}$

Chapter 74   Relative velocity

For relative velocity problems, some fixed datum point needs to be selected. This is often a fixed point on the earth's surface. In any vector equation, only the start and finish points affect the resultant vector of a system. Two different systems are shown in Figure 74.1, but in each of the systems, the resultant vector is **ad**. The vector equation of the system shown in Figure 74.1(a) is:

$$\mathbf{ad = ab + bd}$$

and that for the system shown in Figure 74.1(b) is:

$$\mathbf{ad = ab + bc + cd}$$

Thus in vector equations of this form, only the first and last letters, a and d, respectively, fix the magnitude and direction of the resultant vector.

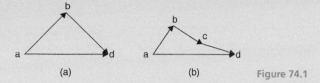

(a)   (b)

Figure 74.1

Application: Two cars, P and Q, are travelling towards the junction of two roads which are at right angles to one another. Car P has a velocity of 45 km/h due east and car Q a velocity of 55 km/h due south. Calculate (i) the velocity of car P relative to car Q, and (ii) the velocity of car Q relative to car P

(i) The directions of the cars are shown in Figure 74.2(a), called a **space diagram**. The velocity diagram is shown in Figure 74.2(b), in which **pe** is taken as the velocity of car P relative to point e on the earth's surface. The velocity of P relative to Q is vector **pq** and the vector equation is **pq = pe + eq**. Hence the vector directions are as shown, **eq** being in the opposite direction to **qe**.

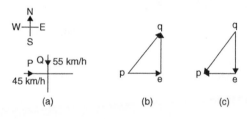

(a)   (b)   (c)

Figure 74.2

From the geometry of the vector triangle,

$$|pq| = \sqrt{45^2 + 55^2} = 71.06 \text{ km/h}$$

and $\quad \arg pq = \tan^{-1}\left(\dfrac{55}{45}\right) = 50.71°$

i.e. **the velocity of car P relative to car Q is 71.06 km/h at 50.71°**

(ii) The velocity of car Q relative to car P is given by the vector equation **qp = qe + ep** and the vector diagram is as shown in Figure 74.2(c), having **ep** opposite in direction to **pe**. From the geometry of this vector triangle:

$$|qp| = \sqrt{45^2 + 55^2} = 71.06 \text{ m/s}$$

and $\quad \arg qp = \tan^{-1}\left(\dfrac{55}{45}\right) = 50.71°$

but must lie in the third quadrant, i.e. the required angle is 180° + 50.71° = 230.71°

Thus the velocity of car Q relative to car P is 71.06 m/s at 230.71°

Chapter 75   i, j, k notation

A method of completely specifying the direction of a vector in space relative to some reference point is to use three unit vectors, **i, j** and **k**, mutually at right angles to each other, as shown in Figure 75.1.

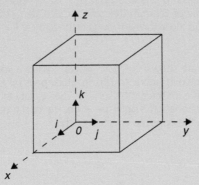

Figure 75.1

Calculations involving vectors given in i, j, k notation are carried out in exactly the same way as standard algebraic calculations, as shown in the applications below.

Application: Determine: $(3\mathbf{i} + 2\mathbf{j} + 2\mathbf{k}) - (4\mathbf{i} - 3\mathbf{j} + 2\mathbf{k})$

$$(3\mathbf{i} + 2\mathbf{j} + 2\mathbf{k}) - (4\mathbf{i} - 3\mathbf{j} + 2\mathbf{k}) = 3\mathbf{i} + 2\mathbf{j} + 2\mathbf{k} - 4\mathbf{i} + 3\mathbf{j} - 2\mathbf{k}$$

$$= -\mathbf{i} + 5\mathbf{j}$$

Application: Given $\mathbf{p} = 3\mathbf{i} + 2\mathbf{k}$, $\mathbf{q} = 4\mathbf{i} - 2\mathbf{j} + 3\mathbf{k}$ and $\mathbf{r} = -3\mathbf{i} + 5\mathbf{j} - 4\mathbf{k}$ determine:

　　(a) $-\mathbf{r}$　　(b) $3\mathbf{p}$　　(c) $2\mathbf{p} + 3\mathbf{q}$　　(d) $-\mathbf{p} + 2\mathbf{r}$　　(e) $0.2\mathbf{p} + 0.6\mathbf{q} - 3.2\mathbf{r}$

(a) $-\mathbf{r} = -(-3\mathbf{i} + 5\mathbf{j} - 4\mathbf{k}) = +3\mathbf{i} - 5\mathbf{j} + 4\mathbf{k}$

(b) $3\mathbf{p} = 3(3\mathbf{i} + 2\mathbf{k}) = 9\mathbf{i} + 6\mathbf{k}$

(c) $2\mathbf{p} + 3\mathbf{q} = 2(3\mathbf{i} + 2\mathbf{k}) + 3(4\mathbf{i} - 2\mathbf{j} + 3\mathbf{k}) = 6\mathbf{i} + 4\mathbf{k} + 12\mathbf{i} - 6\mathbf{j} + 9\mathbf{k}$

$$= 18\mathbf{i} - 6\mathbf{j} + 13\mathbf{k}$$

(d) $-\mathbf{p} + 2\mathbf{r} = -(3\mathbf{i} + 2\mathbf{k}) + 2(-3\mathbf{i} + 5\mathbf{j} - 4\mathbf{k}) = -3\mathbf{i} - 2\mathbf{k} + (-6\mathbf{i} + 10\mathbf{j} - 8\mathbf{k})$

$$= -3\mathbf{i} - 2\mathbf{k} - 6\mathbf{i} + 10\mathbf{j} - 8\mathbf{k}$$

$$= -9\mathbf{i} + 10\mathbf{j} - 10\mathbf{k}$$

(e) $0.2\mathbf{p} + 0.6\mathbf{q} - 3.2\mathbf{r} = 0.2(3\mathbf{i} + 2\mathbf{k}) + 0.6(4\mathbf{i} - 2\mathbf{j} + 3\mathbf{k}) - 3.2(-3\mathbf{i} + 5\mathbf{j} - 4\mathbf{k})$

$$= 0.6\mathbf{i} + 0.4\mathbf{k} + 2.4\mathbf{i} - 1.2\mathbf{j} + 1.8\mathbf{k} + 9.6\mathbf{i} - 16\mathbf{j} + 12.8\mathbf{k}$$

$$= \mathbf{12.6i - 17.2j + 15k}$$

Chapter 76   Combination of two periodic functions

In many engineering situations waveforms have to be combined. There are a number of methods of determining the resultant waveform. These include by:

1. drawing the waveforms and adding graphically
2. drawing the phasors and measuring the resultant
3. using the cosine and sine rules
4. using horizontal and vertical components
5. using complex numbers

Application: Sketch graphs of $y_1 = 4\sin\omega t$ and $y_2 = 3\sin(\omega t - \pi/3)$ on the same axes, over one cycle. Adding ordinates at intervals, obtain a sinusoidal expression for the resultant waveform $y_R = y_1 + y_2$

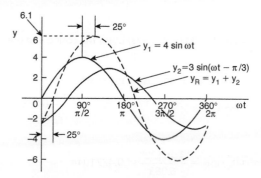

Figure 76.1

$y_1 = 4\sin\omega t$ and $y_2 = 3\sin(\omega t - \pi/3)$ are shown plotted in Figure 76.1.

Ordinates are added at 15° intervals and the resultant is shown by the broken line. The amplitude of the resultant is 6.1 and it **lags** y_1 by 25° or 0.436 rad.

Hence the sinusoidal expression for the resultant waveform is:

$$y_R = \mathbf{6.1}\sin(\omega t - \mathbf{0.436})$$

Application: Determine $4\sin \omega t + 3\sin(\omega t - \pi/3)$ by drawing phasors

The resultant of two periodic functions may be found from their relative positions when the time is zero. $4\sin \omega t$ and $3\sin(\omega t - \pi/3)$ may each be represented as phasors as shown in Figure 76.2, y_1 being 4 units long and drawn horizontally and y_2 being 3 units long, lagging y_1 by $\pi/3$ radians or $60°$. To determine the resultant of $y_1 + y_2$, y_1 is drawn horizontally as shown in Figure 76.3 and y_2 is joined to the end of y_1 at $60°$ to the horizontal. The resultant is given by y_R. This is the same as the diagonal of a parallelogram that is shown completed in Figure 76.4.

The resultant is measured as 6.1 and angle ϕ as $25°$ or 0.436 rad.

Hence, $\mathbf{4\sin \omega t + 3\sin(\omega t - \pi/3) = 6.1\sin(\omega t - 0.436)}$

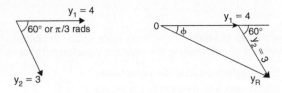

Figure 76.2   Figure 76.3

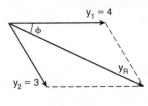

Figure 76.4

Application: Determine $4\sin \omega t + 3\sin(\omega t - \pi/3)$ using the cosine and sine rules

From the phasor diagram of Figure 76.3, and using the cosine rule:

$$y_R^2 = 4^2 + 3^2 - 2(4)(3)\cos 120° = 37 \text{ and } y_R = \sqrt{37} = 6.083$$

Using the sine rule gives:

$$\frac{3}{\sin \phi} = \frac{6.083}{\sin 120°} \text{ from which, } \sin \phi = \frac{3\sin 120°}{6.083} = 0.4271044$$

and $\phi = \sin^{-1} 0.4271044 = 25.28° = 25.28 \times \dfrac{\pi}{180} = 0.441\,\text{rad}$

Hence, by cosine and sine rules,

$$\mathbf{y_R = y_1 + y_2 = 6.083\sin(\omega t - 0.441)}$$

Application: Determine $4\sin \omega t + 3\sin(\omega t - \pi/3)$ using horizontal and vertical components

From the phasors shown in Figure 76.2:

Total horizontal component $= 4\cos 0° + 3\cos 300° = 5.5$

Total vertical component $= 4\sin 0° + 3\sin 300° = -2.598$

By Pythagoras, the resultant, $i_R = \sqrt{[5.5^2 + 2.598^2]} = 6.083$

Phase angle, $\phi = \tan^{-1}\left(\dfrac{2.598}{5.5}\right) = 25.28°$ or $0.441\,\text{rad}$ (ϕ being in the 4th quadrant)

Hence, by using horizontal and vertical components,
$$y_R = y_1 + y_2 = \mathbf{6.083\sin(\omega t - 0.441)}$$

Application: Determine $4\sin \omega t + 3\sin(\omega t - \pi/3)$ using complex numbers

From the phasors shown in Figure 76.2, the resultant may be expressed in polar form (see page 208)

as: $\qquad y_R = 4\angle 0° + 3\angle{-60°}$

i.e. $\qquad y_R = (4 + j0) + (1.5 - j2.598)$

$\qquad\qquad = (5.5 - j2.598) = 6.083\angle{-25.28°}$ A or $6.083\angle{-0.441}\,\text{rad}$ A

Hence, by using complex numbers, the resultant is:
$$y_R = y_1 + y_2 = \mathbf{6.083\sin(\omega t - 0.441)}$$

Chapter 77 The scalar product of two vectors

If $\mathbf{a} = a_1\mathbf{i} + a_2\mathbf{j} + a_3\mathbf{k}$ and $\mathbf{b} = b_1\mathbf{i} + b_2\mathbf{j} + b_3\mathbf{k}$

scalar or dot product: $\qquad \mathbf{a} \bullet \mathbf{b} = a_1 b_1 + a_2 b_2 + a_3 b_3 \qquad (1)$

$|\mathbf{a}| = \sqrt{(a_1^2 + a_2^2 + a_3^2)}$ and $|\mathbf{b}| = \sqrt{(b_1^2 + b_2^2 + b_3^2)} \qquad (2)$

$\cos\theta = \dfrac{\mathbf{a} \bullet \mathbf{b}}{|\mathbf{a}||\mathbf{b}|} = \dfrac{a_1 b_1 + a_2 b_2 + a_3 b_3}{\sqrt{(a_1^2 + a_2^2 + a_3^2)}\sqrt{(b_1^2 + b_2^2 + b_3^2)}} \qquad (3)$

Application: Find vector **a** joining points P and Q where point P has co-ordinates (4, −1, 3) and point Q has co-ordinates (2, 5, 0) and find |**a**|, the magnitude or norm of **a**

Let O be the origin, i.e. its co-ordinates are (0, 0, 0)

The position vector of P and Q are given by **OP** = 4**i** − **j** + 3**k** and **OQ** = 2**i** + 5**j**

By the addition law of vectors **OP** + **PQ** = **OQ**

Hence   **a** = **PQ** = **OQ** − **OP**

i.e.   **a** = **PQ** = (2**i** + 5**j**) − (4**i** − **j** + 3**k**)

$\qquad\qquad$ = −2**i** + 6**j** − 3**k**

From equation (2), the **magnitude** or **norm** of **a**,

$$|a| = \sqrt{(a^2 + b^2 + c^2)} = \sqrt{[(-2)^2 + 6^2 + (-3)^2]} = \sqrt{49} = 7$$

Application: Determine: (i) **p** • **q** (ii) **p** + **q** (iii) |**p** + **q**| and (iv) |**p**|+|**q**| if **p** = 2**i** + **j** − **k** and **q** = **i** − 3**j** + 2**k**

(i) From equation (1), if **p** = a_1**i** + a_2**j** + a_3**k** and **q** = b_1**i** + b_2**j** + b_3**k**

$\quad$ then $\qquad$ **p** • **q** = $a_1 b_1 + a_2 b_2 + a_3 b_3$

$\quad$ When $\qquad$ **p** = 2**i** + **j** − **k**, a_1 = 2, a_2 = 1 and a_3 = −1

$\quad$ and when $\quad$ **q** = **i** − 3**j** + 2**k**, b_1 = 1, b_2 = −3 and b_3 = 2

$\quad$ Hence $\qquad$ **p** • **q** = (2)(1) + (1)(−3) + (−1)(2)

$\quad$ i.e. $\qquad\quad$ **p** • **q** = −3

(ii) **p** + **q** = (2**i** + **j** − **k**) + (**i** − 3**j** + 2**k**) = **3i** − **2j** + **k**

(iii) |**p** + **q**| = |3**i** − 2**j** + **k**|

$\quad$ From equation (2), |**p** + **q**| = $\sqrt{[3^2 + (-2)^2 + 1^2)} = \sqrt{14}$

(iv) From equation (2), |**p**| = |2**i** + **j** − **k**| = $\sqrt{[2^2 + 1^2 + (-1)^2]} = \sqrt{6}$

$\quad$ Similarly, |**q**| = |**i** − 3**j** + 2**k**| = $\sqrt{[1^2 + (-3)^2 + 2^2]} = \sqrt{14}$

$\quad$ Hence |**p**| + |**q**| = $\sqrt{6} + \sqrt{14}$ = 6.191, correct to 3 decimal places

Application: Determine the angle between vectors **oa** and **ob** when **oa** = **i** + 2**j** − 3**k** and **ob** = 2**i** − **j** + 4**k**

From equation (3), $\cos \theta = \dfrac{a_1 b_1 + a_2 b_2 + a_3 b_3}{\sqrt{(a_1^2 + a_2^2 + a_3^2)}\sqrt{(b_1^2 + b_2^2 + b_3^2)}}$

Since **oa** = **i** + 2**j** − 3**k**, a_1 = 1, a_2 = 2 and a_3 = −3

Since **ob** = 2**i** − **j** + 4**k**, b_1 = 2, b_2 = −1 and b_3 = 4

Thus, $\qquad \cos\theta = \dfrac{(1 \times 2) + (2 \times -1) + (-3 \times 4)}{\sqrt{(1^2 + 2^2 + (-3)^2)}\sqrt{(2^2 + (-1)^2 + 4^2)}}$

$$= \dfrac{-12}{\sqrt{14}\sqrt{21}} = -0.6999$$

i.e. $\qquad \theta = \cos^{-1}\theta = 134.4°$ or $225.6°$

By sketching the position of the two vectors, it will be seen that 225.6° is not an acceptable answer.

Thus, the angle between the vectors **oa** and **ob**, $\theta = \mathbf{134.4°}$

Application: A constant force of $\mathbf{F} = 10\mathbf{i} + 2\mathbf{j} - \mathbf{k}$ Newtons displaces an object from $\mathbf{A} = \mathbf{i} + \mathbf{j} + \mathbf{k}$ to $\mathbf{B} = 2\mathbf{i} - \mathbf{j} + 3\mathbf{k}$ (in metres). Find the work done in Newton metres

The work done is the product of the applied force and the distance moved in the direction of the force,

i.e. $\qquad$ **work done = F • d**

From the sketch shown in Figure 77.1, $\mathbf{AB} = \mathbf{AO} + \mathbf{OB} = \mathbf{OB} - \mathbf{OA}$

that is $\mathbf{AB} = (2\mathbf{i} - \mathbf{j} + 3\mathbf{k}) - (\mathbf{i} + \mathbf{j} + \mathbf{k}) = \mathbf{i} - 2\mathbf{j} + 2\mathbf{k}$

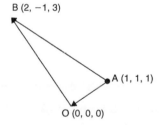

B (2, −1, 3)

A (1, 1, 1)

O (0, 0, 0) $\qquad$ **Figure 77.1**

The work done is **F • d**, that is **F • AB** in this case

i.e. $\qquad$ **work done** $= (10\mathbf{i} + 2\mathbf{j} - \mathbf{k}) • (\mathbf{i} - 2\mathbf{j} + 2\mathbf{k})$

From equation (1), $\qquad \mathbf{a} • \mathbf{b} = a_1b_1 + a_2b_2 + a_3b_3$

Hence, **work done** $= (10 \times 1) + (2 \times (-2)) + ((-1) \times 2) = \mathbf{4\,N\,m}$

Direction cosines

Let **or** $= x\mathbf{i} + y\mathbf{j} + z\mathbf{k}$ and from equation (2),

$$|\mathbf{or}| = \sqrt{x^2 + y^2 + z^2}$$

If **or** makes angles of α, β and γ with the co-ordinate axes i, j and k respectively, then:

$$\cos\alpha = \frac{x}{\sqrt{x^2 + y^2 + z^2}}, \cos\beta = \frac{y}{\sqrt{x^2 + y^2 + z^2}}$$

$$\text{and } \cos\gamma = \frac{z}{\sqrt{x^2 + y^2 + z^2}}$$

The values of $\cos\alpha$, $\cos\beta$ and $\cos\gamma$ are called the **direction cosines of or**

Application: Find the direction cosines of $3\mathbf{i} + 2\mathbf{j} + \mathbf{k}$

$$\sqrt{x^2 + y^2 + z^2} = \sqrt{3^2 + 2^2 + 1^2} = \sqrt{14}$$

The direction cosines are: $\cos\alpha = \dfrac{x}{\sqrt{x^2 + y^2 + z^2}} = \dfrac{3}{\sqrt{14}} = 0.802$

$$\cos\beta = \frac{y}{\sqrt{x^2 + y^2 + z^2}} = \frac{2}{\sqrt{14}} = 0.535$$

and $\qquad\qquad \cos\gamma = \dfrac{z}{\sqrt{x^2 + y^2 + z^2}} = \dfrac{1}{\sqrt{14}} = 0.267$

Chapter 78   Vector products

Let $\mathbf{a} = a_1\mathbf{i} + a_2\mathbf{j} + a_3\mathbf{k}$ and $\mathbf{b} = b_1\mathbf{i} + b_2\mathbf{j} + b_3\mathbf{k}$

Vector or cross product: $\qquad \mathbf{a} \times \mathbf{b} = \begin{vmatrix} \mathbf{i} & \mathbf{j} & \mathbf{k} \\ a_1 & a_2 & a_3 \\ b_1 & b_2 & b_3 \end{vmatrix}$ (1)

$$|\mathbf{a} \times \mathbf{b}| = \sqrt{[(\mathbf{a} \bullet \mathbf{a})(\mathbf{b} \bullet \mathbf{b}) - (\mathbf{a} \bullet \mathbf{b})^2]}$$ (2)

Application: Find (i) $\mathbf{a} \times \mathbf{b}$ and (ii) $|\mathbf{a} \times \mathbf{b}|$ for the vectors $\mathbf{a} = \mathbf{i} + 4\mathbf{j} - 2\mathbf{k}$ and $\mathbf{b} = 2\mathbf{i} - \mathbf{j} + 3\mathbf{k}$

(i) From equation (1),

$$\mathbf{a} \times \mathbf{b} = \begin{vmatrix} \mathbf{i} & \mathbf{j} & \mathbf{k} \\ 1 & 4 & -2 \\ 2 & -1 & 3 \end{vmatrix} = \mathbf{i} \begin{vmatrix} 4 & -2 \\ -1 & 3 \end{vmatrix} - \mathbf{j} \begin{vmatrix} 1 & -2 \\ 2 & 3 \end{vmatrix} + \mathbf{k} \begin{vmatrix} 1 & 4 \\ 2 & -1 \end{vmatrix}$$

$$= i(12 - 2) - j(3 + 4) + k(-1 - 8)$$
$$= 10i - 7j - 9k$$

(ii) From equation (2) $|a \times b| = \sqrt{[(a \bullet a)(b \bullet b) - (a \bullet b)^2]}$

Now $a \bullet a = (1)(1) + (4)(4) + (-2)(-2) = 21$

$b \bullet b = (2)(2) + (-1)(-1) + (3)(3) = 14$

and $a \bullet b = (1)(2) + (4)(-1) + (-2)(3) = -8$

Thus, $|a \times b| = \sqrt{(21 \times 14 - 64)} = \sqrt{230} = 15.17$

Application: Find (a) $(p - 2q) \times r$ (b) $p \times (2r \times 3q)$
if $p = 4i + j - 2k$, $q = 3i - 2j + k$ and $r = i - 2k$

(a) $(p - 2q) \times r = [4i + j - 2k - 2(3i - 2j + k)] \times (i - 2k)$

$$= (-2i + 5j - 4k) \times (i - 2k)$$

$$= \begin{vmatrix} i & j & k \\ -2 & 5 & -4 \\ 1 & 0 & -2 \end{vmatrix} \quad \text{from equation (1)}$$

$$= i \begin{vmatrix} 5 & -4 \\ 0 & -2 \end{vmatrix} - j \begin{vmatrix} -2 & -4 \\ 1 & -2 \end{vmatrix} + k \begin{vmatrix} -2 & 5 \\ 1 & 0 \end{vmatrix}$$

$$= i(-10 - 0) - j(4 + 4) + k(0 - 5)$$

i.e. $(p - 2q) \times r = -10i - 8j - 5k$

(b) $(2r \times 3q) = (2i - 4k) \times (9i - 6j + 3k)$

$$= \begin{vmatrix} i & j & k \\ 2 & 0 & -4 \\ 9 & -6 & 3 \end{vmatrix} = i(0 - 24) - j(6 + 36) + k(-12 - 0)$$

$$= -24i - 42j - 12k$$

Hence, $p \times (2r \times 3q) = (4i + j - 2k) \times (-24i - 42j - 12k)$

$$= \begin{vmatrix} i & j & k \\ 4 & 1 & -2 \\ -24 & -42 & -12 \end{vmatrix}$$

$$= i(-12 - 84) - j(-48 - 48) + k(-168 + 24)$$

$$= -96i + 96j - 144k \quad \text{or} \quad -48(2i - 2j + 3k)$$

Application: Find the moment and the magnitude of the moment of a force of $(i + 2j - 3k)$ Newton's about point B having co-ordinates (0, 1, 1), when the force acts on a line through A whose co-ordinates are (1, 3, 4)

The moment **M** about point B of a force vector **F** that has a position vector of **r** from A is given by:

$$M = r \times F$$

r is the vector from B to A, i.e. **r** = **BA**

But **BA** = **BO** + **OA** = **OA** − **OB**

i.e. **r** = (**i** + 3**j** + 4**k**) − (**j** + **k**) = **i** + 2**j** + 3**k**

Moment, M = **r** × **F** = (**i** + 2**j** + 3**k**) × (**i** + 2**j** − 3**k**)

$$= \begin{vmatrix} i & j & k \\ 1 & 2 & 3 \\ 1 & 2 & -3 \end{vmatrix} = i(-6-6) - j(-3-3) + k(2-2) = -12i + 6j\,\text{N m}$$

The magnitude of **M**, $|M| = |r \times F| = \sqrt{[(r \bullet r)(F \bullet F) - (r \bullet F)^2]}$

$$r \bullet r = (1)(1) + (2)(2) + (3)(3) = 14$$
$$F \bullet F = (1)(1) + (2)(2) + (-3)(-3) = 14$$
$$r \bullet F = (1)(1) + (2)(2) + (3)(-3) = -4$$

i.e. **magnitude,** $|M| = \sqrt{[14 \times 14 - (-4)^2]} = \sqrt{180}\,\text{N m}$

$$= 13.42\,\text{N m}$$

Application: The axis of a circular cylinder coincides with the z-axis and it rotates with an angular velocity of (2**i** − 5**j** + 7**k**) rad/s. Determine the tangential velocity at a point P on the cylinder, whose co-ordinates are (**j** + 3**k**) metres, and the magnitude of the tangential velocity

The velocity v of point P on a body rotating with angular velocity ω about a fixed axis is given by:

$$v = \omega \times r \quad \text{where r is the point on vector P.}$$

Thus, **velocity, v** = (2**i** − 5**j** + 7**k**) × (**j** + 3**k**)

$$= \begin{vmatrix} i & j & k \\ 2 & -5 & 7 \\ 0 & 1 & 3 \end{vmatrix} = i(-15-7) - j(6-0) + k(2-0)$$

$$= (-22i - 6j + 2k)\,\text{m / s}$$

The magnitude of **v**, $|v| = \sqrt{[(\omega \bullet \omega)(r \bullet r) - (\omega \bullet r)^2]}$

$$\omega \bullet \omega = (2)(2) + (-5)(-5) + (7)(7) = 78$$
$$r \bullet r = (0)(0) + (1)(1) + (3)(3) = 10$$
$$\omega \bullet r = (2)(0) + (-5)(1) + (7)(3) = 16$$

Hence, magnitude, $|v| = \sqrt{(78 \times 10 - 16^2)} = \sqrt{524}\,\text{m/s}$

$$= 22.89\,\text{m/s}$$

Matrices and determinants

Why are matrices and determinants important?

Matrices are used to solve problems in electronics, optics, quantum mechanics, statics, robotics, linear programming, optimisation, genetics, and much more. Matrix calculus is a mathematical tool used in connection with linear equations, linear transformations, systems of differential equations, and so on, and is vital for calculating forces, vectors, tensions, masses, loads and a lot of other factors that must be accounted for in engineering and science to ensure safe and resource-efficient structures. Electrical and mechanical engineers, chemists, biologists and scientists all need knowledge of matrices to solve problems. In computer graphics, matrices are used to project a 3-dimensional image on to a 2-dimentional screen, and to create realistic motion. Matrices are therefore very important in solving engineering and science problems.

In the main, matrices and determinants are used to solve a system of simultaneous linear equations. The simultaneous solution of multiple equations finds its way into many common engineering problems. In fact, modern structural engineering analysis techniques are all about solving systems of equations simultaneously. Eigenvalues and eigenvectors, which are based on matrix theory, are very important in engineering and science. For example, car designers analyse eigenvalues in order to damp out the noise in a car, eigenvalue analysis is used in the design of car stereo systems, eigenvalues can be used to test for cracks and deformities in a solid, and oil companies use eigenvalue analysis to explore land for oil.

Chapter 79 Addition, subtraction and multiplication of matrices

If $A = \begin{pmatrix} a & b \\ c & d \end{pmatrix}$ and $B = \begin{pmatrix} e & f \\ g & h \end{pmatrix}$

then $A + B = \begin{pmatrix} a+e & b+f \\ c+g & d+h \end{pmatrix}$ and $A - B = \begin{pmatrix} a-e & b-f \\ c-g & d-h \end{pmatrix}$

and $A \times B = \begin{pmatrix} ae+bg & af+bh \\ ce+dg & cf+dh \end{pmatrix}$

Application: Determine $\begin{pmatrix} 2 & -1 \\ -7 & 4 \end{pmatrix} + \begin{pmatrix} -3 & 0 \\ 7 & -4 \end{pmatrix}$

$$\begin{pmatrix} 2 & -1 \\ -7 & 4 \end{pmatrix} + \begin{pmatrix} -3 & 0 \\ 7 & -4 \end{pmatrix} = \begin{pmatrix} 2+(-3) & -1+0 \\ -7+7 & 4+(-4) \end{pmatrix} = \begin{pmatrix} -1 & -1 \\ 0 & 0 \end{pmatrix}$$

Application: Determine $\begin{pmatrix} 2 & -1 \\ -7 & 4 \end{pmatrix} - \begin{pmatrix} -3 & 0 \\ 7 & -4 \end{pmatrix}$

$$\begin{pmatrix} 2 & -1 \\ -7 & 4 \end{pmatrix} - \begin{pmatrix} -3 & 0 \\ 7 & -4 \end{pmatrix} = \begin{pmatrix} 2-(-3) & -1-0 \\ -7-7 & 4-(-4) \end{pmatrix} = \begin{pmatrix} 5 & -1 \\ -14 & 8 \end{pmatrix}$$

Application:

If $A = \begin{pmatrix} -3 & 0 \\ 7 & -4 \end{pmatrix}$ and $B = \begin{pmatrix} 2 & -1 \\ -7 & 4 \end{pmatrix}$ determine $2A - 3B$

$$2A - 3B = 2\begin{pmatrix} -3 & 0 \\ 7 & -4 \end{pmatrix} - 3\begin{pmatrix} 2 & -1 \\ -7 & 4 \end{pmatrix} = \begin{pmatrix} -6 & 0 \\ 14 & -8 \end{pmatrix} - \begin{pmatrix} 6 & -3 \\ -21 & 12 \end{pmatrix}$$

$$= \begin{pmatrix} -6-6 & 0-(-3) \\ 14-(-21) & -8-12 \end{pmatrix}$$

$$= \begin{pmatrix} -12 & 3 \\ 35 & -20 \end{pmatrix}$$

Application: If $A = \begin{pmatrix} 2 & 3 \\ 1 & -4 \end{pmatrix}$ and $B = \begin{pmatrix} -5 & 7 \\ -3 & 4 \end{pmatrix}$ determine $A \times B$

$$A \times B = \begin{pmatrix} [2 \times -5 + 3 \times -3] & [2 \times 7 + 3 \times 4] \\ [1 \times -5 + -4 \times -3] & [1 \times 7 + -4 \times 4] \end{pmatrix} = \begin{pmatrix} -19 & 26 \\ 7 & -9 \end{pmatrix}$$

Application: Determine $\begin{pmatrix} 3 & 4 & 0 \\ -2 & 6 & -3 \\ 7 & -4 & 1 \end{pmatrix} \times \begin{pmatrix} 2 \\ 5 \\ -1 \end{pmatrix}$

$$\begin{pmatrix} 3 & 4 & 0 \\ -2 & 6 & -3 \\ 7 & -4 & 1 \end{pmatrix} \times \begin{pmatrix} 2 \\ 5 \\ -1 \end{pmatrix} = \begin{pmatrix} (3 \times 2) + (4 \times 5) + (0 \times -1) \\ (-2 \times 2) + (6 \times 5) + (-3 \times -1) \\ (7 \times 2) + (-4 \times 5) + (1 \times -1) \end{pmatrix} = \begin{pmatrix} 26 \\ 29 \\ -7 \end{pmatrix}$$

Chapter 80 The determinant and inverse of a 2 by 2 matrix

If $A = \begin{pmatrix} a & b \\ c & d \end{pmatrix}$ then

the **determinant of A,** $\begin{vmatrix} a & b \\ c & d \end{vmatrix} = a \times d - b \times c$

and the **inverse of A,** $A^{-1} = \dfrac{1}{ad - bc} \begin{vmatrix} d & -b \\ -c & a \end{vmatrix}$

Application: Find the determinant of $\begin{pmatrix} 3 & -4 \\ 1 & 6 \end{pmatrix}$

$$\begin{vmatrix} 3 & -4 \\ 1 & 6 \end{vmatrix} = (3 \times 6) - (-4 \times 1) = 18 - (-4) = 22$$

Application: Find the inverse of $\begin{pmatrix} 3 & -4 \\ 1 & 6 \end{pmatrix}$

Inverse of matrix $\begin{pmatrix} 3 & -4 \\ 1 & 6 \end{pmatrix} = \dfrac{1}{18--4}\begin{pmatrix} 6 & 4 \\ -1 & 3 \end{pmatrix} = \dfrac{1}{22}\begin{pmatrix} 6 & 4 \\ -1 & 3 \end{pmatrix}$

Application: If $A = \begin{pmatrix} 3 & -4 \\ 1 & 6 \end{pmatrix}$ determine $A \times A^{-1}$

From above:
$$A \times A^{-1} = \begin{pmatrix} 3 & -4 \\ 1 & 6 \end{pmatrix} \times \frac{1}{22}\begin{pmatrix} 6 & 4 \\ -1 & 3 \end{pmatrix}$$

$$= \frac{1}{22}\begin{pmatrix} 3 & -4 \\ 1 & 6 \end{pmatrix} \times \begin{pmatrix} 6 & 4 \\ -1 & 3 \end{pmatrix}$$

$$= \frac{1}{22}\begin{pmatrix} 18+4 & 12-12 \\ 6-6 & 4+18 \end{pmatrix}$$

$$= \frac{1}{22}\begin{pmatrix} 22 & 0 \\ 0 & 22 \end{pmatrix} = \begin{pmatrix} 1 & 0 \\ 0 & 1 \end{pmatrix}$$

$\begin{pmatrix} 1 & 0 \\ 0 & 1 \end{pmatrix}$ is called the **unit matrix**; such a matrix has all leading diagonal elements equal to 1 and all other elements equal to 0

Chapter 81   The determinant of a 3 by 3 matrix

(i) The **minor** of an element of a 3 by 3 matrix is the value of the 2 by 2 determinant obtained by covering up the row and column containing that element.

Thus for the matrix $\begin{pmatrix} 1 & 2 & 3 \\ 4 & 5 & 6 \\ 7 & 8 & 9 \end{pmatrix}$ the minor of element 4 is obtained by covering up the row (4 5 6) and the column $\begin{pmatrix} 1 \\ 4 \\ 7 \end{pmatrix}$, leaving the 2 by 2 determinant $\begin{vmatrix} 2 & 3 \\ 8 & 9 \end{vmatrix}$ i.e. the minor of element 4 is $(2 \times 9) - (3 \times 8) = -6$

(ii) The sign of a minor depends on its position within the matrix, the sign pattern being

$$\begin{pmatrix} + & - & + \\ - & + & - \\ + & - & + \end{pmatrix}$$

Thus the signed-minor of element 4 in the matrix $\begin{pmatrix} 1 & 2 & 3 \\ 4 & 5 & 6 \\ 7 & 8 & 9 \end{pmatrix}$ is

$$-\begin{vmatrix} 2 & 3 \\ 8 & 9 \end{vmatrix} = -(-6) = 6$$

The signed-minor of an element is called the **cofactor** of the element.

(iii) **The value of a 3 by 3 determinant is the sum of the products of the elements and their cofactors of any row or any column of the corresponding 3 by 3 matrix.**

There are thus six different ways of evaluating a 3×3 determinant – and all should give the same value.

Using the first row:

$$\begin{vmatrix} a_1 & b_1 & c_1 \\ a_2 & b_2 & c_2 \\ a_3 & b_3 & c_3 \end{vmatrix} = a_1 \begin{vmatrix} b_2 & c_2 \\ b_3 & c_3 \end{vmatrix} - b_1 \begin{vmatrix} a_2 & c_2 \\ a_3 & c_3 \end{vmatrix} + c_1 \begin{vmatrix} a_2 & b_2 \\ a_3 & b_3 \end{vmatrix}$$

Application: Evaluate $\begin{vmatrix} 1 & 4 & -3 \\ -5 & 2 & 6 \\ -1 & -4 & 2 \end{vmatrix}$ using (a) the first row, and (b) the second column

(a) Using the first row:

$$\begin{vmatrix} 1 & 4 & -3 \\ -5 & 2 & 6 \\ -1 & -4 & 2 \end{vmatrix} = 1 \begin{vmatrix} 2 & 6 \\ -4 & 2 \end{vmatrix} - 4 \begin{vmatrix} -5 & 6 \\ -1 & 2 \end{vmatrix} + (-3) \begin{vmatrix} -5 & 2 \\ -1 & -4 \end{vmatrix}$$

$$= (4 + 24) - 4(-10 + 6) - 3(20 + 2)$$
$$= 28 + 16 - 66 = -22$$

(b) Using the second column:

$$\begin{vmatrix} 1 & 4 & -3 \\ -5 & 2 & 6 \\ -1 & -4 & 2 \end{vmatrix} = -4 \begin{vmatrix} -5 & 6 \\ -1 & 2 \end{vmatrix} + 2 \begin{vmatrix} 1 & -3 \\ -1 & 2 \end{vmatrix} - (-4) \begin{vmatrix} 1 & -3 \\ -5 & 6 \end{vmatrix}$$

$$= -4(-10 + 6) + 2(2 - 3) + 4(6 - 15)$$
$$= 16 - 2 - 36 = -22$$

Chapter 82   The inverse of a 3 by 3 matrix

If $A = \begin{pmatrix} a_1 & b_1 & c_1 \\ a_2 & b_2 & c_2 \\ a_3 & b_3 & c_3 \end{pmatrix}$ then the **inverse of matrix A,**

$$A^{-1} = \frac{\text{adj } A}{|A|} \quad \text{where adj } A \text{ is the adjoint}$$

The **adjoint** of a matrix A is obtained by:

(i) forming a matrix B of the cofactors of A, and
(ii) **transposing** matrix B to give B^T, where B^T is the matrix obtained by writing the rows of B as the columns of B^T. Then **adj A = B^T**

Application: Find the inverse of $\begin{pmatrix} 1 & 5 & -2 \\ 3 & -1 & 4 \\ -3 & 6 & -7 \end{pmatrix}$

$$\text{Inverse} = \frac{\text{adjoint}}{\text{determinant}}$$

The matrix of cofactors is $\begin{pmatrix} -17 & 9 & 15 \\ 23 & -13 & -21 \\ 18 & -10 & -16 \end{pmatrix}$

The transpose of the matrix of cofactors

(i.e. the adjoint) is $\begin{pmatrix} -17 & 23 & 18 \\ 9 & -13 & -10 \\ 15 & -21 & -16 \end{pmatrix}$

The determinant of $\begin{pmatrix} 1 & 5 & -2 \\ 3 & -1 & 4 \\ -3 & 6 & -7 \end{pmatrix}$

$$= 1(7 - 24) - 5(-21 + 12) - 2(18 - 3) \text{ using the first row}$$

$$= -17 + 45 - 30 = -2$$

Hence the inverse of $\begin{pmatrix} 1 & 5 & -2 \\ 3 & -1 & 4 \\ -3 & 6 & -7 \end{pmatrix} = \dfrac{\begin{pmatrix} -17 & 23 & 18 \\ 9 & -13 & -10 \\ 15 & -21 & -16 \end{pmatrix}}{-2}$

$$= \begin{pmatrix} 8.5 & -11.5 & -9 \\ -4.5 & 6.5 & 5 \\ -7.5 & 10.5 & 8 \end{pmatrix}$$

Chapter 83   Solution of simultaneous equations by matrices

Two unknowns

The procedure for solving linear simultaneous equations in **two unknowns using matrices** is:

(i) write the equations in the form

$$a_1x + b_1y = c_1$$
$$a_2x + b_2y = c_2$$

(ii) write the matrix equation corresponding to these equations,

i.e. $\begin{pmatrix} a_1 & b_1 \\ a_2 & b_2 \end{pmatrix} \times \begin{pmatrix} x \\ y \end{pmatrix} = \begin{pmatrix} c_1 \\ c_2 \end{pmatrix}$

(iii) determine the inverse matrix of $\begin{pmatrix} a_1 & b_1 \\ a_2 & b_2 \end{pmatrix}$

i.e. $\dfrac{1}{a_1b_2 - b_1a_2} \begin{pmatrix} b_2 & -b_1 \\ -a_2 & a_1 \end{pmatrix}$

(iv) multiply each side of (ii) by the inverse matrix, and

(v) solve for x and y by equating corresponding elements

Applications: Use matrices to solve the simultaneous equations:

$$3x + 5y - 7 = 0 \tag{1}$$
$$4x - 3y - 19 = 0 \tag{2}$$

(i) Writing the equations in the $a_1x + b_1y = c$ form gives:

$$3x + 5y = 7$$

$$4x - 3y = 19$$

(ii) The matrix equation is $\begin{pmatrix} 3 & 5 \\ 4 & -3 \end{pmatrix} \times \begin{pmatrix} x \\ y \end{pmatrix} = \begin{pmatrix} 7 \\ 19 \end{pmatrix}$

(iii) The inverse of matrix $\begin{pmatrix} 3 & 5 \\ 4 & -3 \end{pmatrix}$ is:

$$\frac{1}{3 \times (-3) - 5 \times 4} \begin{pmatrix} -3 & -5 \\ -4 & 3 \end{pmatrix} = \begin{pmatrix} \dfrac{3}{29} & \dfrac{5}{29} \\ \dfrac{4}{29} & \dfrac{-3}{29} \end{pmatrix}$$

(iv) Multiplying each side of (ii) by (iii) and remembering that $A \times A^{-1} = I$, the unit matrix, gives:

$$\begin{pmatrix} 1 & 0 \\ 0 & 1 \end{pmatrix} \begin{pmatrix} x \\ y \end{pmatrix} = \begin{pmatrix} \dfrac{3}{29} & \dfrac{5}{29} \\ \dfrac{4}{29} & \dfrac{-3}{29} \end{pmatrix} \times \begin{pmatrix} 7 \\ 19 \end{pmatrix}$$

Thus $\begin{pmatrix} x \\ y \end{pmatrix} = \begin{pmatrix} \dfrac{21}{29} + \dfrac{95}{29} \\ \dfrac{28}{29} - \dfrac{57}{29} \end{pmatrix}$ i.e. $\begin{pmatrix} x \\ y \end{pmatrix} = \begin{pmatrix} 4 \\ -1 \end{pmatrix}$

(v) By comparing corresponding elements: **x = 4 and y = −1,** which can be checked in the original equations.

Three unknowns

The procedure for solving linear simultaneous equations in **three unknowns using matrices** is:

(i) write the equations in the form

$$a_1x + b_1y + c_1z = d_1$$
$$a_2x + b_2y + c_2z = d_2$$
$$a_3x + b_3y + c_3z = d_3$$

(ii) write the matrix equation corresponding to these equations, i.e.

$$\begin{pmatrix} a_1 & b_1 & c_1 \\ a_2 & b_2 & c_2 \\ a_3 & b_3 & c_3 \end{pmatrix} \times \begin{pmatrix} x \\ y \\ z \end{pmatrix} = \begin{pmatrix} d_1 \\ d_2 \\ d_3 \end{pmatrix}$$

(iii) determine the inverse matrix of $\begin{pmatrix} a_1 & b_1 & c_1 \\ a_2 & b_2 & c_2 \\ a_3 & b_3 & c_3 \end{pmatrix}$

(iv) multiply each side of (ii) by the inverse matrix, and

(v) solve for x, y and z by equating the corresponding elements

Application: Use matrices to solve the simultaneous equations:

$$x + y + z - 4 = 0 \tag{1}$$
$$2x - 3y + 4z - 33 = 0 \tag{2}$$
$$3x - 2y - 2z - 2 = 0 \tag{3}$$

(i) Writing the equations in the $a_1x + b_1y + c_1z = d_1$ form gives:

$$x + y + z = 4$$
$$2x - 3y + 4z = 33$$
$$3x - 2y - 2z = 2$$

(ii) The matrix equation is: $\begin{pmatrix} 1 & 1 & 1 \\ 2 & -3 & 4 \\ 3 & -2 & -2 \end{pmatrix} \times \begin{pmatrix} x \\ y \\ z \end{pmatrix} = \begin{pmatrix} 4 \\ 33 \\ 2 \end{pmatrix}$

(iii) The inverse matrix of $A = \begin{pmatrix} 1 & 1 & 1 \\ 2 & -3 & 4 \\ 3 & -2 & -2 \end{pmatrix}$ is given by $A^{-1} = \dfrac{\text{adj } A}{|A|}$

The adjoint of A is the transpose of the matrix of the cofactors of the elements. The matrix of cofactors is $\begin{pmatrix} 14 & 16 & 5 \\ 0 & -5 & 5 \\ 7 & -2 & -5 \end{pmatrix}$ and the transpose

of this matrix gives: adj $A = \begin{pmatrix} 14 & 0 & 7 \\ 16 & -5 & -2 \\ 5 & 5 & -5 \end{pmatrix}$

The determinant of A, i.e. the sum of the products of elements and their cofactors, using a first row expansion is

$$1 \begin{vmatrix} -3 & 4 \\ -2 & -2 \end{vmatrix} - 1 \begin{vmatrix} 2 & 4 \\ 3 & -2 \end{vmatrix} + 1 \begin{vmatrix} 2 & -3 \\ 3 & -2 \end{vmatrix} = (1 \times 14) - (1 \times -16) + (1 \times 5)$$

$$= 35$$

Hence the inverse of A, $\quad A^{-1} = \dfrac{1}{35}\begin{pmatrix} 14 & 0 & 7 \\ 16 & -5 & -2 \\ 5 & 5 & -5 \end{pmatrix}$

(iv) Multiplying each side of (ii) by (iii), and remembering that $A \times A^{-1} = I$, the unit matrix, gives:

$$\begin{pmatrix} 1 & 0 & 0 \\ 0 & 1 & 0 \\ 0 & 0 & 1 \end{pmatrix} \times \begin{pmatrix} x \\ y \\ z \end{pmatrix} = \frac{1}{35}\begin{pmatrix} 14 & 0 & 7 \\ 16 & -5 & -2 \\ 5 & 5 & -5 \end{pmatrix} \times \begin{pmatrix} 4 \\ 33 \\ 2 \end{pmatrix}$$

$$\begin{pmatrix} x \\ y \\ z \end{pmatrix} = \frac{1}{35}\begin{pmatrix} (14 \times 4) + (0 \times 33) + (7 \times 2) \\ (16 \times 4) + (-5 \times 33) + (-2 \times 2) \\ (5 \times 4) + (5 \times 33) + (-5 \times 2) \end{pmatrix} = \frac{1}{35}\begin{pmatrix} 70 \\ -105 \\ 175 \end{pmatrix} = \begin{pmatrix} 2 \\ -3 \\ 5 \end{pmatrix}$$

(v) By comparing corresponding elements, **x = 2, y = −3, z = 5**, which can be checked in the original equations.

Chapter 84   Solution of simultaneous equations by determinants

Two unknowns

When solving linear simultaneous equations in **two unknowns using determinants:**

(i) write the equations in the form

$$a_1 x + b_1 y + c_1 = 0$$
$$a_2 x + b_2 y + c_2 = 0$$

(ii) the solution is given by: $\dfrac{x}{D_x} = \dfrac{-y}{D_y} = \dfrac{1}{D}$

where $D_x = \begin{vmatrix} b_1 & c_1 \\ b_2 & c_2 \end{vmatrix}$ i.e. the determinant of the coefficients left when the x-column is covered up,

$D_y = \begin{vmatrix} a_1 & c_1 \\ a_2 & c_2 \end{vmatrix}$ i.e. the determinant of the coefficients left when the y-column is covered up,

and $D = \begin{vmatrix} a_1 & b_1 \\ a_2 & b_2 \end{vmatrix}$ i.e. the determinant of the coefficients left when the constants-column is covered up

Application: Solve the following simultaneous equations using determinants:
$$3x - 4y = 12$$
$$7x + 5y = 6.5$$

Following the above procedure:

(i)
$$3x - 4y - 12 = 0$$
$$7x + 5y - 6.5 = 0$$

(ii)
$$\frac{x}{\begin{vmatrix} -4 & -12 \\ 5 & -6.5 \end{vmatrix}} = \frac{-y}{\begin{vmatrix} 3 & -12 \\ 7 & -6.5 \end{vmatrix}} = \frac{1}{\begin{vmatrix} 3 & -4 \\ 7 & 5 \end{vmatrix}}$$

i.e.
$$\frac{x}{(-4)(-6.5) - (-12)(5)} = \frac{-y}{(3)(-6.5) - (-12)(7)} = \frac{1}{(3)(5) - (-4)(7)}$$

i.e.
$$\frac{x}{26 + 60} = \frac{-y}{-19.5 + 84} = \frac{1}{15 + 28}$$

i.e.
$$\frac{x}{86} = \frac{-y}{64.5} = \frac{1}{43}$$

Since $\dfrac{x}{86} = \dfrac{1}{43}$ then $x = \dfrac{86}{43} = 2$

and since $\dfrac{-y}{64.5} = \dfrac{1}{43}$ then $y = -\dfrac{64.5}{43} = -1.5$

Three unknowns

When solving simultaneous equations in **three unknowns using determinants**:

(i) write the equations in the form

$$a_1 x + b_1 y + c_1 z + d_1 = 0$$
$$a_2 x + b_2 y + c_2 z + d_2 = 0$$
$$a_3 x + b_3 y + c_3 z + d_3 = 0$$

(ii) the solution is given by: $\dfrac{x}{D_x} = \dfrac{-y}{D_y} = \dfrac{z}{D_z} = \dfrac{-1}{D}$

where $D_x = \begin{vmatrix} b_1 & c_1 & d_1 \\ b_2 & c_2 & d_2 \\ b_3 & c_3 & d_3 \end{vmatrix}$ i.e. the determinant of the coefficients obtained

by covering up the x-column

$$D_y = \begin{vmatrix} a_1 & c_1 & d_1 \\ a_2 & c_2 & d_2 \\ a_3 & c_3 & d_3 \end{vmatrix} \quad \text{i.e. the determinant of the coefficients obtained by}$$

covering up the y-column

$$D_z = \begin{vmatrix} a_1 & b_1 & d_1 \\ a_2 & b_2 & d_2 \\ a_3 & b_3 & d_3 \end{vmatrix} \quad \text{i.e. the determinant of the coefficients obtained by}$$

covering up the z-column

$$\text{and } D = \begin{vmatrix} a_1 & b_1 & c_1 \\ a_2 & b_2 & c_2 \\ a_3 & b_3 & c_3 \end{vmatrix} \quad \text{i.e. the determinant of the coefficients obtained by}$$

covering up the constants-column

Application: A d.c. circuit comprises three closed loops. Applying Kirchhoff's laws to the closed loops gives the following equations for current flow in milliamperes:

$$2I_1 + 3I_2 - 4I_3 = 26$$

$$I_1 - 5I_2 - 3I_3 = -87$$

$$-7I_1 + 2I_2 + 6I_3 = 12$$

Use determinants to solve for I_1, I_2 and I_3

Following the above procedure:

(i) $2I_1 + 3I_2 - 4I_3 - 26 = 0$

     $I_1 - 5I_2 - 3I_3 + 87 = 0$

     $-7I_1 + 2I_2 + 6I_3 - 12 = 0$

(ii) The solution is given by: $\dfrac{I_1}{D_{I_1}} = \dfrac{-I_2}{D_{I_2}} = \dfrac{I_3}{D_{I_3}} = \dfrac{-1}{D}$, where

$$DI_1 = \begin{vmatrix} 3 & -4 & -26 \\ -5 & -3 & 87 \\ 2 & 6 & -12 \end{vmatrix}$$

$$= (3) \begin{vmatrix} -3 & 87 \\ 6 & -12 \end{vmatrix} - (-4) \begin{vmatrix} -5 & 87 \\ 2 & -12 \end{vmatrix} + (-26) \begin{vmatrix} -5 & -3 \\ 2 & 6 \end{vmatrix}$$

$$= 3(-486) + 4(-114) - 26(-24) = -1290$$

$$DI_2 = \begin{vmatrix} 2 & -4 & -26 \\ 1 & -3 & 87 \\ -7 & 6 & -12 \end{vmatrix}$$

$$= (2)(36 - 522) - (-4)(-12 + 609) + (-26)(6 - 21)$$

$$= -972 + 2388 + 390 = 1806$$

$$DI_3 = \begin{vmatrix} 2 & 3 & -26 \\ 1 & -5 & 87 \\ -7 & 2 & -12 \end{vmatrix}$$

$$= (2)(60 - 174) - (3)(-12 + 609) + (-26)(2 - 35)$$

$$= -228 - 1791 + 858 = -1161$$

and

$$D = \begin{vmatrix} 2 & 3 & -4 \\ 1 & -5 & -3 \\ -7 & 2 & 6 \end{vmatrix}$$

$$= (2)(-30 + 6) - (3)(6 - 21) + (-4)(2 - 35)$$

$$= -48 + 45 + 132 = 129$$

Thus

$$\frac{I_1}{-1290} = \frac{-I_2}{1806} = \frac{I_3}{-1161} = \frac{-1}{129}$$

giving: $I_1 = -\dfrac{-1290}{129} = 10\,\text{mA}, \quad I_2 = \dfrac{1806}{129} = 14\,\text{mA}$ and

$I_3 = -\dfrac{-1161}{129} = 9\,\text{mA}$

Chapter 85 Solution of simultaneous equations using Cramer's rule

Cramer's rule states that if

$$a_{11}x + a_{12}y + a_{13}z = b_1$$
$$a_{21}x + a_{22}y + a_{23}z = b_2$$
$$a_{31}x + a_{32}y + a_{33}z = b_3$$

then

$$x = \frac{D_x}{D}, \quad y = \frac{D_y}{D} \quad \text{and} \quad z = \frac{D_z}{D}$$

where

$$D = \begin{vmatrix} a_{11} & a_{12} & a_{13} \\ a_{21} & a_{22} & a_{23} \\ a_{31} & a_{32} & a_{33} \end{vmatrix}$$

$$D_x = \begin{vmatrix} b_1 & a_{12} & a_{13} \\ b_2 & a_{22} & a_{23} \\ b_3 & a_{32} & a_{33} \end{vmatrix}$$ i.e. the x-column has been replaced by the R.H.S. b column

$$D_y = \begin{vmatrix} a_{11} & b_1 & a_{13} \\ a_{21} & b_2 & a_{23} \\ a_{31} & b_3 & a_{33} \end{vmatrix}$$ i.e. the y-column has been replaced by the R.H.S. b column

$$D_z = \begin{vmatrix} a_{11} & a_{12} & b_1 \\ a_{21} & a_{22} & b_2 \\ a_{31} & a_{32} & b_3 \end{vmatrix}$$ i.e. the z-column has been replaced by the R.H.S. b column

Application: Solve the following simultaneous equations using Cramer's rule

$$x + y + z = 4$$
$$2x - 3y + 4z = 33$$
$$3x - 2y - 2z = 2$$

Following the above method:

$$D = \begin{vmatrix} 1 & 1 & 1 \\ 2 & -3 & 4 \\ 3 & -2 & -2 \end{vmatrix} = 1(6 - -8) - 1(-4 - 12) + 1(-4 - -9)$$
$$= 14 + 16 + 5 = 35$$

$$D_x = \begin{vmatrix} 4 & 1 & 1 \\ 33 & -3 & 4 \\ 2 & -2 & -2 \end{vmatrix} = 4(6 - -8) - 1(-66 - 8) + 1(-66 - -6)$$
$$= 56 + 74 - 60 = 70$$

$$D_y = \begin{vmatrix} 1 & 4 & 1 \\ 2 & 33 & 4 \\ 3 & 2 & -2 \end{vmatrix} = 1(-66 - 8) - 4(-4 - 12) + 1(4 - 99)$$
$$= -74 + 64 - 95 = -105$$

$$D_z = \begin{vmatrix} 1 & 1 & 4 \\ 2 & -3 & 33 \\ 3 & -2 & 2 \end{vmatrix} = 1(-6 - -66) - 1(4 - 99) + 4(-4 - -9)$$
$$= 60 + 95 + 20 = 175$$

Hence, $\quad x = \dfrac{D_x}{D} = \dfrac{70}{35} = 2, \quad y = \dfrac{D_y}{D} = \dfrac{-105}{35} = -3 \quad$ and $\quad z = \dfrac{D_z}{D} = \dfrac{175}{35} = 5$

Chapter 86 Solution of simultaneous equations using Gaussian elimination

If

$$a_{11}x + a_{12}y + a_{13}z = b_1 \tag{1}$$

$$a_{21}x + a_{22}y + a_{23}z = b_2 \tag{2}$$

$$a_{31}x + a_{32}y + a_{33}z = b_3 \tag{3}$$

the three-step **procedure** to solve simultaneous equations in three unknowns using the **Gaussian elimination method** is:

(i) Equation (2) $- \dfrac{a_{21}}{a_{11}} \times$ equation (1) to form equation (2′)

and equation (3) $- \dfrac{a_{31}}{a_{11}} \times$ equation (1) to form equation (3′)

(ii) Equation (3′) $- \dfrac{a_{32} \,(\text{of } 3')}{a_{22} \,(\text{of } 2')} \times$ equation (2′) to form equation (3″)

(iii) Determine z from equation (3″), then y from equation (2′) and finally, x from equation (1)

Application: A d.c. circuit comprises three closed loops. Applying Kirchhoff's laws to the closed loops gives the following equations for current flow in milliamperes:

$$2I_1 + 3I_2 - 4I_3 = 26 \tag{1}$$

$$I_1 - 5I_2 - 3I_3 = -87 \tag{2}$$

$$-7I_1 + 2I_2 + 6I_3 = 12 \tag{3}$$

Use the Gaussian elimination method to solve for I_1, I_2 and I_3

Following the above procedure:

(i) $$2I_1 + 3I_2 - 4I_3 = 26 \tag{1}$$

Equation (2) $- \dfrac{1}{2} \times$ equation (1) gives:

$$0 - 6.5I_2 - I_3 = -100 \tag{2'}$$

Equation (3) $- \dfrac{-7}{2} \times$ equation (1) gives:

$$0 + 12.5l_2 - 8l_3 = 103 \qquad \qquad (3')$$

(ii) $\qquad \qquad 2l_1 + 3l_2 - 4l_3 = 26 \qquad \qquad (1)$

$$0 - 6.5l_2 - l_3 = -100 \qquad \qquad (2')$$

Equation (3') $- \dfrac{12.5}{-6.5} \times$ equation (2') gives:

$$0 + 0 - 9.923l_3 = -89.308 \qquad \qquad (3'')$$

(iii) From equation (3''), $l_3 = \dfrac{-89.308}{-9.923} = 9\,mA$,

from equation (2'), $- 6.5l_2 - 9 = -100$,

from which, $l_2 = \dfrac{-100 + 9}{-6.5} = 14\,mA$

and from equation (1), $2l_1 + 3(14) - 4(9) = 26$,

from which, $l_1 = \dfrac{26 - 42 + 36}{2} = \dfrac{20}{2} = 10\,mA$

Chapter 87   Eigenvalues and eigenvectors

In practical applications, such as coupled oscillations and vibrations, equations occur of the form:

$$\mathbf{A\,x} = \lambda\,\mathbf{x}$$

where **A** is a square matrix and λ (Greek lambda) is a number. Whenever $\mathbf{x} \neq \mathbf{0}$, the values of λ are called the eigenvalues of the matrix A; the corresponding solutions of the equation $\mathbf{A\,x} = \lambda\,\mathbf{x}$ are called the eigenvectors of A.

Sometimes, instead of the term *eigenvalues*, characteristic values or latent roots are used. Also, instead of the term *eigenvectors*, characteristic vectors is used.

From above, if $\mathbf{A\,x} = \lambda\,\mathbf{x}$ then $\mathbf{A\,x} - \lambda\,\mathbf{x} = \mathbf{0}$ i.e. $(\mathbf{A} - \lambda\mathbf{I})\mathbf{x} = \mathbf{0}$ where I is the unit matrix.

If $\mathbf{x} \neq \mathbf{0}$ then $|\mathbf{A} - \lambda\mathbf{I}| = 0$

$|\mathbf{A} - \lambda\mathbf{I}|$ is called the characteristic determinant of **A** and $|\mathbf{A} - \lambda\mathbf{I}| = 0$ is called the characteristic equation. Solving the characteristic equation will give the value(s) of the eigenvalues, as demonstrated in the following applications.

Application: Determine the eigenvalues of the matrix $A = \begin{pmatrix} 3 & 4 \\ 2 & 1 \end{pmatrix}$

The eigenvalue is determined by solving the characteristic equation $|\,\mathbf{A} - \lambda \mathbf{I}\,| = \mathbf{0}$

i.e. $\left|\begin{pmatrix} 3 & 4 \\ 2 & 1 \end{pmatrix} - \lambda \begin{pmatrix} 1 & 0 \\ 0 & 1 \end{pmatrix}\right| = 0$ i.e. $\left|\begin{pmatrix} 3 & 4 \\ 2 & 1 \end{pmatrix} - \begin{pmatrix} \lambda & 0 \\ 0 & \lambda \end{pmatrix}\right| = 0$

i.e. $\begin{vmatrix} 3 - \lambda & 4 \\ 2 & 1 - \lambda \end{vmatrix} = 0$

(Given a square matrix, we can get used to going straight to this characteristic equation)

Hence, $(3 - \lambda)(1 - \lambda) - (2)(4) = 0$

i.e. $3 - 3\lambda - \lambda + \lambda^2 - 8 = 0$

and $\lambda^2 - 4\lambda - 5 = 0$

i.e. $(\lambda - 5)(\lambda + 1) = 0$

from which, $\lambda - 5 = 0$ i.e. $\boldsymbol{\lambda = 5}$ or $\lambda + 1 = 0$ i.e. $\boldsymbol{\lambda = -1}$

(Instead of factorising, the quadratic formula could be used; even electronic calculators can solve quadratic equations).

Hence, the eignevalues of the matrix $\begin{pmatrix} 3 & 4 \\ 2 & 1 \end{pmatrix}$ **are 5 and – 1**

Application: Determine the eigenvectors of the matrix $A = \begin{pmatrix} 3 & 4 \\ 2 & 1 \end{pmatrix}$

From above, the eigenvalues of $\begin{pmatrix} 3 & 4 \\ 2 & 1 \end{pmatrix}$ are $\lambda_1 = 5$ and $\lambda_2 = -1$

Using the equation $(\mathbf{A} - \lambda \mathbf{I})\mathbf{x} = \mathbf{0}$ for $\lambda_1 = 5$

then $\begin{pmatrix} 3 - 5 & 4 \\ 2 & 1 - 5 \end{pmatrix}\begin{pmatrix} x_1 \\ x_2 \end{pmatrix} = \begin{pmatrix} 0 \\ 0 \end{pmatrix}$

i.e. $\begin{pmatrix} -2 & 4 \\ 2 & -4 \end{pmatrix}\begin{pmatrix} x_1 \\ x_2 \end{pmatrix} = \begin{pmatrix} 0 \\ 0 \end{pmatrix}$

from which, $-2x_1 + 4x_2 = 0$

and $-2x_1 - 4x_2 = 0$

From either of these two equations, $x_1 = 2x_2$

Hence, whatever value x_2 is, the value of x_1 will be two times greater. Hence the simplest eigenvector is: $x_1 = \begin{pmatrix} 2 \\ 1 \end{pmatrix}$

Using the equation $(A - \lambda I)x = 0$ for $\lambda_2 = -1$

then $\qquad \begin{pmatrix} 3\frac{1}{2}-1 & 4 \\ 1-\frac{1}{-1} \end{pmatrix}\begin{pmatrix} x_1 \\ x_2 \end{pmatrix} = \begin{pmatrix} 0 \\ 0 \end{pmatrix}$

i.e.

$$\begin{pmatrix} 4 & 4 \\ 2 & 2 \end{pmatrix}\begin{pmatrix} x_1 \\ x_2 \end{pmatrix} = \begin{pmatrix} 0 \\ 0 \end{pmatrix}$$

from which,

$$4x_1 + 4x_2 = 0$$

and

$$2x_1 + 2x_2 = 0$$

From either of these two equations, $x_1 = -x_2$ or $x_2 = -x_1$

Hence, whatever value x_1 is, the value of x_2 will be -1 times greater. Hence the simplest eigenvector is: $x_2 = \begin{pmatrix} 1 \\ -1 \end{pmatrix}$

Summarising, $x_1 = \begin{pmatrix} 2 \\ 1 \end{pmatrix}$ is an eigenvector corresponding to $\lambda_1 = 5$

and $\qquad x_2 = \begin{pmatrix} 1 \\ -1 \end{pmatrix}$ is an eigenvector corresponding to $\lambda_2 = -1$

Application: Determine the eigenvalues of the matrix $A = \begin{pmatrix} 1 & 2 & 1 \\ 6 & -1 & 0 \\ -1 & -2 & -1 \end{pmatrix}$

The eigenvalue is determined by solving the characteristic equation $|A - \lambda I| = 0$

i.e. $\qquad \begin{vmatrix} 1-\lambda & 2 & 1 \\ 6 & -1-\lambda & 0 \\ -1 & -2 & -1-\lambda \end{vmatrix} = 0$

Hence, using the top row:

$$(1 - \lambda)[\,(-1 - \lambda)(-1 - \lambda) - (-2)(0)\,] - 2[\,6(-1 - \lambda) - (-1)(0)\,] \\ + 1[\,(6)(-2) - (-1)(-1 - \lambda)\,] = 0$$

i.e. $\qquad (1 - \lambda)[\,1 + \lambda + \lambda + \lambda^2\,] - 2[\,-6 - 6\lambda\,] + 1[\,-12 - 1 - \lambda\,] = 0$

i.e. $\qquad (1 - \lambda)[\,\lambda^2 + 2\lambda + 1\,] + 12 + 12\lambda - 13 - \lambda = 0$

and $\qquad \lambda^2 + 2\lambda + 1 - \lambda^3 - 2\lambda^2 - \lambda + 12 + 12\lambda - 13 - \lambda = 0$

i.e.
$$-\lambda^3 - \lambda^2 + 12\lambda = 0$$

or
$$\lambda^3 + \lambda^2 - 12\lambda = 0$$

i.e.
$$\lambda(\lambda^2 + \lambda - 12) = 0$$

i.e.
$$\lambda(\lambda - 3)(\lambda + 4) = 0 \quad \text{by factorising}$$

from which, $\quad \lambda = 0, \ \lambda = 3 \ $ or $\ \lambda = -4$

Hence, the eignevalues of the matrix $\begin{pmatrix} 1 & 2 & 1 \\ 6 & -1 & 0 \\ -1 & -2 & -1 \end{pmatrix}$ **are 0, 3 and – 4**

Application: Determine the eigenvectors of the matrix $A = \begin{pmatrix} 1 & 2 & 1 \\ 6 & -1 & 0 \\ -1 & -2 & -1 \end{pmatrix}$

From above, the eigenvalues of $\begin{pmatrix} 1 & 2 & 1 \\ 6 & -1 & 0 \\ -1 & -2 & -1 \end{pmatrix}$ are $\lambda_1 = 0, \lambda_2 = 3$ and $\lambda_3 = -4$

Using the equation $\ (A - \lambda I)x = 0 \ $ for $\lambda_1 = 0$

then
$$\begin{pmatrix} 1-0 & 2 & 1 \\ 6 & -1-0 & 0 \\ -1 & -2 & -1-0 \end{pmatrix} \begin{pmatrix} x_1 \\ x_2 \\ x_3 \end{pmatrix} = \begin{pmatrix} 0 \\ 0 \\ 0 \end{pmatrix}$$

i.e.
$$\begin{pmatrix} 1 & 2 & 1 \\ 6 & -1 & 0 \\ -1 & -2 & -1 \end{pmatrix} \begin{pmatrix} x_1 \\ x_2 \\ x_3 \end{pmatrix} = \begin{pmatrix} 0 \\ 0 \\ 0 \end{pmatrix}$$

from which,

$$x_1 + 2x_2 + x_3 = 0$$

$$6x_1 - x_2 = 0$$

$$-x_1 - 2x_2 - x_3 = 0$$

From the second equation,

$$6x_1 = x_2$$

Substituting in the first equation,

$$x_1 + 12x_2 + x_3 = 0 \text{ i.e.} \qquad -13x_1 = x_3$$

Hence, when $x_1 = 1, x_2 = 6$ and $x_3 = -13$

Hence the simplest eigenvector corresponding to $\lambda_1 = 0$ is: $x_1 = \begin{pmatrix} 1 \\ 6 \\ -13 \end{pmatrix}$

Using the equation $(A - \lambda I)x = 0$ for $\lambda_2 = 3$

then

$$\begin{pmatrix} 1-3 & 2 & 1 \\ 6 & -1-3 & 0 \\ -1 & -2 & -1-3 \end{pmatrix} \begin{pmatrix} x_1 \\ x_2 \\ x_3 \end{pmatrix} = \begin{pmatrix} 0 \\ 0 \\ 0 \end{pmatrix}$$

i.e.

$$\begin{pmatrix} -2 & 2 & 1 \\ 6 & -4 & 0 \\ -1 & -2 & -4 \end{pmatrix} \begin{pmatrix} x_1 \\ x_2 \\ x_3 \end{pmatrix} = \begin{pmatrix} 0 \\ 0 \\ 0 \end{pmatrix}$$

from which,

$$- 2\,x_1 + 2\,x_2 + x_3 = 0$$

$$6\,x_1 - 4\,x_3 = 0$$

$$- x_1 + 2\,x_2 + 4\,x_3 = 0$$

From the second equation,

$$3\,x_1 = 2\,x_2$$

Substituting in the first equation,

$$- 2\,x_1 + 3\,x_1 + x_3 = 0 \qquad \text{i.e.} \qquad x_3 = -\,x_1$$

Hence, if $x_2 = 3$, then $x_1 = 2$ and $x_3 = -2$

Hence the simplest eigenvector corresponding to $\lambda_2 = 3$ is: $x_2 = \begin{pmatrix} 2 \\ 3 \\ -2 \end{pmatrix}$

Using the equation $(A - \lambda I)\,x = 0$ for $\lambda_1 = -4$

then

$$\begin{pmatrix} 1--4 & 2 & 1 \\ 6 & -1--4 & 0 \\ -1 & -2 & -1--4 \end{pmatrix} \begin{pmatrix} x_1 \\ x_2 \\ x_3 \end{pmatrix} = \begin{pmatrix} 0 \\ 0 \\ 0 \end{pmatrix}$$

i.e.

$$\begin{pmatrix} 5 & 2 & 1 \\ 6 & 3 & 0 \\ -1 & -2 & 3 \end{pmatrix} \begin{pmatrix} x_1 \\ x_2 \\ x_3 \end{pmatrix} = \begin{pmatrix} 0 \\ 0 \\ 0 \end{pmatrix}$$

from which,

$$5\,x_1 + 2\,x_2 + x_3 = 0$$

$$6\,x_1 + 3\,x_2 = 0$$

$$- x_1 - 2\,x_2 + 3\,x_3 = 0$$

From the second equation,

$$x_2 = -\,2\,x_1$$

Substituting in the first equation,

$$5\,x_1 - 4\,x_1 + x_3 = 0 \qquad \text{i.e.} \qquad x_3 = -\,x_1$$

Hence, if $x_1 = -1$, then $x_2 = 2$ and $x_3 = 1$

Hence the simplest eigenvector corresponding to $\lambda_2 = -4$ is: $x_3 = \begin{pmatrix} -1 \\ 2 \\ 1 \end{pmatrix}$

Section 10

Boolean algebra and logic circuits

Why are Boolean algebra and logic circuits important?

Logic circuits are the basis for modern digital computer systems; to appreciate how computer systems operate an understanding of digital logic and Boolean algebra is needed. Boolean algebra (named after its developer, George Boole), is the algebra of digital logic circuits all computers use. Boolean algebra is the algebra of binary systems. A logic gate is a physical device implementing a Boolean function, performing a logical operation on one or more logic inputs, and produces a single logic output. Logic gates are implemented using diodes or transistors acting as electronic switches, but can also be constructed using electromagnetic relays, fluidic relays, pneumatic relays, optics, molecules or even mechanical elements. Learning Boolean algebra for logic analysis, learning about gates that process logic signals and learning how to design some smaller logic circuits is clearly of importance to computer engineers.

Chapter 88 Boolean algebra and switching circuits

Function	Boolean expression	Equivalent electrical circuit	Truth Table		

Function	Boolean expression	Equivalent electrical circuit	Truth Table
2-input or-function	A + B (i.e. A, or B, or both A and B)		<table><tr><td>1</td><td>2</td><td>3</td></tr><tr><td colspan="2">Input (switches)</td><td>Output (lamp)</td></tr><tr><td>A</td><td>B</td><td>Z = A + B</td></tr><tr><td>0</td><td>0</td><td>0</td></tr><tr><td>0</td><td>1</td><td>1</td></tr><tr><td>1</td><td>0</td><td>1</td></tr><tr><td>1</td><td>1</td><td>1</td></tr></table>
2-input and-function	A.B (i.e. both A and B)		Input (switches) / Output (lamp); A B Z = A·B; 0 0 0; 0 1 0; 1 0 0; 1 1 1
Not-function	$\overline{A}$		Input A / Output $Z = \overline{A}$; 0 1; 1 0
3-input or-function	A + B + C		Input A B C / Output Z = A + B + C; 0 0 0 → 0; 0 0 1 → 1; 0 1 0 → 1; 0 1 1 → 1; 1 0 0 → 1; 1 0 1 → 1; 1 1 0 → 1; 1 1 1 → 1
3-input and-function	A.B.C		Input A B C / Output Z = A·B·C; 0 0 0 → 0; 0 0 1 → 0; 0 1 0 → 0; 0 1 1 → 0; 1 0 0 → 0; 1 0 1 → 0; 1 1 0 → 0; 1 1 1 → 1

To achieve a given output, it is often necessary to use combinations of switches connected both in series and in parallel. If the output from a switching circuit is given by the Boolean expression: $Z = A.B + \overline{A}.\overline{B}$, the truth table is as shown in Figure 88.1(a). In this table, columns 1 and 2 give all the possible combinations of A and B. Column 3 corresponds to A.B and column 4 to $\overline{A}.\overline{B}$ i.e. a 1 output is obtained when A = 0 and when B − 0. Column 5 is the **or**-function applied to columns 3 and 4 giving an output of $Z = A.B + \overline{A}.\overline{B}$. The corresponding switching circuit is shown in Figure 88.1(b) in which A and B are connected in series to give A.B, $\overline{A}$ and $\overline{B}$ are connected in series to give $\overline{A}.\overline{B}$, and A.B and $\overline{A}.\overline{B}$ are connected in parallel to give $A.B + \overline{A}.\overline{B}$. The circuit symbols used are such that A means the switch is on when A is 1, $\overline{A}$ means the switch is on when A is 0, and so on.

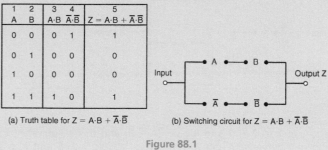

1	2	3	4	5
A	B	A·B	$\overline{A}·\overline{B}$	$Z = A·B + \overline{A}·\overline{B}$
0	0	0	1	1
0	1	0	0	0
1	0	0	0	0
1	1	1	0	1

(a) Truth table for $Z = A·B + \overline{A}·\overline{B}$ (b) Switching circuit for $Z = A·B + \overline{A}·\overline{B}$

Figure 88.1

Application: Derive the Boolean expression and construct a truth table for the switching circuit shown in Figure 88.2.

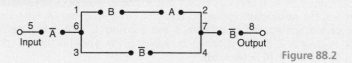

Figure 88.2

The switches between 1 and 2 in Figure 88.2 are in series and have a Boolean expression of B.A. The parallel circuit 1 to 2 and 3 to 4 have a Boolean expression of $(B.A + \overline{B})$. The parallel circuit can be treated as a single switching unit, giving the equivalent of switches 5 to 6, 6 to 7 and 7 to 8 in series. Thus the output is given by: $Z = \overline{A}.(B.A + \overline{B}).\overline{B}$

The truth table is as shown in Table 88.1. Columns 1 and 2 give all the possible combinations of switches A and B. Column 3 is the **and**-function applied to columns 1 and 2, giving B.A. Column 4 is $\overline{B}$, i.e. the opposite to column 2. Column 5 is the **or**-function applied to columns 3 and 4. Column 6 is $\overline{A}$, i.e. the opposite to column 1. The output is column 7 and is obtained by applying the **and**-function to columns 4, 5 and 6.

Table 88.1

1	2	3	4	5	6	7
A	B	B·A	$\bar{B}$	B·A+$\bar{B}$	$\bar{A}$	Z = $\bar{A}$·(B·A+$\bar{B}$)·$\bar{B}$
0	0	0	1	1	1	1
0	1	0	0	0	1	0
1	0	0	1	1	0	0
1	1	1	0	1	0	0

Application: Derive the Boolean expression and construct a truth table for the switching circuit shown in Figure 88.3.

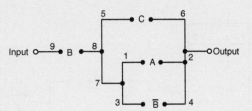

Figure 88.3

The parallel circuit 1 to 2 and 3 to 4 gives $(A + \bar{B})$ and this is equivalent to a single switching unit between 7 and 2. The parallel circuit 5 to 6 and 7 to 2 gives $C + (A + \bar{B})$ and this is equivalent to a single switching unit between 8 and 2. The series circuit 9 to 8 and 8 to 2 gives the output

$$Z = B.[C + (A + \bar{B})]$$

The truth table is shown in Table 88.2. Columns 1, 2 and 3 give all the possible combinations of A, B and C. Column 4 is $\bar{B}$ and is the opposite to column 2. Column 5 is the **or**-function applied to columns 1 and 4, giving $(A + \bar{B})$. Column 6 is the **or**-function applied to columns 3 and 5 giving: $C + (A + \bar{B})$. The output is given in column 7 and is obtained by applying the **and**-function to columns 2 and 6, giving: $Z = B.[C + (A + \bar{B})]$

Table 88.2

1	2	3	4	5	6	7
A	B	C	$\bar{B}$	A + $\bar{B}$	C +(A + $\bar{B}$)	Z = B · [C + (A + $\bar{B}$)]
0	0	0	1	1	1	0
0	0	1	1	1	1	0
0	1	0	0	0	0	0
0	1	1	0	0	1	1
1	0	0	1	1	1	0
1	0	1	1	1	1	0
1	1	0	0	1	1	1
1	1	1	0	1	1	1

Application: Construct a switching circuit to meet the requirements of the Boolean expression:

$$Z = A.\bar{C} + \bar{A}.B + \bar{A}.B.\bar{C}$$

The three terms joined by **or**-functions, (+), indicate three parallel branches, having:

branch 1 A **and** $\bar{C}$ in series
branch 2 $\bar{A}$ **and** B in series
and branch 3 $\bar{A}$ **and** B **and** $\bar{C}$ in series

Hence the required switching circuit is as shown in Figure 88.4.

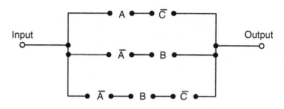

Figure 88.4

The corresponding truth table is shown in Table 88.3.

Column 4 is $\bar{C}$, i.e. the opposite to column 3. Column 5 is A. $\bar{C}$, obtained by applying the **and**-function to columns 1 and 4. Column 6 is $\bar{A}$, the opposite to column 1. Column 7 is $\bar{A}.B$, obtained by applying the **and**-function to columns 2 and 6. Column 8 is $\bar{A}.B.\bar{C}$, obtained by applying the **and**-function to columns 4 and 7. Column 9 is the output, obtained by applying the **or**-function to columns 5, 7 and 8.

Table 88.3

1	2	3	4	5	6	7	8	9
A	B	C	$\bar{C}$	$A \cdot \bar{C}$	$\bar{A}$	$\bar{A} \cdot B$	$\bar{A} \cdot B \cdot \bar{C}$	$Z = A \cdot \bar{C} + \bar{A} \cdot B + \bar{A} \cdot B \cdot \bar{C}$
0	0	0	1	0	1	0	0	0
0	0	1	0	0	1	0	0	0
0	1	0	1	0	1	1	1	1
0	1	1	0	0	1	1	0	1
1	0	0	1	1	0	0	0	1
1	0	1	0	0	0	0	0	0
1	1	0	1	1	0	0	0	1
1	1	1	0	0	0	0	0	0

Chapter 89 Simplifying Boolean expressions

A Boolean expression may be used to describe a complex switching circuit or logic system. If the Boolean expression can be simplified, then the number of switches or logic elements can be reduced resulting in a saving in cost. Three principal ways of simplifying Boolean expressions are:

(a) by using the **laws and rules of Boolean algebra** (see Chapter 90),

(b) by applying **de Morgan's laws** (see Chapter 91), and

(c) by using **Karnaugh maps** (see Chapter 92).

Chapter 90   Laws and rules of Boolean algebra

A summary of the principal laws and rules of Boolean algebra are given in Table 90.1.

Table 90.1

Ref.	Name	Rule or law
1	Commutative laws	$A + B = B + A$
2		$A \cdot B = B \cdot A$
3	Associative laws	$(A + B) + C = A + (B + C)$
4		$(A \cdot B) \cdot C = A \cdot (B \cdot C)$
5	Distributive laws	$A \cdot (B + C) = A \cdot B + A \cdot C$
6		$A + (B \cdot C) = (A + B) \cdot (A + C)$
7	Sum rules	$A + 0 = A$
8		$A + 1 = 1$
9		$A + A = A$
10		$A + \overline{A} = 1$
11	Product rules	$A \cdot 0 = 0$
12		$A \cdot 1 = A$
13		$A \cdot A = A$
14		$A \cdot \overline{A} = 0$
15	Absorption rules	$A + A \cdot B = A$
16		$A \cdot (A + B) = A$
17		$A + \overline{A} \cdot B = A + B$

Application: Simplify the Boolean expression: $\overline{P}.\overline{Q} + \overline{P}.Q + P.\overline{Q}$

With reference to Table 90.1:   **Reference**

$$\overline{P}.\overline{Q} + \overline{P}.Q + P.\overline{Q} = \overline{P}.(\overline{Q} + Q) + P.\overline{Q} \qquad\qquad 5$$
$$= \overline{P}.1 + P.\overline{Q} \qquad\qquad 10$$
$$= \overline{P} + P.\overline{Q} \qquad\qquad 12$$

Application: Simplify $(P + \overline{P}.Q).(Q + \overline{Q}.P)$

With reference to Table 90.1:   **Reference**

$$(P + \overline{P}.Q).(Q + \overline{Q}.P) = P.(Q + \overline{Q}.P) + \overline{P}.Q.(Q + \overline{Q}.P)$$ 5
$$= P.Q + P.\overline{Q}.P + \overline{P}.Q.Q + \overline{P}.Q.\overline{Q}.P$$ 5
$$= P.Q + P.\overline{Q} + \overline{P}.Q + \overline{P}.Q.\overline{Q}.P$$ 13
$$= P.Q + P.\overline{Q} + \overline{P}.Q + 0$$ 14
$$= P.Q + P.\overline{Q} + \overline{P}.Q$$ 7
$$= P.(Q + \overline{Q}) + \overline{P}.Q$$ 5
$$= P.1 + \overline{P}.Q$$ 10
$$= P + \overline{P}.Q$$ 12

Application: Simplify $F.G.\overline{H} + F.G.H + \overline{F}.G.H$

With reference to Table 90.1:   **Reference**

$$F.G.\overline{H} + F.G.H + \overline{F}.G.H = F.G.(\overline{H} + H) + \overline{F}.G.H$$ 5
$$= F.G.1 + \overline{F}.G.H$$ 10
$$= F.G + \overline{F}.G.H$$ 12
$$= G.(F + \overline{F}.H)$$ 5

Application: Simplify $A.\overline{C} + \overline{A}.(B + C) + A.B.(C + \overline{B})$

With reference to Table 90.1:   **Reference**

$$A.\overline{C} + \overline{A}.(B + C) + A.B.(C + \overline{B})$$
$$= A.\overline{C} + \overline{A}.B + \overline{A}.C + A.B.C + A.B.\overline{B}$$ 5
$$= A.\overline{C} + \overline{A}.B + \overline{A}.C + A.B.C + A.0$$ 14
$$= A.\overline{C} + \overline{A}.B + \overline{A}.C + A.B.C$$ 11 and 7
$$= A.(\overline{C} + B.C) + \overline{A}.B + \overline{A}.C$$ 5
$$= A.(\overline{C} + B) + \overline{A}.B + \overline{A}.C$$ 17
$$= A.\overline{C} + A.B + \overline{A}.B + \overline{A}.C$$ 5
$$= A.\overline{C} + B.(A + \overline{A}) + \overline{A}.C$$ 5
$$= A.\overline{C} + B.1 + \overline{A}.C$$ 10
$$= A.\overline{C} + B + \overline{A}.C$$ 12

Chapter 91   De Morgan's laws

De Morgan's laws state that:

$$\overline{A + B} = \overline{A}.\overline{B} \qquad \text{and} \qquad \overline{A.B} = \overline{A} + \overline{B}$$

Application: Simplify the Boolean expression $\overline{\overline{A}.B} + \overline{\overline{A} + B}$ by using de Morgan's laws and the rules of Boolean algebra

Applying de Morgan's law to the first term gives:

$$\overline{\overline{A}.B} = \overline{\overline{A}} + \overline{B} = A + \overline{B} \quad \text{since} \quad \overline{\overline{A}} = A$$

Applying de Morgan's law to the second term gives:

$$\overline{\overline{A} + B} = \overline{\overline{A}}.\overline{B} = A.\overline{B}$$

Thus, $\overline{\overline{A}.B} + \overline{\overline{A} + B} = (A + \overline{B}) + A.\overline{B}$

Removing the bracket and reordering gives: $A + A.\overline{B} + \overline{B}$

But, by rule 15, Table 90.1, $A + A.B = A$. It follows that:

$$A + A.\overline{B} = A$$

Thus: $\overline{\overline{A}.B} + \overline{\overline{A} + B} = A + \overline{B}$

Application: Simplify the Boolean expression $\overline{(A.\overline{B} + C).(\overline{A} + B.\overline{C})}$ by using de Morgan's laws and the rules of Boolean algebra

Applying de Morgan's laws to the first term gives:

$$\overline{(A.\overline{B} + C)} = \overline{A.\overline{B}}.\overline{C} = (\overline{A} + \overline{\overline{B}}).\overline{C} = (\overline{A} + B).\overline{C} = \overline{A}.\overline{C} + B.\overline{C}$$

Applying de Morgan's law to the second term gives:

$$\overline{(\overline{A} + B.\overline{C})} = \overline{\overline{A}} + \overline{(B.\overline{C})} = \overline{A} + (\overline{B} + C)$$

Thus $\overline{(A.\overline{B} + C)}.\overline{(\overline{A} + B.\overline{C})} = (\overline{A}.\overline{C} + B.\overline{C}).(\overline{A} + \overline{B} + C)$
$$= \overline{A}.\overline{A}.\overline{C} + \overline{A}.\overline{B}.\overline{C} + \overline{A}.\overline{C}.C$$
$$+ \overline{A}.B.\overline{C} + B.\overline{B}.\overline{C} + B.\overline{C}.C$$

But from Table 90.1, $\overline{A}.\overline{A} = \overline{A}$ and $\overline{C}.C = B.\overline{B} = 0$

Hence the Boolean expression becomes

$$\overline{A}.\overline{C} + \overline{A}.\overline{B}.\overline{C} + \overline{A}.B.\overline{C} = \overline{A}.\overline{C}(1 + \overline{B} + B)$$
$$= \overline{A}.\overline{C}(1 + B) = \overline{A}.\overline{C}(1) = \overline{A}.\overline{C}$$

Thus: $\overline{(A.\overline{B} + C).(\overline{A} + B.\overline{C})} = \overline{A}.\overline{C}$

Chapter 92   Karnaugh maps

Summary of procedure when simplifying a Boolean expression using a Karnaugh map

1. Draw a four, eight or sixteen-cell matrix, depending on whether there are two, three or four variables.
2. Mark in the Boolean expression by putting 1's in the appropriate cells.
3. Form couples of 8, 4 or 2 cells having common edges, forming the largest groups of cells possible. (Note that a cell containing a 1 may be used more than once when forming a couple. Also note that each cell containing a 1 must be used at least once)
4. The Boolean expression for a couple is given by the variables which are common to all cells in the couple.

(i) Two-variable Karnaugh maps

A truth table for a two-variable expression is shown in Table 92.1(a), the '1' in the third row output showing that $Z = A.\bar{B}$. Each of the four possible Boolean expressions associated with a two-variable function can be depicted as shown in Table 92.1(b) in which one cell is allocated to each row of the truth table. A matrix similar to that shown in Table 92.1(b) can be used to depict $Z = A.\bar{B}$, by putting a 1 in the cell corresponding to $A.\bar{B}$ and 0's in the remaining cells. This method of depicting a Boolean expression is called a two-variable **Karnaugh map**, and is shown in Table 92.1(c).

To simplify a two-variable Boolean expression, the Boolean expression is depicted on a Karnaugh map, as outlined above.

Table 92.1

Inputs		Output	Boolean
A	B	Z	expression
0	0	0	$\bar{A} \cdot \bar{B}$
0	1	0	$\bar{A} \cdot B$
1	0	1	$A \cdot \bar{B}$
1	1	0	$A \cdot B$

(a)

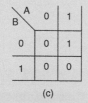

(b)   (c)

Any cells on the map having either a common vertical side or a common horizontal side are grouped together to form a **couple**. (This is a coupling together of cells, not just combining two together). The simplified Boolean expression for a couple is given by those variables common to all cells in the couple.

(ii) Three-variable Karnaugh maps

A truth table for a three-variable expression is shown in Table 92.2(a), the 1's in the output column showing that: $Z = \overline{A}.B.C. + \overline{A}.B.C + A.B.\overline{C}$. Each of the eight possible Boolean expressions associated with a three-variable function can be depicted as shown in Table 92.2(b) in which one cell is allocated to each row of the truth table. A matrix similar to that shown in Table 92.2(b) can be used to depict: $Z = \overline{A}.B.C. + \overline{A}.B.C + A.B.\overline{C}$, by putting 1's in the cells corresponding to the Boolean terms on the right of the Boolean equation and 0's in the remaining cells. This method of depicting a three-variable Boolean expression is called a three-variable Karnaugh map, and is shown in Table 92.2(c).

To simplify a three-variable Boolean expression, the Boolean expression is depicted on a Karnaugh map as outlined above. Any cells on the map having common edges either vertically or horizontally are grouped together to form couples of four cells or two cells. During coupling the horizontal lines at the top and bottom of the cells are taken as a common edge, as are the vertical lines on the left and right of the cells. The simplified Boolean expression for a couple is given by those variables common to all cells in the couple.

Table 92.2

Inputs			Output Z	Boolean expression
A	B	C		
0	0	0	0	$\overline{A}\cdot\overline{B}\cdot\overline{C}$
0	0	1	1	$\overline{A}\cdot\overline{B}\cdot C$
0	1	0	0	$\overline{A}\cdot B\cdot\overline{C}$
0	1	1	1	$\overline{A}\cdot B\cdot C$
1	0	0	0	$A\cdot\overline{B}\cdot\overline{C}$
1	0	1	0	$A\cdot\overline{B}\cdot C$
1	1	0	1	$A\cdot B\cdot\overline{C}$
1	1	1	0	$A\cdot B\cdot C$

(a)

$\begin{array}{c}A\cdot B\\C\end{array}$	00 $(\overline{A}\cdot\overline{B})$	01 $(\overline{A}\cdot B)$	11 $(A\cdot B)$	10 $(A\cdot\overline{B})$
$0(\overline{C})$	$\overline{A}\cdot\overline{B}\cdot\overline{C}$	$\overline{A}\cdot B\cdot\overline{C}$	$A\cdot B\cdot\overline{C}$	$A\cdot\overline{B}\cdot\overline{C}$
$1(C)$	$\overline{A}\cdot\overline{B}\cdot C$	$\overline{A}\cdot B\cdot C$	$A\cdot B\cdot C$	$A\cdot\overline{B}\cdot C$

(b)

$\begin{array}{c}A\cdot B\\C\end{array}$	00	01	11	10
0	0	0	1	0
1	1	1	0	0

(c)

(iii) Four-variable Karnaugh maps

A truth table for a four-variable expression is shown in Table 92.3(a), the 1's in the output column showing that:

$$Z = \overline{A}.\overline{B}.C.\overline{D} + \overline{A}.B.C.\overline{D} + A.\overline{B}.C.\overline{D} + A.B.C.\overline{D}$$

Each of the sixteen possible Boolean expressions associated with a four-variable function can be depicted as shown in Table 92.3(b), in which one cell is allocated to each row of the truth table. A matrix similar to that shown in Table 92.3(b) can be used to depict:

$$Z = \overline{A}.\overline{B}.C.\overline{D} + \overline{A}.\overline{B}.C.D + A.\overline{B}.C.\overline{D} + A.B.C.\overline{D}$$

by putting 1's in the cells corresponding to the Boolean terms on the right of the Boolean equation and 0's in the remaining cells. This method of depicting a four-variable expression is called a four-variable Karnaugh map, and is shown in Table 92.3(c).

Table 92.3

Inputs				Output Z	Boolean expression
A	B	C	D		
0	0	0	0	0	$\overline{A}\cdot\overline{B}\cdot\overline{C}\cdot\overline{D}$
0	0	0	1	0	$\overline{A}\cdot\overline{B}\cdot\overline{C}\cdot D$
0	0	1	0	1	$\overline{A}\cdot\overline{B}\cdot C\cdot\overline{D}$
0	0	1	1	0	$\overline{A}\cdot\overline{B}\cdot C\cdot D$
0	1	0	0	0	$\overline{A}\cdot B\cdot\overline{C}\cdot\overline{D}$
0	1	0	1	0	$\overline{A}\cdot B\cdot\overline{C}\cdot D$
0	1	1	0	1	$\overline{A}\cdot B\cdot C\cdot\overline{D}$
0	1	1	1	0	$\overline{A}\cdot B\cdot C\cdot D$
1	0	0	0	0	$A\cdot\overline{B}\cdot\overline{C}\cdot\overline{D}$
1	0	0	1	0	$A\cdot\overline{B}\cdot\overline{C}\cdot D$
1	0	1	0	1	$A\cdot\overline{B}\cdot C\cdot\overline{D}$
1	0	1	1	0	$A\cdot\overline{B}\cdot C\cdot D$
1	1	0	0	0	$A\cdot B\cdot\overline{C}\cdot\overline{D}$
1	1	0	1	0	$A\cdot B\cdot\overline{C}\cdot D$
1	1	1	0	1	$A\cdot B\cdot C\cdot\overline{D}$
1	1	1	1	0	$A\cdot B\cdot C\cdot D$

(a)

$C\cdot D$ \ $A\cdot B$	00 ($\overline{A}\cdot\overline{B}$)	01 ($\overline{A}\cdot B$)	11 ($A\cdot B$)	10 ($A\cdot\overline{B}$)
00 ($\overline{C}\cdot\overline{D}$)	$\overline{A}\cdot\overline{B}\cdot\overline{C}\cdot\overline{D}$	$\overline{A}\cdot B\cdot\overline{C}\cdot\overline{D}$	$A\cdot B\cdot\overline{C}\cdot\overline{D}$	$A\cdot\overline{B}\cdot\overline{C}\cdot\overline{D}$
01 ($\overline{C}\cdot D$)	$\overline{A}\cdot\overline{B}\cdot\overline{C}\cdot D$	$\overline{A}\cdot B\cdot\overline{C}\cdot D$	$A\cdot B\cdot\overline{C}\cdot D$	$A\cdot\overline{B}\cdot\overline{C}\cdot D$
11 ($C\cdot D$)	$\overline{A}\cdot\overline{B}\cdot C\cdot D$	$\overline{A}\cdot B\cdot C\cdot D$	$A\cdot B\cdot C\cdot D$	$A\cdot\overline{B}\cdot C\cdot D$
10 ($C\cdot\overline{D}$)	$\overline{A}\cdot\overline{B}\cdot C\cdot\overline{D}$	$\overline{A}\cdot B\cdot C\cdot\overline{D}$	$A\cdot B\cdot C\cdot\overline{D}$	$A\cdot\overline{B}\cdot C\cdot\overline{D}$

(b)

$C\cdot D$ \ $A\cdot B$	0.0	0.1	1.1	1.0
0.0	0	0	0	0
0.1	0	0	0	0
1.1	0	0	0	0
1.0	1	1	1	1

(c)

To simplify a four-variable Boolean expression, the Boolean expression is depicted on a Karnaugh map as outlined above. Any cells on the map having common edges either vertically or horizontally are grouped together to form couples of eight cells, four cells or two cells. During coupling, the horizontal lines at the top and bottom of the cells may be considered to be common edges, as are the vertical lines on the left and right of the cells. The simplified Boolean expression for a couple is given by those variables common to all cells in the couple.

Application: Simplify the expression: $\overline{P}.\overline{Q}+\overline{P}.Q$ using Karnaugh map techniques

Using the above procedure:

1. The two-variable matrix is drawn and is shown in Table 92.4.

2. The term $\overline{P}.\overline{Q}$ is marked with a 1 in the top left-hand cell, corresponding to P = 0 and Q = 0; $\overline{P}.Q$ is marked with a 1 in the bottom left-hand cell corresponding to P = 0 and Q = 1.

3. The two cells containing 1's have a common horizontal edge and thus a vertical couple, shown by the broken line, can be formed.

4. The variable common to both cells in the couple is P = 0, i.e. $\overline{P}$ thus

$$\overline{P}.\overline{Q}+\overline{P}.Q = \overline{P}$$

Table 92.4

Q \ P	0	1
0	1 1	0
1	1	0

Application: Simplify $\overline{X}.Y.\overline{Z} + \overline{X}.\overline{Y}.Z + X.Y.\overline{Z} + X.\overline{Y}.Z$ using Karnaugh map techniques

Table 92.5

Z \ X.Y	0.0	0.1	1.1	1.0
0	0	1	1	0
1	1	0	0	1

Using the above procedure:

1. A three-variable matrix is drawn and is shown in Table 92.5.

2. The 1's on the matrix correspond to the expression given, i.e. for $\overline{X}.Y.\overline{Z}$, X = 0, Y = 1 and Z = 0 and hence corresponds to the cell in the top row and second column, and so on.

3. Two couples can be formed, shown by the broken lines. The couple in the bottom row may be formed since the vertical lines on the left and right of the cells are taken as a common edge.

4. The variables common to the couple in the top row are Y = 1 and Z = 0, that is, $Y.\overline{Z}$ and the variables common to the couple in the bottom row are Y = 0, Z = 1, that is, $\overline{Y}.Z$. Hence:

$$\overline{X}.Y.\overline{Z} + \overline{X}.\overline{Y}.Z + X.Y.\overline{Z} + X.\overline{Y}.Z = Y.\overline{Z} + \overline{Y}.Z$$

Application: Simplify $\overline{(P + \overline{Q}.R)} + \overline{(P.Q + \overline{R})}$ using a Karnaugh map technique

The term $(P + \overline{Q}.R)$ corresponds to the cells marked 1 on the matrix in Table 92.6(a), hence $\overline{(P + \overline{Q}.R)}$ corresponds to the cells marked 2. Similarly, $(P.Q + \overline{R})$ corresponds to the cells marked 3 in Table 92.6(a), hence $\overline{(P.Q + \overline{R})}$ corresponds to the cells marked 4. The expression $\overline{(P + \overline{Q}.R)} + \overline{(P.Q + \overline{R})}$ corresponds to cells marked with either a 2 or with a 4 and is shown in Table 92.6(b) by X's. These cells may be coupled as shown by the broken lines. The variables common to the group of four cells is $P = 0$, i.e. $\overline{P}$, and those common to the group of two cells are $Q = 0$, $R = 1$, i.e. $\overline{Q}.R$

Thus: $$\overline{(P + \overline{Q}.R)} + \overline{(P.Q + \overline{R})} = \overline{P} + \overline{Q}.R$$

Table 92.6

P.Q R	0.0	0.1	1.1	1.0
0	3 2	3 2	3 1	3 1
1	4 1	4 2	3 1	4 1

(a)

P.Q R	0.0	0.1	1.1	1.0
0	X	X		
1	X	X		X

(b)

Application: Simplify the expression: $A.B.\overline{C}.\overline{D} + A.B.C.D + \overline{A}.B.C.D + A.B.C.\overline{D} + \overline{A}.B.C.\overline{D}$ using Karnaugh map techniques

Using the procedure, a four-variable matrix is drawn and is shown in Table 92.7. The 1's marked on the matrix correspond to the expression given. Two couples can be formed and are shown by the broken lines. The four-cell couple has $B = 1$,

Table 92.7

A.B C.D	0.0	0.1	1.1	1.0
0.0			1	
0.1				
1.1			1	1
1.0			1	1

$C = 1$, i.e. **B.C** as the common variables to all four cells and the two-cell couple has **A.B.$\overline{\text{D}}$** as the common variables to both cells. Hence, the expression simplifies to: **B.C + A.B.$\overline{\text{D}}$** i.e. **B.(C + A.$\overline{\text{D}}$)**

Chapter 93   Logic circuits and gates

In practice, logic gates are used to perform the **and**, **or** and **not**-functions introduced earlier. Logic gates can be made from switches, magnetic devices or fluidic devices, but most logic gates in use are electronic devices. Various logic gates are available. For example, the Boolean expression (A.B.C) can be produced using a three-input **and**-gate and (C + D) by using a two-input **or**-gate. The principal gates in common use are shown in the table below.

Combinational logic networks

In most logic circuits, more than one gate is needed to give the required output. Except for the **invert**-gate, logic gates generally have two, three or four inputs and are confined to one function only. Thus, for example, a two-input, **or**-gate or a four-input **and**-gate can be used when designing a logic circuit.

Gate type	Traditional symbol	IEC Symbol	Boolean expression	Truth Table			
and-gate			$Z = A.B.C$	INPUTS A B C			OUTPUT $Z = A \cdot B \cdot C$
				0 0 0			0
				0 0 1			0
				0 1 0			0
				0 1 1			0
				1 0 0			0
				1 0 1			0
				1 1 0			0
				1 1 1			1
or-gate			$Z = A + B + C$	INPUTS A B C			OUTPUT $Z = A + B + C$
				0 0 0			0
				0 0 1			1
				0 1 0			1
				0 1 1			1
				1 0 0			1
				1 0 1			1
				1 1 0			1
				1 1 1			1
not-gate or invert-gate			$Z = \overline{A}$	INPUTS A			OUTPUT $Z = \overline{A}$
				0			1
				1			0

Gate type	Traditional symbol	IEC Symbol	Boolean expression	Truth Table
nand-gate			$Z = \overline{A.B.C}$	INPUTS: A B C / $A \cdot B \cdot C$ / OUTPUT $Z = \overline{A \cdot B \cdot C}$ 0 0 0 0 1 0 0 1 0 1 0 1 0 0 1 0 1 1 0 1 1 0 0 0 1 1 0 1 0 1 1 1 0 0 1 1 1 1 1 0
nor-gate			$Z = \overline{A + B + C}$	INPUTS: A B C / $A + B + C$ / OUTPUT $Z = \overline{A + B + C}$ 0 0 0 0 1 0 0 1 1 0 0 1 0 1 0 0 1 1 1 0 1 0 0 1 0 1 0 1 1 0 1 1 0 1 0 1 1 1 1 0
xor-gate			$Z = A \oplus B$	Inputs A B / Output $Z = A\ XOR\ B$ 0 0 0 0 1 1 1 0 1 1 1 0
xnor-gate			$Z = \overline{A \oplus B}$	Inputs A B / Output $Z = A\ XNOR\ B$ 0 0 1 0 1 0 1 0 0 1 1 1

Application: Devise a logic system to meet the requirements of: $Z = A.\overline{B} + C$

With reference to Figure 93.1 an **invert**-gate, shown as (1), gives $\overline{B}$. The **and**-gate, shown as (2), has inputs of A and $\overline{B}$, giving $A.\overline{B}$. The **or**-gate, shown as (3), has inputs of $A.\overline{B}$ and C, giving:

$$Z = A.\ \overline{B} + C$$

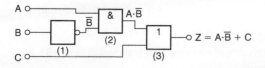

Figure 93.1

Application: Devise a logic system to meet the requirements of $(P + \overline{Q}) . (\overline{R} + S)$

The logic system is shown in Figure 93.2. The given expression shows that two **invert**-functions are needed to give $\overline{Q}$ and $\overline{R}$ and these are shown as gates (1) and (2). Two **or**-gates, shown as (3) and (4), give $(P + \overline{Q})$ and $(\overline{R} + S)$ respectively. Finally, an **and**-gate, shown as (5), gives the required output, $\mathbf{Z = (P + \overline{Q}).(\overline{R} + S)}$

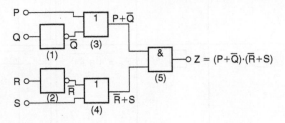

Figure 93.2

Application: Devise a logic circuit to meet the requirements of the output given in Table 93.1, using as few gates as possible

Table 93.1

Inputs			Output
A	B	C	Z
0	0	0	0
0	0	1	0
0	1	0	0
0	1	1	0
1	0	0	0
1	0	1	1
1	1	0	1
1	1	1	1

The '1' outputs in rows 6, 7 and 8 of Table 93.1 show that the Boolean expression is: $Z = A.\overline{B}.C + A.B.\overline{C} + A.B.C$

The logic circuit for this expression can be built using three, 3-input **and**-gates and one, 3-input **or**-gate, together with two **invert**-gates. However, the number of gates required can be reduced by using the techniques introduced earlier, resulting in the cost of the circuit being reduced. Any of the techniques can be used, and in this case, the rules of Boolean algebra (see Table 90.1) are used.

$$Z = A.\overline{B}.C + A.B.\overline{C} + A.B.C = A.[\overline{B}.C + B.\overline{C} + B.C]$$

$$= A.[\overline{B}.C + B(\overline{C} + C)] = A.[\overline{B}.C + B]$$

$$= A.[B + \overline{B}.C] = \mathbf{A.[B + C)]}$$

The logic circuit to give this simplified expression is shown in Figure 93.3.

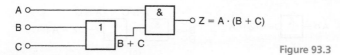

Figure 93.3

Application: Simplify the expression:

$Z = \overline{P}.\overline{Q}.\overline{R}.\overline{S} + \overline{P}.\overline{Q}.\overline{R}.S + \overline{P}.Q.\overline{R}.\overline{S} + \overline{P}.Q.\overline{R}.S + P.\overline{Q}.\overline{R}.\overline{S}$ and devise a logic circuit to give this output

The given expression is simplified using the Karnaugh map techniques introduced earlier. Two couples are formed as shown in Figure 93.4(a) and the simplified expression becomes:

$$Z = \overline{Q}.\overline{R}.\overline{S} + \overline{P}.\overline{R} \text{ i.e. } Z = \overline{R}.(\overline{P} + \overline{Q}.\overline{S})$$

The logic circuit to produce this expression is shown in Figure 93.4(b).

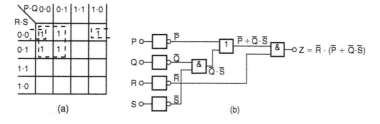

Figure 93.4

Chapter 94   Universal logic gates

The function of any of the five logic gates in common use can be obtained by using either **nand**-gates or **nor**-gates and when used in this manner, the gate selected is called a **universal gate**.

Application: Show how **invert, and, or** and **nor**-functions can be produced using nand-gates only

A single input to a **nand**-gate gives the **invert**-function, as shown in Figure 94.1(a). When two **nand**-gates are connected, as shown in Figure 94.1(b), the output from the first gate is $\overline{A.B.C}$ and this is inverted by the second gate,

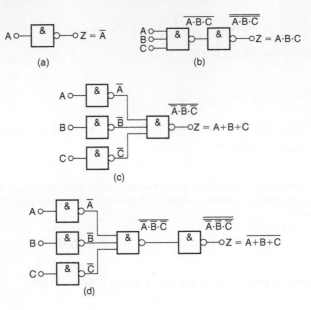

Figure 94.1

giving $Z = \overline{\overline{A.B.C}} = A.B.C$ i.e. the **and**-function is produced. When $\overline{A}$, $\overline{B}$ and $\overline{C}$ are the inputs to a **nand**-gate, the output is $\overline{\overline{A}.\overline{B}.\overline{C}}$

By de Morgan's law, $\overline{\overline{A}.\overline{B}.\overline{C}} = \overline{\overline{A}} + \overline{\overline{B}} + \overline{\overline{C}} = A+B+C$, i.e. a **nand**-gate is used to produce the **or**-function. The logic circuit is shown in Figure 94.1(c). If the output from the logic circuit in Figure 94.1(c) is inverted by adding an additional **nand**-gate, the output becomes the invert of an **or**-function, i.e. the **nor**-function, as shown in Figure 94.1(d).

> **Application:** Show how **invert, or, and** and **nand**-functions can be produced by using **nor**-gates only

A single input to a **nor**-gate gives the **invert**-function, as shown in Figure 94.2(a).

When two **nor**-gates are connected, as shown in Figure 94.2(b), the output from the first gate is $\overline{A + B + C}$ and this is inverted by the second gate, giving $Z = \overline{\overline{A + B + C}} = A + B+C$, i.e. the **or**-function is produced. Inputs of $\overline{A}$, $\overline{B}$ and $\overline{C}$ to a **nor**-gate give an output of $\overline{\overline{A} + \overline{B} + \overline{C}}$

By de Morgan's law, $\overline{\overline{A} + \overline{B} + \overline{C}} = \overline{\overline{A}}.\overline{\overline{B}}.\overline{\overline{C}} = A.B.C$, i.e. the **nor**-gate can be used to produce the **and**-function. The logic circuit is shown in Figure 94.2(c). When the output of the logic circuit, shown in Figure 94.2(c), is inverted by adding an additional **nor**-gate, the output then becomes the invert of an **or**-function, i.e. the **nor**-function as shown in Figure 94.2(d).

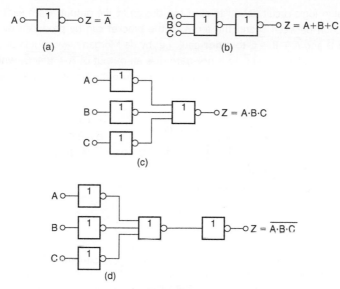

Figure 94.2

> **Application:** Design a logic circuit, using **nand**-gates having not more than three inputs, to meet the requirements of the Boolean expression:
> $Z = \overline{A} + \overline{B} + C + \overline{D}$

When designing logic circuits, it is often easier to start at the output of the circuit. The given expression shows there are four variables joined by **or**-functions. From the principles introduced above, if a four-input **nand**-gate is used to give the expression given, the inputs are $\overline{\overline{A}}, \overline{\overline{B}}, \overline{\overline{C}},$ and $\overline{\overline{D}}$ that is A, B, $\overline{C}$ and D. However, the problem states that three-inputs are not to be exceeded so two of the variables are joined, i.e. the inputs to the three-input **nand**-gate, shown as gate (1) in Figure 94.3, is A.B, $\overline{C}$ and D. From above, the **and**-function is generated by using two **nand**-gates connected in series, as shown by gates (2) and (3) in Figure 94.3. The logic circuit required to produce the given expression is as shown in Figure 94.3.

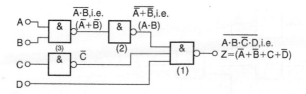

Figure 94.3

> **Application:** Using **nor**-gates only, design a logic circuit to meet the requirements of the expression: $Z = \overline{D}(\overline{A} + B + \overline{C})$

It is usual in logic circuit design to start the design at the output. From earlier, the **and**-function between $\overline{D}$ and the terms in the bracket can be produced by using inputs of $\overline{D}$ and $\overline{A} + B + \overline{C}$ to a **nor**-gate, i.e. by de Morgan's law, inputs of D and $A.\overline{B}.C$. Inputs of $\overline{A}.B$ and $\overline{C}$ to a **nor**-gate give an output of $\overline{\overline{A} + B + \overline{C}}$, which by de Morgan's law is $A.\overline{B}.C$. The logic circuit to produce the required expression is as shown in Figure 94.4.

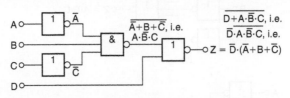

Figure 94.4

Application: An alarm indicator in a grinding mill complex should be activated if (a) the power supply to all mills is off and (b) the hopper feeding the mills is less than 10% full, and (c) if less than two of the three grinding mills are in action. Devise a logic system to meet these requirements

Let variable A represent the power supply on to all the mills, then $\overline{A}$ represents the power supply off. Let B represent the hopper feeding the mills being more than 10% full, then $\overline{B}$ represents the hopper being less than 10% full. Let C, D and E represent the three mills respectively being in action, then $\overline{C}$, $\overline{D}$ and $\overline{E}$ represent the three mills respectively not being in action. The required expression to activate the alarm is: $Z = \overline{A}.\overline{B}(\overline{C} + \overline{D} + \overline{E})$.

There are three variables joined by **and**-functions in the output, indicating that a three-input **and**-gate is required, having inputs of $\overline{A}$, $\overline{B}$ and $(\overline{C} + \overline{D} + \overline{E})$. The term $(\overline{C} + \overline{D} + \overline{E})$ is produced by a three-input **nand**-gate. When variables C, D and E are the inputs to a **nand**-gate, the output is $\overline{C.D.E}$, which, by De Morgan's law is $(\overline{C} + \overline{D} + \overline{E})$. Hence the required logic circuit is as shown in Figure 94.5.

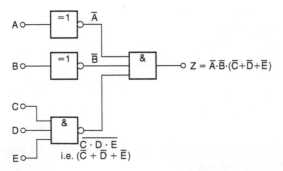

Figure 94.5

Differential calculus and its applications

Why are differential calculus and its applications important?

Calculus is one of the most powerful mathematical tools used by engineers and scientists. Engineers and scientists have to analyse varying quantities, examples including the voltage on a transmission line, the rate of growth of a bacteriological culture, and the rate at which the charge on a capacitor is changing. Rather than using a single equation to define two variables with respect to one another, parametric equations exist as a set that relates the two variables to one another with respect to a third variable.

Parametric equations are useful in defining three-dimensional curves and surfaces, such as determining the velocity or acceleration of a particle following a three-dimensional path. CAD systems use parametric versions of equation. Sometimes in engineering and science, differentiation of parametric equations is necessary, for example, when determining the radius of curvature of part of the surface when finding the surface tension of a liquid.

Engineering applications where implicit differentiation is needed are found in optics, electronics, control, and even some thermodynamics.

Hyperbolic functions have applications in many areas of engineering. For example, the shape formed by a wire freely hanging between two points (known as a catenary curve) is described by the hyperbolic cosine. Hyperbolic functions are also used in electrical engineering applications and for solving differential equations. The hyperbolic sine arises in the gravitational potential of a cylinder, the hyperbolic tangent arises in the calculation of and rapidity of special relativity, the hyperbolic secant arises in the profile of a laminar jet, and the hyperbolic cotangent arises in the Langevin function for magnetic polarisation.

First order partial derivatives can be used for numerous applications, from determining the volume of different shapes to analysing anything from water to heat flow. Second order partial derivatives are used in many fields of engineering and science. One of its applications is in solving problems related to dynamics of rigid bodies and in determination of forces and strength of materials. Partial differentiation is used to estimate errors in calculated quantities that depend on more than one uncertain experimental measurement.

Thermodynamic energy functions (enthalpy, Gibbs free energy, Helmholtz free energy) are functions of two or more variables. Most thermodynamic quantities (temperature, entropy, heat capacity) can be expressed as derivatives of these functions. Many laws of nature are best expressed as relations between the partial derivatives of one or more quantities. Partial differentiation is hence important in many branches of engineering and science.

Chapter 95   Common standard derivatives

A list of some standard derivatives is shown in the table below.

y or f(x)	$\dfrac{dy}{dx}$ or f'(x)
ax^n	anx^{n-1}
$\sin ax$	$a\cos ax$
$\cos ax$	$-a\sin ax$
$\tan ax$	$a\sec^2 ax$
$\sec ax$	$a\sec ax \tan ax$
$\operatorname{cosec} ax$	$-a\operatorname{cosec} ax \cot ax$
$\cot ax$	$-a\operatorname{cosec}^2 ax$
e^{ax}	ae^{ax}
$\ln ax$	$\dfrac{1}{x}$

Application: Differentiate $y = 5x^4 + 4x - \dfrac{1}{2x^2} + \dfrac{1}{\sqrt{x}} - 3$ with respect to x

$y = 5x^4 + 4x - \dfrac{1}{2x^2} + \dfrac{1}{\sqrt{x}} - 3$ is rewritten as: $y = 5x^4 + 4x - \dfrac{1}{2}x^{-2} + x^{-\frac{1}{2}} - 3$

Thus $\dfrac{dy}{dx} = (5)(4)x^{4-1} + (4)(1)x^{1-1} - \dfrac{1}{2}(-2)x^{-2-1} + (1)\left(-\dfrac{1}{2}\right)x^{-\frac{1}{2}-1} - 0$

$= 20x^3 + 4 + x^{-3} - \dfrac{1}{2}x^{-\frac{3}{2}}$

i.e. $\dfrac{dy}{dx} = 20x^3 + 4 + \dfrac{1}{x^3} - \dfrac{1}{2\sqrt{x^3}}$

Application: Find the differential coefficient of $y = 3\sin 4x - 2\cos 3x$

When $y = 3\sin 4x - 2\cos 3x$

then $\dfrac{dy}{dx} = (3)(4\cos 4x) - (2)(-3\sin 3x) = \mathbf{12\cos 4x + 6\sin 3x}$

Application: Determine the derivative of $f(\theta) = \dfrac{2}{e^{3\theta}} + 6\ln 2\theta$

$$f(\theta) = \frac{2}{e^{3\theta}} + 6\ln 2\theta = 2e^{-3\theta} + 6\ln 2\theta$$

Hence, $f'(\theta) = (2)(-3)e^{-3\theta} + 6\left(\dfrac{1}{\theta}\right) = -6e^{-3\theta} + \dfrac{6}{\theta} = \dfrac{-6}{e^{3\theta}} + \dfrac{6}{\theta}$

Application: Find the gradient of the curve $y = 3x^4 - 2x^2 + 5x - 2$ at the point $(1, 4)$

The gradient of a curve at a given point is given by the corresponding value of the derivative.

Thus, since $y = 3x^4 - 2x^2 + 5x - 2$,

then $\quad$ **gradient** $= \dfrac{dy}{dx} = \mathbf{12x^3 - 4x + 5}$

At the point $(1, 4)$, $x = 1$, thus the **gradient** $= 12(1)^3 - 4(1) + 5 = \mathbf{13}$

Chapter 96   Products and quotients

When $y = uv$, and u and v are both functions of x, then:

$$\frac{dy}{dx} = u\frac{dv}{dx} + v\frac{du}{dx}$$

This is called the product rule of differentiation

When $y = \dfrac{u}{v}$, and u and v are both functions of x then:

$$\frac{dy}{dx} = \frac{v\dfrac{du}{dx} - u\dfrac{dv}{dx}}{v^2}$$

This is called the quotient rule of differentiation

Application: Find the differential coefficient of $y = 3x^2 \sin 2x$

$3x^2 \sin 2x$ is a product of two terms $3x^2$ and $\sin 2x$

Let $u = 3x^2$ and $v = \sin 2x$

Using the product rule: $\quad \dfrac{dy}{dx} = u \quad \dfrac{dv}{dx} + v \quad \dfrac{du}{dx}$

$$\qquad\qquad\qquad \downarrow \quad\quad \downarrow \qquad\quad \downarrow \quad \downarrow$$

gives: $\qquad\qquad \dfrac{dy}{dx} = (3x^2)(2\cos 2x) + (\sin 2x)(6x)$

i.e. $\qquad\qquad \dfrac{dy}{dx} = 6x^2\cos 2x + 6x\sin 2x$

$$\qquad\qquad\qquad = 6x\,(x\cos 2x + \sin 2x)$$

Application: Find the differential coefficient of $y = \dfrac{4\sin 5x}{5x^4}$

$\dfrac{4\sin 5x}{5x^4}$ is a quotient. Let $u = 4\sin 5x$ and $v = 5x^4$

$$\frac{dy}{dx} = \frac{v\dfrac{du}{dx} - u\dfrac{dv}{dx}}{v^2} = \frac{(5x^4)(20\cos 5x) - (4\sin 5x)(20x^3)}{(5x^4)^2}$$

$$= \frac{100x^4\cos 5x - 80x^3\sin 5x}{25x^8}$$

$$= \frac{20x^3[5x\cos 5x - 4\sin 5x]}{25x^8}$$

i.e. $\qquad \dfrac{dy}{dx} = \dfrac{4}{5x^5}(5x\cos 5x - 4\sin 5x)$

Application: Determine the differential coefficient of $y = \tan ax$

$y = \tan ax = \dfrac{\sin ax}{\cos ax}$. Differentiation of $\tan ax$ is thus treated as a quotient with $u = \sin ax$ and $v = \cos ax$

$$\frac{dy}{dx} = \frac{v\dfrac{du}{dx} - u\dfrac{dv}{dx}}{v^2} = \frac{(\cos ax)(a\cos ax) - (\sin ax)(-a\sin ax)}{(\cos ax)^2}$$

$$= \frac{a\cos^2 ax + a\sin^2 ax}{(\cos ax)^2} = \frac{a(\cos^2 ax + \sin^2 ax)}{\cos^2 ax}$$

$$= \frac{a}{\cos^2 ax} \qquad \text{since } \cos^2 ax + \sin^2 ax = 1$$

Hence, $\dfrac{dy}{dx} = \mathbf{a\,sec^2\,ax} \qquad$ since $\sec^2 ax = \dfrac{1}{\cos^2 ax}$

Chapter 97   Function of a function

It is often easier to make a substitution before differentiating.

If y is a function of x then: $\dfrac{dy}{dx} = \dfrac{dy}{du} \times \dfrac{du}{dx}$

This is known as the **'function of a function'** rule (or sometimes the **chain rule**).

Application: Differentiate $y = (3x - 1)^9$

If $y = (3x - 1)^9$ then, by making the substitution $u = (3x - 1)$, $y = u^9$, which is of the 'standard' form.

Hence, $\dfrac{dy}{du} = 9u^8$ and $\dfrac{du}{dx} = 3$

Then $\dfrac{dy}{dx} = \dfrac{dy}{du} \times \dfrac{du}{dx} = (9u^8)(3) = 27u^8$

Rewriting u as $(3x - 1)$ gives: $\dfrac{dy}{dx} = \mathbf{27(3x - 1)^8}$

Application: Determine the differential coefficient of $y = \sqrt{3x^2 + 4x - 1}$

$y = \sqrt{3x^2 + 4x - 1} = (3x^2 + 4x - 1)^{1/2}$

Let $u = 3x^2 + 4x - 1$ then $y = u^{1/2}$

Hence $\dfrac{du}{dx} = 6x + 4$ and $\dfrac{dy}{du} = \dfrac{1}{2}u^{-1/2} = \dfrac{1}{2\sqrt{u}}$

Using the function of a function rule,

$$\frac{dy}{dx} = \frac{dy}{du} \times \frac{du}{dx} = \left(\frac{1}{2\sqrt{u}}\right)(6x + 4) = \frac{3x + 2}{\sqrt{u}}$$

i.e. $\dfrac{dy}{dx} = \dfrac{3x + 2}{\sqrt{(3x^2 + 4x - 1)}}$

Chapter 98 Successive differentiation

When a function $y = f(x)$ is differentiated with respect to x the differential coefficient is written as $\dfrac{dy}{dx}$ or $f'(x)$. If the expression is differentiated again, the second differential coefficient is obtained and is written as $\dfrac{d^2y}{dx^2}$ or $f''(x)$. By successive differentiation further higher derivatives such as $\dfrac{d^3y}{dx^3}$ and $\dfrac{d^4y}{dx^4}$ may be obtained.

Thus, if $y = 3x^4$, $\dfrac{dy}{dx} = 12x^3$, $\dfrac{d^2y}{dx^2} = 36x^2$, $\dfrac{d^3y}{dx^3} = 72x$,

$\dfrac{d^4y}{dx^4} = 72$ and $\dfrac{d^5y}{dx^5} = 0$

Application: If $f(x) = 2x^5 - 4x^3 + 3x - 5$ determine $f''(x)$

If $f(x) = 2x^5 - 4x^3 + 3x - 5$

then $\qquad f'(x) = 10x^4 - 12x^2 + 3$

and $\qquad \mathbf{f''(x) = 40x^3 - 24x = 4x(10x^2 - 6)}$

Application: Evaluate $\dfrac{d^2y}{d\theta^2}$ when $\theta = 0$ given $y = 4\sec 2\theta$

Since $y = 4\sec 2\theta$, then $\dfrac{dy}{d\theta} = (4)(2)\sec 2\theta \tan 2\theta$

$$= 8\sec 2\theta \tan 2\theta \qquad \text{(i.e. a product)}$$

$$\frac{d^2y}{d\theta^2} = (8\sec 2\theta)(2\sec^2 2\theta) + (\tan 2\theta)[(8)(2)\sec 2\theta \tan 2\theta]$$

$$= 16\sec^3 2\theta + 16\sec 2\theta \tan^2 2\theta$$

When $\theta = 0$, $\qquad \dfrac{d^2y}{d\theta^2} = 16\sec^3 0 + 16\sec 0 \tan^2 0$

$$= 16(1) + 16(1)(0) = \mathbf{16}$$

Chapter 99   Differentiation of hyperbolic functions

A list of standard derivatives for hyperbolic functions is shown in the table below.

y or f(x)	$\dfrac{dy}{dx}$ or f'(x)
$\sinh ax$	$a\cosh ax$
$\cosh ax$	$a\sinh ax$
$\tanh ax$	$a\,\mathrm{sech}^2 ax$
$\mathrm{sech}\, ax$	$-a\,\mathrm{sech}\, ax \tanh ax$
$\mathrm{cosech}\, ax$	$-a\,\mathrm{cosech}\, ax \coth ax$
$\coth ax$	$-a\,\mathrm{cosech}^2 ax$

Application: Differentiate the following with respect to x:

(a) $y = 4\,\mathrm{sh}\,2x - \dfrac{3}{7}\,\mathrm{ch}\,3x$ (b) $y = 5\,\mathrm{th}\,\dfrac{x}{2} - 2\coth 4x$

(a) $\dfrac{dy}{dx} = 4(2\cosh 2x) - \dfrac{3}{7}(3\sinh 3x) = \mathbf{8\cosh 2x} - \dfrac{9}{7}\sinh 3x$

(b) $\dfrac{dy}{dx} = 5\left(\dfrac{1}{2}\sec h^2 \dfrac{x}{2}\right) - 2(-4\operatorname{cosech}^2 4x)$

$\qquad = \dfrac{5}{2}\sec h^2 \dfrac{x}{2} + 8\operatorname{cosech}^2 4x$

Application: Differentiate the following with respect to the variable:

$\qquad$ (a) $y = 4\sin 3t\, ch\, 4t$ $\qquad$ (b) $y = \ln(sh\, 3\theta) - 4\, ch^2 3\theta$

(a) $\quad y = 4\sin 3t\, ch\, 4t$ $\qquad$ (i.e. a product)

$\dfrac{dy}{dt} = (4\sin 3t)(4\, sh\, 4t) + (ch\, 4t)(4)(3\cos 3t)$

$\qquad = 16\sin 3t\, sh\, 4t + 12\, ch\, 4t\cos 3t$

$\qquad = 4(4\sin 3t\, sh\, 4t + 3\cos 3t\, ch\, 4t)$

(b) $\quad y = \ln(sh\, 3\theta) - 4\, ch^2 3\theta$ $\qquad$ (i.e. a function of a function)

$\dfrac{dy}{d\theta} = \left(\dfrac{1}{sh\, 3\theta}\right)(3\, ch\, 3\theta) - (4)(2\, ch\, 3\theta)(3\, sh\, 3\theta)$

$\qquad = 3\coth 3\theta - 24\, ch\, 3\theta\, sh\, 3\theta = 3(\coth 3\theta - 8\, ch\, 3\theta\, sh\, 3\theta)$

Chapter 100 Rates of change using differentiation

If a quantity y depends on and varies with a quantity x then the rate of change of y with respect to x is $\dfrac{dy}{dx}$. Thus, for example, the rate of change of pressure p with height h is $\dfrac{dp}{dh}$.

A rate of change with respect to time is usually just called 'the rate of change', the 'with respect to time' being assumed. Thus, for example, a rate of change of current, i, is $\dfrac{di}{dt}$ and a rate of change of temperature, θ, is $\dfrac{d\theta}{dt}$, and so on.

Application: Newton's law of cooling is given by $\theta = \theta_0 e^{-kt}$, where the excess of temperature at zero time is $\theta_0\,°C$ and at time t seconds is $\theta\,°C$. Determine the rate of change of temperature after 40 s, given that $\theta_0 = 16°C$ and $k = 0.03$

The rate of change of temperature is $\dfrac{d\theta}{dt}$

Since $\theta = \theta_0 e^{-kt}$ $\quad$ then $\quad \dfrac{d\theta}{dt} = (\theta_0)(-k)e^{-kt} = -k\theta_0 e^{-kt}$

When $\theta_0 = 16$, $k = 0.03$ and $t = 40$

then $\dfrac{d\theta}{dt} = -(0.03)(16)e^{-(0.03)(40)}$

$= -0.48\,e^{-1.2} = \mathbf{-0.145°C/s}$

Application: The luminous intensity I candelas of a lamp at varying voltage V is given by: $I = 4 \times 10^{-4}V^2$. Determine the voltage at which the light is increasing at a rate of 0.6 candelas per volt

---

The rate of change of light with respect to voltage is given by $\dfrac{dI}{dV}$

Since $I = 4 \times 10^{-4}V^2$, $\dfrac{dI}{dV} = (4 \times 10^{-4})(2)V = 8 \times 10^{-4}\,V$

When the light is increasing at 0.6 candelas per volt then

$+ 0.6 = 8 \times 10^{-4}V$, from which,

$$\textbf{voltage V} = \frac{0.6}{8 \times 10^{-4}} = 0.075 \times 10^{+4} = 750 \text{ volts}$$

Chapter 101 Velocity and acceleration

If a body moves a distance x metres in a time t seconds then:

(i) distance, x = f(t)

(ii) velocity, v = f'(t) or $\dfrac{dx}{dt}$, which is the gradient of the distance/time graph

(iii) acceleration, a = $\dfrac{dv}{dt}$ = f''(x) or $\dfrac{d^2x}{dt^2}$, which is the gradient of the velocity/time graph.

Application: The distance x metres travelled by a vehicle in time t seconds after the brakes are applied is given by $x = 20t - \dfrac{5}{3}t^2$. Determine (a) the speed of the vehicle (in km/h) at the instant the brakes are applied, and (b) the distance the car travels before it stops

---

(a) Distance, $x = 20t - \dfrac{5}{3}t^2$. Hence velocity, $v = \dfrac{dx}{dt} = 20 - \dfrac{10}{3}t$

At the instant the brakes are applied, time $= 0$

Hence **velocity, v** $= 20\,\text{m/s} = \dfrac{20 \times 60 \times 60}{1000}$ km/h $= \mathbf{72\,km/h}$

(Note: changing from m/s to km/h merely involves multiplying by 3.6)

(b) When the car finally stops, the velocity is zero,

i.e. $v = 20 - \dfrac{10}{3}t = 0$, from which, $20 = \dfrac{10}{3}t$, giving $t = 6\,s$.

Hence the distance travelled before the car stops is given by:

$x = 20t - \dfrac{5}{3}t^2 = 20(6) - \dfrac{5}{3}(6)^2 = 120 - 60 = 60\,m$

Application: The angular displacement θ radians of a flywheel varies with time t seconds and follows the equation $\theta = 9t^2 - 2t^3$. Determine (a) the angular velocity and acceleration of the flywheel when time, $t = 1\,s$, and (b) the time when the angular acceleration is zero.

(a) Angular displacement $\theta = 9t^2 - 2t^3$ rad

Angular velocity, $\omega = \dfrac{d\theta}{dt} = 18t - 6t^2$ rad/s

When time $t = 1\,s$, $\omega = 18(1) - 6(1)^2 = \mathbf{12\ rad/s}$

Angular acceleration, $\alpha = \dfrac{d^2\theta}{dt^2} = 18 - 12t$ rad/s

When time $t = 1\,s$, $\alpha = 18 - 12(1) = \mathbf{6\ rad/s^2}$

(b) When the angular acceleration is zero, $18 - 12t = 0$, from which, $18 = 12t$, giving time, $\mathbf{t = 1.5\ s}$

Application: The displacement x cm of the slide valve of an engine is given by: $x = 2.2\cos 5\pi t + 3.6\sin 5\pi t$. Evaluate the velocity (in m/s) when time $t = 30\,ms$.

Displacement $x = 2.2\cos 5\pi t + 3.6\sin 5\pi t$

Velocity $v = \dfrac{dx}{dt} = (2.2)(-5\pi)\sin 5\pi t + (3.6)(5\pi)\cos 5\pi t$
$= -11\pi\sin 5\pi t + 18\pi\cos 5\pi t$ cm/s

When time $t = 30\,ms$,

velocity $= -11\pi\sin(5\pi \times 30 \times 10^{-3}) + 18\pi\cos(5\pi \times 30 \times 10^{-3})$

$= -11\pi\sin 0.4712 + 18\pi\cos 0.4712$

$= -11\pi\sin 27° + 18\pi\cos 27°$

$= -15.69 + 50.39$

$= 34.7\,cm/s = \mathbf{0.347\ m/s}$

Chapter 102   Turning points

Procedure for finding and distinguishing between stationary points

(i) Given $y = f(x)$, determine $\dfrac{dy}{dx}$ (i.e. $f'(x)$)

(ii) Let $\frac{dy}{dx} = 0$ and solve for the values of x

(iii) Substitute the values of x into the original equation, y = f(x), to find the corresponding y-ordinate values. This establishes the co-ordinates of the stationary points.

To determine the nature of the stationary points:
Either

(iv) Find $\frac{d^2y}{dx^2}$ and substitute into it the values of x found in (ii).

    If the result is: (a) positive – the point is a minimum one,

                       (b) negative – the point is a maximum one,

                       (c) zero – the point is a point of inflexion

or

(v) Determine the sign of the gradient of the curve just before and just after the stationary points. If the sign change for the gradient of the curve is:

        (a) positive to negative – the point is a maximum one

        (b) negative to positive – the point is a minimum one

        (c) positive to positive or negative to negative – the point is a point of inflexion

Application: Find the maximum and minimum values of the curve $y = x^3 - 3x + 5$

Since $y = x^3 - 3x + 5$ then $\frac{dy}{dx} = 3x^2 - 3$

For a maximum or minimum value $\frac{dy}{dx} = 0$

Hence, $3x^2 - 3 = 0$, from which, $3x^2 = 3$ and $x = \pm 1$

When $x = 1$, $y = (1)^3 - 3(1) + 5 = 3$

When $x = -1$, $y = (-1)^3 - 3(-1) + 5 = 7$

Hence, **(1, 3) and (−1, 7) are the co-ordinates of the turning points**.

Considering the point (1, 3):

If x is slightly less than 1, say 0.9, then $\frac{dy}{dx} = 3(0.9)^2 - 3$, which is negative.

If x is slightly more than 1, say 1.1, then $\frac{dy}{dx} = 3(1.1)^2 - 3$, which is positive.

Since the gradient changes from negative to positive, **the point (1, 3) is a minimum point**.

Considering the point (−1, 7):

If x is slightly less than −1, say −1.1, then $\frac{dy}{dx} = 3(-1.1)^2 - 3$, which is positive.

If x is slightly more than −1, say −0.9, then $\frac{dy}{dx} = 3(-0.9)^2 - 3$, which is negative.

Since the gradient changes from positive to negative, **the point (−1, 7) is a maximum point**.

Since $\dfrac{dy}{dx} = 3x^2 - 3$, then $\dfrac{d^2y}{dx^2} = 6x$

When x = 1, $\dfrac{d^2y}{dx^2}$ is positive, hence (1, 3) is a **minimum value**.

When x = −1, $\dfrac{d^2y}{dx^2}$ is negative, hence (−1, 7) is a **maximum value**.

Thus the maximum value is 7 and the minimum value is 3.

It can be seen that the second differential method of determining the nature of the turning points is, in this case, quicker than investigating the gradient.

Application: Determine the area of the largest piece of rectangular ground that can be enclosed by 100 m of fencing, if part of an existing straight wall is used as one side

Let the dimensions of the rectangle be x and y as shown in Figure 102.1, where PQ represents the straight wall.

From Figure 102.1, $x + 2y = 100$ (1)

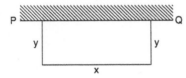

Figure 102.1

Area of rectangle, $A = xy$ (2)

Since the maximum area is required, a formula for area A is needed in terms of one variable only. From equation (1), $x = 100 - 2y$

Hence, area, $A = xy = (100 - 2y)y = 100y - 2y^2$

$\dfrac{dA}{dy} = 100 - 4y = 0$, for a turning point, from which, y = 25 m.

$\dfrac{d^2A}{dy^2} = -4$, which is negative, giving a maximum value.

When y = 25 m, x = 50 m from equation (1).

Hence, the **maximum possible area** = xy = (50)(25) = **1250 m²**

Application: An open rectangular box with square ends is fitted with an overlapping lid which covers the top and the front face. Determine the maximum volume of the box if 6 m² of metal are used in its construction

A rectangular box having square ends of side x and length y is shown in Figure 102.2.

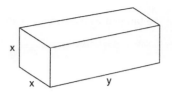

Figure 102.2

Surface area of box, A, consists of two ends and five faces (since the lid also covers the front face).

Hence, $\qquad A = 2x^2 + 5xy = 6$ (1)

Since it is the maximum volume required, a formula for the volume in terms of one variable only is needed. Volume of box, $V = x^2y$

From equation (1), $\qquad y = \dfrac{6 - 2x^2}{5x} = \dfrac{6}{5x} - \dfrac{2x}{5}$ (2)

Hence, volume $V = x^2y = x^2\left(\dfrac{6}{5x} - \dfrac{2x}{5}\right) = \dfrac{6x}{5} - \dfrac{2x^3}{5}$

$\dfrac{dV}{dx} = \dfrac{6}{5} - \dfrac{6x^2}{5} = 0$ for a maximum or minimum value.

Hence, $6 = 6x^2$, giving $x = 1\,\text{m}$ $\qquad$ ($x = -1$ is not possible, and is thus neglected).

$\dfrac{d^2V}{dx^2} = \dfrac{-12x}{5}$. When $x = 1$, $\dfrac{d^2V}{dx^2}$ is negative, giving a maximum value.

From equation (2), $\qquad$ when $x = 1$, $y = \dfrac{6}{5} - \dfrac{2}{5} = \dfrac{4}{5}$

Hence, **the maximum volume of the box,**

$$\mathbf{V} = x^2y = (1)^2\left(\dfrac{4}{5}\right) = \dfrac{4}{5}\,\text{m}^3$$

Chapter 103   Tangents and normals

The **equation of the tangent** to a curve $y = f(x)$ at the point (x_1, y_1) is given by:
$$y - y_1 = m(x - x_1)$$

where $m = \dfrac{dy}{dx} = $ gradient of the curve at (x_1, y_1).

The **equation of the normal** to a curve at the point (x_1, y_1) is given by:
$$y - y_1 = -\dfrac{1}{m}(x - x_1)$$

Application: Find the equation of the tangent to the curve $y = x^2 - x - 2$ at the point $(1, -2)$

Gradient, $m = \dfrac{dy}{dx} = 2x - 1$

At the point $(1, -2)$, $x = 1$ and $m = 2(1) - 1 = 1$

Hence the equation of the tangent is: $y - y_1 = m(x - x_1)$

i.e. $y - -2 = 1(x - 1)$

i.e. $y + 2 = x - 1$

or $\mathbf{y = x - 3}$

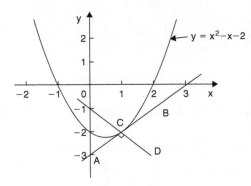

Figure 103.1

The graph of $y = x^2 - x - 2$ is shown in Figure 103.1. The line AB is the tangent to the curve at the point C, i.e. $(1, -2)$, and the equation of this line is $y = x - 3$.

Application: Find the equation of the normal to the curve $y = x^2 - x - 2$ at the point $(1, -2)$

$m = 1$ from above, hence the equation of the normal is:

$$y - -2 = -\frac{1}{1}(x - 1)$$

i.e. $y + 2 = -x + 1$ or $\mathbf{y = -x - 1}$

Thus the line CD in Figure 103.1 has the equation $y = -x - 1$

Chapter 104 Small changes using differentiation

If y is a function of x, i.e. y = f(x), and the approximate change in y corresponding to a small change δx in x is required, then:

$$\frac{\delta y}{\delta x} \approx \frac{dy}{dx}$$

and $\qquad \delta y \approx \dfrac{dy}{dx} \cdot \delta x \qquad$ or $\qquad \delta y \approx f'(x) \cdot \delta x$

Application: The time of swing T of a pendulum is given by $T = k\sqrt{\ell}$, where ℓ is a constant. Determine the percentage change in the time of swing if the length of the pendulum ℓ changes from 32.1 cm to 32.0 cm

If $T = k\sqrt{\ell} = k\ell^{\frac{1}{2}}$, then $\dfrac{dT}{d\ell} = k\left(\dfrac{1}{2}\ell^{-\frac{1}{2}}\right) = \dfrac{k}{2\sqrt{\ell}}$

Approximate change in T, $\delta t \approx \dfrac{dT}{d\ell}\delta\ell \approx \left(\dfrac{k}{2\sqrt{\ell}}\right)\delta\ell$

$$\approx \left(\frac{k}{2\sqrt{\ell}}\right)(-0.1) \qquad \text{(negative}$$
$$\text{since } \ell \text{ decreases)}$$

Percentage error $= \left(\dfrac{\text{approximate change in T}}{\text{original value of T}}\right)100\%$

$$= \frac{\left(\dfrac{k}{2\sqrt{\ell}}\right)(-0.1)}{k\sqrt{\ell}} \times 100\% = \left(\frac{-0.1}{2\ell}\right)100\%$$

$$= \left(\frac{-0.1}{2(32.1)}\right)100\% = -0.156\%$$

Hence, the change in the time of swing is a decrease of 0.156%

Chapter 105   Parametric equations

The following are some of the more **common parametric equations**, and Figure 105.1 shows typical shapes of these curves.

(a) Ellipse $\qquad\qquad x = a\cos\theta, y = b\sin\theta$

(b) Parabola $\qquad\qquad x = at^2, y = 2at$

(c) Hyperbola $\qquad\qquad x = a\sec\theta, y = b\tan\theta$

(d) Rectangular hyperbola $\quad x = ct, y = \dfrac{c}{t}$

(e) Cardioid $\qquad\qquad x = a(2\cos\theta - \cos 2\theta),$
$\qquad\qquad\qquad\qquad y = a(2\sin\theta - \sin 2\theta)$

(f) Astroid $\qquad\qquad x = a\cos^3\theta, y = a\sin^3\theta$

(g) Cycloid $\qquad\qquad x = a(\theta - \sin\theta), y = a(1 - \cos\theta)$

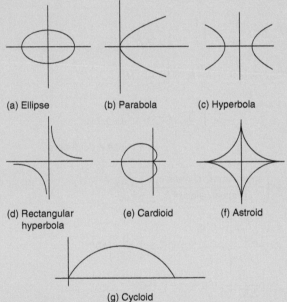

(a) Ellipse $\qquad$ (b) Parabola $\qquad$ (c) Hyperbola

(d) Rectangular $\qquad$ (e) Cardioid $\qquad$ (f) Astroid
hyperbola

(g) Cycloid $\qquad\qquad\qquad\qquad\qquad$ Figure 105.1

Differentiation in parameters

When x and y are both functions of θ, then:

$$\frac{dy}{dx} = \frac{\dfrac{dy}{d\theta}}{\dfrac{dx}{d\theta}} \qquad\qquad (1)$$

and

$$\frac{d^2y}{dx^2} = \frac{\dfrac{d}{d\theta}\left(\dfrac{dy}{dx}\right)}{\dfrac{dx}{d\theta}} \tag{2}$$

Application: Given $x = 5\theta - 1$ and $y = 2\theta(\theta - 1)$, determine $\dfrac{dy}{dx}$ in terms of θ.

$x = 5\theta - 1$, hence $\dfrac{dx}{d\theta} = 5$

$y = 2\theta(\theta - 1) = 2\theta^2 - 2\theta$, hence $\dfrac{dy}{d\theta} = 4\theta - 2 = 2(2\theta - 1)$

From equation (1), $\quad \dfrac{dy}{dx} = \dfrac{\dfrac{dy}{d\theta}}{\dfrac{dx}{d\theta}} = \dfrac{2(2\theta - 1)}{5} \quad$ or $\quad \dfrac{2}{5}(2\theta - 1)$

Application: The parametric equations of a cycloid are $x = 4(\theta - \sin\ \theta)$, $y = 4(1 - \cos\theta)$. Determine

$$\text{(a)}\ \frac{dy}{dx} \qquad \text{(b)}\ \frac{d^2y}{dx^2}$$

(a) $x = 4(\theta - \sin\theta)$, hence $\dfrac{dx}{d\theta} = 4 - 4\cos\theta = 4(1 - \cos\theta)$

$y = 4(1 - \cos\theta)$, hence $\dfrac{dy}{d\theta} = 4\sin\theta$

From equation (1), $\quad \dfrac{dy}{dx} = \dfrac{\dfrac{dy}{d\theta}}{\dfrac{dx}{d\theta}} = \dfrac{4\sin\theta}{4(1 - \cos\theta)} = \dfrac{\sin\theta}{(1 - \cos\theta)}$

(b) From equation (2),

$$\frac{d^2y}{dx^2} = \frac{\dfrac{d}{d\theta}\left(\dfrac{\sin\theta}{1 - \cos\theta}\right)}{4(1 - \cos\theta)} = \frac{\dfrac{(1 - \cos\theta)(\cos\theta) - (\sin\theta)(\sin\theta)}{(1 - \cos\theta)^2}}{4(1 - \cos\theta)}$$

$$= \frac{\cos\theta - \cos^2\theta - \sin^2\theta}{4(1 - \cos\theta)^3} = \frac{\cos\theta - (\cos^2\theta + \sin^2\theta)}{4(1 - \cos\theta)^3}$$

$$= \frac{\cos\theta - 1}{4(1 - \cos\theta)^3} = \frac{-(1 - \cos\theta)}{4(1 - \cos\theta)^3} = \frac{-1}{4(1 - \cos\theta)^2}$$

Application: When determining the surface tension of a liquid, the radius of curvature ρ, of part of the surface is given by:

$$\rho = \frac{\sqrt{\left[1+\left(\dfrac{dy}{dx}\right)^2\right]^3}}{\dfrac{d^2y}{dx^2}}$$

Find the radius of curvature of the part of the surface having the parametric equations

$$x = 3t^2, y = 6t \text{ at the point } t = 2$$

$x = 3t^2$, hence $\dfrac{dx}{dt} = 6t$ and $y = 6t$, hence $\dfrac{dy}{dt} = 6$

From equation (1),

$$\frac{dy}{dx} = \frac{\dfrac{dy}{dt}}{\dfrac{dx}{dt}} = \frac{6}{6t} = \frac{1}{t}$$

From equation (2),

$$\frac{d^2y}{dt^2} = \frac{\dfrac{d}{dt}\left(\dfrac{dy}{dx}\right)}{\dfrac{dx}{dt}} = \frac{\dfrac{d}{dt}\left(\dfrac{1}{t}\right)}{6t} = \frac{\dfrac{d}{dt}(t^{-1})}{6t} = \frac{-t^{-2}}{6t} = \frac{-\dfrac{1}{t^2}}{6t} = \frac{-1}{6t^3}$$

Hence radius of curvature, $\rho = \dfrac{\sqrt{\left[1+\left(\dfrac{dy}{dx}\right)^2\right]^3}}{\dfrac{d^2y}{dx^2}} = \dfrac{\sqrt{\left[1+\left(\dfrac{1}{t}\right)^2\right]^3}}{\dfrac{-1}{6t^3}}$

When t = 2,

$$\rho = \frac{\sqrt{\left[1+\left(\dfrac{1}{2}\right)^2\right]^3}}{-\dfrac{1}{6(2)^3}} = \frac{\sqrt{(1.25)^3}}{-\dfrac{1}{48}} = -48\sqrt{1.25^3} = -67.08$$

Chapter 106   Differentiating implicit functions

$$\frac{d}{dx}[f(y)] = \frac{d}{dy}[f(y)] \times \frac{dy}{dx} \qquad (1)$$

Sometimes with equations involving, say, y and x, it is impossible to make y the subject of the formula. The equation is then called an **implicit function** and examples of such functions include $y^3 + 2x^2 = y^2 - x$ and $\sin y = x^2 + 2xy$

Application: Differentiate $u = \sin 3t$ with respect to x

$$\frac{du}{dx} = \frac{du}{dt} \times \frac{dt}{dx} = \frac{d}{dt}(\sin 3t) \times \frac{dt}{dx} = 3\cos 3t \frac{dt}{dx}$$

Application: Differentiate $u = 4\ln 5y$ with respect to t

$$\frac{du}{dt} = \frac{du}{dy} \times \frac{dy}{dt} = \frac{d}{dy}(4\ln 5y) \times \frac{dy}{dt} = \left(\frac{4}{y}\right)\frac{dy}{dt}$$

Application: Determine $\frac{d}{dx}(x^2 y)$

$$\frac{d}{dx}(x^2 y) = (x^2)\frac{d}{dx}(y) + (y)\frac{d}{dx}(x^2), \qquad \text{by the product rule}$$

$$= (x^2)\left(1\frac{dy}{dx}\right) + y(2x), \qquad \text{by using equation (1)}$$

$$= x^2\frac{dy}{dx} + 2xy$$

Application: Determine $\frac{d}{dx}\left(\frac{3y}{2x}\right)$

$$\frac{d}{dx}\left(\frac{3y}{2x}\right) = \frac{(2x)\dfrac{d}{dx}(3y) - (3y)\dfrac{d}{dx}(2x)}{(2x)^2} = \frac{(2x)\left(3\dfrac{dy}{dx}\right) - (3y)(2)}{4x^2}$$

$$= \frac{6x\dfrac{dy}{dx} - 6y}{4x^2} = \frac{3}{2x^2}\left(x\frac{dy}{dx} - y\right)$$

Application: Given $3x^2 + y^2 - 5x + y = 2$ determine $\dfrac{dy}{dx}$

Differentiating term by term with respect to x gives:

$$\frac{d}{dx}(3x^2) + \frac{d}{dx}(y^2) - \frac{d}{dx}(5x) + \frac{d}{dx}(y) = \frac{d}{dx}(2)$$

i.e. $6x + 2y\dfrac{dy}{dx} - 5 + 1\dfrac{dy}{dx} = 0$ using equation (1) and standard derivatives.

Rearranging gives: $(2y + 1)\dfrac{dy}{dx} = 5 - 6x$

from which, $\dfrac{dy}{dx} = \dfrac{5 - 6x}{2y + 1}$

Application: Determine the values of $\dfrac{dy}{dx}$ when $x = 4$ given that $x^2 + y^2 = 25$

Differentiating each term in turn with respect to x gives:

$$\frac{d}{dx}(x^2) + \frac{d}{dx}(y^2) = \frac{d}{dx}(25) \quad \text{i.e. } 2x + 2y\frac{dy}{dx} = 0$$

Hence $\dfrac{dy}{dx} = -\dfrac{2x}{2y} = -\dfrac{x}{y}$

Since $x^2 + y^2 = 25$, when $x = 4$, $y = \sqrt{(25 - 4^2)} = \pm 3$

Thus, when $x = 4$ and $y = \pm 3$, $\dfrac{dy}{dx} = -\dfrac{4}{\pm 3} = \pm\dfrac{4}{3}$

$x^2 + y^2 = 25$ is the equation of a circle, centre at the origin and radius 5, as shown in Figure 106.1. At $x = 4$, the two gradients are shown.

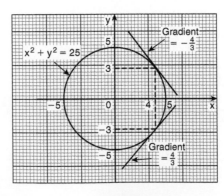

Figure 106.1

Above, $x^2 + y^2 = 25$ was differentiated implicitly; actually, the equation could be transposed to $y = \sqrt{(25 - x^2)}$ and differentiated using the function of a function rule. This gives

$$\frac{dy}{dx} = \frac{1}{2}(25 - x^2)^{-\frac{1}{2}}(-2x) = -\frac{x}{\sqrt{(25 - x^2)}}$$

and when $x = 4$, $\dfrac{dy}{dx} = -\dfrac{4}{\sqrt{(25 - 4^2)}} = \pm\dfrac{4}{3}$ as obtained above.

Chapter 107 Differentiation of logarithmic functions

Logarithmic differentiation is achieved with knowledge of (i) the laws of logarithms, (ii) the differential coefficients of logarithmic functions, and (iii) the differentiation of implicit functions.

(i) **The laws of logarithms** are: 1. $\log(A \times B) = \log A + \log B$

2. $\log\left(\dfrac{A}{B}\right) = \log A - \log B$

3. $\log A^n = n\log A$

(ii) The differential coefficient of the logarithmic function $\ln x$ is given by:

$$\frac{d}{dx}(\ln x) = \frac{1}{x}$$

More generally, it may be shown that:

$$\frac{d}{dx}[\ln f(x)] = \frac{f'(x)}{f(x)} \tag{1}$$

(iii) Differentiation of implicit functions is obtained using:

$$\frac{d}{dx}[f(y)] = \frac{d}{dy}[f(y)] \times \frac{dy}{dx} \tag{2}$$

Application: If $y = \ln(3x^2 + 2x - 1)$ determine $\dfrac{dy}{dx}$

If $y = \ln(3x^2 + 2x - 1)$ then $\dfrac{dy}{dx} = \dfrac{6x + 2}{3x^2 + 2x - 1}$

Application: If $y = \ln(\sin 3x)$ determine $\dfrac{dy}{dx}$

If $y = \ln(\sin 3x)$ then $\dfrac{dy}{dx} = \dfrac{3 \cos 3x}{\sin 3x} = 3 \cot 3x$

Application: Differentiate $y = \dfrac{(1 + x)^2 \sqrt{(x - 1)}}{x\sqrt{(x + 2)}}$ with respect to x

(i) Taking Napierian logarithms of both sides of the equation gives:

$$\ln y = \ln \left\{ \frac{(1 + x)^2 \sqrt{(x - 1)}}{x\sqrt{(x + 2)}} \right\} = \ln \left\{ \frac{(1 + x)^2 (x - 1)^{1/2}}{x(x + 2)^{1/2}} \right\}$$

(ii) Applying the laws of logarithms gives:

$$\ln y = \ln(1 + x)^2 + \ln(x - 1)^{1/2} - \ln x - \ln(x + 2)^{1/2} \text{ by laws 1 and 2}$$

i.e. $\ln y = 2 \ln(1 + x) + \dfrac{1}{2} \ln(x - 1) - \ln x - \dfrac{1}{2} \ln(x + 2)$ by law 3

(iii) Differentiating each term in turn with respect to x using equations (1) and (2) gives:

$$\frac{1}{y} \frac{dy}{dx} = \frac{2}{(1 + x)} + \frac{\frac{1}{2}}{(x - 1)} - \frac{1}{x} - \frac{\frac{1}{2}}{(x + 2)}$$

(iv) Rearranging the equation to make $\dfrac{dy}{dx}$ the subject gives:

$$\frac{dy}{dx} = y \left\{ \frac{2}{(1 + x)} + \frac{1}{2(x - 1)} - \frac{1}{x} - \frac{1}{2(x + 2)} \right\}$$

(v) Substituting for y in terms of x gives:

$$\frac{dy}{dx} = \frac{(1 + x)^2 \sqrt{(x - 1)}}{x\sqrt{(x + 2)}} \left\{ \frac{2}{(1 + x)} + \frac{1}{2(x - 1)} - \frac{1}{x} - \frac{1}{2(x + 2)} \right\}$$

Application: Determine $\dfrac{dy}{dx}$ given $y = x^x$

Taking Napierian logarithms of both sides of $y = x^x$ gives:

$$\ln y = \ln x^x = x \ln x \qquad \text{by law 3}$$

Differentiating both sides with respect to x gives:

$$\frac{1}{y} \frac{dy}{dx} = (x)\left(\frac{1}{x}\right) + (\ln x)(1), \text{ using the product rule}$$

i.e. $\dfrac{1}{y}\dfrac{dy}{dx} = 1 + \ln x$ from which, $\dfrac{dy}{dx} = y(1 + \ln x)$

i.e. $\dfrac{dy}{dx} = x^x(1 + \ln x)$

Application: Determine the differential coefficient of $y = \sqrt[x]{(x-1)}$ and evaluate $\dfrac{dy}{dx}$ when $x = 2$.

$y = \sqrt[x]{(x-1)} = (x-1)^{1/x}$ since by the laws of indices: $\sqrt[n]{a^m} = a^{\frac{m}{n}}$

Taking Napierian logarithms of both sides gives:

$$\ln y = \ln(x-1)^{1/x} = \frac{1}{x}\ln(x-1) \qquad \text{by law 3}$$

Differentiating each side with respect to x gives:

$$\frac{1}{y}\frac{dy}{dx} = \left(\frac{1}{x}\right)\left(\frac{1}{x-1}\right) + [\ln(x-1)]\left(\frac{-1}{x^2}\right) \qquad \text{by the product rule}$$

Hence $\dfrac{dy}{dx} = y\left\{\dfrac{1}{x(x-1)} - \dfrac{\ln(x-1)}{x^2}\right\}$

i.e. $\dfrac{dy}{dx} = \sqrt[x]{(x-1)}\left\{\dfrac{1}{x(x-1)} - \dfrac{\ln(x-1)}{x^2}\right\}$

When $x = 2$, $\dfrac{dy}{dx} = \sqrt[2]{(1)}\left\{\dfrac{1}{2(1)} - \dfrac{\ln(1)}{4}\right\} = \pm1\left\{\dfrac{1}{2} - 0\right\} = \pm\dfrac{1}{2}$

Chapter 108   Differentiation of inverse trigonometric functions

If $y = 3x - 2$, then by transposition, $x = \dfrac{y+2}{3}$. Interchanging x and y terms gives $y = \dfrac{x+2}{3}$ and this is called the **inverse function** of $y = 3x - 2$.

Inverse trigonometric functions are denoted by prefixing the function with $^{-1}$ or 'arc'. For example, if $y = \sin x$, then $x = \sin^{-1}y$ or $x = \arcsin y$. Similarly, if $y = \cos x$, then $y = \cos^{-1}y$ or $x = \arccos y$, and so on. A sketch of each of the inverse trigonometric functions is shown in Figure 108.1.

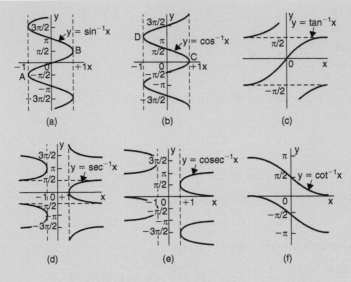

Figure 108.1

Table 108.1   **Differential coefficients of inverse trigonometric functions**

	y or f(x)	$\dfrac{dy}{dx}$ or f'(x)
(i)	$\sin^{-1}\dfrac{x}{a}$	$\dfrac{1}{\sqrt{a^2 - x^2}}$
	$\sin^{-1} f(x)$	$\dfrac{f'(x)}{\sqrt{1 - [f(x)]^2}}$
(ii)	$\cos^{-1}\dfrac{x}{a}$	$\dfrac{-1}{\sqrt{a^2 - x^2}}$
	$\cos^{-1} f(x)$	$\dfrac{-f'(x)}{\sqrt{1 - [f(x)]^2}}$
(iii)	$\tan^{-1}\dfrac{x}{a}$	$\dfrac{a}{a^2 + x^2}$
	$\tan^{-1} f(x)$	$\dfrac{f'(x)}{1 + [f(x)]^2}$
(iv)	$\sec^{-1}\dfrac{x}{a}$	$\dfrac{a}{x\sqrt{x^2 - a^2}}$

$\sec^{-1} f(x)$	$\dfrac{f'(x)}{f(x)\sqrt{[f(x)]^2 - 1}}$
(v) $\operatorname{cosec}^{-1}\dfrac{x}{a}$	$\dfrac{-a}{x\sqrt{x^2 - a^2}}$
$\operatorname{cosec}^{-1} f(x)$	$\dfrac{-f'(x)}{f(x)\sqrt{[f(x)]^2 - 1}}$
(vi) $\cot^{-1}\dfrac{x}{a}$	$\dfrac{-a}{a^2 + x^2}$
$\cot^{-1} f(x)$	$\dfrac{-f'(x)}{1 + [f(x)]^2}$

Application: Find $\dfrac{dy}{dx}$ given $y = \sin^{-1} 5x^2$

From Table 108.1(i), if $y = \sin^{-1} f(x)$ then $\dfrac{dy}{dx} = \dfrac{f'(x)}{\sqrt{1 - [f(x)]^2}}$

Hence, if $y = \sin^{-1} 5x^2$ then $f(x) = 5x^2$ and $f'(x) = 10x$

Thus, $\dfrac{dy}{dx} = \dfrac{10x}{\sqrt{1 - (5x^2)^2}} = \dfrac{10x}{\sqrt{1 - 25x^4}}$

Application: Find the differential coefficient of $y = \ln(\cos^{-1} 3x)$

Let $u = \cos^{-1} 3x$ then $y = \ln u$

By the function of a function rule,

$$\frac{dy}{dx} = \frac{dy}{du} \cdot \frac{du}{dx} = \frac{1}{u} \times \frac{d}{dx}(\cos^{-1} 3x)$$

$$= \frac{1}{\cos^{-1} 3x}\left\{\frac{-3}{\sqrt{1 - (3x)^2}}\right\}$$

i.e. $\dfrac{d}{dx}[\ln(\cos^{-1} 3x)] = \dfrac{-3}{\sqrt{1 - 9x^2}\,\cos^{-1} 3x}$

Application: Find $\dfrac{dy}{dt}$ given $y = \tan^{-1}\dfrac{3}{t^2}$

Using the general form from Table 108.1(iii),

$$f(t) = \frac{3}{t^2} = 3t^{-2} \text{ from which, } f'(t) = \frac{-6}{t^3}$$

Hence, $$\frac{d}{dt}\left(\tan^{-1}\frac{3}{t^2}\right) = \frac{f'(t)}{1+[f(t)]^2} = \frac{-\dfrac{6}{t^3}}{\left\{1+\left(\dfrac{3}{t^2}\right)^2\right\}} = \frac{-\dfrac{6}{t^3}}{\dfrac{t^4+9}{t^4}}$$

$$= \left(-\frac{6}{t^3}\right)\left(\frac{t^4}{t^4+9}\right) = -\frac{6t}{t^4+9}$$

Chapter 109   Differentiation of inverse hyperbolic functions

Inverse hyperbolic functions are denoted by prefixing the function with $^{-1}$ or 'ar'. For example, if $y = \sinh x$, then $x = \sinh^{-1}y$ or $x = \operatorname{arsinh} y$. Similarly, if $y = \operatorname{sech} x$, then $x = \operatorname{sech}^{-1}y$ or $x = \operatorname{arsech} y$, and so on. A sketch of each of the inverse hyperbolic functions is shown in Figure 109.1.

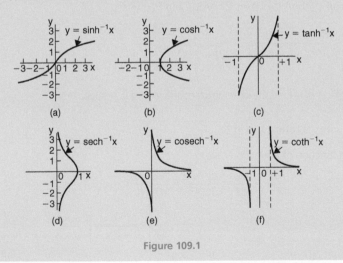

Figure 109.1

Table 109.1   **Differential coefficients of inverse hyperbolic functions**

y or f(x)	$\dfrac{dy}{dx}$ or f'(x)
(i) $\sinh^{-1}\dfrac{x}{a}$	$\dfrac{1}{\sqrt{x^2 + a^2}}$
$\sinh^{-1} f(x)$	$\dfrac{f'(x)}{\sqrt{[f(x)]^2 + 1}}$
(ii) $\cosh^{-1}\dfrac{x}{a}$	$\dfrac{1}{\sqrt{x^2 - a^1}}$
$\cosh^{-1} f(x)$	$\dfrac{f'(x)}{\sqrt{[f(x)]^2 - 1}}$
(iii) $\tanh^{-1}\dfrac{x}{a}$	$\dfrac{a}{a^2 - x^2}$
$\tanh^{-1} f(x)$	$\dfrac{f'(x)}{1 - [f(x)]^2}$
(iv) $\text{sech}^{-1}\dfrac{x}{a}$	$\dfrac{-a}{x\sqrt{a^2 - x^2}}$
$\text{sech}^{-1} f(x)$	$\dfrac{-f'(x)}{f(x)\sqrt{1 - [f(x)]^2}}$
(v) $\text{cosech}^{-1}\dfrac{x}{a}$	$\dfrac{-a}{x\sqrt{x^2 + a^2}}$
$\text{cosech}^{-1} f(x)$	$\dfrac{-f'(x)}{f(x)\sqrt{[f(x)]^2 + 1}}$
(vi) $\coth^{-1}\dfrac{x}{a}$	$\dfrac{a}{a^2 - x^2}$
$\coth^{-1} f(x)$	$\dfrac{f'(x)}{1 - [f(x)]^2}$

Application: Find the differential coefficient of $y = \sinh^{-1} 2x$

From Table 109.1(i), $\dfrac{d}{dx}[\sinh^{-1} f(x)] = \dfrac{f'(x)}{\sqrt{[f(x)]^2 + 1}}$

Hence $\dfrac{d}{dx}(\sinh^{-1}2x) = \dfrac{2}{\sqrt{[(2x)^2+1]}} = \dfrac{2}{\sqrt{[4x^2+1]}}$

Application: Determine $\dfrac{d}{dx}\left[\cosh^{-1}\sqrt{(x^2+1)}\,\right]$

If $y = \cosh^{-1}f(x)$, $\dfrac{dy}{dx} = \dfrac{f'(x)}{\sqrt{\left\{\,[f(x)]^2-1\,\right\}}}$

If $y = \cosh^{-1}\sqrt{(x^2+1)}$, then $f(x) = \sqrt{(x^2+1)}$ and

$f'(x) = \dfrac{1}{2}(x+1)^{-1/2}(2x) = \dfrac{x}{\sqrt{(x^2+1)}}$

Hence $\dfrac{d}{dx}\left[\cosh^{-1}\sqrt{(x^2+1)}\right] = \dfrac{\dfrac{x}{\sqrt{(x^2+1)}}}{\sqrt{\left\{[\sqrt{(x^2+1)}]^2-1\right\}}} = \dfrac{\dfrac{x}{\sqrt{(x^2+1)}}}{\sqrt{(x^2+1-1)}}$

$= \dfrac{\dfrac{x}{\sqrt{(x^2+1)}}}{x} = \dfrac{1}{\sqrt{(x^2+1)}}$

Application: Find the differential coefficient of $y = \text{sech}^{-1}(2x-1)$

From Table 109.1(iv), $\dfrac{d}{dx}[\text{sech}^{-1}f(x)] = \dfrac{-f'(x)}{f(x)\sqrt{1-\left[f(x)\right]^2}}$

Hence

$\dfrac{d}{dx}[\text{sech}^{-1}(2x-1)] = \dfrac{-2}{(2x-1)\sqrt{[1-(2x-1)^2]}}$

$= \dfrac{-2}{(2x-1)\sqrt{[1-(4x^2-4x+1)]}}$

$= \dfrac{-2}{(2x-1)\sqrt{(4x-4x^2)}} = \dfrac{-2}{(2x-1)\sqrt{[4x(1-x)]}}$

$= \dfrac{-2}{(2x-1)\,2\sqrt{[x(1-x)]}} = \dfrac{-1}{(2x-1)\sqrt{[x(1-x)]}}$

Logarithmic forms of the inverse hyperbolic functions

Inverse hyperbolic functions may be evaluated most conveniently when expressed in a **logarithmic form**.

$$\sinh^{-1}\frac{x}{a} = \ln\left\{\frac{x + \sqrt{a^2 + x^2}}{a}\right\} \tag{1}$$

$$\cosh^{-1}\frac{x}{a} = \ln\left\{\frac{x + \sqrt{x^2 - a^2}}{a}\right\} \tag{2}$$

and

$$\tanh^{-1}\frac{x}{a} = \frac{1}{2}\ln\left(\frac{a + x}{a - x}\right) \tag{3}$$

A calculator with inverse hyperbolic functions may also be used to evaluate such functions.

Application: Evaluate $\sinh^{-1}\frac{3}{4}$

To evaluate $\sinh^{-1}\frac{3}{4}$ let $x = 3$ and $a = 4$ in equation (1).

Then, $\sinh^{-1}\frac{3}{4} = \ln\left\{\frac{3 + \sqrt{4^2 + 3^2}}{4}\right\} = \ln\left(\frac{3 + 5}{4}\right) = \ln 2 = 0.6931$

Application: Evaluate, correct to 4 decimal places, $\sinh^{-1}2$

From equation (1), with $x = 2$ and $a = 1$,

$$\sinh^{-1}2 = \ln\left\{\frac{2 + \sqrt{1^2 + 2^2}}{1}\right\} = \ln\left(2 + \sqrt{5}\right) = \ln 4.2361$$

$$= 1.4436, \text{ correct to 4 decimal places}$$

Application: Evaluate $\cosh^{-1}1.4$, correct to 3 decimal places

From equation (2), $\cosh^{-1}\dfrac{x}{a} = \ln\left\{\dfrac{x \pm \sqrt{x^2 - a^2}}{a}\right\}$

and $\cosh^{-1}1.4 = \cosh^{-1}\dfrac{14}{10} = \cosh^{-1}\dfrac{7}{5}$ hence, $x = 7$ and $a = 5$
Then,

$$\cosh^{-1}\dfrac{7}{5} = \ln\left\{\dfrac{7 + \sqrt{7^2 - 5^2}}{5}\right\} = \ln 2.3798$$

$$= \mathbf{0.867}, \text{ correct to 3 decimal places}$$

Application: Evaluate $\tanh^{-1}\dfrac{3}{5}$, correct to 4 decimal places

---

From equation (3),

$$\tanh^{-1}\dfrac{x}{a} = \dfrac{1}{2}\ln\left(\dfrac{a + x}{a - x}\right); \quad \text{substituting } x = 3 \text{ and } a = 5 \text{ gives:}$$

$$\tanh^{-1}\dfrac{3}{5} = \dfrac{1}{2}\ln\left(\dfrac{5 + 3}{5 - 3}\right) = \dfrac{1}{2}\ln 4$$

$$= \mathbf{0.6931}, \text{ correct to 4 decimal places}$$

Each of the above Applications could have been determined using a scientific notation calculator. For example, to evaluate $\tan^{-1}\dfrac{3}{5}$

1. Press 'hyp' **2.** Select 6 **3.** Enter $\dfrac{3}{5}$ **4.** Press '=' and **0.6931** .... appears.

Chapter 110   Partial differentiation

When differentiating a function having two variables, one variable is kept constant and the differential coefficient of the other variable is found with respect to that variable. The differential coefficient obtained is called a **partial derivative** of the function.

First order partial derivatives

If $V = \pi r^2 h$ then $\dfrac{\partial V}{\partial r}$ means 'the partial derivative of V with respect to r, with h remaining constant'

Thus $\dfrac{\partial V}{\partial r} = (\pi h)\dfrac{\mathrm{d}}{\mathrm{d}r}(r^2) = (\pi h)(2r) = 2\pi r h$

Similarly, $\dfrac{\partial V}{\partial h}$ means 'the partial derivative of V with respect to h, with r remaining constant'

Thus $\dfrac{\partial V}{\partial h} = (\pi r^2)\dfrac{d}{dh}(h) = (\pi r^2)(1) = \pi r^2$

Second order partial derivatives

(i) Differentiating $\dfrac{\partial V}{\partial r}$ with respect to r, keeping h constant, gives

$$\dfrac{\partial}{\partial r}\left(\dfrac{\partial V}{\partial r}\right), \text{ which is written as } \dfrac{\partial^2 V}{\partial r^2}$$

Thus if $V = \pi r^2 h$ then $\dfrac{\partial^2 V}{\partial r^2} = \dfrac{\partial}{\partial r}(2\pi rh) = 2\pi h$

(ii) Differentiating $\dfrac{\partial V}{\partial h}$ with respect to h, keeping r constant, gives

$$\dfrac{\partial}{\partial h}\left(\dfrac{\partial V}{\partial h}\right), \text{ which is written as } \dfrac{\partial^2 V}{\partial h^2}$$

Thus $\dfrac{\partial^2 V}{\partial h^2} = \dfrac{\partial}{\partial h}(\pi r^2) = 0$

(iii) Differentiating $\dfrac{\partial V}{\partial h}$ with respect to r, keeping h constant, gives

$$\dfrac{\partial}{\partial r}\left(\dfrac{\partial V}{\partial h}\right), \text{ which is written as } \dfrac{\partial^2 V}{\partial r\,\partial h}$$

Thus $\dfrac{\partial^2 V}{\partial r\,\partial h} = \dfrac{\partial}{\partial r}\left(\dfrac{\partial V}{\partial h}\right) = \dfrac{\partial}{\partial r}(\pi r^2) = 2\pi r$

(iv) Differentiating $\dfrac{\partial V}{\partial r}$ with respect to h, keeping r constant, gives

$$\dfrac{\partial}{\partial h}\left(\dfrac{\partial V}{\partial r}\right), \text{ which is written as } \dfrac{\partial^2 V}{\partial h\,\partial r}$$

Thus $\dfrac{\partial^2 V}{\partial h\,\partial r} = \dfrac{\partial}{\partial h}\left(\dfrac{\partial V}{\partial r}\right) = \dfrac{\partial}{\partial h}(2\pi rh) = 2\pi r$

$\dfrac{\partial^2 V}{\partial r^2}, \dfrac{\partial^2 V}{\partial h^2}, \dfrac{\partial^2 V}{\partial r\,\partial h}$ and $\dfrac{\partial^2 V}{\partial h\,\partial r}$ are examples of **second order partial derivatives**. It is seen from (iii) and (iv) that $\dfrac{\partial^2 V}{\partial r\,\partial h} = \dfrac{\partial^2 V}{\partial h\,\partial r}$ and such a result is always true for continuous functions.

Application: If $Z = 5x^4 + 2x^3y^2 - 3y$ determine $\dfrac{\partial Z}{\partial x}$ and $\dfrac{\partial Z}{\partial y}$

If $Z = 5x^4 + 2x^3y^2 - 3y$

then $\dfrac{\partial Z}{\partial x} = \dfrac{d}{dx}(5x^4) + (2y^2)\dfrac{d}{dx}(x^3) - (3y)\dfrac{d}{dx}(1)$

$= 20x^3 + (2y^2)(3x^2) - (3y)(0) = 20x^3 + 6x^2y^2$

and $\dfrac{\partial Z}{\partial y} = (5x^4)\dfrac{d}{dy}(1) + (2x^3)\dfrac{d}{dy}(y^2) - 3\dfrac{d}{dy}(y)$

$= 0 + (2x^3)(2y) - 3 = 4x^3y - 3$

Application: The time of oscillation, t, of a pendulum is given by: $t = 2\pi\sqrt{\dfrac{l}{g}}$

where l is the length of the pendulum and g the free fall acceleration due to gravity. Find $\dfrac{\partial t}{\partial l}$ and $\dfrac{\partial t}{\partial g}$

To find $\dfrac{\partial t}{\partial l}$, g is kept constant.

$t = 2\pi\sqrt{\dfrac{l}{g}} = \left(\dfrac{2\pi}{\sqrt{g}}\right)\sqrt{l} = \left(\dfrac{2\pi}{\sqrt{g}}\right)l^{1/2}$

Hence, $\dfrac{\partial t}{\partial l} = \left(\dfrac{2\pi}{\sqrt{g}}\right)\dfrac{d}{dl}(l^{1/2}) = \left(\dfrac{2\pi}{\sqrt{g}}\right)\left(\dfrac{1}{2}l^{-1/2}\right) = \left(\dfrac{2\pi}{\sqrt{g}}\right)\left(\dfrac{1}{2\sqrt{l}}\right) = \dfrac{\pi}{\sqrt{lg}}$

To find $\dfrac{\partial t}{\partial g}$, l is kept constant.

$t = 2\pi\sqrt{\dfrac{l}{g}} = (2\pi\sqrt{l})\left(\sqrt{\dfrac{1}{g}}\right) = (2\pi\sqrt{l})g^{-1/2}$

Hence

$$\dfrac{\partial t}{\partial g} = (2\pi\sqrt{l})\left(-\dfrac{1}{2}g^{-3/2}\right) = (2\pi\sqrt{l})\left(\dfrac{-1}{2\sqrt{g^3}}\right) = \dfrac{-\pi\sqrt{l}}{\sqrt{g^3}} = -\pi\sqrt{\dfrac{l}{g^3}}$$

Application: Given $Z = 4x^2y^3 - 2x^3 + 7y^2$ find

(a) $\dfrac{\partial^2 Z}{\partial x^2}$ (b) $\dfrac{\partial^2 Z}{\partial y^2}$ (c) $\dfrac{\partial^2 Z}{\partial x\,\partial y}$ (d) $\dfrac{\partial^2 Z}{\partial y\,\partial x}$

(a) $\dfrac{\partial Z}{\partial x} = 8xy^3 - 6x^2$

$$\dfrac{\partial^2 Z}{\partial x^2} = \dfrac{\partial}{\partial x}\left(\dfrac{\partial Z}{\partial x}\right) = \dfrac{\partial}{\partial x}(8xy^3 - 6x^2) = 8y^3 - 12x$$

(b) $\dfrac{\partial Z}{\partial y} = 12x^2y^2 + 14y$

$$\dfrac{\partial^2 Z}{\partial y^2} = \dfrac{\partial}{\partial y}\left(\dfrac{\partial Z}{\partial y}\right) = \dfrac{\partial}{\partial y}(12x^2y^2 + 14y) = 24x^2y + 14$$

(c) $\dfrac{\partial^2 Z}{\partial x\,\partial y} = \dfrac{\partial}{\partial x}\left(\dfrac{\partial Z}{\partial y}\right) = \dfrac{\partial}{\partial x}(12x^2y^2 + 14y) = 24xy^2$

(d) $\dfrac{\partial^2 Z}{\partial y\,\partial x} = \dfrac{\partial}{\partial y}\left(\dfrac{\partial Z}{\partial x}\right) = \dfrac{\partial}{\partial y}(8xy^3 - 6x^2) = 24xy^2$

Chapter 111   Total differential

If $Z = f(u, v, w, \dots)$, then the **total differential, dZ**, is given by:

$$dZ = \dfrac{\partial Z}{\partial u}\,du + \dfrac{\partial Z}{\partial v}\,dv + \dfrac{\partial Z}{\partial w}\,dw + \dots$$

Application: If $Z = f(u, v, w)$ and $Z = 3u^2 - 2v + 4w^3v^2$ determine the total differential dZ

Total differential, $dZ = \dfrac{\partial Z}{\partial u}\,du + \dfrac{\partial Z}{\partial v}\,dv + \dfrac{\partial Z}{\partial w}\,dw$

$\dfrac{\partial Z}{\partial u} = 6u$ (i.e. v and w are kept constant)

$\dfrac{\partial Z}{\partial v} = -2 + 8w^3v$ (i.e. u and w are kept constant)

$\dfrac{\partial Z}{\partial w} = 12w^2v^2$ (i.e. u and v are kept constant)

Hence, $\mathbf{dZ = 6u\,du + (8vw^3 - 2)dv + (12v^2w^2)dw}$

Application: If $z = f(x, y)$ and $z = x^2y^3 + \dfrac{2x}{y} + 1$, determine the total differential, dz

The total differential is the sum of the partial differentials, i.e. $dz = \dfrac{\partial z}{\partial x} dx + \dfrac{\partial z}{\partial z} dy$

$$\frac{\partial z}{\partial x} = 2xy^3 + \frac{2}{y} \quad \text{(i.e. y is kept constant)}$$

Since $z = x^2 y^3 + \dfrac{2x}{y} + 1 = x^2 y^3 + 2xy^{-1} + 1$

then $\dfrac{\partial z}{\partial x} = x^2 (3y^2) + 2x (-y^2) = 3x^2 y^2 - \dfrac{2x}{y^2}$ (i.e. x is kept constant)

Hence $\mathbf{dz = \left(2xy^3 + \dfrac{2}{y}\right)dx + \left(3x^2y^2 - \dfrac{2x}{y^2} dy\right)}$

Chapter 112   Rates of change using partial differentiation

If $Z = f(u, v, w, \ldots)$ and $\dfrac{du}{dt}, \dfrac{dv}{dt}, \dfrac{dw}{dt}, \ldots$ denote the rate of change of u, v, w,

....respectively, then the rate of change of Z, $\dfrac{dZ}{dt}$, is given by:

$$\frac{dZ}{dt} = \frac{\partial Z}{\partial u}\frac{du}{dt} + \frac{\partial Z}{\partial v}\frac{dv}{dt} + \frac{\partial Z}{\partial w}\frac{dw}{dt} + .. \tag{1}$$

Application: If the height of a right circular cone is increasing at 3 mm/s and its radius is decreasing at 2 mm/s, find the rate at which the volume is changing (in cm^3/s) when the height is 3.2 cm and the radius is 1.5 cm.

Volume of a right circular cone, $V = \dfrac{1}{3}\pi r^2 h$

Using equation (1), the rate of change of volume,

$$\frac{dV}{dt} = \frac{\partial V}{\partial r}\frac{dr}{dt} + \frac{\partial V}{\partial h}\frac{dh}{dt}$$

$\dfrac{\partial V}{\partial r} = \dfrac{2}{3}\pi rh$ and $\dfrac{\partial V}{\partial h} = \dfrac{1}{3}\pi r^2$

Since the height is increasing at 3 mm/s, i.e. 0.3 cm/s, then $\dfrac{dh}{dt} = +0.3$ and since the radius is decreasing at 2 mm/s, i.e. 0.2 cm/s, then $\dfrac{dr}{dt} = -0.2$

Hence, $\dfrac{dV}{dt} = \left(\dfrac{2}{3}\pi rh\right)(-0.2) + \left(\dfrac{1}{3}\pi r^2\right)(+0.3) = \dfrac{-0.4}{3}\pi rh + 0.1\pi r^2$

However, $h = 3.2$ cm and $r = 1.5$ cm.

Hence $\dfrac{dV}{dt} = \dfrac{-0.4}{3}\pi(1.5)(3.2) + (0.1)\pi(1.5)^2$

$= -2.011 + 0.707 = -1.304 \text{ cm}^3/\text{s}$

Thus, the rate of change of volume is 1.30 cm³/s decreasing

Chapter 113　Small changes using partial differentiation

If $Z = f(u, v, w, \ldots)$ and $\delta u, \delta v, \delta w, \ldots$ denote **small changes** in $u, v, w, \ldots$ respectively, then the corresponding approximate change δZ in Z is given by:

$$\delta Z \approx \frac{\partial Z}{\partial u}\,\delta u + \frac{\partial Z}{\partial v}\,\delta v + \frac{\partial Z}{\partial w}\,\delta w + \ldots \qquad (1)$$

Application: If the modulus of rigidity $G = (R^4\theta)/L$, where R is the radius, θ the angle of twist and L the length, find the approximate percentage error in G when R is increased by 2%, θ is reduced by 5% and L is increased by 4%

From equation (1), $\quad \delta G \approx \dfrac{\partial G}{\partial R}\,\delta R + \dfrac{\partial G}{\partial \theta}\,\delta \theta + \dfrac{\partial G}{\partial L}\,\delta L$

Since $\quad G = \dfrac{R^4\theta}{L}, \dfrac{\partial G}{\partial R} = \dfrac{4R^3\theta}{L}, \dfrac{\partial G}{\partial \theta} = \dfrac{R^4}{L} \quad$ and $\quad \dfrac{\partial G}{\partial L} = \dfrac{-R^4\theta}{L^2}$

Since R is increased by 2%, $\delta R = \dfrac{2}{100}R = 0.02\,R$

Similarly, $\quad \delta\theta = -0.05\,\theta \quad$ and $\quad \delta L = 0.04\,L$

Hence $\quad \delta G \approx \left(\dfrac{4R^3\theta}{L}\right)(0.02\,R) + \left(\dfrac{R^4}{L}\right)(-0.05\,\theta) + \left(-\dfrac{R^4\theta}{L^2}\right)(0.04\,L)$

$\approx \dfrac{R^4\theta}{L}[0.08 - 0.05 - 0.04] \approx -0.01\dfrac{R^4\theta}{L}$

i.e. $\quad \delta G \approx -\dfrac{1}{100}G$

Hence the approximate percentage error in G is a 1% decrease.

Application: If the second moment of area I of a rectangle is given by $I = \dfrac{bl^3}{3}$, find the approximate error in the calculated value of I, if b and I are measured as 40 mm and 90 mm respectively and the measurement errors are −5 mm in b and +8 mm in I.

Using equation (1), the approximate error in I, $\delta I \approx \dfrac{\partial I}{\partial b}\,\delta b + \dfrac{\partial I}{\partial I}\,\delta I$

$\dfrac{\partial I}{\partial b} = \dfrac{I^3}{3}$ and $\dfrac{\partial I}{\partial I} = \dfrac{3bI^2}{3} = bI^2$

$\delta b = -5\,mm$ and $\delta I = +8\,mm$

Hence $\delta I \approx \left(\dfrac{I^3}{3}\right)(-5) + (bI^2)(+8)$

Since $b = 40\,mm$ and $I = 90\,mm$ then

$$\delta I \approx \left(\dfrac{90^3}{3}\right)(-5) + 40(90)^2(8)$$

$$\approx -1{,}215{,}000 + 2{,}592{,}000$$

$$\approx 1{,}377{,}000\,mm^4 \approx 137.7\,cm^4$$

Hence, the approximate error in the calculated value of I is a 137.7 cm⁴ increase.

Chapter 114   Maxima, minima and saddle points of functions of two variables

Procedure to determine maxima, minima and saddle points for functions of two variables

Given $z = f(x, y)$:

(i) determine $\dfrac{\partial z}{\partial x}$ and $\dfrac{\partial z}{\partial y}$

(ii) for stationary points, $\dfrac{\partial z}{\partial x} = 0$ and $\dfrac{\partial z}{\partial y} = 0$,

(iii) solve the simultaneous equations $\dfrac{\partial z}{\partial x} = 0$ and $\dfrac{\partial z}{\partial y} = 0$ for x and y, which gives the co-ordinates of the stationary points,

(iv) determine $\dfrac{\partial^2 z}{\partial x^2}, \dfrac{\partial^2 z}{\partial y^2}$ and $\dfrac{\partial^2 z}{\partial x\,\partial y}$

(v) for each of the co-ordinates of the stationary points, substitute values of x and y into $\dfrac{\partial^2 z}{\partial x^2}, \dfrac{\partial^2 z}{\partial y^2}$ and $\dfrac{\partial^2 z}{\partial x\,\partial y}$ and evaluate each,

(vi) evaluate $\left(\dfrac{\partial^2 z}{\partial x\,\partial y}\right)^2$ for each stationary point,

(vii) substitute the values of $\dfrac{\partial^2 z}{\partial x^2}, \dfrac{\partial^2 z}{\partial y^2}$ and $\dfrac{\partial^2 z}{\partial x \, \partial y}$ into the equation

$$\Delta = \left(\frac{\partial^2 z}{\partial x \, \partial y}\right)^2 - \left(\frac{\partial^2 z}{\partial x^2}\right)\left(\frac{\partial^2 z}{\partial y^2}\right) \text{ and evaluate,}$$

(viii) (a) if $\Delta > 0$ then the stationary point is a **saddle point**

(b) if $\Delta < 0$ and $\dfrac{\partial^2 z}{\partial x^2} < 0,$ then the stationary point is a **maximum point**, and

(c) if $\Delta < 0$ and $\dfrac{\partial^2 z}{\partial x^2} < 0,$ then the stationary point is a **minimum point**

Application: Determine the co-ordinates of the stationary point and its nature for the function $z = (x - 1)^2 + (y - 2)^2$

Following the above procedure:

(i) $\dfrac{\partial z}{\partial x} = 2(x - 1)$ and $\dfrac{\partial z}{\partial y} = 2(y - 2)$

(ii) $2(x - 1) = 0$ ⟶ (1)

$2(y - 2) = 0$ ⟶ (2)

(iii) From equations (1) and (2), $x = 1$ and $y = 2$, thus the only stationary point exists at (1, 2)

(iv) Since $\dfrac{\partial z}{\partial x} = 2(x - 1) = 2x - 2,\ \dfrac{\partial^2 z}{\partial x^2} = 2$

and since $\dfrac{\partial z}{\partial y} = 2(y - 2) = 2y - 4,\ \dfrac{\partial^2 z}{\partial y^2} = 2$

and $\dfrac{\partial^2 z}{\partial x \, \partial y} = \dfrac{\partial}{\partial x}\left(\dfrac{\partial z}{\partial y}\right) = \dfrac{\partial}{\partial x}\,(2y - 4) = 0$

(v) $\dfrac{\partial^2 z}{\partial x^2} = \dfrac{\partial^2 z}{\partial y^2} = 2$ and $\dfrac{\partial^2 z}{\partial x \, \partial y} = 0$

(vi) $\left(\dfrac{\partial^2 z}{\partial x \, \partial y}\right)^2 = 0$

(vii) $\Delta = (0)^2 - (2)(2) = -4$

(viii) Since $\Delta < 0$ and $\dfrac{\partial^2 z}{\partial x^2} > 0$, **the stationary point (1, 2) is a minimum.**

The surface $z = (x - 1)^2 + (y - 2)^2$ is shown in three dimensions in Figure 114.1. Looking down towards the x–y plane from above, it is possible to produce a **contour map**. A contour is a line on a map that gives places having the same vertical height above a datum line (usually the mean sea-level on a geographical map). A contour map for $z = (x - 1)^2 + (y - 2)^2$ is shown in Figure 114.2. The values of z are shown on the map and these give an indication of the rise and fall to a stationary point.

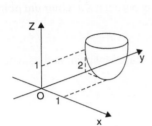

Figure 114.1   Figure 114.2

Application: Find the co-ordinates of the stationary points on the surface $z = (x^2 + y^2)^2 - 8(x^2 - y^2)$, and distinguish between them

Following the procedure:

(i) $\dfrac{\partial z}{\partial x} = 2(x^2 + y^2)2x - 16x$ and $\dfrac{\partial z}{\partial y} = 2(x^2 + y^2)2y + 16y$

(ii) for stationary points,

$$2(x^2 + y^2)2x - 16x = 0$$

i.e. $$4x^3 + 4xy^2 - 16x = 0 \qquad (1)$$

and $$2(x^2 + y^2)2y + 16y = 0$$

i.e. $$4y(x^2 + y^2 + 4) = 0 \qquad (2)$$

(iii) From equation (1), $y^2 = \dfrac{16x - 4x^3}{4x} = 4 - x^2$

Substituting $y^2 = 4 - x^2$ in equation (2) gives

$$4y(x^2 + 4 - x^2 + 4) = 0$$

i.e. $32y = 0$ and $y = 0$

When $y = 0$ in equation (1), $4x^3 - 16x = 0$

i.e. $4x(x^2 - 4) = 0$

from which, $x = 0$ or $x = \pm 2$

The co-ordinates of the stationary points are (0, 0), (2, 0) and (−2, 0)

(iv) $\dfrac{\partial^2 z}{\partial x^2} = 12x^2 + 4y^2 - 16$, $\dfrac{\partial^2 z}{\partial y^2} = 4x^2 + 12y^2 + 16$

and $\dfrac{\partial^2 z}{\partial x\,\partial y} = 8xy$

(v) For the point (0, 0), $\dfrac{\partial^2 z}{\partial x^2} = -16$, $\dfrac{\partial^2 z}{\partial y^2} = 16$ and $\dfrac{\partial^2 z}{\partial x\,\partial y} = 0$

For the point (2, 0), $\dfrac{\partial^2 z}{\partial x^2} = 32$, $\dfrac{\partial^2 z}{\partial y^2} = 32$ and $\dfrac{\partial^2 z}{\partial x\,\partial y} = 0$

For the point (−2, 0), $\dfrac{\partial^2 z}{\partial x^2} = 32$, $\dfrac{\partial^2 z}{\partial y^2} = 32$ and $\dfrac{\partial^2 z}{\partial x\,\partial y} = 0$

(vi) $\left(\dfrac{\partial^2 z}{\partial x\,\partial y}\right)^2 = 0$ for each stationary point

(vii) $\Delta_{(0,0)} = (0)^2 - (-16)(16) = 256$
$\Delta_{(2,0)} = (0)^2 - (32)(32) = -1024$
$\Delta_{(-2,0)} = (0)^2 - (32)(32) = -1024$

(viii) Since $\Delta_{(0,0)} > 0$, **the point (0, 0) is a saddle point**

Since $\Delta_{(2,0)} < 0$ and $\left.\left(\dfrac{\partial^2 z}{\partial x^2}\right)\right|_{(2,0)} > 0$, **the point (2, 0) is a minimum point**

Since $\Delta_{(-2,0)} < 0$ and $\left.\left(\dfrac{\partial^2 z}{\partial x^2}\right)\right|_{(-2,0)} > 0$, **the point (−2, 0) is a minimum point**

Looking down towards the x–y plane from above, an approximate contour map can be constructed to represent the value of z. Such a map is shown in Figure 114.3. To produce a contour map requires a large number of x–y co-ordinates to be chosen and the values of z at each co-ordinate calculated. Here are a few examples of points used to construct the contour map.

When z = 0, $0 = (x^2 + y^2)^2 - 8(x^2 - y^2)$
In addition, when, say, $y = 0$ (i.e. on the x-axis)
$$0 = x^4 - 8x^2 \text{ i.e. } x^2(x^2 - 8) = 0$$
from which, $x = 0$ or $x = \pm\sqrt{8}$

Hence the contour $z = 0$ crosses the x-axis at 0 and $\pm\sqrt{8}$, i.e. at co-ordinates (0, 0), (2.83, 0) and (−2.83, 0) shown as points S, a and b respectively.

When z = 0 and x = 2 then $0 = (4 + y^2)^2 - 8(4 - y^2)$

i.e. $0 = 16 + 8y^2 + y^4 - 32 + 8y^2$

i.e. $0 = y^4 + 16y^2 - 16$

Let $y^2 = p$, then $p^2 + 16p - 16 = 0$

and $p = \dfrac{-16 \pm \sqrt{16^2 - 4(1)(-16)}}{2} = \dfrac{-16 \pm 17.89}{2}$

$= 0.945$ or -16.945

Hence $y = \sqrt{p} = \sqrt{(0.945)}$ or $\sqrt{(-16.945)} = \pm 0.97$ or complex roots

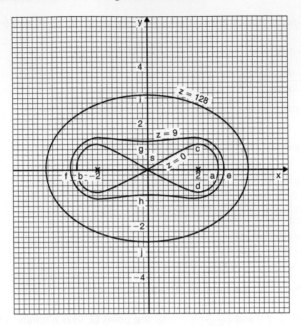

Figure 114.3

Hence the $z = 0$ contour passes through the co-ordinates (2, 0.97) and (2, −0.97) shown as c and d in Figure 114.3.

Similarly, for the **$z = 9$** contour, when $y = 0$,

$$9 = (x^2 + 0^2)^2 - 8(x^2 - 0^2)$$

i.e. $9 = x^4 - 8x^2$

i.e. $x^4 - 8x^2 - 9 = 0$

Hence $(x^2 - 9)(x^2 + 1) = 0$ from which, $x = \pm 3$ or complex roots Thus the $z = 9$ contour passes through (3, 0) and (−3, 0), shown as e and f in Figure 114.3.

If $z = 9$ and $x = 0$, $9 = y^4 + 8y^2$

i.e. $y^4 + 8y^2 - 9 = 0$

i.e. $(y^2 + 9)(y^2 - 1) = 0$

from which, $y = \pm 1$ or complex roots

Thus the $z = 9$ contour also passes through (0, 1) and (0, −1), shown as g and h in Figure 114.3.

When, say, $x = 4$ and $y = 0$, $z = (4^2)^2 - 8(4^2) = 128$

When $z = 128$ and $x = 0$, $128 = y^4 + 8y^2$

i.e. $y^4 + 8y^2 - 128 = 0$

i.e. $(y^2 + 16)(y^2 - 8) = 0$

from which, $y = \pm\sqrt{8}$ or complex roots

Thus the z = 128 contour passes through (0, 2.83) and (0, −2.83), shown as i and j in Figure 114.3.

In a similar manner many other points may be calculated with the resulting approximate contour map shown in Figure 114.3. It is seen that two 'hollows' occur at the minimum points, and a 'cross-over' occurs at the saddle point S, which is typical of such contour maps.

Application: An open rectangular container is to have a volume of $62.5\,m^3$. Find the least surface area of material required

Let the dimensions of the container be x, y and z as shown in Figure 114.4.

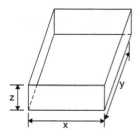

Figure 114.4

$$\text{Volume } V = xyz = 62.5 \tag{1}$$

$$\text{Surface area, } S = xy + 2yz + 2xz \tag{2}$$

From equation (1), $\quad z = \dfrac{62.5}{xy}$

Substituting in equation (2) gives:

$$S = xy + 2y\left(\frac{62.5}{xy}\right) + 2x\left(\frac{62.5}{xy}\right)$$

i.e. $\quad S = xy + \dfrac{125}{x} + \dfrac{125}{y}\quad$ which is a function of two variables

$\dfrac{\partial S}{\partial x} = y - \dfrac{125}{x^2} = 0$ for a stationary point, $\quad$ hence $x^2y = 125 \tag{3}$

$\dfrac{\partial S}{\partial y} = x - \dfrac{125}{y^2} = 0$ for a stationary point, $\quad$ hence $xy^2 = 125 \tag{4}$

Dividing equation (3) by (4) gives: $\dfrac{x^2y}{xy^2} = 1$ i.e. $\dfrac{x}{y} = 1$ $\quad$ i.e. $\quad x = y$

Substituting y = x in equation (3) gives $x^3 = 125$, from which, x = 5 m.

Hence y = 5 m also.

From equation (1), $(5)(5)(z) = 62.5$ from which, $z = \dfrac{62.5}{25} = 2.5\,m$

$$\frac{\partial^2 S}{\partial x^2} = \frac{250}{x^3}, \quad \frac{\partial^2 S}{\partial y^2} = \frac{250}{y^3} \quad \text{and} \quad \frac{\partial^2 S}{\partial x\,\partial y} = 1$$

When $x = y = 5$, $\dfrac{\partial^2 S}{\partial x^2} = 2$, $\dfrac{\partial^2 S}{\partial y^2} = 2$ and $\dfrac{\partial^2 S}{\partial x\,\partial y} = 1$

$\Delta = (1)^2 - (2)(2) = -3$

Since $\Delta < 0$ and $\dfrac{\partial^2 S}{\partial x^2} > 0$, then the surface area S is a **minimum**

Hence the minimum dimensions of the container to have a volume of $62.5\,m^3$ are **5 m by 5 m by 2.5 m**

From equation (2), **minimum surface area,**

$$S = (5)(5) + 2(5)(2.5) + 2(5)(2.5) = \mathbf{75\,m^2}$$

Integral calculus and its applications

Why are integral calculus and its applications important?

Engineering is all about problem solving and many problems in engineering can be solved using calculus. Physicists, chemists, engineers, and many other scientific and technical specialists use calculus in their everyday work; it is a technique of fundamental importance. Both integration and differentiation have numerous applications in engineering and science and some typical examples include determining areas, mean and r.m.s. values, volumes of solids of revolution, centroids, second moments of area, differential equations and Fourier series. Most complex engineering problems cannot be solved without calculus. Calculus can solve many problems for which algebra alone is insufficient. For example, calculus is needed to calculate the force exerted on a particle a specific distance from an electrically charged wire, and is needed for computations involving arc length, centre of mass, work and pressure. Sometimes the integral is not a standard one; in these cases, it may be possible to replace the variable of integration by a function of a new variable. A change in variable can reduce an integral to a standard form.

The algebraic technique of resolving a complicated fraction into partial fractions is often needed by electrical and mechanical engineers for not only determining certain integrals in calculus, but for determining inverse Laplace transforms, and for analysing linear differential equations like resonant circuits and feedback control systems.

Integration by parts is a very important technique that is used often in engineering and science. It is the foundation for the theory of differential equations and is used with Fourier series.

Double and triple integrals have engineering applications in finding areas, masses and forces of two-dimensional regions, and in determining volumes, average values of functions, centres of mass, moments of inertia, and surface areas. A multiple integral is a type of definite integral extended to functions of more than one real variable.

There are two main reasons for why there is a need to do numerical integration – analytical integration may be impossible or infeasible, or it may be necessary to integrate tabulated data rather than known functions. Integration is involved in practically every physical theory in some way – vibration, distortion under weight, or one of many types of fluid flow – be it heat flow, air flow (over a wing), or water flow (over a ship's hull, through a pipe, or perhaps even groundwater flow regarding a contaminant), and so on; all these things can be either directly solved by integration (for simple systems), or some type of numerical integration (for complex systems). Numerical integration is also essential for the evaluation of integrals of functions available only at discrete points; such functions often arise in the numerical solution of differential equations or from experimental data taken at discrete intervals.

One of the important applications of integration is to find the area bounded by a curve. Often such an area can have a physical significance like the work done by a motor, or the distance travelled by a vehicle. Other examples where finding the area under a curve is important can involve position, velocity, force, charge density, resistivity and current density.

Electrical currents and voltages often vary with time and engineers may wish to know the average or mean value of such a current or voltage over some particular time interval. The mean value of a time-varying function is defined in terms of an integral. An associated quantity is the root mean square (r.m.s.) value of a current which is used, for example, in the calculation of the power dissipated by a resistor. Mean and r.m.s. values are required with alternating currents and voltages, pressure of sound waves, and much more.

Revolving a plane figure about an axis generates a volume. The solid generated by rotating a plane area about an axis in its plane is called a solid of revolution, and integration may be used to calculate such a volume. There are many applications of volumes of solids of revolution in engineering and particularly in manufacturing.

The centroid of an area is similar to the centre of mass of a body. Calculating the centroid involves only the geometrical shape of the area; the centre of gravity will equal the centroid if the body has constant density. Centroids of basic shapes can be intuitive – such as the centre of a circle; centroids of more complex shapes can be found using integral calculus – as long as the area, volume or line of an object can be described by a mathematical equation. Centroids are of considerable importance in manufacturing, and in mechanical, civil and structural design engineering.

The second moment of area is a property of a cross-section that can be used to predict the resistance of a beam to bending and deflection around an axis that lies in the cross-sectional plane. The stress in, and deflection of, a beam under load depends not only on the load but also on the geometry of the beam's cross-section; larger values of second moment cause smaller values of stress and deflection. This is why beams with larger second moments of area, such as I-beams, are used in building construction in preference to other beams with the same cross-sectional area. The second moment of area has applications in many scientific disciplines including fluid mechanics, engineering mechanics, and biomechanics – for example to study the structural properties of bone during bending. The static roll stability of a ship depends on the second moment of area of the waterline section – short fat ships are stable, long thin ones are not. Calculations involving the second moment of area are very important in many areas of engineering.

Chapter 115 Standard integrals

Table 115.1 A list of standard integrals

$$\int ax^n \ dx \qquad = \frac{ax^{n+1}}{n+1} + c \qquad \text{(except when } n = -1\text{)}$$

$$\int \cos ax \ dx \qquad = \frac{1}{a}\sin ax + c$$

$$\int \sin ax \ dx \qquad = -\frac{1}{a}\cos ax + c$$

$$\int \sec^2 ax \ dx \qquad = \frac{1}{a}\tan ax + c$$

$$\int \csc^2 ax \ dx \qquad = -\frac{1}{a}\cot ax + c$$

$$\int \csc ax \cot ax \ dx \qquad = -\frac{1}{a}\csc ax + c$$

$$\int \sec ax \tan ax \ dx \qquad = \frac{1}{a}\sec ax + c$$

$$\int e^{ax} \ dx \qquad = \frac{1}{a}e^{ax} + c$$

$$\int \frac{1}{x} dx \qquad = \ln x + c$$

Application: Find $\int 3x^4 \ dx$

$$\int 3x^4 \ dx = \frac{3x^{4+1}}{4+1} + c = \frac{3}{5}x^5 + c$$

Application: Find $\int \frac{2}{x^2} dx$

$$\int \frac{2}{x^2} dx = \int 2x^{-2} \ dx = \frac{2x^{-2+1}}{-2+1} + c = \frac{2x^{-1}}{-1} + c = \frac{-2}{x} + c$$

Application: Find $\int \sqrt{x} \ dx$

$$\int \sqrt{x}\, dx = \int x^{\frac{1}{2}}\, dx = \frac{x^{\frac{1}{2}+1}}{\frac{1}{2}+1} + c = \frac{x^{\frac{3}{2}}}{\frac{3}{2}} + c = \frac{2}{3}\sqrt{x^3} + c$$

Application: Find $\int \dfrac{-5}{9\sqrt[4]{t^3}}\, dt$

$$\int \frac{-5}{9\sqrt[4]{t^3}}\, dt = \int \frac{-5}{9t^{\frac{3}{4}}}\, dt = -\frac{5}{9}\int t^{-\frac{3}{4}}\, dt$$

$$= -\frac{5}{9}\left[\frac{t^{-\frac{3}{4}+1}}{-\frac{3}{4}+1}\right] + c = -\frac{5}{9}\left(\frac{t^{\frac{1}{4}}}{\frac{1}{4}}\right) + c$$

$$= -\left(\frac{5}{9}\right)\left(\frac{4}{1}\right)t^{\frac{1}{4}} + c = -\frac{20}{9}\sqrt[4]{t} + c$$

Application: Find $\int \dfrac{(1+\theta)^2}{\sqrt{\theta}}\, d\theta$

$$\int \frac{(1+\theta)^2}{\sqrt{\theta}}\, d\theta = \int \frac{1 + 2\theta + \theta^2}{\sqrt{\theta}}\, d\theta$$

$$= \int \left(\frac{1}{\theta^{\frac{1}{2}}} + \frac{2\theta}{\theta^{\frac{1}{2}}} + \frac{\theta^2}{\theta^{\frac{1}{2}}}\right) d\theta = \int \left(\theta^{-\frac{1}{2}} + 2\theta^{1-\frac{1}{2}} + \theta^{2-\frac{1}{2}}\right) d\theta$$

$$= \int \left(\theta^{-\frac{1}{2}} + 2\theta^{\frac{1}{2}} + \theta^{\frac{3}{2}}\right) d\theta$$

$$= \frac{\theta^{-\frac{1}{2}+1}}{-\frac{1}{2}+1} + \frac{2\theta^{\frac{1}{2}+1}}{\frac{1}{2}+1} + \frac{\theta^{\frac{3}{2}+1}}{\frac{3}{2}+1} + c = \frac{\theta^{\frac{1}{2}}}{\frac{1}{2}} + \frac{2\theta^{\frac{3}{2}}}{\frac{3}{2}} + \frac{\theta^{\frac{5}{2}}}{\frac{5}{2}} + c$$

$$= 2\theta^{\frac{1}{2}} + \frac{4}{3}\theta^{\frac{3}{2}} + \frac{2}{5}\theta^{\frac{5}{2}} + c = 2\sqrt{\theta} + \frac{4}{3}\sqrt{\theta^3} + \frac{2}{5}\sqrt{\theta^5} + c$$

Application: Find $\int (4\cos 3x - 5\sin 2x)\, dx$

$$\int (4\cos 3x - 5\sin 2x)\, dx = (4)\left(\frac{1}{3}\right)\sin 3x - (5)\left(-\frac{1}{2}\right)\cos 2x$$

$$= \frac{4}{3}\sin 3x + \frac{5}{2}\cos 2x + c$$

Application: Find $\int (7 \sec^2 4t + 3 \cosec^2 2t)\, dt$

$$\int (7 \sec^2 4t + 3 \cosec^2 2t)\, dt = (7)\left(\frac{1}{4}\right) \tan 4t + (3)\left(-\frac{1}{2}\right) \cot 2t + c$$

$$= \frac{7}{4} \tan 4t - \frac{3}{2} \cot 2t + c$$

Application: Find $\int \dfrac{2}{3e^{4t}}\, dt$

$$\int \frac{2}{3e^{4t}}\, dt = \frac{2}{3} \int e^{-4t}\, dt = \left(\frac{2}{3}\right)\left(-\frac{1}{4}\right) e^{-4t} + c$$

$$= -\frac{1}{6} e^{-4t} + c = -\frac{1}{6e^{4t}} + c$$

Application: Find $\int \dfrac{3}{5x}\, dx$

$$\int \frac{3}{5x}\, dx = \frac{3}{5} \int \frac{1}{x}\, dx = \frac{3}{5} \ln x + c$$

Definite Integrals

Application: Evaluate $\int_{-2}^{3} (4 - x^2)\, dx$

$$\int_{-2}^{3} (4 - x^2)\, dx = \left[4x - \frac{x^3}{3} \right]_{-2}^{3} = \left(4(3) - \frac{3^3}{3} \right) - \left(4(-2) - \frac{(-2)^3}{3} \right)$$

$$= (12 - 9) - \left(-8 - \frac{-8}{3} \right) = (3) - \left(-5\frac{1}{3} \right) = 8\frac{1}{3}$$

Application: Evaluate $\int_{0}^{\pi/2} 3 \sin 2x\, dx$

$$\int_{0}^{\pi/2} 3 \sin 2x\, dx = \left[(3)\left(-\frac{1}{2}\right) \cos 2x \right]_{0}^{\pi/2} = \left[-\frac{3}{2} \cos 2x \right]_{0}^{\pi/2}$$

$$= \left\{ -\frac{3}{2} \cos 2\left(\frac{\pi}{2}\right) \right\} - \left\{ -\frac{3}{2} \cos 2(0) \right\}$$

$$= \left\{ -\frac{3}{2}(-1) \right\} - \left\{ -\frac{3}{2}(1) \right\} = \frac{3}{2} + \frac{3}{2} = 3$$

Application: Evaluate $\int_1^2 4\cos 3t\, dt$

$$\int_1^2 4\cos 3t\, dt = \left[(4)\left(\frac{1}{3}\right)\sin 3t\right]_1^2 = \left[\frac{4}{3}\sin 3t\right]_1^2 = \left\{\frac{4}{3}\sin 6\right\} - \left\{\frac{4}{3}\sin 3\right\}$$

(Note that limits of trigonometric functions are always expressed in **radians**, thus, for example, sin 6 means the sine of 6 radians = −0.279415..)

Hence, $\int_1^2 4\cos 3t\, dt = \left\{\frac{4}{3}(-0.279415..)\right\} - \left\{\frac{4}{3}(0.141120..)\right\}$
$$= (-0.37255) - (0.18816) = -0.5607$$

Application: Evaluate $\int_1^2 4e^{2x}\, dx$

$$\int_1^2 4e^{2x}\, dx = \left[\frac{4}{2}e^{2x}\right]_1^2 = 2[e^{2x}]_1^2 = 2[e^4 - e^2]$$
$$= 2[54.5982 - 7.3891] = 94.42$$

Application: Evaluate $\int_1^4 \frac{3}{4u}\, du$

$$\int_1^4 \frac{3}{4u}\, du = \left[\frac{3}{4}\ln u\right]_1^4 = \frac{3}{4}[\ln 4 - \ln 1] = \frac{3}{4}[1.3863 - 0] = 1.040$$

Chapter 116 Non-standard integrals

Functions that require integrating are not always in the 'standard form' shown in Chapter 115. However, it is often possible to change a function into a form that can be integrated by using either:

1. an algebraic substitution – see Chapter 117,
2. trigonometric and hyperbolic substitutions – see Chapter 118,
3. partial fractions – see Chapter 119
4. $t = \tan\frac{\theta}{2}$ substitution – see Chapter 120
5. integration by parts – see Chapter 121, or
6. reduction formulae – see Chapter 122.

Chapter 117 Integration using algebraic substitutions

Algebraic substitutions are demonstrated in the following Applications.

Application: Determine $\int \cos(3x + 7)\, dx$

$\int \cos(3x + 7)\, dx$ is not a standard integral of the form shown in Table 115.1, page 317, thus an algebraic substitution is made.

Let $u = 3x + 7$ then $\dfrac{du}{dx} = 3$ and rearranging gives: $dx = \dfrac{du}{3}$

Hence $\displaystyle\int \cos(3x + 7)\, dx = \int (\cos u)\, \dfrac{du}{3}$

$$= \int \frac{1}{3}\cos u \, du, \text{ which is a standard integral}$$

$$= \frac{1}{3}\sin u + c$$

Rewriting u as $(3x + 7)$ gives: $\displaystyle\int \cos(3x + 7)\, dx = \frac{1}{3}\sin(3x + 7) + c$, which may be checked by differentiating it.

Application: Find $\int (2x - 5)^7\, dx$

Let $u = (2x - 5)$ then $\dfrac{du}{dx} = 2$ and $dx = \dfrac{du}{2}$

Hence,

$$\int (2x - 5)^7\, dx = \int u^7\, \frac{du}{2} = \frac{1}{2}\int u^7\, du = \frac{1}{2}\left(\frac{u^8}{8}\right) + c = \frac{1}{16}u^8 + c$$

Rewriting u as $(2x - 5)$ gives: $\displaystyle\int (2x - 5)^7\, dx = \frac{1}{16}(2x - 5)^8 + c$

Application: Evaluate $\displaystyle\int_0^{\pi/6} 24 \sin^5 \theta \cos \theta \, d\theta$

Let $u = \sin \theta$ then $\dfrac{du}{d\theta} = \cos \theta$ and $d\theta = \dfrac{du}{\cos \theta}$

Hence, $\displaystyle\int 24 \sin^5 \theta \cos \theta \, d\theta = \int 24 u^5 \cos \theta \, \frac{du}{\cos \theta}$

$$= 24 \int u^5 \, du, \text{ by cancelling}$$

$$= 24 \frac{u^6}{6} + c = 4u^6 + c = 4(\sin \theta)^6 + c$$

$$= 4 \sin^6 \theta + c$$

Thus, $\displaystyle\int_0^{\pi/6} 24 \sin^5 \theta \cos \theta \, d\theta = \left[4 \sin^6 \theta\right]_0^{\pi/6}$

$$= 4\left[\left(\sin \frac{\pi}{6}\right)^6 - (\sin 0)^6\right]$$

$$= 4\left[\left(\frac{1}{2}\right)^6 - 0\right] = \frac{1}{16} \text{ or } 0.0625$$

Application: Determine $\displaystyle\int \frac{2x}{\sqrt{(4x^2 - 1)}} \, dx$

Let $u = 4x^2 - 1$ then $\dfrac{du}{dx} = 8x$ and $dx = \dfrac{du}{8x}$

Hence $\displaystyle\int \frac{2x}{\sqrt{(4x^2 - 1)}} \, dx = \int \frac{2x}{\sqrt{u}} \frac{du}{8x} = \frac{1}{4} \int \frac{1}{\sqrt{u}} \, du$, by cancelling

$$= \frac{1}{4} \int u^{-\frac{1}{2}} = \frac{1}{4} \left[\frac{u^{-\frac{1}{2}+1}}{-\frac{1}{2}+1}\right] + c = \frac{1}{4}\left[\frac{u^{\frac{1}{2}}}{\frac{1}{2}}\right] + c$$

$$= \frac{1}{2} \sqrt{u} + c = \frac{1}{2} \sqrt{4x^2 - 1} + c$$

Change of limits

When evaluating definite integrals involving substitutions it is sometimes more convenient to **change the limits** of the integral.

Application: Evaluate $\displaystyle\int_1^3 5x\sqrt{2x^2 + 7} \, dx$, taking positive values of square roots only:

Let $u = 2x^2 + 7$, then $\dfrac{du}{dx} = 4x$ and $dx = \dfrac{du}{4x}$

When $x = 3$, $u = 2(3)^2 + 7 = 25$ and when $x = 1$, $u = 2(1)^2 + 7 = 9$

Hence, $\displaystyle\int_{x=1}^{x=3} 5x\sqrt{2x^2 + 7} \, dx = \int_{u=9}^{u=25} 5x\sqrt{u} \frac{du}{4x}$

$$= \frac{5}{4} \int_9^{25} \sqrt{u} \, du = \frac{5}{4} \int_9^{25} u^{\frac{1}{2}} du$$

Thus the limits have been changed, and it is unnecessary to change the integral back in terms of x.

Thus, $\displaystyle\int_{x=1}^{x=3} 5x\sqrt{2x^2+7}\ dx = \frac{5}{4}\left[\frac{u^{3/2}}{3/2}\right]_9^{25} = \frac{5}{6}\left[\sqrt{u^3}\right]_9^{25}$

$$= \frac{5}{6}\left[\sqrt{25^3} - \sqrt{9^3}\right] = \frac{5}{6}(125 - 27)$$

$$= 81\frac{2}{3}$$

Chapter 118   Integration using trigonometric and hyperbolic substitutions

Table 118.1   **Integrals using trigonometric substitutions**

f(x)	$\int$ f(x) dx	Method
1. $\cos^2 x$	$\frac{1}{2}\left(x + \dfrac{\sin 2x}{2}\right) + c$	Use $\cos 2x = 2\cos^2 x - 1$
2. $\sin^2 x$	$\frac{1}{2}\left(x - \dfrac{\sin 2x}{2}\right) + c$	Use $\cos 2x = 1 - 2\sin^2 x$
3. $\tan^2 x$	$\tan x - x + c$	Use $1 + \tan^2 x = \sec^2 x$
4. $\cot^2 x$	$-\cot x - x + c$	Use $\cot^2 x + 1 = \operatorname{cosec}^2 x$
5. $\cos^m x \sin^n x$	(a) If either m or n is odd (but not both), use $\cos^2 x + \sin^2 x = 1$ (b) If both m and n are even, use either $\cos 2x = 2\cos^2 x - 1$ or $\cos 2x = 1 - 2\sin^2 x$	
6. $\sin A \cos B$		Use $\dfrac{1}{2}\,[\sin(A+B) + \sin(A-B)]$
7. $\cos A \sin B$		Use $\dfrac{1}{2}\,[\sin(A+B) - \sin(A-B)]$
8. $\cos A \cos B$		Use $\dfrac{1}{2}\,[\cos(A+B) + \cos(A-B)]$

Table 118.1 **Continued**

f(x)	$\int f(x)\,dx$	Method
9. sin A sin B		Use $-\dfrac{1}{2}\,[\cos(A + B) - \cos(A - B)]$
10. $\dfrac{1}{\sqrt{(a^2 - x^2)}}$	$\sin^{-1}\dfrac{x}{a} + c$	Use $x = a\sin\theta$ substitution
11. $\sqrt{a^2 - x^2}$	$\dfrac{a^2}{2}\sin^{-1}\dfrac{x}{a} + \dfrac{x}{2}\sqrt{a^2 - x^2} + c$	
12. $\dfrac{1}{a^2 + x^2}$	$\dfrac{1}{a}\tan^{-1}\dfrac{x}{a} + c$	Use $x = a\tan\theta$ substitution
13. $\dfrac{1}{\sqrt{(x^2 + a^2)}}$	$\sinh^{-1}\dfrac{x}{a} + c$ or $\ln\left\{\dfrac{x + \sqrt{(x^2 + a^2)}}{a}\right\} + c$	Use $x = a\sinh\theta$ substitution
14. $\sqrt{(x^2 + a^2)}$	$\dfrac{a^2}{2}\sinh^{-1}\dfrac{x}{a} + \dfrac{x}{2}\sqrt{(x^2 + a^2)} + c$	
15. $\dfrac{1}{\sqrt{(x^2 - a^2)}}$	$\cosh^{-1}\dfrac{x}{a} + c$ or $\ln\left\{\dfrac{x + \sqrt{(x^2 - a^2)}}{a}\right\} + c$	Use $x = a\cosh\theta$ substitution
16. $\sqrt{(x^2 - a^2)}$	$\dfrac{x}{2}\sqrt{(x^2 - a^2)} - \dfrac{a^2}{2}\cosh^{-1}\dfrac{x}{a} + c$	

Application: Evaluate $\displaystyle\int_0^{\pi/4} 2\cos^2 4t\, dt$

Since $\cos 2t = 2\cos^2 t - 1$ (from Chapter 52),

then $\cos^2 t = \dfrac{1}{2}(1 + \cos 2t)$ and $\cos^2 4t = \dfrac{1}{2}(1 + \cos 8t)$

Hence $\displaystyle\int_0^{\pi/4} 2\cos^2 4t\, dt = 2\int_0^{\pi/4} \dfrac{1}{2}(1 + \cos 8t)\, dt = \left[t + \dfrac{\sin 8t}{8}\right]_0^{\pi/4}$

$$= \left[\frac{\pi}{4} + \frac{\sin 8\left(\frac{\pi}{4}\right)}{8} \right] - \left[0 + \frac{\sin 0}{8} \right]$$

$$= \frac{\pi}{4} \text{ or } 0.7854$$

Application: Find $3\int \tan^2 4x \, dx$

Since $\qquad 1 + \tan^2 x = \sec^2 x$, then $\tan^2 x = \sec^2 x - 1$ and
$$\tan^2 4x = \sec^2 4x - 1$$

Hence, $\quad 3\int \tan^2 4x \, dx = 3\int (\sec^2 4x - 1) \, dx = 3\left(\frac{\tan 4x}{4} - x \right) + c$

Application: Determine $\int \sin^5\theta \, d\theta$

Since $\cos^2\theta + \sin^2\theta = 1$ then $\sin^2\theta = (1 - \cos^2\theta)$

Hence, $\int \sin^5\theta \, d\theta = \int \sin\theta(\sin^2\theta)^2 \, d\theta = \int \sin\theta(1 - \cos^2\theta)^2 \, d\theta$

$$= \int \sin\theta(1 - 2\cos^2\theta + \cos^4\theta) \, d\theta$$

$$= \int (\sin\theta - 2\sin\theta\cos^2\theta + \sin\theta\cos^4\theta) \, d\theta$$

$$= -\cos\theta + \frac{2\cos^3\theta}{3} - \frac{\cos^5\theta}{5} + c$$

[Whenever a power of a cosine is multiplied by a sine of power 1, or vice-versa, the integral may be determined by inspection as follows.

In general, $\quad \int \cos^n\theta \sin\theta \, d\theta = \frac{-\cos^{n+1}\theta}{(n+1)} + c$

and $\qquad \int \sin^n\theta \cos\theta \, d\theta = \frac{\sin^{n+1}\theta}{(n+1)} + c$

Alternatively, an algebraic substitution may be used.]

Application: Evaluate $\int_0^{\pi/2} \sin^2 x \cos^3 x \, dx$

$$\int_0^{\pi/2} \sin^2 x \cos^3 x \, dx = \int_0^{\pi/2} \sin^2 x \cos^2 x \cos x \, dx$$

$$= \int_0^{\pi/2} \sin^2 x (1 - \sin^2 x) \cos x \, dx$$

$$= \int_0^{\pi/2} (\sin^2 x \cos x - \sin^4 \cos x) dx$$

$$= \left[\frac{\sin^3 x}{3} - \frac{\sin^5 x}{5} \right]_0^{\pi/2}$$

$$= \left[\frac{\left(\sin \dfrac{\pi}{2}\right)^3}{3} - \frac{\left(\sin \dfrac{\pi}{2}\right)^5}{5} \right] - [0 - 0]$$

$$= \frac{1}{3} - \frac{1}{5} = \frac{2}{15} \text{ or } 0.1333$$

Application: Find $\int \sin^2 t \cos^4 t \, dt$

$$\int \sin^2 t \cos^4 t \, dt = \int \sin^2 t (\cos^2 t)^2 \, dt = \int \left(\frac{1 - \cos 2t}{2} \right) \left(\frac{1 + \cos 2t}{2} \right)^2 dt$$

$$= \frac{1}{8} \int (1 - \cos 2t)(1 + 2\cos 2t + \cos^2 2t) \, dt$$

$$= \frac{1}{8} \int (1 + 2\cos 2t + \cos^2 2t - \cos 2t - 2\cos^2 2t - \cos^3 2t) \, dt$$

$$= \frac{1}{8} \int (1 + \cos 2t - \cos^2 2t - \cos^3 2t) \, dt$$

$$= \frac{1}{8} \int \left[1 + \cos 2t - \left(\frac{1 + \cos 4t}{2} \right) - \cos 2t(1 - \sin^2 2t) \right] dt$$

$$= \frac{1}{8} \int \left(\frac{1}{2} - \frac{\cos 4t}{2} + \cos 2t \sin^2 2t \right) dt$$

$$= \frac{1}{8} \left(\frac{t}{2} - \frac{\sin 4t}{8} + \frac{\sin^3 2t}{6} \right) + c$$

Application: Determine $\int \sin 3t \cos 2t \, dt$

$$\int \sin 3t \cos 2t \, dt = \int \frac{1}{2}[\sin(3t + 2t) + \sin(3t - 2t)]\, dt,$$

from 6 of Table 118.1,

$$= \frac{1}{2} \int (\sin 5t + \sin t)\, dt = \frac{1}{2}\left(\frac{-\cos 5t}{5} - \cos t \right) + c$$

Application: Evaluate $\int_0^1 2 \cos 6\theta \cos \theta \, d\theta$,   correct to 4 decimal places

$$\int_0^1 2 \cos 6\theta \cos \theta \, d\theta = 2 \int_0^1 \frac{1}{2}[\cos(6\theta + \theta) + \cos(6\theta - \theta)]\, d\theta,$$

from 8 of Table 118.1

$$= \int_0^1 (\cos 7\theta + \cos 5\theta)\, d\theta$$

$$= \left[\frac{\sin 7\theta}{7} + \frac{\sin 5\theta}{5} \right]_0^1$$

$$= \left(\frac{\sin 7}{7} + \frac{\sin 5}{5} \right) - \left(\frac{\sin 0}{7} + \frac{\sin 0}{5} \right)$$

'sin 7' means 'the sine of 7 radians'

Hence, $\int_0^1 2 \cos 6\theta \cos \theta \, d\theta = (0.09386 + -0.19178) - (0)$

$$= -\textbf{0.0979}, \text{ correct to 4 decimal places}$$

Application: Evaluate $\int_0^4 \sqrt{16 - x^2}\, dx$

From 11 of Table 118.1,

$$\int_0^4 \sqrt{16 - x^2}\, dx = \left[\frac{16}{2} \sin^{-1}\frac{x}{4} + \frac{x}{2}\sqrt{(16 - x^2)} \right]_0^4$$

$$= \left[8 \sin^{-1}1 + 2\sqrt{0} \right] - [8 \sin^{-1}0 + 0]$$

$$= 8 \sin^{-1}1 = 8\left(\frac{\pi}{2} \right) = 4\pi \text{ or } 12.57$$

Application: Evaluate $\int_0^2 \frac{1}{(4 + x^2)}\, dx$

From 12 of Table 118.1, $\int_0^2 \frac{1}{(4 + x^2)}\, dx = \frac{1}{2}\left[\tan^{-1}\frac{x}{2} \right]_0^2$ since $a = 2$

$$= \frac{1}{2}(\tan^{-1}1 - \tan^{-1}0)$$

$$= \frac{1}{2}\left(\frac{\pi}{4} - 0\right)$$

$$= \frac{\pi}{8} \text{ or } 0.3927$$

Application: Evaluate $\int_0^2 \frac{1}{\sqrt{(x^2 + 4)}} dx$, correct to 4 decimal places

$$\int_0^2 \frac{1}{\sqrt{(x^2 + 4)}} dx = \left[\sinh^{-1}\frac{x}{2}\right]_0^2 \quad \text{or} \quad \left[\ln\left\{\frac{x + \sqrt{(x^2 + 4)}}{2}\right\}\right]_0^2$$

from 13 of Table 118.1, where a = 2

Using the logarithmic form,

$$\int_0^2 \frac{1}{\sqrt{(x^2 + 4)}} dx = \left[\ln\left(\frac{2 + \sqrt{8}}{2}\right) - \ln\left(\frac{0 + \sqrt{4}}{2}\right)\right]$$

$$= \ln 2.4142 - \ln 1$$

$$= 0.8814, \text{ correct to 4 decimal places}$$

Application: Determine $\int \frac{2x - 3}{\sqrt{(x^2 - 9)}} dx$

$$\int \frac{2x - 3}{\sqrt{(x^2 + 9)}} dx = \int \frac{2x}{\sqrt{(x^2 - 9)}} dx - \int \frac{3}{\sqrt{(x^2 - 9)}} dx$$

The first integral is determined using the algebraic substitution $u = (x^2 - 9)$, and the second integral is of the form $\int \frac{1}{\sqrt{(x^2 - a^2)}} dx$ (see 15 of Table 118.1)

Hence,

$$\int \frac{2x}{\sqrt{(x^2 - 9)}} dx - \int \frac{3}{\sqrt{(x^2 - 9)}} dx = 2\sqrt{(x^2 - 9)} - 3\cosh^{-1}\frac{x}{3} + c$$

Application: Evaluate $\int_2^3 \sqrt{(x^2 - 4)} dx$

$$\int_2^3 \sqrt{(x^2 - 4)} dx = \left[\frac{x}{2}\sqrt{(x^2 - 4)} - \frac{4}{2}\cosh^{-1}\frac{x}{2}\right]_2^3$$

from 16 of Table 118.1, when a = 2,

$$= \left(\frac{3}{2}\sqrt{5} - 2\cosh^{-1}\frac{3}{2}\right) - \left(0 - 2\cosh^{-1}1\right)$$

$$= 1.429, \text{ by calculator}$$

or since $\cosh^{-1}\frac{x}{a} = \ln\left\{\frac{x + \sqrt{(x^2 - a^2)}}{a}\right\}$

then $\cosh^{-1}\frac{3}{2} = \ln\left\{\frac{3 + \sqrt{(3^2 - 2^2)}}{2}\right\}$

$$= \ln 2.6180 = 0.9624$$

Similarly, $\cosh^{-1}1 = 0$

Hence, $\displaystyle\int_2^3 \sqrt{(x^2 - 4)}\, dx = \left|\frac{3}{2}\sqrt{5} - 2(0.9624)\right| - [0]$

$$= \mathbf{1.429}, \text{ correct to 4 significant figures}$$

Chapter 119 Integration using partial fractions

1. Linear factors

Application: Determine $\displaystyle\int \frac{11 - 3x}{x^2 + 2x - 3}\, dx$

As shown on page 63: $\dfrac{11 - 3x}{x^2 + 2x - 3} \equiv \dfrac{2}{(x - 1)} - \dfrac{5}{(x + 3)}$

Hence $\displaystyle\int \frac{11 - 3x}{x^2 + 2x - 3}\, dx = \int\left\{\frac{2}{(x - 1)} - \frac{5}{(x + 3)}\right\} dx$

$$= 2\ln(x - 1) - 5\ln(x + 3) + c$$

(by algebraic substitutions (see chapter 117))

or $\ln\left\{\dfrac{(x - 1)^2}{(x + 3)^5}\right\} + c$ by the laws of logarithms

Application: Evaluate $\displaystyle\int_2^3 \frac{x^3 - 2x^2 - 4x - 4}{x^2 + x - 2}\, dx$, correct to 4 significant figures

By dividing out and resolving into partial fractions, it was shown on page chapter 64:

$$\frac{x^3 - 2x^2 - 4x - 4}{x^2 + x - 2} \equiv x - 3 + \frac{4}{(x + 2)} - \frac{3}{(x - 1)}$$

Hence,

$$\int_2^3 \frac{x^3 - 2x^2 - 4x - 4}{x^2 + x - 2}\ dx \equiv \int_2^3 \left\{ x - 3 + \frac{4}{(x+2)} - \frac{3}{(x-1)} \right\} dx$$

$$= \left[\frac{x^2}{2} - 3x + 4\ \ln(x + 2) - 3\ \ln(x - 1) \right]_2^3$$

$$= \left(\frac{9}{2} - 9 + 4\ln 5 - 3\ln 2 \right) - (2 - 6 + 4\ln 4 - 3\ln 1)$$

$$= \mathbf{-1.687}, \text{ correct to 4 significant figures}$$

2. Repeated linear factors

Application: Find $\int \dfrac{5x^2 - 2x - 19}{(x+3)(x-1)^2}\ dx$

It was shown on page 65:

$$\frac{5x^2 - 2x - 19}{(x+3)(x-1)^2} \equiv \frac{2}{(x+3)} + \frac{2}{(x-1)} - \frac{4}{(x-1)^2}$$

Hence, $\int \dfrac{5x^2 - 2x - 19}{(x+3)(x-1)^2}\ dx \equiv \int \left\{ \dfrac{2}{(x+3)} + \dfrac{3}{(x-1)} - \dfrac{4}{(x-1)^2} \right\}\ dx$

$$= 2\ \ln(x + 3) + 3\ \ln(x - 1) + \frac{4}{(x-1)} + c$$

$$\text{or }\ \ln\{(x + 3)^2\ (x - 1)^3\} + \frac{4}{(x-1)} + c$$

3. Quadratic factors

Application: Find $\int \dfrac{3 + 6x + 4x^2 - 2x^3}{x^2(x^2 + 3)}\ dx$

It was shown on page 65: $\dfrac{3 + 6x + 4x^2 - 2x^2}{x^2(x^2 + 3)} \equiv \dfrac{2}{x} + \dfrac{1}{x^2} + \dfrac{3 - 4x}{(x^2 + 3)}$

Thus,

$$\int \frac{3 + 6x + 4x^2 - 2x^3}{x^2(x^2 + 3)}\, dx = \int \left(\frac{2}{x} + \frac{1}{x^2} + \frac{3 - 4x}{(x^2 + 3)} \right) dx$$

$$= \int \left\{ \frac{2}{x} + \frac{1}{x^2} + \frac{3}{(x^2 + 3)} - \frac{4x}{(x^2 + 3)} \right\} dx$$

$$\int \frac{3}{(x^2 + 3)}\, dx = 3 \int \frac{1}{x^2 + \left(\sqrt{3}\right)^2}\, dx = \frac{3}{\sqrt{3}} \tan^{-1} \frac{x}{\sqrt{3}},$$

from 12, Table 118.1, page 324.

$\int \frac{4x}{x^2 + 3}\, dx$ is determined using the algebraic substitution $u = (x^2 + 3)$

Hence, $\int \left\{ \frac{2}{x} + \frac{1}{x^2} + \frac{3}{(x^2 + 3)} - \frac{4x}{(x^2 + 3)} \right\} dx$

$$= 2 \ln x - \frac{1}{x} + \frac{3}{\sqrt{3}} \tan^{-1} \frac{x}{\sqrt{3}} - 2 \ln(x^2 + 3) + c$$

$$\text{or } \ln \left(\frac{x}{x^2 + 3} \right)^2 - \frac{1}{x} + \sqrt{3} \tan^{-1} \frac{x}{\sqrt{3}} + c$$

Chapter 120 The $t = \tan \frac{\theta}{2}$ substitution

To determine $\int \frac{1}{a \cos \theta + b \sin \theta + c}\, d\theta$, where a, b and c are constants, if $t = \tan \frac{\theta}{2}$ then:

$$\sin \theta = \frac{2t}{(1 + t^2)} \qquad (1)$$

$$\cos \theta = \frac{1 - t^2}{1 + t^2} \qquad (2)$$

$$d\theta = \frac{2dt}{1 + t^2} \qquad (3)$$

Application: Determine $\int \frac{d\theta}{\sin \theta}$

If $t = \tan \frac{\theta}{2}$ then $\sin \theta = \frac{2t}{1 + t^2}$ and $d\theta = \frac{2dt}{1 + t^2}$ from equations (1) and (3).

Thus, $\displaystyle\int \frac{d\theta}{\sin\theta} = \int \frac{1}{\dfrac{2t}{1+t^2}}\left(\frac{2dt}{1+t^2}\right) = \int \frac{1}{t}dt = \ln t + c$

Hence, $\displaystyle\int \frac{d\theta}{\sin\theta} = \ln\left|\tan\frac{\theta}{2}\right| + c$

Application: Determine $\displaystyle\int \frac{dx}{\cos x}$

If $t = \tan\dfrac{x}{2}$ then $\cos x = \dfrac{1-t^2}{1+t^2}$ and $dx = \dfrac{2dt}{1+t^2}$ from equations (2) and (3).

Thus $\displaystyle\int \frac{dx}{\cos x} = \int \frac{1}{\dfrac{1-t^2}{1+t^2}}\left(\frac{2dt}{1+t^2}\right) = \int \frac{2}{1-t^2}dt$

$\dfrac{2}{1-t^2}$ may be resolved into partial fractions (see Chapter 21)

Let $\dfrac{2}{1-t^2} = \dfrac{2}{(1-t)(1+t)} = \dfrac{A}{(1-t)} + \dfrac{B}{(1+t)} = \dfrac{A(1+t) + B(1-t)}{(1-t)(1+t)}$

Hence $2 = A(1+t) + B(1-t)$

When $t = 1$, $2 = 2A$, from which, $A = 1$

When $t = -1$, $2 = 2B$, from which, $B = 1$

Hence, $\displaystyle\int \frac{2}{1-t^2} = \int \left(\frac{1}{1-t} + \frac{1}{1+t}\right)dt = -\ln(1-t) + \ln(1+t) + c$

$$= \ln\left\{\frac{(1+t)}{(1-t)}\right\} + c$$

Thus, $\displaystyle\int \frac{dx}{\cos x} = \ln\left\{\frac{1+\tan\dfrac{x}{2}}{1-\tan\dfrac{x}{2}}\right\} + c$

Note that since $\tan\dfrac{\pi}{4} = 1$, the above result may be written as:

$$\int \frac{dx}{\cos x} = \ln\left\{\frac{\tan\dfrac{\pi}{4} + \tan\dfrac{x}{2}}{\tan\dfrac{\pi}{4} - \tan\dfrac{x}{2}}\right\} + c = \ln\left\{\tan\left(\frac{\pi}{4} + \frac{x}{2}\right)\right\} + c$$

from compound angles, chapter 54.

Application: Determine $\displaystyle\int \frac{d\theta}{5 + 4\cos\theta}$

If $t = \tan\dfrac{\theta}{2}$ then $\cos\theta = \dfrac{1 - t^2}{1 + t^2}$ and $d\theta = \dfrac{2\,dt}{1 + t^2}$ from equations (2) and (3).

Thus, $\displaystyle\int \dfrac{d\theta}{5 + 4\cos\theta} = \int \dfrac{1}{5 + 4\left(\dfrac{1 - t^2}{1 + t^2}\right)}\left(\dfrac{2\,dt}{1 + t^2}\right)$

$$= \int \dfrac{1}{\dfrac{5(1 + t^2) + 4(1 - t^2)}{1 + t^2}}\left(\dfrac{2\,dt}{(1 + t^2)}\right)$$

$$= 2\int \dfrac{dt}{t^2 + 9} = 2\int \dfrac{dt}{t^2 + 3^2} = 2\left(\dfrac{1}{3}\tan^{-1}\dfrac{t}{3}\right) + c$$

Hence, $\displaystyle\int \dfrac{d\theta}{5 + 4\cos\theta} = \dfrac{2}{3}\tan^{-1}\left(\dfrac{1}{3}\tan\dfrac{\theta}{2}\right) + c$

Application: Determine $\displaystyle\int \dfrac{dx}{\sin x + \cos x}$

If $t = \tan\dfrac{x}{2}$ then $\sin x = \dfrac{2t}{1 + t^2}$, $\cos x = \dfrac{1 - t^2}{1 + t^2}$ and $dx = \dfrac{2\,dt}{1 + t^2}$ from equations (1), (2) and (3).

Thus, $\displaystyle\int \dfrac{dx}{\sin x + \cos x} = \int \dfrac{\dfrac{2\,dt}{1 + t^2}}{\left(\dfrac{2t}{1 + t^2}\right) + \left(\dfrac{1 - t^2}{1 + t^2}\right)} = \int \dfrac{\dfrac{2\,dt}{1 + t^2}}{\dfrac{2t + 1 - t^2}{1 + t^2}}$

$$= \int \dfrac{2\,dt}{1 + 2t - t^2} = \int \dfrac{-2\,dt}{t^2 - 2t - 1}$$

$$= \int \dfrac{-2\,dt}{(t - 1)^2 - 2} = \int \dfrac{2\,dt}{\left(\sqrt{2}\right)^2 - (t - 1)^2}$$

$$= 2\left[\dfrac{1}{2\sqrt{2}}\ln\left\{\dfrac{\sqrt{2} + (t - 1)}{\sqrt{2} - (t - 1)}\right\}\right] + c$$

by using partial fractions $\displaystyle\int \dfrac{1}{a^2 - x^2}\,dx = \dfrac{1}{2a}\ln\left(\dfrac{a + x}{a - x}\right)$

i.e. $\displaystyle\int \dfrac{dx}{\sin x + \cos x} = \dfrac{1}{\sqrt{2}}\ln\left\{\dfrac{\sqrt{2} - 1 + \tan\dfrac{x}{2}}{\sqrt{2} + 1 - \tan\dfrac{x}{2}}\right\} + c$

Application: Determine $\displaystyle\int \frac{dx}{7 - 3\sin x + 6\cos x}$

From equations (1) to (3),

$$\int \frac{dx}{7 - 3\sin x + 6\cos x} = \int \frac{\dfrac{2\,dt}{1 + t^2}}{7 - 3\left(\dfrac{2t}{1 + t^2}\right) + 6\left(\dfrac{1 - t^2}{1 + t^2}\right)}$$

$$= \int \frac{\dfrac{2\,dt}{1 + t^2}}{\dfrac{7(1 + t^2) - 3(2t) + 6(1 - t^2)}{1 + t^2}}$$

$$= \int \frac{2\,dt}{7 + 7t^2 - 6t + 6 - 6t^2}$$

$$= \int \frac{2\,dt}{t^2 - 6t + 13} = \int \frac{2\,dt}{(t - 3)^2 + 2^2}$$

$$= 2\left[\frac{1}{2}\tan^{-1}\left(\frac{t - 3}{2}\right)\right] + c$$

from 12 of Table 118.1, page 324.

Hence, $\displaystyle\int \frac{dx}{7 - 3\sin x + 6\cos x} = \tan^{-1}\left(\frac{\tan\dfrac{x}{2} - 3}{2}\right) + c$

121 Integration by parts

If u and v are both functions of x, then:

$$\int u\frac{dv}{dx}\,dx = \int uv - \int v\frac{du}{dx}\,dx$$

or $$\int u\,dv = uv - \int v\,du$$

This is known as the **integration by parts formula**.

Application: Determine $\displaystyle\int x\cos x\,dx$

From the integration by parts formula, $\displaystyle\int u\,dv = uv - \int v\,du$

Let $u = x$, from which $\dfrac{du}{dx} = 1$,   i.e. $du = dx$

and let $dv = \cos x \, dx$, from which $v = \displaystyle\int \cos x \, dx = \sin x$

Expressions for u, du, v and dv are now substituted into the 'by parts' formula as shown below.

$\displaystyle\int$	u	dv	=	u	v	$- \displaystyle\int$	v	du
$\displaystyle\int$	x	$\cos x \, dx$	=	(x)	(sin x)	$- \displaystyle\int$	(sin x)	(dx)

i.e. $\displaystyle\int x \cos x \, dx = x \sin x - (-\cos x) + c = x \sin x + \cos x + c$

[This result may be checked by differentiating the right-hand side, i.e.

$\dfrac{d}{dx}(x \sin x + \cos x + c) = [(x)(\cos x) + (\sin x)(1)] - \sin x + 0$ using the product rule

$$= x \cos x, \qquad \text{which is the function being integrated}]$$

Application: Find $\displaystyle\int 3t \, e^{2t} \, dt$

Let $u = 3t$, from which, $\dfrac{du}{dt} = 3$, i.e. $du = 3dt$

and let $dv = e^{2t} \, dt$, from which, $v = \displaystyle\int e^{2t} dt = \dfrac{1}{2} e^{2t}$

Substituting into $\displaystyle\int u \, dv = uv - \displaystyle\int v \, du$ gives:

$$\int 3t \, e^{2t} \, dt = (3t)\left(\frac{1}{2} e^{2t}\right) - \int \left(\frac{1}{2} e^{2t}\right)(3 \, dt) = \frac{3}{2} t \, e^{2t} - \frac{3}{2} \int e^{2t} \, dt$$

$$= \frac{3}{2} t \, e^{2t} - \frac{3}{2}\left(\frac{e^{2t}}{2}\right) + c$$

Hence, $\displaystyle\int 3t \, e^{2t} \, dt = \dfrac{3}{2} e^{2t}\left(t - \dfrac{1}{2}\right) + c$, which may be checked by differentiating

Application: Evaluate $\displaystyle\int_0^{\pi/2} 2\theta \sin \theta \, d\theta$

Let $u = 2\theta$, from which, $\dfrac{du}{d\theta} = 2$, i.e. $du = 2d\theta$

and let $dv = \sin \theta \, d\theta$, from which, $v = \displaystyle\int \sin \theta \, d\theta = -\cos \theta$

Substituting into $\displaystyle\int u \, dv = uv - \displaystyle\int v \, du$ gives:

$$\int 2\theta \sin \theta \, d\theta = (2\theta)(-\cos \theta) - \int (-\cos \theta)(2 \, d\theta)$$

$$= -2\theta \cos \theta + 2 \int \cos \theta \, d\theta = -2\theta \cos \theta + 2 \sin \theta + c$$

Hence, $\displaystyle\int_0^{\pi/2} 2\theta \sin \theta \, d\theta = \left[-2\theta \cos \theta + 2 \sin \theta\right]_0^{\pi/2}$

$$= \left[-2\left(\frac{\pi}{2}\right)\cos \frac{\pi}{2} + 2 \sin \frac{\pi}{2}\right] - [0 + 2 \sin 0]$$

$$= (-0 + 2) - (0 + 0) = \mathbf{2}$$

since $\cos \dfrac{\pi}{2} = 0$ and $\sin \dfrac{\pi}{2} = 1$

Application: Determine $\displaystyle\int x^2 \sin x \, dx$

Let $u = x^2$, from which, $\dfrac{du}{dx} = 2x$, i.e. $du = 2x \, dx$,

and let $dv = \sin x \, dx$, from which, $v = \displaystyle\int \sin x \, dx = -\cos x$

Substituting into $\displaystyle\int u \, dv = uv - \int v \, du$ gives:

$$\int x^2 \sin x \, dx = (x^2)(-\cos x) - \int (-\cos x)(2x \, dx)$$

$$= -x^2 \cos x + 2\left[\int x \cos x \, dx\right]$$

The integral, $\displaystyle\int x \cos x \, dx$, is not a 'standard integral' and it can only be determined by using the integration by parts formula again.

From the application, on page 334, $\displaystyle\int x \cos x \, dx \, x = x \sin x + \cos x$

Hence, $\displaystyle\int x^2 \sin x \, dx = -x^2 \cos x + 2\{x \sin x + \cos x\} + c$

$$= -x^2 \cos x + 2x \sin x + 2 \cos x + c$$

$$= (2 - x^2) \cos x + 2x \sin x + c$$

In general, if the algebraic term of a product is of power n, then the integration by parts formula is applied n times.

Application: Find $\displaystyle\int x \ln x \, dx$

The logarithmic function is chosen as the 'u part'

Thus, when $u = \ln x$, then $\dfrac{du}{dx} = \dfrac{1}{x}$, i.e. $du = \dfrac{dx}{x}$

Letting $dv = x \, dx$ gives $v = \displaystyle\int x \, dx = \dfrac{x^2}{2}$

Substituting into $\int u\,dv = uv - \int v\,du$ gives:

$$\int x \ln x\,dx = (\ln x)\left(\frac{x^2}{2}\right) - \int \left(\frac{x^2}{2}\right)\frac{dx}{x}$$

$$= \frac{x^2}{2}\ln x - \frac{1}{2}\int x\,dx = \frac{x^2}{2}\ln x - \frac{1}{2}\left(\frac{x^2}{2}\right) + c$$

Hence, $\int x \ln x\,dx = \frac{x^2}{2}\left(\ln x - \frac{1}{2}\right) + c$ or $\frac{x^2}{4}(2\ln x - 1) + c$

Chapter 122 Reduction formulae

When using integration by parts, an integral such as $\int x^2 e^x\,dx$ requires integration by parts twice. Similarly, $\int x^3 e^x\,dx$ requires integration by parts three times. Thus, integrals such as $\int x^5 e^x\,dx$, $\int x^6 \cos x\,dx$ and $\int x^8 \sin 2x\,dx$ or example, would take a long time to determine using integration by parts. Reduction formulae provide a quicker method for determining such integrals.

Below is a list of the most common reduction formulae.

$$\int x^n e^x\,dx = I_n = x^n e^x - n I_{n-1} \tag{1}$$

$$\int x^n \cos x\,dx = I_n = x^n \sin x + nx^{n-1}\cos x - n(n-1)I_{n-2} \tag{2}$$

$$\int_0^\pi x^n \cos x\,dx = I_n = -n\pi^{n-1} - n(n-1)I_{n-2} \tag{3}$$

$$\int x^n \sin x\,dx = I_n = -x^n \cos x + nx^{n-1}\sin x - n(n-1)I_{n-2} \tag{4}$$

$$\int \sin^n x\,dx = I_n = -\frac{1}{n}\sin^{n-1} x \cos x + \frac{n-1}{n}I_{n-2} \tag{5}$$

$$\int \cos^n x\,dx = I_n = \frac{1}{n}\cos^{n-1} \sin x + \frac{n-1}{n}I_{n-2} \tag{6}$$

$$\int_0^{\pi/2} \sin^n x\,dx = \int_0^{\pi/2} \cos^n x\,dx = I_n = \frac{n-1}{n}I_{n-2} \tag{7}$$

$$\int \tan^n x\,dx = I_n = \frac{\tan^{n-1}x}{n-1} - I_{n-2} \tag{8}$$

$$\int (\ln x)^n\,dx = I_n = x(\ln x)^n - n I_{n-1} \tag{9}$$

Integrals of the form $\int x^n e^x \, dx$

Application: Determine $\int x^3 e^x \, dx$ using a reduction formula

From equation (1), $I_n = x^n e^x - nI_{n-1}$

Hence $\int x^3 e^x \, dx = I_3 = x^3 e^x - 3I_2$

$$I_2 = x^2 e^x - 2I_1$$
$$I_1 = x^1 e^x - 1I_0$$

and $I_0 = \int x^0 e^x \, dx = \int e^x \, dx = e^x$

Thus $\int x^3 e^x \, dx = x^3 e^x - 3[x^2 e^x - 2I_1]$

$$= x^3 e^x - 3[x^2 e^x - 2(xe^x - I_0)]$$
$$= x^3 e^x - 3[x^2 e^x - 2(xe^x - e^x)]$$
$$= x^3 e^x - 3x^2 e^x + 6(xe^x - e^x)$$
$$= x^3 e^x - 3x^2 e^x + 6xe^x - 6e^x$$

i.e. $\int x^3 e^x \, dx = e^x(x^3 - 3x^2 + 6x - 6) + c$

Integrals of the form $\int x^n \cos x \, dx$

Application: Determine $\int x^2 \cos x \, dx$ using a reduction formula

Using the reduction formula of equation (2):
$$\int x^2 \cos x \, dx = I_2 = x^2 \sin x + 2x^1 \cos x - 2(1)I_0$$

and $I_0 = \int x^0 \cos x \, dx = \int \cos x \, dx = \sin x$

Hence $\int x^2 \cos x \, dx = x^2 \sin x + 2x \cos x - 2\sin x + c$

Application: Evaluate $\int_1^2 4t^3 \cos t \, dt$, correct to 4 significant figures

From equation (2),
$$\int t^3 \cos t \, dt = I_3 = t^3 \sin t + 3t^2 \cos t - 3(2)I_1$$
and
$$I_1 = t^1 \sin t + 1t^0 \cos t - 1(0)I_{n-2}$$
$$= t \sin t + \cos t$$

Hence $\int t^3 \cos t\, dt = t^3 \sin t + 3t^2 \cos t - 3(2)[t \sin t + \cos t]$

$$= t^3 \sin t + 3t^2 \cos t - 6t \sin t - 6 \cos t$$

Thus, $\int_1^2 4t^3 \cos t\, dt = \left[4(t^3 \sin t + 3t^2 \cos t - 6t \sin t - 6 \cos t)\right]_1^2$

$$= [4(8 \sin 2 + 12 \cos 2 - 12 \sin 2 - 6 \cos 2)]$$

$$- [4(\sin 1 + 3 \cos 1 - 6 \sin 1 - 6 \cos 1)]$$

$$= (-24.53628) - (-23.31305)$$

$$= -1.223$$

Integrals of the form $\int x^n \sin x\, dx$

Application: Determine $\int x^3 \sin x\, dx$ using a reduction formula

Using equation (4),

$$\int x^3 \sin x\, dx = I_3 = -x^3 \cos x + 3x^2 \sin x - 3(2)I_1$$

and $I_1 = -x^1 \cos x + 1x^0 \sin x = -x \cos x + \sin x$

Hence,

$$\int x^3 \sin x\, dx = -x^3 \cos x + 3x^2 \sin x - 6[-x \cos x + \sin x]$$

$$= \mathbf{-x^3 \cos x + 3x^2 \sin x + 6x \cos x - 6 \sin x + c}$$

Integrals of the form $\int \sin^n x\, dx$

Application: Determine $\int \sin^4 x\, dx$ using a reduction formula

Using equation (5), $\int \sin^4 x\, dx = I_4 = -\dfrac{1}{4} \sin^3 x \cos x + \dfrac{3}{4} I_2$

$I_2 = -\dfrac{1}{2} \sin^1 x \cos x + \dfrac{1}{2} I_0$ and $I_0 = \int \sin^0 x\, dx = \int 1\, dx = x$

Hence

$$\int \sin^4 x\, dx = I_4 = -\frac{1}{4} \sin^3 x \cos x + \frac{3}{4}\left[-\frac{1}{2} \sin x \cos x + \frac{1}{2}(x)\right]$$

$$= -\frac{1}{4}\sin^3 x \cos x - \frac{3}{8}\sin x \cos x + \frac{3}{8}x + c$$

Integrals of the form $\int \cos^n x\, dx$

Application: Determine $\int \cos^4 x\, dx$ using a reduction formula

Using equation (6), $\int \cos^4 x\, dx = I_4 = \frac{1}{4}\cos^3 x \sin x + \frac{3}{4}I_2$

and $I_2 = \frac{1}{2}\cos x \sin x + \frac{1}{2}I_0$ and $I_0 = \int \cos^0 x\, dx = \int 1\, dx = x$

Hence, $\int \cos^4 x\, dx = \frac{1}{4}\cos^3 x \sin x + \frac{3}{4}\left(\frac{1}{2}\cos x \sin x + \frac{1}{2}x\right)$

$$= \frac{1}{4}\cos^3 x \sin x + \frac{3}{8}\cos x \sin x + \frac{3}{8}x + c$$

Application: Evaluate $\int_0^{\pi/2} \cos^5 x\, dx$

From equation (7), $\int_0^{\pi/2} \cos^n x\, dx = I_n = \frac{n-1}{n}I_{n-2}$ (This is usually known as

Wallis's formula)

Thus, $\int_0^{\pi/2} \cos^5 x\, dx = \frac{4}{5}I_3$

$I_3 = \frac{2}{3}I_1$ and $I_1 = \int_0^{\pi/2} \cos^1 x\, dx = \left[\sin x\right]_0^{\pi/2} = (1-0) = 1$

Hence $\int_0^{\pi/2} \cos^5 x\, dx = \frac{4}{5}I_3 = \frac{4}{5}\left[\frac{2}{3}I_1\right] = \frac{4}{5}\left[\frac{2}{3}(1)\right] = \frac{8}{15}$

Further reduction formulae

Application: Determine $\int \tan^7 x\, dx$

From equation (8), $I_n = \frac{\tan^{n-1} x}{n-1} - I_{n-2}$

When $n = 7$, $I_7 = \int \tan^7 x\, dx = \dfrac{\tan^6 x}{6} - I_5$

$I_5 = \dfrac{\tan^4 x}{4} - I_3$ and $I_3 = \dfrac{\tan^2 x}{2} - I_1$

$I_1 = \int \tan x\, dx = \ln(\sec x)$ using $\tan x = \dfrac{\sin x}{\cos x}$ and letting $u = \cos x$

Thus $\int \tan^7 x\, dx = \dfrac{\tan^6 x}{6} - \left[\dfrac{\tan^4 x}{4} - \left(\dfrac{\tan^2 x}{2} - (\ln(\sec x)) \right) \right]$

Hence,

$$\int \tan^7 x\, dx = \frac{1}{6} \tan^6 x - \frac{1}{4} \tan^4 x + \frac{1}{2} \tan^2 x - \ln(\sec x) + c$$

Application: Evaluate $\displaystyle\int_0^{\pi/2} \sin^2 t \cos^6 t\, dt$ using a reduction formula

$$\int_0^{\pi/2} \sin^2 t \cos^6 t\, dt = \int_0^{\pi/2} (1 - \cos^2 t)\cos^6 t\, dt$$
$$= \int_0^{\pi/2} \cos^6 t\, dt - \int_0^{\pi/2} \cos^8 t\, dt$$

If $I_n = \displaystyle\int_0^{\pi/2} \cos^n t\, dt$ then $\displaystyle\int_0^{\pi/2} \sin^2 t \cos^6 t\, dt = I_6 - I_8$

and from equation (7), $I_6 = \dfrac{5}{6} I_4 = \dfrac{5}{6}\left[\dfrac{3}{4} I_2 \right] = \dfrac{5}{6}\left[\dfrac{3}{4}\left(\dfrac{1}{2} I_0 \right) \right]$

and $I_0 = \displaystyle\int_0^{\pi/2} \cos^0 t\, dt = \int_0^{\pi/2} 1\, dt = [x]_0^{\pi/2} = \dfrac{\pi}{2}$

Hence $I_6 = \dfrac{5}{6} \cdot \dfrac{3}{4} \cdot \dfrac{1}{2} \cdot \dfrac{\pi}{2} = \dfrac{15\pi}{96}$ or $\dfrac{5\pi}{32}$

Similarly, $I_8 = \dfrac{7}{8} I_6 = \dfrac{7}{8} \cdot \dfrac{5\pi}{32}$

Thus $\displaystyle\int_0^{\pi/2} \sin^2 t \cos^6 t\, dt = I_6 - I_8 = \dfrac{5\pi}{32} - \dfrac{7}{8} \cdot \dfrac{5\pi}{32} = \dfrac{1}{8} \cdot \dfrac{5\pi}{32} = \dfrac{5\pi}{256}$

Chapter 123 Double and triple integrals

Double integrals

The procedure to determine a double integral of the form:

$\int_{y_1}^{y_2} \int_{x_1}^{x_2} f(x, y)\, dx\, dy$ is:

(i) integrate $f(x, y)$ with respect to x between the limits of $x = x_1$ and $x = x_2$ (where y is regarded as being a constant), and

(ii) integrate the result in (i) with respect to y between the limits of $y = y_1$ and $y = y_2$

It is seen from this procedure that to determine a double integral we start with the innermost integral and then work outwards.

Double integrals may be used to determine areas under curves, second moments of area, centroids and moments of inertia.

(Sometimes $\int_{y_1}^{y_2} \int_{x_1}^{x_2} f(x, y)\, dx\, dy$ is written as: $\int_{y_1}^{y_2} dy \int_{x_1}^{x_2} f(x, y)\, dx$. All this means is that the right hand side integral is determined first).

Application Evaluate $\int_{1}^{3} \int_{2}^{5} (2x - 3y)\, dx\, dy$

Following the above procedure:

(i) $(2x - 3y)$ is integrated with respect to x between $x = 2$ and $x = 5$, with y regarded as a constant

i.e. $\int_{2}^{5} (2x - 3y)\, dx = \left[\dfrac{2x^2}{2} - (3y)x \right]_{2}^{5} = \left[x^2 - 3xy \right]_{2}^{5} = \left[\left(5^2 - 3(5)y \right) - \left(2^2 - 3(2)y \right) \right]$

$= (25 - 15y) - (4 - 6y) = 25 - 15y - 4 + 6y$

$= 21 - 9y$

(ii) $\int_{1}^{3} \int_{2}^{5} (2x - 3y)\, dx\, dy = \int_{1}^{3} (21 - 9y)\, dy = \left[21y - \dfrac{9y^2}{2} \right]_{1}^{3}$

$= \left[\left(21(3) - \dfrac{9(3)^2}{2} \right) - \left(21(1) - \dfrac{9(1)^2}{2} \right) \right]$

$= (63 - 40.5) - (21 - 4.5)$

$= 63 - 40.5 - 21 + 4.5 = 6$

Hence, $\int_{1}^{3} \int_{2}^{5} (2x - 3y)\, dx\, dy = 6$

Application: Evaluate $\int_0^4 \int_1^2 (3x^2 - 2)\, dx\, dy$

Following the above procedure:

(i) $(3x^2 - 2)$ is integrated with respect to x between $x = 1$ and $x = 2$,

i.e. $\displaystyle \int_1^2 (3x^2 - 2)\, dx = \left[\frac{3x^3}{3} - 2x \right]_1^2 = \left[\left(2^3 - 2(2) \right) - \left(1^3 - 2(1) \right) \right]$

$$= (8 - 4) - (1 - 2) = 8 - 4 - 1 + 2 = 5$$

(ii) $\displaystyle \int_0^4 \int_1^2 (3x^2 - 2)\, dx\, dy = \int_0^4 (5)\, dy = \left[5y \right]_0^4 = \left[\left(5(4) \right) - \left(5(0) \right) \right] = 20 - 0 = 20$

Hence, $\displaystyle \int_0^4 \int_1^2 (3x^2 - 2)\, dx\, dy = 20$

Application: Evaluate $\int_1^3 \int_0^2 (2x^2 y)\, dx\, dy$

Following the above procedure:

(i) $(2x^2 y)$ is integrated with respect to x between $x = 0$ and $x = 2$,

i.e. $\displaystyle \int_0^2 (2x^2 y)\, dx = \left[\frac{2x^3 y}{3} \right]_0^2 = \left[\left(\frac{2(2)^3 y}{3} \right) - (0) \right] = \frac{16}{3} y$

(ii) $\displaystyle \int_1^3 \int_0^2 (2x^2 y)\, dx\, dy = \int_1^3 \left(\frac{16}{3} y \right) dy = \left[\frac{16 y^2}{6} \right]_1^3 = \left[\left(\frac{16(3)^2}{6} \right) - \left(\frac{16(1)^2}{6} \right) \right]$

$$= 24 - 2.67 = 21.33$$

Hence, $\displaystyle \int_1^3 \int_0^2 (2x^2 y)\, dx\, dy = 21.33$

Application: Evaluate $\int_1^4 \int_0^\pi (2 + \sin 2\theta)\, d\theta\, dr$

Following the above procedure:

(i) $(2 + \sin 2\theta)$ is integrated with respect to θ between $\theta = 0$ and $\theta = \pi$,

i.e. $\displaystyle \int_0^\pi (2 + \sin 2\theta)\, dx = \left[2\theta - \frac{1}{2} \cos 2\theta \right]_0^\pi = \left[\left(2\pi - \frac{1}{2} \cos 2\pi \right) - \left(0 - \frac{1}{2} \cos 0 \right) \right]$

$$= (2\pi - 0.5) - (0 - 0.5) = 2\pi$$

(ii) $\displaystyle \int_1^4 \int_0^\pi (2 + \sin 2\theta)\, d\theta\, dr = \int_1^4 \left(2\pi \right) dr = \left[2\pi r \right]_1^4 = \left[\left(2\pi(4) \right) - \left(2\pi(1) \right) \right]$

$$= 8\pi - 2\pi = \mathbf{6\pi \text{ or } 18.85}$$

Hence, $\int_1^4 \int_0^\pi \left(2 + \sin 2\theta\right) d\theta \, dr = 18.85$

Triple integrals

The procedure to determine a triple integral of the form: $\int_{z_1}^{z_2} \int_{y_1}^{y_2} \int_{x_1}^{x_2} f(x, y, z) \, dx \, dy \, dz$ is:

(i) integrate $f(x, y, z)$ with respect to x between the limits of $x = x_1$ and $x = x_2$ (where y and z are regarded as being constants),

(ii) integrate the result in (i) with respect to y between the limits of $y = y_1$ and $y = y_1$, and

(iii) integrate the result in (ii) with respect to z between the limits of $z = z_1$ and $z = z_2$

It is seen from this procedure that to determine a triple integral we start with the innermost integral and then work outwards.

Application: Evaluate $\int_1^2 \int_{-1}^3 \int_0^2 (x - 3y + z) \, dx \, dy \, dz$

Following the above procedure:

(i) $(x - 3y + z)$ is integrated with respect to x between $x = 0$ and $x = 2$, with y and z regarded as constants,

i.e. $\int_0^2 (x - 3y + z) \, dx = \left[\dfrac{x^2}{2} - (3y)x + (z)x\right]_0^2 = \left[\left(\dfrac{2^2}{2} - (3y)(2) + (z)(2)\right) - (0)\right]$

$$= 2 - 6y + 2z$$

(ii) $(2 - 6y + 2z)$ is integrated with respect to y between $y = -1$ and $y = 3$, with z regarded as a constant, i.e.

$\int_{-1}^3 (2 - 6y + 2z) \, dy = \left[2y - \dfrac{6y^2}{2} + (2z)y\right]_{-1}^3$

$$= \left[\left(2(3) - \dfrac{6(3)^2}{2} + (2z)(3)\right) - \left(2(-1) - \dfrac{6(-1)^2}{2} + (2z)(-1)\right)\right]$$

$$= [(6 - 27 + 6z) - (-2 - 3 - 2z)] = 6 - 27 + 6z + 2 + 3 + 2z$$
$$= 8z - 16$$

(iii) $(8z - 16)$ is integrated with respect to z between z = 1 and z = 2

i.e. $\displaystyle\int_1^2 (8z - 16)\, dz = \left[\frac{8z^2}{2} - 16z\right]_1^2 = \left[\left(\frac{8(2)^2}{2} - 16(2)\right) - \left(\frac{8(1)^2}{2} - 16(1)\right)\right]$

$= [(16 - 32) - (4 - 16)]$

$= 16 - 32 - 4 + 16 = -4$

Hence, $\displaystyle\int_1^2 \int_{-1}^3 \int_0^2 (x - 3y + z)\, dx\, dy\, dz = -4$

Application: Evaluate $\displaystyle\int_1^3 \int_0^2 \int_0^1 (2a^2 - b^2 + 3c^2)\, da\, db\, dc$

Following the above procedure:

(i) $(2a^2 - b^2 + 3c^2)$ is integrated with respect to 'a' between a = 0 and a = 1, with 'b' and 'c' regarded as constants,

i.e. $\displaystyle\int_0^1 (2a^2 - b^2 + 3c^2)\, da = \left[\frac{2a^3}{3} - (b^2)a + (3c^2)a\right]_0^1$

$= \left[\left(\frac{2}{3} - (b^2) + (3c^2)\right) - (0)\right]$

$= \frac{2}{3} - b^2 + 3c^2$

(ii) $\left(\frac{2}{3} - b^2 + 3c^2\right)$ is integrated with respect to 'b' between b = 0 and b = 2, with c regarded as a constant, i.e.

$\displaystyle\int_0^2 \left(\frac{2}{3} - b^2 + 3c^2\right) db = \left[\frac{2}{3}b - \frac{b^3}{3} + (3c^2)b\right]_0^2 = \left[\left(\frac{2}{3}(2) - \frac{(2)^3}{3} + (3c^2)(2)\right) - (0)\right]$

$= \left(\frac{4}{3} - \frac{8}{3} + 6c^2\right) - (0) = 6c^2 - \frac{4}{3}$

(iii) $\left(6c^2 - \frac{4}{3}\right)$ is integrated with respect to 'c' between c = 1 and c = 3

i.e. $\displaystyle\int_1^3 \left(6c^2 - \frac{4}{3}\right) dc = \left[\frac{6c^3}{3} - \frac{4}{3}c\right]_1^3 = \left[(54 - 4) - \left(2 - \frac{4}{3}\right)\right]$

$= [(50) - (0.67)] = \mathbf{49.33}$

Hence, $\displaystyle\int_1^3 \int_0^2 \int_0^1 (2a^2 - b^2 + 3c^2)\, da\, db\, dc = \mathbf{49.33}$

Chapter 124   Numerical integration

The trapezoidal rule states:

$$\int_a^b y\, dx \approx \left(\begin{array}{c}\text{width of}\\\text{interval}\end{array}\right)\left\{\frac{1}{2}\left(\begin{array}{c}\text{first} + \text{last}\\\text{ordinate}\end{array}\right) + \left(\begin{array}{c}\text{sum of remaining}\\\text{ordinates}\end{array}\right)\right\} \tag{1}$$

The mid-ordinate rule states:

$$\int_a^b y\, dx \approx (\text{width of interval})(\text{sum of mid-ordinates}) \tag{2}$$

Simpson's rule states:

$$\int_a^b y\, dx \approx \frac{1}{3}\left(\begin{array}{c}\text{width of}\\\text{interval}\end{array}\right)\left[\begin{array}{c}\left(\begin{array}{c}\text{first} + \text{last}\\\text{ordinate}\end{array}\right) + 4\left(\begin{array}{c}\text{sum of even}\\\text{ordinates}\end{array}\right)\\ + 2\left(\begin{array}{c}\text{sum of remaining}\\\text{odd ordinates}\end{array}\right)\end{array}\right] \tag{3}$$

Application: Using the trapezoidal rule with 8 intervals, evaluate $\int_1^3 \frac{2}{\sqrt{x}}\, dx$, correct to 3 decimal places

With 8 intervals, the width of each is $\frac{3-1}{8}$ i.e. 0.25 giving ordinates at 1.00, 1.25, 1.50, 1.75, 2.00, 2.25, 2.50, 2.75 and 3.00. Corresponding values of $\frac{2}{\sqrt{x}}$ are shown in the table below.

x	1.00	1.25	1.50	1.75	2.00	2.25	2.50	2.75	3.00
$\frac{2}{\sqrt{x}}$	2.0000	1.7889	1.6330	1.5119	1.4142	1.3333	1.2649	1.2060	1.1547

From equation (1):

$$\int_1^3 \frac{2}{\sqrt{x}}\, dx \approx (0.25)\left\{\begin{array}{l}\frac{1}{2}(2.000 + 1.1547) + 1.7889\\ + 1.6330 + 1.5119 + 1.4142\\ + 1.3333 + 1.2649 + 1.2060\end{array}\right\}$$

$$= 2.932, \text{ correct to 3 decimal places}$$

The greater the number of intervals chosen (i.e. the smaller the interval width) the more accurate will be the value of the definite integral. The exact value is found when the number of intervals is infinite, which is, of course, what the process of integration is based upon. Using integration (see Chapter 115):

$$\int_1^3 \frac{2}{\sqrt{x}}\,dx = \int_1^3 2x^{-\frac{1}{2}}\,dx$$

$$= \left[\frac{2x^{(-1/2)+1}}{-\frac{1}{2}+1}\right]_1^3 = \left[4x^{1/2}\right]_1^3$$

$$= 4\left[\sqrt{x}\right]_1^3 = 4\left[\sqrt{3} - \sqrt{1}\right]$$

$$= 2.928, \text{ correct to 3 decimal places}$$

Application: Using the trapezoidal rule, evaluate $\int_0^{\pi/2} \frac{1}{1+\sin x}\,dx$ using 6 intervals

With 6 intervals, each will have a width of $\dfrac{\frac{\pi}{2}-0}{6}$, i.e. $\dfrac{\pi}{12}$ rad

(or 15°) and the ordinates occur at 0, $\dfrac{\pi}{12}, \dfrac{\pi}{6}, \dfrac{\pi}{4}, \dfrac{\pi}{3}, \dfrac{5\pi}{12}$ and $\dfrac{\pi}{2}$

Corresponding values of $\dfrac{1}{1+\sin x}$ are shown in the table below.

x	0	$\dfrac{\pi}{12}$ (or 15°)	$\dfrac{\pi}{6}$ (or 30°)	$\dfrac{\pi}{4}$ (or 45°)
$\dfrac{1}{1+\sin x}$	1.0000	0.79440	0.66667	0.58579

x	$\dfrac{\pi}{3}$ (or 60°)	$\dfrac{5\pi}{12}$ (or 75°)	$\dfrac{\pi}{2}$ (or 90°)
$\dfrac{1}{1+\sin x}$	0.53590	0.50867	0.50000

From equation (1):

$$\int_0^{\pi/2} \frac{1}{1+\sin x}\, dx \approx \left(\frac{\pi}{12}\right) \begin{bmatrix} \frac{1}{2}(1.00000 + 0.50000) + 0.79440 \\ + 0.66667 + 0.58579 \\ + 0.53590 + 0.50867 \end{bmatrix}$$

$$= 1.006, \text{correct to 4 significant figures}$$

Application: Using the mid-ordinate rule with 8 intervals, evaluate $\int_1^3 \frac{2}{\sqrt{x}}\, dx$, correct to 3 decimal places

With 8 intervals, each will have a width of 0.25 and the ordinates will occur at 1.00, 1.25, 1.50, 1.75, ... and thus mid-ordinates at 1.125, 1.375, 1.625, 1.875......
Corresponding values of $\frac{2}{\sqrt{x}}$ are shown in the following table.

x	1.125	1.375	1.625	1.875	2.125	2.375	2.625	2.875
$\frac{2}{\sqrt{x}}$	1.8856	1.7056	1.5689	1.4606	1.3720	1.2978	1.2344	1.1795

From equation (2):

$$\int_1^3 \frac{2}{\sqrt{x}}\, dx \approx (0.25)[1.8856 + 1.7056 + 1.5689 + 1.4606 + 1.3720$$
$$+ 1.2978 + 1.2344 + 1.1795]$$
$$= 2.926, \text{correct to 3 decimal places}$$

As previously, the greater the number of intervals the nearer the result is to the true value (of 2.928, correct to 3 decimal places).

Application: Using Simpson's rule with 8 intervals, evaluate $\int_1^3 \frac{2}{\sqrt{x}}\, dx$, correct to 3 decimal places:

With 8 intervals, each will have a width of $\frac{3-1}{8}$, i.e. 0.25 and the ordinates occur at 1.00, 1.25, 1.50, 1.75, ..., 3.0. The values of the ordinates are as shown in the table above

Thus, from equation (3):

$$\int_1^3 \frac{2}{\sqrt{x}}\, dx \approx \frac{1}{3}(0.25)[(2.0000 + 1.1547) + 4(1.7889 + 1.5119$$
$$+ 1.3333 + 1.2060) + 2(1.6330 + 1.4142 + 1.2649)]$$

$$= \frac{1}{3}(0.25)[3.1547 + 23.3604 + 8.6242]$$
$$= 2.928, \text{ correct to 3 decimal places}$$

It is noted that the latter answer is exactly the same as that obtained by integration. In general, Simpson's rule is regarded as the most accurate of the three approximate methods used in numerical integration.

Application: An alternating current i has the following values at equal intervals of 2.0 milliseconds.

Time (ms)	0	2.0	4.0	6.0	8.0	10.0	12.0
Current i (A)	0	3.5	8.2	10.0	7.3	2.0	0

Charge, q, in millicoulombs, is given by $q = \int_0^{12.0} i\,dt$. Use Simpson's rule to determine the approximate charge in the 12 millisecond period

From equation (3):

$$\text{Charge, } q = \int_0^{12.0} i\,dt \approx \frac{1}{3}(2.0)[(0+0) + 4(3.5 + 10.0 + 2.0)$$
$$+ 2(8.2 + 7.3)]$$
$$= 62 \text{ mC}$$

Chapter 125 Area under and between curves

The area shown shaded in Figure 125.1 is given by:

$$\text{total shaded area} = \int_a^b f(x)\,dx - \int_b^c f(x)\,dx + \int_c^d f(x)\,dx$$

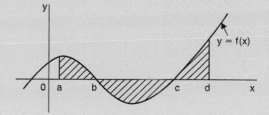

Figure 125.1

The area shown shaded in Figure 125.2, is given by:

$$\text{shaded area} = \int_a^b [f_2(x) - f_1(x)]\,dx$$

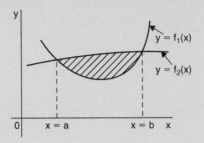

Figure 125.2

Application: The velocity v of a body t seconds after a certain instant is $(2t^2 + 5)$m/s. Find by integration how far it moves in the interval from $t = 0$ to $t = 4$s

Since $2t^2 + 5$ is a quadratic expression, the curve $v = 2t^2 + 5$ is a parabola cutting the v-axis at $v = 5$, as shown in Figure 125.3.

The distance travelled is given by the area under the v/t curve, shown shaded in Figure 125.3.

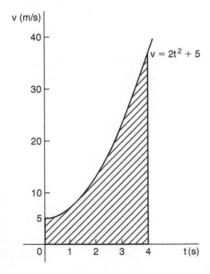

Figure 125.3

By integration,

$$\text{shaded area} = \int_0^4 v\,dt = \int_0^4 (2t^2 + 5)\,dt = \left[\frac{2t^3}{3} + 5t\right]_0^4$$

i.e.   **distance travelled = 62.67 m**

Application: Determine the area enclosed by the curve $y = x^3 + 2x^2 - 5x - 6$ and the x-axis between $x = -3$ and $x = 2$

A table of values is produced and the graph sketched as shown in Figure 125.4 where the area enclosed by the curve and the x-axis is shown shaded.

x	−3	−2	−1	0	1	2
x^3	−27	−8	−1	0	1	8
$2x^2$	18	8	2	0	2	8
$-5x$	15	10	5	0	−5	−10
−6	−6	−6	−6	−6	−6	−6
y	0	4	0	−6	−8	0

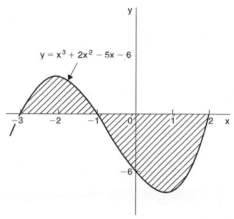

$y = x^3 + 2x^2 - 5x - 6$

Figure 125.4

Shaded area $= \displaystyle\int_{-3}^{-1} y\,dx - \int_{-1}^{2} y\,dx$, the minus sign before the second integral being necessary since the enclosed area is below the x-axis.

Hence, shaded area

$$= \int_{-3}^{-1} (x^3 + 2x^2 - 5x - 6)\,dx - \int_{-1}^{2}(x^3 + 2x^2 - 5x - 6)\,dx$$

$$= \left[\frac{x^4}{4} + \frac{2x^3}{3} - \frac{5x^2}{2} - 6x\right]_{-3}^{-1} - \left[\frac{x^4}{4} + \frac{2x^3}{3} - \frac{5x^2}{2} - 6x\right]_{-1}^{2}$$

$$= \left[\left\{\frac{1}{4} - \frac{2}{3} - \frac{5}{2} + 6\right\} - \left\{\frac{81}{4} - 18 - \frac{45}{2} + 18\right\}\right]$$

$$\quad - \left[\left\{4 + \frac{16}{3} - 10 - 12\right\} - \left\{\frac{1}{4} - \frac{2}{3} - \frac{5}{2} + 6\right\}\right]$$

$$= \left[\left\{ 3\frac{1}{12} \right\} - \left\{ -2\frac{1}{4} \right\} \right] - \left[\left\{ -12\frac{2}{3} \right\} - \left\{ 3\frac{1}{12} \right\} \right] = \left[5\frac{1}{3} \right] - \left[-15\frac{3}{4} \right]$$

$= 21\dfrac{1}{12}$ or 21.083 square units

Application: Find the area enclosed by the curve y = sin 2x, the x-axis and the ordinates x = 0 and $x = \dfrac{\pi}{3}$

A sketch of y = sin 2x is shown in Figure 125.5.

(Note that y = sin 2x has a period of $\dfrac{2\pi}{2}$, i.e. π radians)

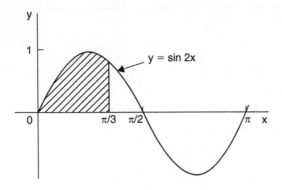

Figure 125.5

Shaded area $= \displaystyle\int_{0}^{\pi/3} y \, dx = \int_{0}^{\pi/3} \sin 2x \, dx$

$$= \left[-\frac{1}{2} \cos 2x \right]_{0}^{\pi/3} = \left\{ -\frac{1}{2} \cos \frac{2\pi}{3} \right\} - \left\{ -\frac{1}{2} \cos 0 \right\}$$

$$= \left\{ -\frac{1}{2} \left(-\frac{1}{2} \right) \right\} - \left\{ -\frac{1}{2} (1) \right\} = \frac{1}{4} + \frac{1}{2} = \frac{3}{4} \text{ square units}$$

Application: Determine the area between the curve y = x³ − 2x² − 8x and the x-axis

y = x³ − 2x² − 8x = x(x² − 2x − 8) = x(x + 2)(x − 4)

When y = 0, x = 0 or (x + 2) = 0 or (x − 4) = 0, i.e. when y = 0, x = 0 or −2 or 4, which means that the curve crosses the x-axis at 0, −2, and 4. Since the curve is a continuous function, only one other co-ordinate value needs to be calculated before a sketch of the curve can be produced. When x = 1, y = −9, showing that the part of the curve between x = 0 and x = 4 is negative. A sketch of y = x³ − 2x² − 8x is shown in Figure 125.6. (Another method of sketching Figure 125.6 would have been to draw up a table of values.)

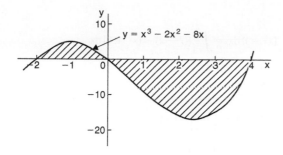

Figure 125.6

Shaded area $= \displaystyle\int_{-2}^{0} (x^3 - 2x^2 - 8x)\, dx - \int_{0}^{4} (x^3 - 2x^2 - 8x)\, dx$

$$= \left[\frac{x^4}{4} - \frac{2x^3}{3} - \frac{8x^2}{2} \right]_{-2}^{0} - \left[\frac{x^4}{4} - \frac{2x^3}{3} - \frac{8x^2}{2} \right]_{0}^{4}$$

$$= \left(6\frac{2}{3} \right) - \left(-42\frac{2}{3} \right)$$

$$= 49\frac{1}{3} \text{ square units}$$

Application: Determine the area enclosed between the curves $y = x^2 + 1$ and $y = 7 - x$

At the points of intersection the curves are equal. Thus, equating the y values of each curve gives:

$$x^2 + 1 = 7 - x$$

from which, $x^2 + x - 6 = 0$

Factorising gives: $(x - 2)(x + 3) = 0$

from which $x = 2$ and $x = -3$

By firstly determining the points of intersection the range of x-values has been found. Tables of values are produced as shown below.

x	−3	−2	−1	0	1	2
$y = x^2 + 1$	10	5	2	1	2	5

x	−3	0	2
$y = 7 - x$	10	7	5

A sketch of the two curves is shown in Figure 125.7.

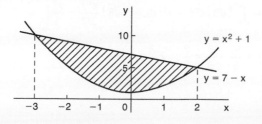

Figure 125.7

Shaded area

$$= \int_{-3}^{2} (7 - x)\, dx - \int_{-3}^{2} (x^2 + 1)\, dx = \int_{-3}^{2} [(7 - x) - (x^2 + 1)]\, dx$$

$$= \int_{-3}^{2} (6 - x - x^2)\, dx = \left[6x - \frac{x^2}{2} - \frac{x^3}{3} \right]_{-3}^{2}$$

$$= \left(12 - 2 - \frac{8}{3} \right) - \left(-18 - \frac{9}{2} + 9 \right) = \left(7\frac{1}{3} \right) - \left(-13\frac{1}{2} \right)$$

$$= 20\frac{5}{6} \text{ sq. units}$$

Application: Calculate the area enclosed by the curves $y = x^2$ and $y^2 = 8x$

At the points of intersection the co-ordinates of the curves are equal.

When $y = x^2$ then $y^2 = x^4$

Hence, at the points of intersection $x^4 = 8x$, by equating the y^2 values.

Thus $x^4 - 8x = 0$, from which $x(x^3 - 8) = 0$, i.e. $x = 0$ or $(x^3 - 8) = 0$

Hence at the points of intersection $x = 0$ or $x = 2$.

When $x = 0$, $y = 0$ and when $x = 2$, $y = 2^2 = 4$

Hence the points of intersection of the curves $y = x^2$ and $y^2 = 8x$ are (0, 0) and (2, 4).

A sketch of $y = x^2$ and $y^2 = 8x$ is shown in Figure 125.8.

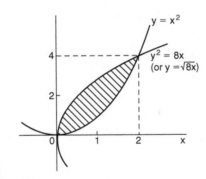

Figure 125.8

Shaded area

$$= \int_{0}^{2} \left(\sqrt{8x} - x^2 \right) dx = \int_{0}^{2} \left(\sqrt{8} \right) x^{1/2} - x^2)\, dx = \left[\left(\sqrt{8} \right) \frac{x^{3/2}}{3/2} - \frac{x^3}{3} \right]_{0}^{2}$$

$$= \left\{ \frac{\sqrt{8}\sqrt{8}}{3/2} - \frac{8}{3} \right\} - \{0\} = \frac{16}{3} - \frac{8}{3} = \frac{8}{3} = 2\frac{2}{3} \text{ sq. units}$$

Application: Determine by integration the area bounded by the three straight lines $y = 4 - x$, $y = 3x$ and $3y = x$

Each of the straight lines are shown sketched in Figure 125.9.

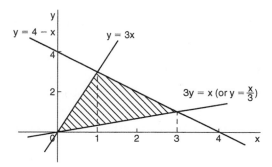

Figure 125.9

$$\text{Shaded area} = \int_0^1 \left(3x - \frac{x}{3}\right) dx + \int_1^3 \left[(4 - x) - \frac{x}{3}\right] dx$$

$$= \left[\frac{3x^2}{2} - \frac{x^2}{6}\right]_0^1 + \left[4x - \frac{x^2}{2} - \frac{x^2}{6}\right]_1^3$$

$$= \left[\left(\frac{3}{2} - \frac{1}{6}\right) - (0)\right] + \left[\left(12 - \frac{9}{2} - \frac{9}{6}\right) - \left(4 - \frac{1}{2} - \frac{1}{6}\right)\right]$$

$$= \left(1\frac{1}{3}\right) + \left(6 - 3\frac{1}{3}\right)$$

$$= 4 \text{ square units}$$

126 Mean or average values

The mean or average value of the curve shown in Figure 126.1, between $x = a$ and $x = b$, is given by:

$$\text{mean or average value, } \bar{y} = \frac{1}{b - a} \int_a^b f(x)\, dx$$

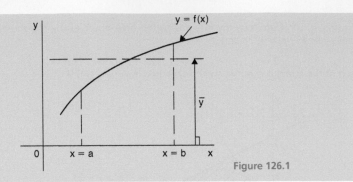

Figure 126.1

Application: Determine the mean value of $y = 5x^2$ between $x = 1$ and $x = 4$

Mean value, $\bar{y} = \dfrac{1}{4-1} \displaystyle\int_1^4 y\,dx = \dfrac{1}{3} \displaystyle\int_1^4 5x^2\,dx$

$$= \dfrac{1}{3}\left[\dfrac{5x^3}{3}\right]_1^4 = \dfrac{5}{9}\left[x^3\right]_1^4 = \dfrac{5}{9}(64-1)$$

$$= 35$$

Application: A sinusoidal voltage is given by $v = 100\sin \omega t$ volts. Determine the mean value of the voltage over half a cycle using integration

Half a cycle means the limits are 0 to π radians.

Mean value, $\bar{v} = \dfrac{1}{\pi - 0} \displaystyle\int_0^\pi v\,d(\omega t) = \dfrac{1}{\pi} \displaystyle\int_0^\pi 100\sin \omega t\,d(\omega t)$

$$= \dfrac{100}{\pi}\left[-\cos \omega t\right]_0^\pi = \dfrac{100}{\pi}\left[(-\cos \pi) - (-\cos 0)\right]$$

$$= \dfrac{100}{\pi}\left[(+1) - (-1)\right] = \dfrac{200}{\pi} = 63.66 \text{ volts}$$

[Note that for a sine wave, **mean value $= \dfrac{2}{\pi} \times$ maximum value**

In this case, mean value $= \dfrac{2}{\pi} \times 100 = 63.66\,\text{V}$]

Application: The number of atoms, N, remaining in a mass of material during radioactive decay after time t seconds is given by $N = N_0 e^{-\lambda t}$, where N_0 and λ are constants. Determine the mean number of atoms in the mass of material for the time period $t = 0$ and $t = \dfrac{1}{\lambda}$

Mean number of atoms

$$= \frac{1}{\frac{1}{\lambda} - 0} \int_0^{1/\lambda} N \, dt = \frac{1}{\frac{1}{\lambda}} \int_0^{1/\lambda} N_0 e^{-\lambda t} \, dt = \lambda N_0 \int_0^{1/\lambda} e^{-\lambda t} \, dt$$

$$= \lambda N_0 \left[\frac{e^{-\lambda t}}{-\lambda} \right]_0^{1/\lambda} = -N_0 \left[e^{-\lambda(1/\lambda)} - e^0 \right] = -N_0 \left[e^{-1} - e^0 \right]$$

$$= +N_0 \left[e^0 - e^{-1} \right] = N_0 \left[1 - e^{-1} \right] = 0.632 \, N_0$$

Chapter 127 Root mean square values

With reference to Figure 126.1, page 356, the r.m.s. value of $y = f(x)$ over the range $x = a$ to $x = b$ is given by:

$$\text{r.m.s. value} = \sqrt{\left\{ \frac{1}{b-a} \int_a^b y^2 \, dx \right\}}$$

The r.m.s. value of an alternating current is defined as 'that current which will give the same heating effect as the equivalent direct current'.

Application: Determine the r.m.s. value of $y = 2x^2$ between $x = 1$ and $x = 4$

$$\text{R.m.s. value} = \sqrt{\left\{ \frac{1}{4-1} \int_1^4 y^2 \, dx \right\}} = \sqrt{\left\{ \frac{1}{3} \int_1^4 (2x^2)^2 \, dx \right\}}$$

$$= \sqrt{\left\{ \frac{1}{3} \int_1^4 4x^4 \, dx \right\}}$$

$$= \sqrt{\left\{ \frac{4}{3} \left[\frac{x^5}{5} \right]_1^4 \right\}} = \sqrt{\left\{ \frac{4}{15} (1024 - 1) \right\}} = \sqrt{272.8} = 16.5$$

Application: A sinusoidal voltage has a maximum value of 100 V. Calculate its r.m.s. value

A sinusoidal voltage v having a maximum value of 10 V may be written as $v = 10 \sin \theta$. Over the range $\theta = 0$ to $\theta = \pi$,

$$\text{r.m.s. value} = \sqrt{\left\{ \frac{1}{\pi - 0} \int_0^\pi v^2 \, d\theta \right\}} = \sqrt{\left\{ \frac{1}{\pi} \int_0^\pi (100 \sin \theta)^2 \, d\theta \right\}}$$

$$= \sqrt{\left\{\frac{10000}{\pi} \int_0^\pi \sin^2 \theta \, d\theta\right\}} \quad \text{which is not a 'standard' integral}$$

It is shown in chapter 54 that $\cos 2A = 1 - 2\sin^2 A$

Rearranging $\cos 2A = 1 - 2\sin^2 A$ gives $\sin^2 A = \frac{1}{2}(1 - \cos 2A)$

Hence,

$$\sqrt{\left\{\frac{10000}{\pi} \int_0^\pi \sin^2 \theta \, d\theta\right\}} = \sqrt{\left\{\frac{10000}{\pi} \int_0^\pi \frac{1}{2}(1 - \cos 2\theta) \, d\theta\right\}}$$

$$= \sqrt{\left\{\frac{10000}{\pi} \frac{1}{2}\left[\theta - \frac{\sin 2\theta}{2}\right]_0^\pi\right\}}$$

$$= \sqrt{\left\{\frac{10000}{\pi} \frac{1}{2}\left[\left(\pi - \frac{\sin 2\pi}{2}\right) - \left(0 - \frac{\sin 0}{2}\right)\right]\right\}}$$

$$= \sqrt{\left\{\frac{10000}{\pi} \frac{1}{2}[\pi]\right\}} = \sqrt{\left\{\frac{10000}{2}\right\}} = \frac{100}{\sqrt{2}}$$

$$= 70.71 \text{ volts}$$

[Note that for a sine wave, **r.m.s. value** $= \dfrac{1}{\sqrt{2}} \times$ **maximum value**.

In this case, r.m.s. value $= \dfrac{1}{\sqrt{2}} \times 100 = 70.71\text{V}$]

Application: In a frequency distribution the average distance from the mean, y, is related to the variable, x, by the equation $y = 2x^2 - 1$. Determine, correct to 3 significant figures, the r.m.s. deviation from the mean for values of x from -1 to $+4$

R.m.s. deviation

$$= \sqrt{\left\{\frac{1}{4--1}\int_{-1}^4 y^2 \, dx\right\}} = \sqrt{\left\{\frac{1}{5}\int_{-1}^4 (2x^2 - 1)^2 \, dx\right\}}$$

$$= \sqrt{\left\{\frac{1}{5}\int_{-1}^4 (4x^4 - 4x^2 + 1) \, dx\right\}}$$

$$= \sqrt{\left\{\frac{1}{5}\left[\frac{4x^5}{5} - \frac{4x^3}{3} + x\right]_{-1}^4\right\}}$$

$$= \sqrt{\left\{\frac{1}{5}\left[\left(\frac{4}{5}(4)^5 - \frac{4}{3}(4)^3 + 4\right) - \left(\frac{4}{5}(-1)^5 - \frac{4}{3}(-1)^3 + (-1)\right)\right]\right\}}$$

$$= \sqrt{\left\{\frac{1}{5}[(737.87) - (-0.467)]\right\}} = \sqrt{\left\{\frac{1}{5}[738.34]\right\}}$$

$= \sqrt{147.67} = 12.152 = 12.2$, correct to 3 significant figures.

Chapter 128 Volumes of solids of revolution

With reference to Figure 128.1, the volume of solid of revolution V obtained by rotating the shaded area through one revolution is given by:

$$V = \int_a^b \pi y^2 \, dx \quad \text{about the x-axis}$$

$$V = \int_c^d \pi x^2 \, dy \quad \text{about the y-axis}$$

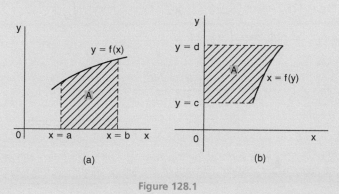

Figure 128.1

Application: The curve $y = x^2 + 4$ is rotated one revolution about (a) the x-axis, and (b) the y-axis, between the limits $x = 1$ and $x = 4$. Determine the volume of the solid of revolution produced in each case

(a) Revolving the shaded area shown in Figure 128.2 about the x-axis 360° produces a solid of revolution given by:

$$\text{Volume} = \int_1^4 \pi y^2 \, dx = \int_1^4 \pi (x^2 + 4)^2 \, dx$$

$$= \int_1^4 \pi (x^4 + 8x^2 + 16) \, dx = \pi \left[\frac{x^5}{5} + \frac{8x^3}{3} + 16x \right]_1^4$$

$$= \pi[(204.8 + 170.67 + 64) - (0.2 + 2.67 + 16)]$$

$$= \mathbf{420.6\pi \ cubic \ units}$$

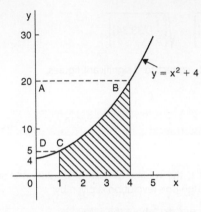

Figure 128.2

(b) The volume produced when the curve $y = x^2 + 4$ is rotated about the y-axis between $y = 5$ (when $x = 1$) and $y = 20$ (when $x = 4$), i.e. rotating area ABCD of Figure 128.2 about the y-axis is given by: volume $= \int_5^{20} \pi x^2 \, dy$

Since $y = x^2 + 4$, then $x^2 = y - 4$

Hence,

$$\text{volume} = \int_5^{20} \pi(y - 4) \, dy = \pi \left[\frac{y^2}{2} - 4y \right]_5^{20} = \pi[(120) - (-7.5)]$$

$$= \textbf{127.5}\boldsymbol{\pi} \textbf{ cubic units}$$

Application: The area enclosed by the curve $y = 3e^{\frac{x}{3}}$, the x-axis and ordinates $x = -1$ and $x = 3$ is rotated 360° about the x-axis. Determine the volume generated.

A sketch of $y = 3e^{\frac{x}{3}}$ is shown in Figure 128.3.

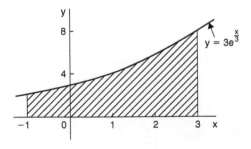

Figure 128.3

When the shaded area is rotated 360° about the x-axis then:

$$\text{volume generated} = \int_{-1}^{3} \pi y^2 \, dx = \int_{-1}^{3} \pi \left(3e^{\frac{x}{3}} \right)^2 dx = 9\pi \int_{-1}^{3} e^{\frac{2x}{3}} \, dx$$

$$= 9\pi \left[\frac{e^{\frac{2x}{3}}}{\frac{2}{3}}\right]_{-1}^{3} = \frac{27\pi}{2}\left(e^2 - e^{-\frac{2}{3}}\right)$$

$= 92.82\pi$ cubic units

Application: Calculate the volume of a frustum of a sphere of radius 4 cm that lies between two parallel planes at 1 cm and 3 cm from the centre and on the same side of it

The volume of a frustum of a sphere may be determined by integration by rotating the curve $x^2 + y^2 = 4^2$ (i.e. a circle, centre 0, radius 4) one revolution about the x-axis, between the limits $x = 1$ and $x = 3$ (i.e. rotating the shaded area of Figure 128.4).

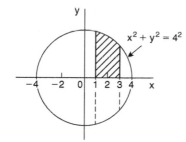

Figure 128.4

$$\text{Volume of frustum} = \int_{1}^{3} \pi y^2 \, dx = \int_{1}^{3} \pi(4^2 - x^2) \, dx$$

$$= \pi\left[16x - \frac{x^3}{3}\right]_{1}^{3} = \pi\left[(39) - \left(15\frac{2}{3}\right)\right]$$

$$= 23\frac{1}{3}\pi \text{ cubic units}$$

Chapter 129 Centroids

Centroid of area between a curve and the x-axis

If $\bar{x}$ and $\bar{y}$ denote the co-ordinates of the centroid C of area A in Figure 129.1 then:

$$\bar{x} = \frac{\int_{a}^{b} xy \, dx}{\int_{a}^{b} y \, dx} \qquad \text{and} \qquad \bar{y} = \frac{\frac{1}{2}\int_{a}^{b} y^2 \, dx}{\int_{a}^{b} y \, dx}$$

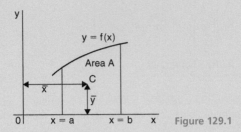

Figure 129.1

Centroid of area between a curve and the y-axis

If $\bar{x}$ and $\bar{y}$ denote the co-ordinates of the centroid C of area A in Figure 129.2 then:

$$\bar{x} = \frac{\frac{1}{2}\int_c^d x^2\, dy}{\int_c^d x\, dy} \qquad \text{and} \qquad \bar{y} = \frac{\int_c^d xy\, dy}{\int_c^d x\, dy}$$

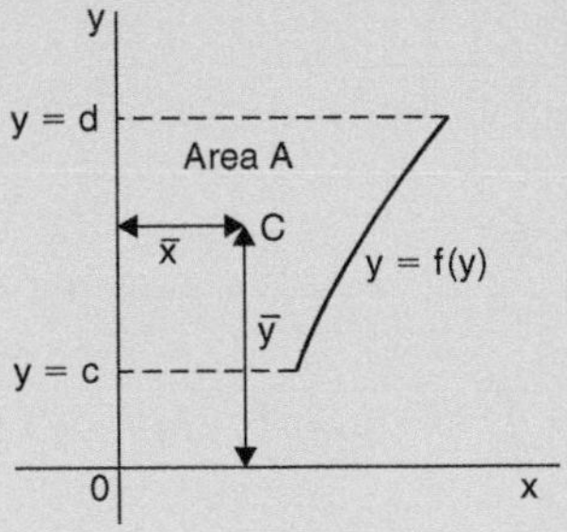

Figure 129.2

Application: Find the position of the centroid of the area bounded by the curve $y = 3x^2$, the x-axis and the ordinates $x = 0$ and $x = 2$

If $(\bar{x}, \bar{y})$ are the co-ordinates of the centroid of the given area then:

$$\bar{x} = \frac{\int_0^2 xy\, dx}{\int_0^2 y\, dx} = \frac{\int_0^2 x(3x^2)\, dx}{\int_0^2 3x^2\, dx} = \frac{\int_0^2 3x^3\, dx}{\int_0^2 3x^2\, dx}$$

$$= \frac{\left[\dfrac{3x^4}{4}\right]_0^2}{\left[x^3\right]_0^2} = \frac{12}{8} = 1.5$$

$$\bar{y} = \frac{\frac{1}{2}\int_0^2 y^2\, dx}{\int_0^2 y\, dx} = \frac{\frac{1}{2}\int_0^2 (3x^2)^2\, dx}{8} = \frac{\frac{1}{2}\int_0^2 9x^4\, dx}{8}$$

$$= \frac{\dfrac{9}{2}\left[\dfrac{x^5}{5}\right]_0^2}{8} = \frac{\dfrac{9}{2}\left(\dfrac{32}{5}\right)}{8} = \frac{18}{5} = 3.6$$

Hence the centroid lies at (1.5, 3.6)

Application: Locate the position of the centroid enclosed by the curves $y = x^2$ and $y^2 = 8x$

Figure 129.3 shows the two curves intersecting at (0, 0) and (2, 4). These are the same curves as used in the application on page 354, where the shaded area was calculated as $2\frac{2}{3}$ square units. Let the co-ordinates of centroid C be $\overline{x}$ and $\overline{y}$

By integration, $\qquad \overline{x} = \dfrac{\displaystyle\int_0^2 xy\,dx}{\displaystyle\int_0^2 y\,dx}$

The value of y is given by the height of the typical strip shown in Figure 129.3, i.e. $y = \sqrt{8x} - x^2$

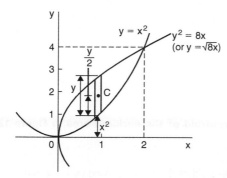

Figure 129.3

Hence,

$$\overline{x} = \frac{\displaystyle\int_0^2 x\left(\sqrt{8x} - x^2\right)dx}{2\frac{2}{3}} = \frac{\displaystyle\int_0^2 \left(\sqrt{8}\,x^{3/2} - x^3\right)}{2\frac{2}{3}} = \frac{\left[\sqrt{8}\,\dfrac{x^{5/2}}{\frac{5}{2}} - \dfrac{x^4}{4}\right]_0^2}{2\frac{2}{3}}$$

$$= \frac{\left(\sqrt{8}\,\dfrac{\sqrt{2^5}}{\frac{5}{2}} - 4\right)}{2\frac{2}{3}} = \frac{2\frac{2}{5}}{2\frac{2}{3}} = 0.9$$

Care needs to be taken when finding $\bar{y}$ in such examples as this.

From Figure 129.3, $y = \sqrt{8x} - x^2$ and $\dfrac{y}{2} = \dfrac{1}{2}\left(\sqrt{8x} - x^2\right)$

The perpendicular distance from centroid C of the strip to Ox is $\dfrac{1}{2}\left(\sqrt{8x} - x^2\right) + x^2$

Taking moments about Ox gives:

$$(\text{total area}) \, (\bar{y}) = \sum_{x=0}^{x=2} (\text{area of strip})(\text{perpendicular distance of centroid of strip to Ox})$$

Hence, $(\text{area}) \, (\bar{y}) = \displaystyle\int_0^2 \left[\sqrt{8x} - x^2\right]\left[\dfrac{1}{2}\left(\sqrt{8x} - x^2\right) + x^2\right] dx$

i.e. $\left(2\dfrac{2}{3}\right)(\bar{y}) = \displaystyle\int_0^2 \left[\sqrt{8x} - x^2\right]\left[\dfrac{\sqrt{8x}}{2} + \dfrac{x^2}{2}\right] dx$

$$= \int_0^2 \left(\dfrac{8x}{2} - \dfrac{x^4}{2}\right) dx = \left[\dfrac{8x^2}{4} - \dfrac{x^5}{10}\right]_0^2$$

$$= \left(8 - 3\dfrac{1}{5}\right) - (0) = 4\dfrac{4}{5}$$

Hence $\bar{y} = \dfrac{4\dfrac{4}{5}}{2\dfrac{2}{3}} = 1.8$

Thus the position of the centroid of the enclosed area in Figure 129.3 is at (0.9, 1.8)

Application: Locate the centroid of the area enclosed by the curve $y = 2x^2$, the y-axis and ordinates $y = 1$ and $y = 4$, correct to 3 decimal places

$$\bar{x} = \dfrac{\dfrac{1}{2}\displaystyle\int_1^4 x^2 \, dy}{\displaystyle\int_1^4 x \, dy} = \dfrac{\dfrac{1}{2}\displaystyle\int_1^4 \dfrac{y}{2} \, dy}{\displaystyle\int_1^4 \sqrt{\dfrac{y}{2}} \, dy} = \dfrac{\dfrac{1}{2}\left[\dfrac{y^2}{4}\right]_1^4}{\left[\dfrac{2y^{3/2}}{3\sqrt{2}}\right]_1^4} = \dfrac{\dfrac{15}{8}}{\dfrac{14}{3\sqrt{2}}} = 0.568$$

$$\bar{y} = \dfrac{\displaystyle\int_1^4 xy \, dy}{\displaystyle\int_1^4 x \, dy} = \dfrac{\displaystyle\int_1^4 \sqrt{\dfrac{y}{2}} \, (y) \, dy}{\dfrac{14}{3\sqrt{2}}} = \dfrac{\displaystyle\int_1^4 \dfrac{y^{3/2}}{\sqrt{2}} \, dy}{\dfrac{14}{3\sqrt{2}}}$$

$$= \frac{\dfrac{1}{\sqrt{2}} \left[\dfrac{y^{5/2}}{\dfrac{5}{2}} \right]_1^4}{\dfrac{14}{3\sqrt{2}}} = \frac{\dfrac{2}{5\sqrt{2}}(31)}{\dfrac{14}{3\sqrt{2}}} = 2.657$$

Hence the position of the centroid is at (0.568, 2.657)

Chapter 130 Theorem of Pappus

A theorem of Pappus states:

'If a plane area is rotated about an axis in its own plane but not intersecting it, the volume of the solid formed is given by the product of the area and the distance moved by the centroid of the area'.

With reference to Figure 130.1, when the curve $y = f(x)$ is rotated one revolution about the x-axis between the limits $x = a$ and $x = b$, the volume V generated is given by:

$$\text{volume } V = (A)(2\pi\bar{y}), \text{ from which, } \bar{y} = \frac{V}{2\pi A}$$

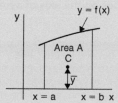

Figure 130.1

Application: Determine the position of the centroid of a semicircle of radius r by using the theorem of Pappus

A semicircle is shown in Figure 130.2 with its diameter lying on the x-axis and its centre at the origin.

Area of semicircle $= \dfrac{\pi r^2}{2}$. When the area is rotated about the x-axis one revolution a sphere is generated of volume $\dfrac{4}{3}\pi r^3$

Let centroid C be at a distance $\bar{y}$ from the origin as shown in Figure 130.2.

From the theorem of Pappus, volume generated = area × distance moved through by centroid

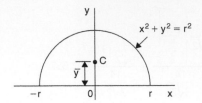

Figure 130.2

i.e.　　　$\dfrac{4}{3}\pi r^3 = \left(\dfrac{\pi r^2}{2}\right)(2\pi\bar{y})$

Hence　　　$\bar{y} = \dfrac{\dfrac{4}{3}\pi r^3}{\pi^2 r^2} = \dfrac{4r}{3\pi}$

[By integration,

$$\bar{y} = \dfrac{\dfrac{1}{2}\displaystyle\int_{-r}^{r} y^2\,dx}{\text{area}} = \dfrac{\dfrac{1}{2}\displaystyle\int_{-r}^{r}(r^2 - x^2)\,dx}{\dfrac{\pi r^2}{2}} = \dfrac{\dfrac{1}{2}\left[r^2 x - \dfrac{x^3}{3}\right]_{-r}^{r}}{\dfrac{\pi r^2}{2}}$$

$$= \dfrac{\dfrac{1}{2}\left[\left(r^3 - \dfrac{r^3}{3}\right) - \left(-r^3 + \dfrac{r^3}{3}\right)\right]}{\dfrac{\pi r^2}{2}} = \dfrac{4r}{3\pi}$$

Hence the centroid of a semicircle lies on the axis of symmetry, distance $\dfrac{4r}{3\pi}$ (or $0.424\,r$) from its diameter.

Application: (a) Calculate the area bounded by the curve $y = 2x^2$, the x-axis and ordinates $x = 0$ and $x = 3$ (b) If the area in part (a) is revolved (i) about the x-axis and (ii) about the y-axis, find the volumes of the solids produced, and (c) locate the position of the centroid using (i) integration, and (ii) the theorem of Pappus

(a)　The required area is shown shaded in Figure 130.3.

$$\text{Area} = \int_0^3 y\,dx = \int_0^3 2x^2\,dx = \left[\dfrac{2x^3}{3}\right]_0^3 = 18\ \text{square units}$$

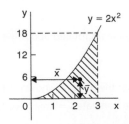

Figure 130.3

(b) (i)   When the shaded area of Figure 130.3 is revolved 360° about the x-axis, the volume generated

$$= \int_0^3 \pi y^2 \, dx = \int_0^3 \pi (2x^2)^2 \, dx = \int_0^3 4\pi x^4 \, dx$$

$$= 4\pi \left[\frac{x^5}{5} \right]_0^3 = 4\pi \left(\frac{243}{5} \right)$$

$$= \textbf{194.4}\boldsymbol{\pi} \textbf{ cubic units}$$

(ii)   When the shaded area of Figure 130.3 is revolved 360° about the y-axis, the volume generated = (volume generated by x = 3) − (volume generated by y = 2x²)

$$= \int_0^{18} \pi (3)^2 \, dy - \int_0^{18} \pi \left(\frac{y}{2} \right) dy = \pi \int_0^{18} \left(9 - \frac{y}{2} \right) dy$$

$$= \pi \left[9y - \frac{y^2}{4} \right]_0^{18}$$

$$= \textbf{81}\boldsymbol{\pi} \textbf{ cubic units}$$

(c)   If the co-ordinates of the centroid of the shaded area in Figure 130.3 are $(\overline{x}, \overline{y})$ then:

(i) by integration,

$$\overline{x} = \frac{\int_0^3 xy \, dx}{\int_0^3 y \, dx} = \frac{\int_0^3 x(2x^2) \, dx}{18} = \frac{\int_0^3 2x^3 \, dx}{18} = \frac{\left[\frac{2x^4}{4} \right]_0^3}{18}$$

$$= \frac{81}{36} = 2.25$$

$$\overline{y} = \frac{\frac{1}{2} \int_0^3 y^2 \, dx}{\int_0^3 y \, dx} = \frac{\frac{1}{2} \int_0^3 (2x^2)^2 \, dx}{18} = \frac{\frac{1}{2} \int_0^3 4x^4 \, dx}{18}$$

$$= \frac{\frac{1}{2} \left[\frac{4x^5}{5} \right]_0^3}{18} = 5.4$$

(ii)   using the theorem of Pappus:
Volume generated when shaded area is revolved about 0y = (area)(2π$\overline{x}$)

i.e.   81π = (18)(2π$\overline{x}$), from which, $\overline{x} = \dfrac{81\pi}{36\pi} = 2.25$

Volume generated when shaded area is revolved about 0x = (area)(2π$\overline{y}$)

i.e.   194.4π = (18)(2π$\overline{y}$), from which, $\overline{y} = \dfrac{194.4\pi}{36\pi} = 5.4$

Hence, the centroid of the shaded area in Figure 130.3 is at (2.25, 5.4)

Application: A metal disc has a radius of 5.0 cm and is of thickness 2.0 cm. A semicircular groove of diameter 2.0 cm is machined centrally around the rim to form a pulley. Using Pappus' theorem, determine the volume and mass of metal removed and the volume and mass of the pulley if the density of the metal is 8000 kg/m³

A side view of the rim of the disc is shown in Figure 130.4.

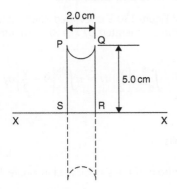

Figure 130.4

When area PQRS is rotated about axis XX the volume generated is that of the pulley.

The centroid of the semicircular area removed is at a distance of $\dfrac{4r}{3\pi}$ from its diameter (see earlier example), i.e. $\dfrac{4(1.0)}{3\pi}$, i.e. 0.424 cm from PQ. Thus the distance of the centroid from XX is (5.0 − 0.424), i.e. 4.576 cm.

The distance moved through in one revolution by the centroid is 2π(4.576) cm.

Area of semicircle $= \dfrac{\pi r^2}{2} = \dfrac{\pi(1.0)^2}{2} = \dfrac{\pi}{2}\text{ cm}^2$.

By the theorem of Pappus,

volume generated = area × distance moved by centroid

$$= \left(\frac{\pi}{2}\right)(2\pi)(4.576)$$

i.e.   **volume of metal removed = 45.16 cm³**

Mass of metal removed = density × volume

$$= 8000\,\text{kg/m}^3 \times \frac{45.16}{10^6}\,\text{m}^3$$
$$= \textbf{0.361 kg or 361 g}$$

Volume of pulley = volume of cylindrical disc − volume of metal removed

$$= \pi(5.0)^2(2.0) - 45.16 = \textbf{111.9 cm}^3$$

Mass of pulley = density × volume

$$= 8000\,\text{kg/m}^3 \times \frac{111.9}{10^6}\,\text{m}^3 = \textbf{0.895 kg or 895 g}$$

131 Second moments of area

Table 131.1 **Summary of standard results of the second moments of areas of regular sections**

Shape	Position of axis	Second moment of area, I	Radius of gyration, k
Rectangle	(1) Coinciding with b	$\dfrac{bl^3}{3}$	$\dfrac{l}{\sqrt{3}}$
length l	(2) Coinciding with l	$\dfrac{lb^3}{3}$	$\dfrac{b}{\sqrt{3}}$
breadth b	(3) Through centroid, parallel to b	$\dfrac{bl^3}{12}$	$\dfrac{l}{\sqrt{12}}$
	(4) Through centroid, parallel to l	$\dfrac{lb^3}{12}$	$\dfrac{b}{\sqrt{12}}$
Triangle	(1) Coinciding with b	$\dfrac{bh^3}{12}$	$\dfrac{h}{\sqrt{6}}$
Perpendicular height h	(2) Through centroid, parallel to base	$\dfrac{bh^3}{36}$	$\dfrac{h}{\sqrt{18}}$
base b	(3) Through vertex, parallel to base	$\dfrac{bh^3}{4}$	$\dfrac{h}{\sqrt{2}}$
Circle radius r	(1) Through centre, perpendicular to plane (i.e. polar axis)	$\dfrac{\pi r^4}{2}$	$\dfrac{r}{\sqrt{2}}$
	(2) Coinciding with diameter	$\dfrac{\pi r^4}{4}$	$\dfrac{r}{2}$
	(3) About a tangent	$\dfrac{5\pi r^4}{4}$	$\dfrac{\sqrt{5}}{2}r$
Semicircle radius r	Coinciding with diameter	$\dfrac{\pi r^4}{8}$	$\dfrac{r}{2}$

Parallel axis theorem

If C is the centroid of area A in Figure 131.1, then:

$$I_{DD} = I_{GG} + Ad^2$$

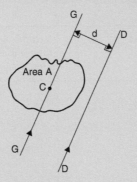

Figure 131.1

Perpendicular axis theorem

If OX and OY lie in the plane of area A in Figure 131.2, then:

$$I_{OZ} = I_{OX} + I_{OY}$$

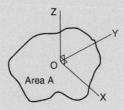

Figure 131.2

Application: Determine the second moment of area and the radius of gyration about axes AA, BB and CC for the rectangle shown in Figure 131.3

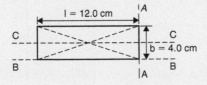

Figure 131.3

From Table 131.1, the second moment of area about axis AA,

$$I_{AA} = \frac{bl^3}{3} = \frac{(4.0)(12.0)^3}{3} = 2304 \text{ cm}^4$$

Radius of gyration, $k_{AA} = \frac{l}{\sqrt{3}} = \frac{12.0}{\sqrt{3}} = 6.93 \text{ cm}$

Similarly, $\quad I_{BB} = \dfrac{lb^3}{3} = \dfrac{(12.0)(4.0)^3}{3} = 256 \text{ cm}^4$

and $\quad k_{BB} = \dfrac{b}{\sqrt{3}} = \dfrac{4.0}{\sqrt{3}} = 2.31 \text{ cm}$

The second moment of area about the centroid of a rectangle is $\dfrac{bl^3}{12}$ when the axis through the centroid is parallel with the breadth b. In this case, the axis CC is parallel with the length l

Hence $\quad I_{CC} = \dfrac{lb^3}{12} = \dfrac{(12.0)(4.0)^3}{12} = 64 \text{ cm}^4$

and $\quad k_{CC} = \dfrac{b}{\sqrt{12}} = \dfrac{4.0}{\sqrt{12}} = 1.15 \text{ cm}$

Application: Find the second moment of area and the radius of gyration about axis PP for the rectangle shown in Figure 131.4

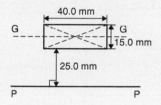

Figure 131.4

$I_{GG} = \dfrac{lb^3}{12}$ where $l = 40.0\,\text{mm}$ and $b = 15.0\,\text{mm}$

Hence $I_{GG} = \dfrac{(40.0)(15.0)^3}{12} = 11250 \text{ mm}^4$

From the parallel axis theorem, $I_{PP} = I_{GG} + Ad^2$, where $A = 40.0 \times 15.0 = 600\,\text{mm}^2$ and $d = 25.0 + 7.5 = 32.5\,\text{mm}$, the perpendicular distance between GG and PP.

Hence, $\mathbf{I_{PP}} = 11250 + (600)(32.5)^2 = \mathbf{645000\,mm^4}$

$I_{PP} = Ak_{PP}^2$, from which, $k_{PP} = \sqrt{\dfrac{I_{PP}}{\text{area}}} = \sqrt{\left(\dfrac{645000}{600}\right)} = 32.79 \text{ mm}$

Application: Determine the second moment of area and radius of gyration about axis QQ of the triangle BCD shown in Figure 131.5

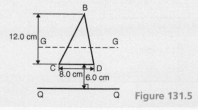

Figure 131.5

372 *Mathematics Pocket Book for Engineers and Scientists*

Using the parallel axis theorem: $I_{QQ} = I_{GG} + Ad^2$, where I_{GG} is the second moment

of area about the centroid of the triangle, i.e. $\dfrac{bh^3}{36} = \dfrac{(8.0)(12.0)^3}{36} = 384\,cm^4$,

A is the area of the triangle $= \dfrac{1}{2}bh = \dfrac{1}{2}(8.0)(12.0) = 48\,cm^2$ and d is the distance

between axes GG and $QQ = 6.0 + \dfrac{1}{3}(12.0) = 10\,cm$

Hence the second moment of area about axis QQ,

$$I_{QQ} = 384 + (48)(10)^2 = \mathbf{5184\,cm^4}$$

Radius of gyration, $K_{QQ} = \sqrt{\dfrac{I_{QQ}}{\text{area}}} = \sqrt{\left(\dfrac{5184}{48}\right)} = 10.4\,cm$

Application: Determine the second moment of area and radius of gyration of the circle shown in Figure 131.6 about axis YY

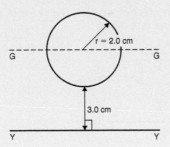

Figure 131.6

In Figure 131.6, $I_{GG} = \dfrac{\pi r^4}{4} = \dfrac{\pi}{4}(2.0)^4 = 4\pi\,cm^4$

Using the parallel axis theorem, $I_{YY} = I_{GG} + Ad^2$, where $d = 3.0 + 2.0 = 5.0\,cm$.

Hence $\mathbf{I_{YY}} = 4\pi + [\pi(2.0)^2](5.0)^2 = 4\pi + 100\pi = 104\pi = \mathbf{327\,cm^4}$

Radius of gyration, $k_{YY} = \sqrt{\dfrac{I_{YY}}{\text{area}}} = \sqrt{\left(\dfrac{104\pi}{\pi(2.0)^2}\right)} = \sqrt{26} = 5.10\,cm$

Application: Determine the second moment of area and radius of gyration for the semicircle shown in Figure 131.7 about axis XX

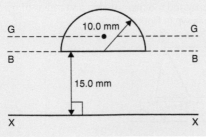

Figure 131.7

The centroid of a semicircle lies at $\dfrac{4r}{3\pi}$ from its diameter

Using the parallel axis theorem: $I_{BB} = I_{GG} + Ad^2$,

where $I_{BB} = \dfrac{\pi r^4}{8}$ (from Table 131.1) $= \dfrac{\pi(10.0)^4}{8} = 3927\,\text{mm}^4$,

$A = \dfrac{\pi r^2}{2} = \dfrac{\pi(10.0)^2}{2} = 157.1\,\text{mm}^2$

and $\quad d = \dfrac{4r}{3\pi} = \dfrac{4(10.0)}{3\pi} = 4.244\,\text{mm}$

Hence, $3927 = I_{GG} + (157.1)(4.244)^2$

i.e. $3927 = I_{GG} + 2830$, from which, $I_{GG} = 3927 - 2830 = 1097\,\text{mm}^4$

Using the parallel axis theorem again: $I_{xx} = I_{GG} + A(15.0 + 4.244)^2$

i.e. $\mathbf{I_{xx}} = 1097 + (157.1)(19.244)^2 = 1097 + 58179 = 59276\,\text{mm}^4$ or $\mathbf{59280\,mm^4}$, correct to 4 significant figures.

Radius of gyration, $k_{xx} = \sqrt{\dfrac{I_{xx}}{\text{area}}} = \sqrt{\left(\dfrac{59276}{157.1}\right)} = 19.42\,\text{mm}$

Application: Determine the polar second moment of area of the propeller shaft cross-section shown in Figure 131.8

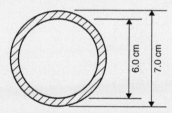

6.0 cm

7.0 cm

Figure 131.8

The polar second moment of area of a circle $= \dfrac{\pi r^4}{2}$

The polar second moment of area of the shaded area is given by the polar second moment of area of the 7.0 cm diameter circle minus the polar second moment of area of the 6.0 cm diameter circle.

Hence the polar second moment of area of the cross-section shown

$$= \frac{\pi}{2}\left(\frac{7.0}{2}\right)^4 - \frac{\pi}{2}\left(\frac{6.0}{2}\right)^4 = 235.7 - 127.2 = 108.5\,\text{cm}^4$$

Application: Determine the second moment of area and radius of gyration of a rectangular lamina of length 40 mm and width 15 mm about an axis through one corner, perpendicular to the plane of the lamina

The lamina is shown in Figure 131.9.

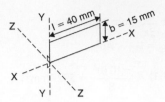

Figure 131.9

From the perpendicular axis theorem: $I_{ZZ} = I_{XX} + I_{YY}$

$$I_{XX} = \frac{lb^3}{3} = \frac{(40)(15)^3}{3} = 45000\,mm^4$$

and $$I_{YY} = \frac{bl^3}{3} = \frac{(15)(40)^3}{3} = 320000\,mm^4$$

Hence $I_{ZZ} = 45000 + 320000 = \mathbf{365000\,mm^4}$ or $\mathbf{36.5\,cm^4}$

Radius of gyration, $k_{ZZ} = \sqrt{\dfrac{I_{ZZ}}{area}} = \sqrt{\left(\dfrac{365000}{(40)(15)}\right)}$

$$= 24.7\,mm \text{ or } 2.47\,cm$$

Application: Determine correct to 3 significant figures, the second moment of area about axis XX for the composite area shown in Figure 131.10.

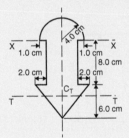

Figure 131.10

For the semicircle, $I_{XX} = \dfrac{\pi r^4}{8} = \dfrac{\pi(4.0)^4}{8} = 100.5\,cm^4$

For the rectangle $I_{XX} = \dfrac{bl^3}{3} = \dfrac{(6.0)(8.0)^3}{3} = 1024\,cm^4$

For the triangle, about axis TT through centroid C_T,

$$I_{TT} = \frac{bh^3}{36} = \frac{(10)(6.0)^3}{36} = 60\,cm^4$$

By the parallel axis theorem, the second moment of area of the triangle about axis XX

$$= 60 + \left[\frac{1}{2}(10)(6.0)\right]\left[8.0 + \frac{1}{3}(6.0)\right]^2 = 3060\,cm^4$$

Total second moment of area about XX $= 100.5 + 1024 + 3060$

$$= 4184.5 = \mathbf{4180\,cm^4}, \text{ correct to 3 significant figures}$$

Differential equations

Why are differential equations important?

Differential equations play an important role in modelling virtually every physical, technical, or biological process, from celestial motion, to bridge design, to interactions between neurons. Further applications are found in fluid dynamics with the design of containers and funnels, in heat conduction analysis with the design of heat spreaders in microelectronics, in rigid-body dynamic analysis, with falling objects, in exponential growth of current in an R-L circuit, in fluid mechanics analysis, in heat transfer analysis, in kinematic analysis of rigid body dynamics, with exponential decay in radioactive material, Newton's law of cooling and in mechanical oscillations.

First-order differential equations model phenomena of cooling, population growth, radioactive decay, mixture of salt solutions, series circuits, survivability with AIDS, draining a tank, economics and finance, drug distribution, pursuit problem and harvesting of renewable natural resources.

However, differential equations such as those used to solve real-life problems may not necessarily be directly solvable, i.e. do not have closed form solutions. Instead, solutions can be approximated using numerical methods, and in science and engineering a numeric approximation to the solution is often good enough to solve a problem.

Second order differential equations have many engineering and scientific applications. These include free vibration analysis with simple and damped mass-spring systems, resonant and non-resonant vibration analysis, with modal analysis, time-varying mechanical forces or pressure, fluid induced vibration such as intermittent wind, forced electrical and mechanical oscillations, tidal waves, acoustics, ultrasonic and random movements of support.

Differential equations govern the fundamental operation of important areas such as automobile dynamics, tyre dynamics, aerodynamics, acoustics, active control systems, including speed control, engine performance and emissions control, climate control, ABS control systems, airbag deployment systems, structural dynamics of buildings, bridges and dams, for example, earthquake and wind engineering, industrial process control, control and operation of automation (robotic) systems, the operation of the electric power grid, electric power generation, orbital dynamics of satellite systems, heat transfer from electrical equipment (including computer chips), economic systems, biological systems, chemical systems, and so on.

In engineering, physics and economics, quantities are frequently encountered – for example energy – that depends on many variables, such as position, velocity and temperature. Usually this dependency is expressed through a partial differential equation, and solving these equations is important for understanding these complex relationships. Solving ordinary differential equations involves finding a function (or a set of functions) of one independent variable, but partial differential equations are for functions of two or more variables. Examples of physical models using partial differential equations are the heat equation for the evolution of the temperature distribution in a body, the wave equation for the motion of a wave front, the flow equation for the flow of fluids and Laplace's equation for an electrostatic potential or elastic strain field. In such cases, not only are the initial conditions needed, but also boundary conditions for the region in which the model applies.

Chapter 132 The solution of equations of the form $\frac{dy}{dx} = f(x)$

A differential equation of the form $\frac{dy}{dx} = f(x)$ is solved by direct integration, i.e.

$$y = \int f(x)\, dx$$

Application: Find the particular solution of the differential equation $5\frac{dy}{dx} + 2x = 3$, given the boundary conditions $y = 1\frac{2}{5}$ when $x = 2$

Since $5\frac{dy}{dx} + 2x = 3$ then $\frac{dy}{dx} = \frac{3 - 2x}{5} = \frac{3}{5} - \frac{2x}{5}$

Hence, $y = \int \left(\frac{3}{5} - \frac{2x}{5}\right) dx$

i.e. $y = \frac{3x}{5} - \frac{x^2}{5} + c$, which is the general solution.

Substituting the boundary conditions $y = 1\frac{2}{5}$ and $x = 2$ to evaluate c gives:

$1\frac{2}{5} = \frac{6}{5} - \frac{4}{5} + c$, from which, $c = 1$.

Hence the particular solution is $y = \frac{3x}{5} - \frac{x^2}{5} + 1$

Chapter 133 The solution of equations of the form $\frac{dy}{dx} = f(y)$

A differential equation of the form $\frac{dy}{dx} = f(y)$ is initially rearranged to give $dx = \frac{dy}{f(y)}$ and then the solution is obtained by direct integration, i.e.

$$\int dx = \int \frac{dy}{f(y)}$$

Application:

(a) The variation of resistance, R ohms, of an aluminium conductor with tempera-
ture $\theta°C$ is given by $\dfrac{dR}{d\theta} = \alpha R$, where α is the temperature coefficient
of resistance of aluminium. If $R = R_0$ when $\theta = 0°C$, solve the equation for R.

(b) If $\alpha = 38 \times 10^{-4}/°C$, determine the resistance of an aluminium conductor at
50°C, correct to 3 significant figures, when its resistance at 0°C is 24.0 Ω

(a) $\dfrac{dR}{d\theta} = \alpha R$ is of the form $\dfrac{dy}{dx} = f(y)$

Rearranging gives: $d\theta = \dfrac{dR}{\alpha R}$

Integrating both sides gives: $\displaystyle\int d\theta = \int \dfrac{dR}{\alpha R}$

i.e. $\theta = \dfrac{1}{\alpha} \ln R + c$, which is the general solution

Substituting the boundary conditions $R = R_0$ when $\theta = 0$ gives:

$$0 = \frac{1}{\alpha} \ln R_0 + c \text{ from which } c = -\frac{1}{\alpha} \ln R_0$$

Hence the particular solution is

$\theta = \dfrac{1}{\alpha} \ln R - \dfrac{1}{\alpha} \ln R_0 = \dfrac{1}{\alpha} (\ln R - \ln R_0)$

i.e. $\theta = \dfrac{1}{\alpha} \ln \left(\dfrac{R}{R_0} \right)$ or $\alpha\theta = \ln \left(\dfrac{R}{R_0} \right)$

Hence $e^{\alpha\theta} = \dfrac{R}{R_0}$ from which, $\mathbf{R = R_0\, e^{\alpha\theta}}$

(b) Substituting $\alpha = 38 \times 10^{-4}$, $R_0 = 24.0$ and $\theta = 50$ into $R = R_0\, e^{\alpha\theta}$ gives the resistance at 50°C, i.e.

$$R_{50} = 24.0\ e^{(38\times10^{-4}\times50)} = 29.0 \text{ ohms}$$

Chapter 134 The solution of equations of the form $\frac{dy}{dx} = f(x) \cdot f(y)$

A differential equation of the form $\frac{dy}{dx} = f(x).f(y)$, where $f(x)$ is a function of x only and $f(y)$ is a function of y only, may be rearranged as $\frac{dy}{f(y)} = f(x)dx$, and then the solution is obtained by direct integration, i.e.

$$\int \frac{dy}{f(y)} = \int f(x)\, dx$$

Application: Solve the equation $4xy \frac{dy}{dx} = y^2 - 1$

Separating the variables gives: $\left(\frac{4y}{y^2 - 1}\right) dy = \frac{1}{x} dx$

Integrating both sides gives: $\int \left(\frac{4y}{y^2 - 1}\right) dy = \int \left(\frac{1}{x}\right) dx$

Using the substitution $u = y^2 - 1$, the general solution is:

$$2 \ln (y^2 - 1) = \ln x + c \tag{1}$$

or $\ln (y^2 - 1)^2 - \ln x = c$

from which, $\ln \left\{ \frac{(y^2 - 1)^2}{x} \right\} = c$

and $\frac{(y^2 - 1)^2}{x} = e^c \tag{2}$

If in equation (1), $c = \ln A$, where A is a different constant,

then $\ln (y^2 - 1)^2 = \ln x + \ln A$

i.e. $\ln (y^2 - 1)^2 = \ln Ax$

i.e. $(y^2 - 1)^2 = Ax \tag{3}$

Equations (1) to (3) are thus three valid solutions of the differential equations

$$4xy\frac{dy}{dx} = y^2 - 1$$

Application: The current i in an electric circuit containing resistance R and inductance L in series with a constant voltage source E is given by the differential equation $E - L\left(\dfrac{di}{dt}\right) = Ri$.

Solve the equation to find i in terms of time t, given that when t = 0, i = 0

In the R–L series circuit shown in Figure 134.1, the supply p.d., E, is given by

$$E = V_R + V_L$$

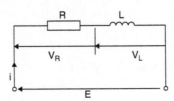

Figure 134.1

$V_R = iR$ and $V_L = L\dfrac{di}{dt}$

Hence $E = iR + L\dfrac{di}{dt}$ from which $E - L\dfrac{di}{dt} = Ri$

Most electrical circuits can be reduced to a differential equation.

Rearranging $E - L\dfrac{di}{dt} = Ri$ gives: $\qquad \dfrac{di}{dt} = \dfrac{E - Ri}{L}$

and separating the variables gives: $\qquad \dfrac{di}{E - Ri} = \dfrac{dt}{L}$

Integrating both sides gives: $\qquad \displaystyle\int \dfrac{di}{E - Ri} = \int \dfrac{dt}{L}$

Hence the general solution is: $-\dfrac{1}{R}\ln(E - Ri) = \dfrac{t}{L} + c$

$\qquad\qquad$ (by making a substitution u = E − Ri, see chapter 117)

When t = 0, i = 0, thus $-\dfrac{1}{R}\ln E = c$

Thus the particular solution is: $-\dfrac{1}{R}\ln(E - Ri) = \dfrac{t}{L} - \dfrac{1}{R}\ln E$

Transposing gives: $-\dfrac{1}{R}\ln(E-Ri)+\dfrac{1}{R}\ln E=\dfrac{t}{L}$

$$\dfrac{1}{R}[\ln E-\ln(E-Ri)]=\dfrac{t}{L}$$

$$\ln\left(\dfrac{E}{E-Ri}\right)=\dfrac{Rt}{L}\quad\text{from which}\quad\dfrac{E}{E-Ri}=e^{\frac{Rt}{L}}$$

Hence $\dfrac{E-Ri}{E}=e^{-\frac{Rt}{L}}$ and $E-Ri=Ee^{-\frac{Rt}{L}}$ and $Ri=E-Ee^{-\frac{Rt}{L}}$

Hence current, $i=\dfrac{E}{R}\left(1-e^{-\frac{Rt}{L}}\right)$ which represents the law of growth

of current in an inductive circuit as shown in Figure 134.2.

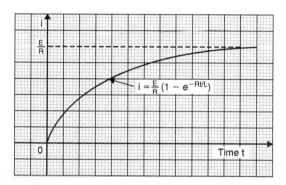

Figure 134.2

Chapter 135 Homogeneous first order differential equations

Procedure to solve differential equations of the form $P\dfrac{dy}{dx}=Q$

1. Rearrange $P\dfrac{dy}{dx}=Q$ into the form $\dfrac{dy}{dx}=\dfrac{P}{Q}$.

2. Make the substitution $y=vx$ (where v is a function of x), from which, $\dfrac{dy}{dx}=v(1)+x\dfrac{dv}{dx}$ by the product rule.

3. Substitute for both y and $\dfrac{dy}{dx}$ in the equation $\dfrac{dy}{dx} = \dfrac{P}{Q}$.

    Simplify, by cancelling, and an equation results in which the variables are separable.
4. Separate the variables and solve.
5. Substitute $v = \dfrac{y}{x}$ to solve in terms of the original variables.

Application: Determine the particular solution of the equation $x\dfrac{dy}{dx} = \dfrac{x^2 + y^2}{y}$, given the boundary conditions that $x = 1$ when $y = 4$

Using the above procedure:

1. Rearranging $x\dfrac{dy}{dx} = \dfrac{x^2 + y^2}{y}$ gives $\dfrac{dy}{dx} = \dfrac{x^2 + y^2}{xy}$ which is homogeneous in x
    and y since each of the three terms on the right hand side are of the same degree (i.e. degree 2).
2. Let $y = vx$ then $\dfrac{dy}{dx} = v(1) + x\dfrac{dv}{dx}$
3. Substituting for y and $\dfrac{dy}{dx}$ in the equation $\dfrac{dy}{dx} = \dfrac{x^2 + y^2}{xy}$ gives:

$$v + x\dfrac{dv}{dx} = \dfrac{x^2 + (vx)^2}{x(vx)} = \dfrac{x^2 + v^2x^2}{vx^2} = \dfrac{1 + v^2}{v}$$

4. Separating the variables give:

$$x\dfrac{dv}{dx} = \dfrac{1 + v^2}{v} - v = \dfrac{1 + v^2 - v^2}{v} = \dfrac{1}{v}$$

Hence, $vdv = \dfrac{1}{x}dx$

Integrating both sides gives:

$$\int v\,dv = \int \dfrac{1}{x}dx \text{ i.e. } \dfrac{v^2}{2} = \ln x + c$$

5. Replacing v by $\dfrac{y}{x}$ gives: $\dfrac{y^2}{2x^2} = \ln x + c$, which is the general solution.

When $x = 1$, $y = 4$, thus: $\dfrac{16}{2} = \ln 1 + c$, from which, $c = 8$

Hence, **the particular solution is:** $\dfrac{y^2}{2x^2} = \ln x + 8$ or $y^2 = 2x^2 (\ln x + 8)$

Chapter 136 Linear first order differential equations

Procedure to solve differential equations of the form $\frac{dy}{dx} + Py = Q$

1. Rearrange the differential equation into the form

   $\frac{dy}{dx} + Py = Q$, where P and Q are functions of x

2. Determine $\int P\,dx$

3. Determine the integrating factor $e^{\int P\,dx}$

4. Substitute $e^{\int P\,dx}$ into the equation:

$$y\,e^{\int P\,dx} = \int e^{\int P\,dx}\,Q\,dx \qquad\qquad (1)$$

5. Integrate the right hand side of equation (1) to give the general solution of the differential equation. Given boundary conditions, the particular solution may be determined.

Application: Solve the differential equation $\dfrac{1}{x}\dfrac{dy}{dx} + 4y = 2$, given the boundary conditions $x = 0$ when $y = 4$

Using the above procedure:

1. Rearranging gives $\dfrac{dy}{dx} + 4xy = 2x$, which is of the form

   $\dfrac{dy}{dx} + Py = Q$ where $P = 4x$ and $Q = 2x$

2. $\displaystyle\int P\,dx = \int 4x\,dx = 2x^2$

3. Integrating factor, $e^{\int P\,dx} = e^{2x^2}$

4. Substituting into equation (1) gives:

$$y\,e^{2x^2} = \int e^{2x^2}(2x)\,dx$$

5. Hence the general solution is: $y\,e^{2x^2} = \dfrac{1}{2}e^{2x^2} + c$, by using the substitution $u = 2x^2$

When $x = 0$, $y = 4$, thus $4e^0 = \dfrac{1}{2}e^0 + c$, from which, $c = \dfrac{7}{2}$

Hence the particular solution is:

$$y\,e^{2x^2} = \dfrac{1}{2}e^{2x^2} + \dfrac{7}{2}$$

i.e. $y = \dfrac{1}{2} + \dfrac{7}{2}e^{-2x^2}$ or $y = \dfrac{1}{2}(1 + 7e^{-2x^2})$

Chapter 137   Numerical methods for first order differential equations (1) – Euler's method

Euler's method

$$y_1 = y_0 + h(y')_0 \tag{1}$$

Application: Obtain a numerical solution of the differential equation $\dfrac{dy}{dx} = 3(1 + x) - y$ given the initial conditions that $x = 1$ when $y = 4$, for the range $x = 1.0$ to $x = 2.0$ with intervals of 0.2

$\dfrac{dy}{dx} = y' = 3(1 + x) - y$

With $x_0 = 1$ and $y_0 = 4$, $(y')_0 = 3(1 + 1) - 4 = \mathbf{2}$

By Euler's method: $y_1 = y_0 + h(y')_0$, from equation (1)

Hence $\mathbf{y_1} = 4 + (0.2)(2) = \mathbf{4.4}$, since $h = 0.2$

At point Q in Figure 137.1, $x_1 = 1.2$, $y_1 = 4.4$

and $(y')_1 = 3(1 + x_1) - y_1$

i.e. $\mathbf{(y')_1} = 3(1 + 1.2) - 4.4 = \mathbf{2.2}$

If the values of x, y and y' found for point Q are regarded as new starting values of x_0, y_0 and $(y')_0$, the above process can be repeated and values found for the point R shown in Figure 137.2.

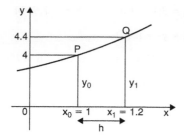

Figure 137.1

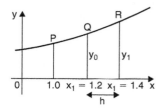

Figure 137.2

Thus at point R, $y_1 = y_0 + h(y')_0$ from equation (1)

$$= 4.4 + (0.2)(2.2) = \textbf{4.84}$$

When $x_1 = 1.4$ and $y_1 = 4.84$, $(y')_0 = 3(1 + 1.4) - 4.84 = \textbf{2.36}$

This step by step Euler's method can be continued and it is easiest to list the results in a table, as shown in Table 137.1. The results for lines 1 to 3 have been produced above.

Table 137.1

	X_0	y_0	$(y')_0$
1.	1	4	2
2.	1.2	4.4	2.2
3.	1.4	4.84	2.36
4.	1.6	5.312	2.488
5.	1.8	5.8096	2.5904
6.	2.0	6.32768	

For line 4, where $x_0 = 1.6$: $y_1 = y_0 + h(y')_0$

$$= 4.84 + (0.2)(2.36) = \textbf{5.312}$$

and $(y')_0 = 3(1 + 1.6) - 5.312 = \textbf{2.488}$

For line 5, where $x_0 = 1.8$: $y_1 = y_0 + h(y')_0$

$$= 5.312 + (0.2)(2.488) = \textbf{5.8096}$$

and $\qquad\qquad$ $(y')_0 = 3(1 + 1.8) - 5.8096 = \mathbf{2.5904}$

For line 6, where $x_0 = 2.0$: $\quad \mathbf{y_1} = y_0 + h\,(y')_0$

$$= 5.8096 + (0.2)(2.5904) = \mathbf{6.32768}$$

(As the range is 1.0 to 2.0 there is no need to calculate $(y')_0$ in line 6)

The particular solution is given by the value of y against x.

A graph of the solution of $\dfrac{dy}{dx} = 3(1 + x) - y$ with initial conditions $x = 1$ and $y = 4$ is shown in Figure 137.3.

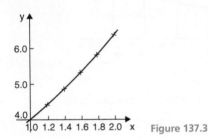

Figure 137.3

In practice it is probably best to plot the graph as each calculation is made, which checks that there is a smooth progression and that no calculation errors have occurred.

Chapter 138   Numerical methods for first order differential equations (2) – the Euler-Cauchy method

Euler-Cauchy method

$$y_{P_1} = y_0 + h(y')_0 \qquad\qquad (1)$$

$$y_{c_1} = y_0 + \frac{1}{2}h[\,(y')_0 + f(x_1, y_{P_1})\,] \qquad\qquad (2)$$

Application: Applying the Euler-Cauchy method, solve the differential equation $\dfrac{dy}{dx} = y - x$ in the range 0(0.1)0.5, given the initial conditions that at $x = 0$, $y = 2$

$$\frac{dy}{dx} = y' = y - x$$

Since the initial conditions are $x_0 = 0$ and $y_0 = 2$ then $(y')_0 = 2 - 0 = 2$

Interval $h = 0.1$, hence $x_1 = x_0 + h = 0 + 0.1 = 0.1$

From equation (1), $y_{P_1} = y_0 + h(y')_0 = 2 + (0.1)(2) = 2.2$

From equation (2), $y_{C_1} = y_0 + \dfrac{1}{2} h[(y')_0 + f(x_1, y_{P_1})]$

$$= y_0 + \frac{1}{2} h[(y')_0 + (y_{P_1} - x_1)] \text{ in this case}$$

$$= 2 + \frac{1}{2}(0.1)[2 + (2.2 - 0.1)] = 2.205$$

$(y')_0 = y_{C_1} - x_1 = 2.205 - 0.1 = 2.105$

If a table of values is produced, as in Euler's method, lines 1 and 2 have so far been determined for Table 138.1.

Table 138.1

	X	y	$(y')_0$
1.	0	2	2
2.	0.1	2.205	2.105
3.	0.2	2.421025	2.221025
4.	0.3	2.649232625	2.349232625
5.	0.4	2.890902051	2.490902051
6.	0.5	3.147446766	

The results in line 2 are now taken as x_0, y_0 and $(y')_0$ for the next interval and the process is repeated.

For line 3, $x_1 = 0.2$

$$y_{P_1} = y_0 + h(y')_0 = 2.205 + (0.1)(2.105) = 2.4155$$

$$y_{C_1} = y_0 + \frac{1}{2} h[(y')_0 + f(x_1, y_{P_1})]$$

$$= 2.205 + \frac{1}{2}(0.1)[2.105 + (2.4155 - 0.2)] = 2.421025$$

$(y')_0 = y_{C_1} - x_1 = 2.421025 - 0.2 = 2.221025$

For line 4, $x_1 = 0.3$

$$y_{P_1} = y_0 + h(y')_0 = 2.421025 + (0.1)(2.221025) = 2.6431275$$

$$y_{C_1} = y_0 + \frac{1}{2} h[(y')_0 + f(x_1, y_{P_1})]$$

$$= 2.421025 + \frac{1}{2}(0.1)\,[2.221025 + (2.6431275 - 0.3)]$$

$$= 2.649232625$$

$$(y')_0 = y_{C_1} - x_1 = 2.649232625 - 0.3 = 2.349232625$$

For line 5, $x_1 = 0.4$

$$y_{P_1} = y_0 + h(y')_0 = 2.649232625 + (0.1)(2.349232625)$$

$$= 2.884155888$$

$$y_{C_1} = y_0 + \frac{1}{2}h[(y')_0 + f(x_1, y_{P_1})]$$

$$= 2.649232625 + \frac{1}{2}(0.1)\,[2.349232625$$

$$+ (2.884155888 - 0.4)] = 2.890902051$$

$$(y')_0 = y_{C_1} - x_1 = 2.890902051 - 0.4 = 2.490902051$$

For line 6, $x_1 = 0.5$

$$y_{P_1} = y_0 + h(y')_0 = 2.890902051 + (0.1)(2.490902051)$$

$$= 3.139992256$$

$$y_{C_1} = y_0 + \frac{1}{2}h[(y')_0 + f(x_1, y_{P_1})]$$

$$= 2.890902051 + \frac{1}{2}(0.1)\,[2.490902051$$

$$+ (3.139992256 - 0.5)] = 3.147446766$$

Chapter 139 Numerical methods for first order differential equations (3) – the Runge-Kutta method

Runge-Kutta method

To solve the differential equation $\frac{dy}{dx} = f(x, y)$ given the initial condition $y = y_0$ at $x = x_0$ for a range of values of $x = x_0(h)x_n$:

1. Identify x_0, y_0 and h, and values of $x_1, x_2, x_3, \ldots$

2. Evaluate $k_1 = f(x_n, y_n)$ starting with $n = 0$

3. Evaluate $k_2 = f\left(x_n + \frac{h}{2}, y_n + \frac{h}{2}k_1\right)$

4. Evaluate $k_3 = f\left(x_n + \frac{h}{2}, y_n + \frac{h}{2}k_2\right)$

5. Evaluate $k_4 = f\left(x_n + h, y_n + hk_3\right)$

6. Use the values determined from steps 2 to 5 to evaluate:

$$y_{n+1} = y_n + \frac{h}{6}\{k_1 + 2k_2 + 2k_3 + k_4\}$$

7. Repeat steps 2 to 6 for n = 1, 2, 3, ...

Application: Use the Runge-Kutta method to solve the differential equation: $\frac{dy}{dx} = y - x$ in the range 0(0.1)0.5, given the initial conditions that at x = 0, y = 2

Using the above procedure:

1. $x_0 = 0$, $y_0 = 2$ and since h = 0.1, and the range is from x = 0 to x = 0.5, then

    $x_1 = 0.1$, $x_2 = 0.2$, $x_3 = 0.3$, $x_4 = 0.4$, and $x_5 = 0.5$

Let n = 0 to determine y_1:

2. $k_1 = f(x_0, y_0) = f(0, 2)$; since $\frac{dy}{dx} = y - x$, $f(0, 2) = 2 - 0 = \mathbf{2}$

3. $k_2 = f\left(x_0 + \frac{h}{2}, y_0 + \frac{h}{2}k_1\right) = f\left(0 + \frac{0.1}{2}, 2 + \frac{0.1}{2}(2)\right) = f(0.05, 2.1)$

    $= 2.1 - 0.05 = \mathbf{2.05}$

4. $k_3 = f\left(x_0 + \frac{h}{2}, y_0 + \frac{h}{2}k_2\right) = f\left(0 + \frac{0.1}{2}, 2 + \frac{0.1}{2}(2.05)\right)$

    $= f(0.05, 2.1025)$

    $= 2.1025 - 0.05 = \mathbf{2.0525}$

5. $k_4 = f(x_0 + h, y_0 + hk_3) = f(0 + 0.1, 2 + 0.1(2.0525))$

    $= f(0.1, 2.20525)$

    $= 2.20525 - 0.1 = \mathbf{2.10525}$

6. $y_{n+1} = y_n + \frac{h}{6}\{k_1 + 2k_2 + 2k_3 + k_4\}$ and when n = 0:

    $y_1 = y_0 + \frac{h}{6}\{k_1 + 2k_2 + 2k_3 + k_4\}$

    $= 2 + \frac{0.1}{6}\{2 + 2(2.05) + 2(2.0525) + 2.10525\}$

    $= 2 + \frac{0.1}{6}\{12.31025\} = 2.205171$

A table of values may be constructed as shown in Table 139.1. The working has been shown for the first two rows.

Table 139.1

n	x_n	k_1	k_2	k_3	k_4	y_n
0	0					**2**
1	0.1	2.0	2.05	2.0525	2.10525	**2.205171**
2	0.2	2.105171	2.160430	2.163193	2.221490	**2.421403**
3	0.3	2.221403	2.282473	2.285527	2.349956	**2.649859**
4	0.4	2.349859	2.417352	2.420727	2.491932	**2.891925**
5	0.5	2.491825	2.566416	2.570146	2.648840	**3.148721**

Let n = 1 to determine y_2:

2. $k_1 = f(x_1, y_1) = f(0.1, 2.205171)$; since $\dfrac{dy}{dx} = y - x$,

$$f(0.1, 2.205171) = 2.205171 - 0.1 = \textbf{2.105171}$$

3. $k_2 = f\left(x_1 + \dfrac{h}{2}, y_1 + \dfrac{h}{2}k_1\right)$

$$= f\left(0.1 + \dfrac{0.1}{2}, 2.205171 + \dfrac{0.1}{2}(2.105171)\right)$$

$$= f(0.15, 2.31042955)$$

$$= 2.31042955 - 0.15 = \textbf{2.160430}$$

4. $k_3 = f\left(x_1 + \dfrac{h}{2}, y_1 + \dfrac{h}{2}k_2\right)$

$$= f\left(0.1 + \dfrac{0.1}{2}, 2.205171 + \dfrac{0.1}{2}(2.160430)\right)$$

$$= f(0.15, 2.3131925) = 2.3131925 - 0.15 = \textbf{2.163193}$$

5. $k_4 = f(x_1 + h, y_1 + hk_3) = f(0.1 + 0.1, 2.205171 + 0.1(2.163193))$

$$= f(0.2, 2.421490) = 2.421490 - 0.2$$

$$= \textbf{2.221490}$$

6. $y_{n+1} = y_n + \dfrac{h}{6}\{k_1 + 2k_2 + 2k_3 + k_4\}$ and when n = 1:

$$y_2 = y_1 + \dfrac{h}{6}\{k_1 + 2k_2 + 2k_3 + k_4\}$$

$$= 2.205171 + \dfrac{0.1}{6}\{2.105171 + 2(2.160430) + 2(2.163193) + 2.221490\}$$

$$= 2.205171 + \dfrac{0.1}{6}\{12.973907\} = 2.421403$$

This completes the third row of Table 139.1. In a similar manner y_3, y_4 and y_5 can be calculated and the results are as shown in Table 139.1.

This problem is the same as the application on page 386 which used the Euler-Cauchy method, and a comparison of results can be made.

The differential equation $\dfrac{dy}{dx} = y - x$ may be solved analyticall using the y integrating factor method shown on page 383, with the solution: $\mathbf{y = x + 1 + e^x}$

Substituting values of x of 0, 0.1, 0.2, ..., 0.5 will give the exact values. A comparison of the results obtained by Euler's method (which is left to the reader to produce), the Euler-Cauchy method and the Runga-Kutta method, together with the exact values is shown in Table 139.2.

Table 139.2

x	Euler's method y	Euler-Cauchy method Y	Runge-Kutta method y	Exact value $y = x + 1 + e^x$
0	2	2	2	2
0.1	2.2	2.205	2.205171	2.205170918
0.2	2.41	2.421025	2.421403	2.421402758
0.3	2.631	2.649232625	2.649859	2.649858808
0.4	2.8641	2.2890902051	2.891825	2.891824698
0.5	3.11051	3.147446766	3.148721	3.148721271

It is seen from Table 139.2 that **the Runge-Kutta method is exact, correct to 5 decimal places**.

Chapter 140 Second order differential equations of the form $a\dfrac{d^2y}{dx^2} + b\dfrac{dy}{dx} + cy = 0$

Procedure to solve differential equations of the form

$a\dfrac{d^2y}{dx^2} + b\dfrac{dy}{dx} + cy = 0$

1. Rewrite the differential equation $a\dfrac{d^2y}{dx^2} + b\dfrac{dy}{dx} + cy = 0$ as

$(aD^2 + bD + c)y = 0$

2. Substitute m for D and solve the auxiliary equation $am^2 + bm + c = 0$ for m

3. If the roots of the auxiliary equation are:
   (a) **real and different,** say $m = \alpha$ and $m = \beta$, then the general solution is:
      $y = Ae^{\alpha x} + Be^{\beta x}$
   (b) **real and equal,** say $m = \alpha$ twice, then the general solution is:
      $y = (Ax + B)e^{\alpha x}$
   (c) **complex,** say $m = \alpha \pm j\beta$, then the general solution is:
      $y = e^{\alpha x}\{A\cos \beta x + B\sin \beta x\}$

4. Given boundary conditions, constants A and B may be determined and the **particular solution** of the differential equation obtained. The particular solution obtained with differential equations may be verified by substituting expressions for y, $\dfrac{dy}{dx}$ and $\dfrac{d^2y}{dx^2}$ into the original equation.

Application: The oscillations of a heavily damped pendulum satisfy the differential equation $\dfrac{d^2x}{dt^2} + 6\dfrac{dx}{dt} + 8x = 0$, where x cm is the displacement of the bob at time t seconds.
The initial displacement is equal to $+ 4$ cm and the initial velocity $\left(\text{i.e. } \dfrac{dx}{dt}\right)$ is 8 cm/s. Solve the equation for x.

Using the above procedure:

1. $\dfrac{d^2x}{dt^2} + 6\dfrac{dx}{dt} + 8x = 0$ in D-operator form is $(D^2 + 6D + 8)x = 0$,

   where $D \equiv \dfrac{d}{dt}$

2. The auxiliary equation is $m^2 + 6m + 8 = 0$

   Factorising gives: $(m + 2)(m + 4) = 0$, from which, $m = -2$ or $m = -4$

3. Since the roots are real and different, **the general solution is:**

   $x = Ae^{-2t} + Be^{-4t}$

4. Initial displacement means that time $t = 0$. At this instant, $x = 4$

   Thus $4 = A + B$ (1)

   Velocity, $\dfrac{dx}{dt} = -2Ae^{-2t} - 4Be^{-4t}$

   $\dfrac{dx}{dt} = 8$ cm/s when $t = 0$, thus $8 = -2A - 4B$ (2)

From equations (1) and (2), A = 12 and B = −8

Hence the particular solution is: $x = 12e^{-2t} - 8e^{-4t}$

i.e. **displacement, $x = 4(3e^{-2t} - 2e^{-4t})$ cm**

Application: The equation $\dfrac{d^2i}{dt^2} + \dfrac{R}{L}\dfrac{di}{dt} + \dfrac{1}{LC}i = 0$ represents a current i flowing in an electrical circuit containing resistance R, inductance L and capacitance C connected in series. If R = 200 ohms, L = 0.20 henry and C = 20 × 10⁻⁶ farads, solve the equation for i given the boundary conditions that when t = 0, i = 0 and $\dfrac{di}{dt} = 100$

Using the procedure:

1. $\dfrac{d^2i}{dt^2} + \dfrac{R}{L}\dfrac{di}{dt} + \dfrac{1}{LC}i = 0$ in D-operator form is

    $\left(D^2 + \dfrac{R}{L}D + \dfrac{1}{LC}\right)i = 0$ where $D \equiv \dfrac{d}{dt}$

2. The auxiliary equation is $m^2 + \dfrac{R}{L}m + \dfrac{1}{LC} = 0$

    Hence, $m = \dfrac{-\dfrac{R}{L} \pm \sqrt{\left[\left(\dfrac{R}{L}\right)^2 - 4(1)\left(\dfrac{1}{LC}\right)\right]}}{2}$

    When R = 200, L = 0.20 and C = 20 × 10⁻⁶,

    then $m = \dfrac{-\dfrac{200}{0.20} \pm \sqrt{\left[\left(\dfrac{200}{0.20}\right)^2 - \dfrac{4}{(0.20)(20 \times 10^{-6})}\right]}}{2}$

    $= \dfrac{-1000 \pm \sqrt{0}}{2} = -500$

3. Since the two roots are real and equal (i.e. −500 twice, since for a second order differential equation there must be two solutions), **the general solution is: $i = (At + B)e^{-500t}$**

4. When t = 0, i = 0, hence B = 0

    $\dfrac{di}{dt} = (At + B)(-500e^{-500t}) + (e^{-500t})(A)$ by the product rule

When $t = 0$, $\dfrac{di}{dt} = 100$, thus $100 = -500B + A$

i.e. $A = 100$, since $B = 0$

Hence the particular solution is: $i = 100\,te^{-500t}$

Application: The equation of motion of a body oscillating on the end of a spring is $\dfrac{d^2x}{dt^2} + 100x = 0$, where x is the displacement in metres of the body from its equilibrium position after time t seconds. Determine x in terms of t given that at time $t = 0$, $x = 2\,\text{m}$ and $\dfrac{dx}{dt} = 0$

An equation of the form $\dfrac{d^2x}{dt^2} + m^2x = 0$ is a differential equation representing simple harmonic motion (S.H.M.). Using the procedure:

1. $\dfrac{d^2x}{dt^2} + 100x = 0$ in D-operator form is $(D^2 + 100)x = 0$

2. The auxiliary equation is $m^2 + 100 = 0$, i.e. $m^2 = -100$ and

$$m = \sqrt{-100}$$

i.e. $$m = \pm j10$$

3. Since the roots are complex, the **general solution** is:

$$x = e^0(A\cos 10t + B\sin 10t),$$

i.e. **$x = (A\cos 10t + B\sin 10t)$ metres**

4. When $t = 0$, $x = 2$, thus $2 = A$

$$\dfrac{dx}{dt} = -10A\,\sin 10t + 10B\,\cos 10t$$

When $t = 0$, $\dfrac{dx}{dt} = 0$ thus $0 = -10A\sin 0 + 10B\cos 0$　i.e.　$B = 0$

Hence the particular solution is: $x = 2\cos 10t$ metres

Chapter 141 Second order differential equations of the form $a \dfrac{d^2y}{dx^2} + b \dfrac{dy}{dx} + cy = f(x)$

Procedure to solve differential equations of the form $a \dfrac{d^2y}{dx^2} + b \dfrac{dy}{dx} + cy = f(x)$

1. Rewrite the given differential equation as $(aD^2 + bD + c)y = f(x)$
2. Substitute m for D, and solve the auxiliary equation $am^2 + bm + c = 0$ for m
3. Obtain the **complementary function**, u, which is achieved using the same procedure as on page 391–92
4. To determine the **particular integral**, v , firstly assume a particular integral which is suggested by f(x), but which contains undetermined coefficients. Table 141.1 gives some suggested substitutions for different functions f(x).
5. Substitute the suggested P.I. into the differential equation $(aD^2 + bD + c)v = f(x)$ and equate relevant coefficients to find the constants introduced.
6. The general solution is given by y = C.F. + P.I. i.e. y = u + v
7. Given boundary conditions, arbitrary constants in the C.F. may be determined and the particular solution of the differential equation obtained.

Table 141.1 **Form of particular integral for different functions**

Type	Straightforward cases Try as particular integral:	'Snag' cases Try as particular integral:
(a) f(x) = a constant	v = k	v = kx (used when C.F. contains a constant)
(b) f(x) = polynomial (i.e. f(x) = L + Mx + Nx² + .. where any of the coefficients may be zero)	$v = a + bx + cx^2 + ..$	
(c) f(x) = an exponential function (i.e. $f(x) = Ae^{ax}$)	$v = ke^{ax}$	(i) $v = kxe^{ax}$ (used when e^{ax} appears in the C.F.) (ii) $v = kx^2 e^{ax}$ (used when e^{ax} **and** xe^{ax} both appear in the C.F.)
(d) f(x) a sine or cosine function (i.e. f(x) a sin px + b cos px where a or b may be zero)	v = A sin px + B cos px	v = x(A sin px + B cos px) (used when sin px and/or cos px appears in the C.F.)

Table 141.1   Continued

Type	Straightforward cases Try as particular integral:	'Snag' cases Try as particular integral:

(e) f(x) = a sum e.g.
    (i) f(x) = 4x² − 3 sin 2x
    (ii) f(x) = 2 − x + e³ˣ
	(i) v = ax² + bx
	        + c+d sin 2x
	        + e cos 2x
	    (ii) v = ax + b+ce³ˣ	

(f) f(x) = a product
    e.g. f(x) = 2eˣ cos 2x
	v = eˣ(A sin 2x
	        + B cos 2x)	

Application: In a galvanometer the deflection θ satisfies the differential equation $\dfrac{d^2\theta}{dt^2} + 4\dfrac{d\theta}{dt} + 4\theta = 8$. Solve the equation for θ given that when t = 0, $\theta = \dfrac{d\theta}{dt} = 2$

1. $\dfrac{d^2\theta}{dt^2} + 4\dfrac{d\theta}{dt} + 4\theta = 8$ in D-operator form is: $(D^2 + 4D + 4)\theta = 8$

2. Auxiliary equation is: $m^2 + 4m + 4 = 0$
    i.e. $(m + 2)(m + 2) = 0$
    from which, $m = -2$ twice

3. Hence, C.F., $u = (At + B)e^{-2t}$

4. Let the particular integral, P.I., $v = k$

5. Substituting $v = k$ gives: $(D^2 + 4D + 4)k = 8$
    $D(k) = 0$ and $D^2(k) = D(0) = 0$
    Hence, $4k = 8$ from which, $k = 2$
    Hence, P.I., $v = 2$

6. The general solution, $\theta = u + v = (At + B)e^{-2t} + 2$

7. t = 0 and $\theta = 2$, hence, $2 = B + 2$ from which, B = 0

$$\frac{d\theta}{dt} = (At + B)(-2e^{-2t}) + (e^{-2t})(A)$$

    x = 0 and $\dfrac{d\theta}{dt} = 2$, hence, $2 = -2B + A$ from which, A = 2
    Hence, $\theta = 2te^{-2t} + 2$

    i.e. $\theta = 2(te^{-2t} + 1)$

Application: Solve $2\dfrac{d^2y}{dx^2} - 11\dfrac{dy}{dx} + 12y = 3x - 2$

1. $2\dfrac{d^2y}{dx^2} - 11\dfrac{dy}{dx} + 12y = 3x - 2$ in D-operator form is

   $(2D^2 - 11D + 12)y = 3x - 2$

2. Substituting m for D gives the auxiliary equation $2m^2 - 11m + 12 = 0$

   Factorising gives: $(2m - 3)(m - 4) = 0$, from which, $m = \dfrac{3}{2}$ or $m = 4$

3. Since the roots are real and different, the C.F.,

$$u = Ae^{\frac{3}{2}x} + Be^{4x}$$

4. Since $f(x) = 3x - 2$ is a polynomial, let the P.I., $v = ax + b$ (see Table 141.1(b))

5. Substituting $v = ax + b$ into $(2D^2 - 11D + 12)v = 3x - 2$ gives:

$$(2D^2 - 11D + 12)(ax + b) = 3x - 2,$$

   i.e. $\quad 2D^2(ax + b) - 11D(ax + b) + 12(ax + b) = 3x - 2$

   i.e. $\quad\quad\quad\quad\quad 0 - 11a + 12ax + 12b = 3x - 2$

   Equating the coefficients of x gives: $12a = 3$, from which, $a = \dfrac{1}{4}$

   Equating the constant terms gives: $-11a + 12b = -2$

   i.e. $-11\left(\dfrac{1}{4}\right) + 12b = -2$ from which, $12b = -2 + \dfrac{11}{4} = \dfrac{3}{4}$ i.e. $b = \dfrac{1}{16}$

   Hence the P.I., $v = ax + b = \dfrac{1}{4}x + \dfrac{1}{16}$

6. The general solution is given by $y = u + v$

$$\text{i.e. } y = Ae^{\frac{3}{2}x} + Be^{4x} + \dfrac{1}{4}x + \dfrac{1}{16}$$

Application: Solve $\dfrac{d^2y}{dx^2} - 2\dfrac{dy}{dx} + y = 3e^{4x}$ given that when $x = 0$, $y = -\dfrac{2}{3}$

and $\dfrac{dy}{dx} = 4\dfrac{1}{3}$

1. $\dfrac{d^2y}{dx^2} - 2\dfrac{dy}{dx} + y = 3e^{4x}$ in D-operator form is

   $(D^2 - 2D + 1)y = 3e^{4x}$

2. Substituting m for D gives the auxiliary equation $m^2 - 2m + 1 = 0$
   Factorising gives: $(m - 1)(m - 1) = 0$, from which, $m = 1$ twice

3. Since the roots are real and equal the C.F., $u = (Ax + B)e^x$

4. Let the particular integral, $v = ke^{4x}$ (see Table 141.1(c))

5. Substituting $v = ke^{4x}$ into $(D^2 - 2D + 1)v = 3e^{4x}$ gives:

$$(D^2 - 2D + 1)ke^{4x} = 3e^{4x}$$

i.e. $D^2(ke^{4x}) - 2D(ke^{4x}) + 1(ke^{4x}) = 3e^{4x}$

i.e. $16ke^{4x} - 8ke^{4x} + ke^{4x} = 3e^{4x}$

Hence $9ke^{4x} = 3e^{4x}$, from which, $k = \dfrac{1}{3}$

Hence the P.I., $\mathbf{v = ke^{4x}} = \dfrac{1}{3}e^{4x}$

6. The general solution is given by $y = u + v$, i.e.

$$y = (Ax + B)e^x + \dfrac{1}{3}e^{4x}$$

7. When $x = 0$, $y = -\dfrac{2}{3}$ thus $-\dfrac{2}{3} = (0 + B)e^0 + \dfrac{1}{3}e^0$,   from which, $B = -1$

$\dfrac{dy}{dx} = (Ax + B)e^x + e^x(A) + \dfrac{4}{3}e^{4x}$

When $x = 0$, $\dfrac{dy}{dx} = 4\dfrac{1}{3}$,   thus $\dfrac{13}{3} = B + A + \dfrac{4}{3}$ from which, $A = 4$,

since $B = -1$

Hence the particular solution is: $y = (4x - 1)e^x + \dfrac{1}{3}e^{4x}$

Application: $L\dfrac{d^2q}{dt^2} + R\dfrac{dq}{dt} + \dfrac{1}{C}q = V_0 \sin wt$ represents the variation of capacitor charge in an electric circuit. Determine an expression for q at time t seconds given that $R = 40\Omega$, $L = 0.02H$, $C = 50 \times 10^{-6}F$, $V_0 = 540.8V$ and $\omega = 200\,rad/s$ and given the boundary conditions that when $t = 0$, $q = 0$ and $\dfrac{dq}{dt} = 4.8$

$L\dfrac{d^2q}{dt^2} + R\dfrac{dq}{dt} + \dfrac{1}{C}q = V_0 \sin \omega t$ in D-operator form is:

$$\left(LD^2 + RD + \dfrac{1}{C}\right)q = V_0 \sin \omega t$$

The auxiliary equation is: $Lm^2 + Rm + \dfrac{1}{C} = 0$

and

$$m = \frac{-R \pm \sqrt{R^2 - \frac{4L}{C}}}{2L} = \frac{-40 \pm \sqrt{40^2 - \frac{4(0.02)}{50 \times 10^{-6}}}}{2(0.02)} = \frac{-40 \pm \sqrt{0}}{0.04} = -1000$$

Hence, C.F., $u = (At + B)e^{-1000t}$

Let P.I., $v = A \sin \omega t + B \cos \omega t$

$$\left(L D^2 + R D + \frac{1}{C}\right)[A \sin \omega t + B \cos \omega t] = V_0 \sin \omega t$$

$D(v) = A\omega \cos \omega t - B\omega \sin \omega t$ and $D^2(v) = -A\omega^2 \sin \omega t - B\omega^2 \cos \omega t$

Thus,

$$\left(L D^2 + R D + \frac{1}{C}\right) v = 0.02(-A\omega^2 \sin \omega t - B\omega^2 \cos \omega t)$$

$$+ 40(A\omega \cos \omega t - B\omega \sin \omega t)$$

$$+ \frac{1}{50 \times 10^{-6}} (A \sin \omega t + B \cos \omega t) = V_0 \sin \omega t$$

i.e. $-800A \sin 200t - 800B \cos 200t + 8000A \cos 200t$
$-8000B \sin 200t + 20000A \sin 200t + 20000B \cos 200t = 540.8 \sin 200t$

Hence, $-800A - 8000B + 20000A = 540.8$
and $-800B + 8000A + 20000B = 0$
i.e. $19200A - 8000B = 540.8$ (1)
and $8000A + 19200B = 0$ (2)
$8 \times$ (1) gives: $153600A - 64000B = 4326.4$ (3)
$19.2 \times$ (2) gives: $153600A + 368640B = 0$ (4)
(3)–(4) gives: $-432640B = 4326.4$

from which, $B = \dfrac{4326.4}{432640} = -0.01$

Substituting in (1) gives: $19200A - 8000(-0.01) = 540.8$

i.e. $19200A + 80 = 540.8$

and $A = \dfrac{540.8 - 80}{19200} = \dfrac{460.8}{19200} = 0.024$

Hence, P.I., $v = 0.024 \sin 200t - 0.01 \cos 200t$

Thus, $q = u + v = (At + B)e^{-1000t} + 0.024 \sin 200t - 0.01 \cos 200t$

When $t = 0$, $q = 0$, hence, $0 = B - 0.01$ from which, $B = 0.01$

$$\frac{dq}{dt} = (At + B)\left(-1000e^{-1000t}\right) + Ae^{-1000t} + (0.024)(200) \cos 200t$$

$$+ (0.01)(200) \sin 200t$$

When t = 0, $\dfrac{dq}{dt}$ = 4.8,   hence, 4.8 = −1000B + A + 4.8

i.e. A = 1000B = 1000(0.01) = 10

Thus,   **q = (10t + 0.01)e^{-1000t} + 0.024 sin 200t − 0.010 cos 200t**

Chapter 142   Power series methods of solving ordinary differential equations (1) – Leibniz theorem

Leibniz's theorem

To find the n'th derivative of a product y = uv:

$$y^{(n)} = (uv)^{(n)} = u^{(n)}v + nu^{(n-1)}v^{(1)} + \frac{n(n-1)}{2!}u^{(n-2)}v^{(2)}$$
$$+ \frac{n(n-1)(n-2)}{3!}u^{(n-3)}v^{(3)} + \cdots \qquad (1)$$

Application: Find the 5th derivative of y = x^4 sin x

If y = x^4 sin x, then using Leibniz's equation with u = sin x and v = x^4 gives:

$$y^{(n)} = \left[\sin\left(x + \frac{n\pi}{2}\right)x^4\right] + n\left[\sin\left(x + \frac{(n-1)\pi}{2}\right)4x^3\right]$$
$$+ \frac{n(n-1)}{2!}\left[\sin\left(x + \frac{(n-2)\pi}{2}\right)12x^2\right]$$
$$+ \frac{n(n-1)(n-2)}{3!}\left[\sin\left(x + \frac{(n-3)\pi}{2}\right)24x\right]$$
$$+ \frac{n(n-1)(n-2)(n-3)}{4!}\left[\sin\left(x + \frac{(n-4)\pi}{2}\right)24\right]$$

and

$$y^{(5)} = x^4\sin\left(x + \frac{5\pi}{2}\right) + 20x^3\sin(x + 2\pi) + \frac{(5)(4)}{2}(12x^2)\sin\left(x + \frac{3\pi}{2}\right)$$
$$+ \frac{(5)(4)(3)}{(3)(2)}(24x)\sin(x + \pi) + \frac{(5)(4)(3)(2)}{(4)(3)(2)}(24)\sin\left(x + \frac{\pi}{2}\right)$$

Since $\sin\left(x + \frac{5\pi}{2}\right) \equiv \sin\left(x + \frac{\pi}{2}\right) \equiv \cos x$, $\sin(x + 2\pi) \equiv \sin x$,

$$\sin\left(x + \frac{3\pi}{2}\right) \equiv -\cos x, \quad \text{and } \sin(x + \pi) \equiv -\sin x,$$

then $y^{(5)} = x^4 \cos x + 20x^3 \sin x + 120x^2 (-\cos x) + 240x (-\sin x) + 120 \cos x$

i.e. $y^{(5)} = (x^4 - 120x^2 + 120) \cos x + (20x^3 - 240x) \sin x$

Chapter 143 Power series methods of solving ordinary differential equations (2) – Leibniz-Maclaurin method

Leibniz-Maclaurin method

(i) Differentiate the given equation n times, using the Leibniz theorem of equation (1), page 400

(ii) rearrange the result to obtain the recurrence relation at x = 0,

(iii) determine the values of the derivatives at x = 0, i.e. find $(y)_0$ and $(y')_0$,

(iv) substitute in the Maclaurin expansion for y = f(x) (see page 84),

(v) simplify the result where possible and apply boundary condition (if given).

Application: Determine the power series solution of the differential equation:
$\frac{d^2y}{dx^2} + x\frac{dy}{dx} + 2y = 0$ using Leibniz-Maclaurin's method, given the boundary conditions $x = 0$, $y = 1$ and $\frac{dy}{dx} = 2$

Following the above procedure:

(i) The differential equation is rewritten as: $y'' + xy' + 2y = 0$ and from the Leibniz theorem of equation (1), each term is differentiated n times, which gives:

$$y^{(n+2)} + \left\{y^{(n+1)}(x) + ny^{(n)}(1) + 0\right\} + 2y^{(n)} = 0$$

i.e. $\qquad y^{(n+2)} + xy^{(n+1)} + (n + 2)y^{(n)} = 0 \qquad\qquad (2)$

(ii) At x = 0, equation (2) becomes:

$$y^{(n+2)} + (n + 2)y^{(n)} = 0$$

from which, $\qquad y^{(n+2)} = -(n + 2)y^{(n)}$

This equation is called a **recurrence relation** or **recurrence formula**, because each recurring term depends on a previous term.

(iii) Substituting n = 0, 1, 2, 3, ... will produce a set of relationships between the various coefficients. For

n = 0, $(y'')_0 = -2(y)_0$

$n = 1$, $(y''')_0 = -3\,(y')_0$

$n = 2$, $(y^{(4)})_0 = -4(y'')_0 = -4\left\{-2(y)_0\right\} = 2 \times 4(y)_0$

$n = 3$, $(y^{(5)})_0 = -5(y''')_0 = -5\left\{-3(y')_0\right\} = 3 \times 5(y')_0$

$n = 4$, $(y^{(6)})_0 = -6(y^{(4)})_0 = -6\left\{2 \times 4(y)_0\right\} = -2 \times 4 \times 6(y)_0$

$n = 5$, $(y^{(7)})_0 = -7(y^{(5)})_0 = -7\left\{3 \times 5(y')_0\right\} = -3 \times 5 \times 7(y')_0$

$n = 6$, $(y^{(8)})_0 = -8(y^{(6)})_0 = -8\left\{-2 \times 4 \times 6\left(y\right)_0\right\}$

$$= 2 \times 4 \times 6 \times 8(y)_0$$

(iv) Maclaurin's theorem from page 84 may be written as:

$$y = (y)_0 + x\,(y')_0 + \frac{x^2}{2!}\,(y'')_0 + \frac{x^3}{3!}\,(y''')_0 + \frac{x^4}{4!}\,(y^{(4)})_0 + \dots.$$

Substituting the above values into Maclaurin's theorem gives:

$$y = (y)_0 + x\,(y')_0 + \frac{x^2}{2!}\left\{-2(y)_0\right\} + \frac{x^3}{3!}\left\{-3\,(y')_0\right\}$$

$$+ \frac{x^4}{4!}\left\{2 \times 4\,(y)_0\right\} + \frac{x^5}{5!}\left\{3 \times 5\,(y')_0\right\} + \frac{x^6}{6!}\left\{-2 \times 4 \times 6\,(y)_0\right\}$$

$$+ \frac{x^7}{7!}\left\{-3 \times 5 \times 7(y')_0\right\} + \frac{x^8}{8!}\left\{2 \times 4 \times 6 \times 8\,(y)_0\right\}$$

(v) Collecting similar terms together gives:

$$y = (y)_0\left\{1 - \frac{2x^2}{2!} + \frac{2 \times 4x^4}{4!} - \frac{2 \times 4 \times 6x^6}{6!} + \frac{2 \times 4 \times 6 \times 8x^8}{8!} - \dots\right\}$$

$$+ (y')_0\left\{x - \frac{3x^3}{3!} + \frac{3 \times 5x^5}{5!} - \frac{3 \times 5 \times 7x^7}{7!} + \dots\right\}$$

i.e. $$y = (y)_0\left\{1 - \frac{x^2}{1} + \frac{x^4}{1 \times 3} - \frac{x^6}{3 \times 5} + \frac{x^8}{3 \times 5 \times 7} - \dots\right\}$$

$$+ (y')_0\left\{\frac{x}{1} - \frac{x^3}{1 \times 2} + \frac{x^5}{2 \times 4} - \frac{x^7}{2 \times 4 \times 6} + \dots\right\}$$

The boundary conditions are that at $x = 0$, $y = 1$ and $\dfrac{dy}{dx} = 2$, i.e. $(y)_0 = 1$ and $(y')_0 = 2$.

Hence,   the   power   series   solution   of   the   differential   equation:

$\dfrac{d^2y}{dx^2} + x\dfrac{dy}{dx} + 2y = 0$ is:

$$y = \left\{1 - \frac{x^2}{1} + \frac{x^4}{1 \times 3} - \frac{x^6}{3 \times 5} + \frac{x^8}{3 \times 5 \times 7} - \dots\right\}$$

$$+ 2\left\{\frac{x}{1} - \frac{x^3}{1 \times 2} + \frac{x^5}{2 \times 4} - \frac{x^7}{2 \times 4 \times 6} + \dots\right\}$$

Chapter 144 Power series methods of solving ordinary differential equations (3) – Frobenius method

Frobenius method

A differential equation of the form $y'' + Py' + Qy = 0$, where P and Q are both functions of x, can be represented by a power series as follows:

(i) Assume a trial solution of the form

$$y = x^c\{a_0 + a_1x + a_2x^2 + a_3x^3 + \cdots + a_rx^r + \cdots\}$$

(ii) differentiate the trial series,

(iii) substitute the results in the given differential equation,

(iv) equate coefficients of corresponding powers of the variable on each side of the equation; this enables index c and coefficients a_1, a_2, a_3, ... from the trial solution, to be determined.

Application: Determine, using the Frobenius method, the general power series solution of the differential equation: $3x\dfrac{d^2y}{dx^2} + \dfrac{dy}{dx} - y = 0$

The differential equation may be rewritten as: $3xy'' + y' - y = 0$

(i) Let a trial solution be of the form

$$y = x^c\{a_0 + a_1x + a_2x^2 + a_3x^3 + \cdots + a_rx^r + \cdots\} \tag{1}$$

where $a_0 \neq 0$,

i.e. $y = a_0x^c + a_1x^{c+1} + a_2x^{c+2} + a_3x^{c+3} + \cdots + a_rx^{c+r} + \cdots \tag{2}$

(ii) Differentiating equation (2) gives:

$$y' = a_0cx^{c-1} + a_1(c + 1)x^c + a_2(c + 2)x^{c+1} + \cdots + a_r(c + r)x^{c+r-1} + \cdots$$

and

$$y'' = a_0c(c - 1)x^{c-2} + a_1c(c + 1)x^{c-1} + a_2(c + 1)(c + 2)x^c + \cdots$$
$$+ a_r(c + r - 1)(c + r)x^{c+r-2} + \cdots$$

(iii) Substituting y, y' and y'' into each term of the given equation $3xy'' + y' - y = 0$ gives:

$$3xy'' = 3a_0c(c - 1)x^{c-1} + 3a_1c(c + 1)x^c + 3a_2(c + 1)(c + 2)x^{c+1}$$
$$+ \cdots + 3a_r(c + r - 1)(c + r)x^{c+r-1} + \cdots \tag{a}$$

$$y' = a_0cx^{c-1} + a_1(c + 1)x^c + a_2(c + 2)x^{c+1} + \cdots$$
$$+ a_r(c + r)x^{c+r-1} + \cdots \qquad \text{(b)}$$

$$-y = -a_0x^c - a_1x^{c+1} - a_2x^{c+2} - a_3x^{c+3} - \cdots - a_rx^{c+r} - \cdots \qquad \text{(c)}$$

(iv) The sum of these three terms forms the left-hand side of the equation. Since the right-hand side is zero, the coefficients of each power of x can be equated to zero.

For example, the coefficient of x^{c-1} is equated to zero giving:

$$3a_0c(c - 1) + a_0c = 0$$

or $a_0c[3c - 3 + 1] = a_0\,c(3c - 2) = 0 \qquad \text{(3)}$

The coefficient of x^c is equated to zero giving:

$$3a_1c(c + 1) + a_1(c + 1) - a_0 = 0$$

i.e. $a_1(3c^2 + 3c + c + 1) - a_0 = a_1(3c^2 + 4c + 1) - a_0 = 0$

or $a_1(3c + 1)(c + 1) - a_0 = 0 \qquad \text{(4)}$

In each of series (a), (b) and (c) an x^c term is involved, after which, a general relationship can be obtained for x^{c+r}, where $r \geq 0$.

In series (a) and (b), terms in x^{c+r-1} are present; replacing r by $(r + 1)$ will give the corresponding terms in x^{c+r}, which occurs in all three equations, i.e.

in series (a), $3a_{r+1}(c + r)(c + r+1)x^{c+r}$

in series (b), $a_{r+1}(c + r+1)x^{c+r}$

in series (c), $-a_rx^{c+r}$

Equating the total coefficients of x^{c+r} to zero gives:

$$3a_{r+1}(c + r)(c + r + 1) + a_{r+1}(c + r + 1) - a_r = 0$$

which simplifies to:

$$a_{r+1}\{(c + r + 1)(3c + 3r + 1)\} - a_r = 0 \qquad \text{(5)}$$

Equation (3), which was formed from the coefficients of the lowest power of x, i.e. x^{c-1}, is called the **indicial equation**, from which the value of c is obtained. From equation (3), since $a_0 \neq 0$, then

$$c = 0 \text{ or } c = \frac{2}{3}$$

(a) When c = 0:

From equation (4), if c = 0, $a_1(1 \times 1) - a_0 = 0$, i.e. $\mathbf{a_1 = a_0}$

From equation (5), if c = 0, $a_{r+1}(r + 1)(3r + 1) - a_r = 0$,

i.e. $a_{r+1} = \dfrac{a_r}{(r + 1)(3r + 1)} \qquad r \geq 0$

Thus,

when $r = 1$, $a_2 = \dfrac{a_1}{(2 \times 4)} = \dfrac{a_0}{(2 \times 4)}$ since $a_1 = a_0$

when $r = 2$, $a_3 = \dfrac{a_2}{(3 \times 7)} = \dfrac{a_0}{(2 \times 4)(3 \times 7)}$ or $\dfrac{a_0}{(2 \times 3)(4 \times 7)}$

when $r = 3$, $a_4 = \dfrac{a_3}{(4 \times 10)} = \dfrac{a_0}{(2 \times 3 \times 4)(4 \times 7 \times 10)}$ and so on.

From equation (1), the trial solution was:

$$y = x^c\{a_0 + a_1 x + a_2 x^2 + a_3 x^3 + \cdots + a_r x^r + \cdots\}$$

Substituting $c = 0$ and the above values of $a_1, a_2, a_3, \ldots$ into the trial solution gives:

$$y = x^0 \left[\begin{array}{l} a_0 + a_0 x + \left(\dfrac{a_0}{(2 \times 4)}\right) x^2 + \left(\dfrac{a_0}{(2 \times 3)(4 \times 7)}\right) x^3 \\[2mm] + \left(\dfrac{a_0}{(2 \times 3 \times 4)(4 \times 7 \times 10)}\right) x^4 + \cdots \end{array} \right]$$

i.e.
$$y = a_0 \left[\begin{array}{l} 1 + x + \dfrac{x^2}{(2 \times 4)} + \dfrac{x^3}{(2 \times 3)(4 \times 7)} \\[2mm] + \dfrac{x^4}{(2 \times 3 \times 4)(4 \times 7 \times 10)} + \cdots \end{array} \right] \qquad (6)$$

(b) When $c = \dfrac{2}{3}$:

From equation (4), if $c = \dfrac{2}{3}$, $a_1 (3)\left(\dfrac{5}{3}\right) - a_0 = 0$, i.e. $a_1 = \dfrac{a_0}{5}$

From equation (5), if $c = \dfrac{2}{3}$, $a_{r+1}\left(\dfrac{2}{3} + r + 1\right)(2 + 3r + 1) - a_r = 0$,

i.e. $a_{r+1}\left(r + \dfrac{5}{3}\right)(3r + 3) - a_r = a_{r+1}(3r^2 + 8r + 5) - a_r = 0$,

i.e. $a_{r+1} = \dfrac{a_r}{(r + 1)(3r + 5)} \qquad r \geq 0$

Thus, when $r = 1$, $a_2 = \dfrac{a_1}{(2 \times 8)} = \dfrac{a_0}{(2 \times 5 \times 8)}$ since $a_1 = \dfrac{a_0}{5}$

when $r = 2$, $a_3 = \dfrac{a_2}{(3 \times 11)} = \dfrac{a_0}{(2 \times 3)(5 \times 8 \times 11)}$

when $r = 3$, $a_4 = \dfrac{a_3}{(4 \times 14)} = \dfrac{a_0}{(2 \times 3 \times 4)(5 \times 8 \times 11 \times 14)}$ and so on.

From equation (1), the trial solution was:

$$y = x^c\{a_0 + a_1 x + a_2 x^2 + a_3 x^3 + \cdots + a_r x^r + \cdots\}$$

Substituting $c = \dfrac{2}{3}$ and the above values of $a_1, a_2, a_3, \ldots$ into the trial solution gives:

$$y = x^{\frac{2}{3}} \left\{ \begin{array}{l} a_0 + \left(\dfrac{a_0}{5}\right)x + \left(\dfrac{a_0}{2 \times 5 \times 8}\right)x^2 + \left(\dfrac{a_0}{(2 \times 3)(5 \times 8 \times 11)}\right)x^3 \\ + \left(\dfrac{a_0}{(2 \times 3 \times 4)(5 \times 8 \times 11 \times 14)}\right)x^4 + \cdots \end{array} \right\}$$

i.e.

$$y = a_0 x^{\frac{2}{3}} \left\{ \begin{array}{l} 1 + \dfrac{x}{5} + \dfrac{x^2}{(2 \times 5 \times 8)} + \dfrac{x^3}{(2 \times 3)(5 \times 8 \times 11)} \\ + \dfrac{x^4}{(2 \times 3 \times 4)(5 \times 8 \times 11 \times 14)} + \cdots \end{array} \right\} \tag{7}$$

Since a_0 is an arbitrary (non-zero) constant in each solution, its value could well be different.

Let $a_0 = A$ in equation (6), and $a_0 = B$ in equation (7). Also, if the first solution is denoted by $u(x)$ and the second by $v(x)$, then the general solution of the given differential equation is $y = u(x) + v(x)$. Hence,

$$y = A \left\{ \begin{array}{l} 1 + x + \dfrac{x^2}{(2 \times 4)} + \dfrac{x^3}{(2 \times 3)(4 \times 7)} \\ + \dfrac{x^4}{(2 \times 3 \times 4)(4 \times 7 \times 10)} + \cdots \end{array} \right\}$$

$$+ B x^{\frac{2}{3}} \left\{ \begin{array}{l} 1 + \dfrac{x}{5} + \dfrac{x^2}{(2 \times 5 \times 8)} + \dfrac{x^3}{(2 \times 3)(5 \times 8 \times 11)} \\ + \dfrac{x^4}{(2 \times 3 \times 4)(5 \times 8 \times 11 \times 14)} + \cdots \end{array} \right\}$$

Chapter 145   Power series methods of solving ordinary differential equations (4) – Bessel's equation

Bessel's equation

The solution of $x^2 \dfrac{d^2 y}{dx^2} + x \dfrac{dy}{dx} + (x^2 - v^2)y = 0$

is: $$y = A x^v \left\{ \begin{array}{l} 1 - \dfrac{x^2}{2^2(v + 1)} + \dfrac{x^4}{2^4 \times 2!(v + 1)(v + 2)} \\ - \dfrac{x^6}{2^6 \times 3!(v + 1)(v + 2)(v + 3)} + \cdots \end{array} \right\}$$

$$+ Bx^{-v} \left\{ \begin{array}{l} 1 + \dfrac{x^2}{2^2(v-1)} + \dfrac{x^4}{2^4 \times 2!(v-1)(v-2)} \\[3mm] + \dfrac{x^6}{2^6 \times 3!(v-1)(v-2)(v-3)} + \cdots \end{array} \right\}$$

or, in terms of Bessel functions and gamma functions:

$$y = A\,J_v(x) + B\,J_{-v}(x)$$

$$= A\left(\frac{x}{2}\right)^v \left\{ \frac{1}{\Gamma(v+1)} - \frac{x^2}{2^2(1!)\Gamma(v+2)} + \frac{x^4}{2^4(2!)\Gamma(v+4)} - \cdots \right\}$$

$$+ B\left(\frac{x}{2}\right)^{-v} \left\{ \frac{1}{\Gamma(1-v)} - \frac{x^2}{2^2(1!)\Gamma(2-v)} + \frac{x^4}{2^4(2!)\Gamma(3-v)} - \cdots \right\}$$

In general terms: $J_v(x) = \left(\dfrac{x}{2}\right)^v \displaystyle\sum_{k=0}^{\infty} \dfrac{(-1)^k x^{2k}}{2^{2k}(k!)\Gamma(v+k+1)}$

and $\qquad J_{-v}(x) = \left(\dfrac{x}{2}\right)^{-v} \displaystyle\sum_{k=0}^{\infty} \dfrac{(-1)^k x^{2k}}{2^{2k}(k!)\Gamma(k-v+1)}$

and in particular:

$$J_n(x) = \left(\frac{x}{2}\right)^n \left\{ \frac{1}{n!} - \frac{1}{(n+1)!}\left(\frac{x}{2}\right)^2 + \frac{1}{(2!)(n+2)!}\left(\frac{x}{2}\right)^4 - \cdots \right\}$$

$$J_0(x) = 1 - \frac{x^2}{2^2(1!)^2} + \frac{x^4}{2^4(2!)^2} - \frac{x^6}{2^6(3!)^2} + \cdots$$

and $\quad J_1(x) = \dfrac{x}{2} - \dfrac{x^3}{2^3(1!)(2!)} + \dfrac{x^5}{2^5(2!)(3!)} - \dfrac{x^7}{2^7(3!)(4!)} + \cdots$

Chapter 146 Power series methods of solving ordinary differential equations (5) – Legendre's equation and Legendre's polynomials

Legendre's equation

The solution of $(1 - x^2)\dfrac{d^2y}{dx^2} - 2x\dfrac{dy}{dx} + k(k+1)y = 0$

is: $y = a_0 \left\{ 1 - \dfrac{k(k+1)}{2!}x^2 + \dfrac{k(k+1)(k-2)(k+3)}{4!}x^4 - \cdots \right\}$

$$+ a_1 \left\{ \begin{array}{l} x - \dfrac{(k-1)(k+2)}{3!} x^3 \\ + \dfrac{(k-1)(k-3)(k+2)(k+4)}{5!} x^5 - \cdots \end{array} \right\} \tag{1}$$

Legendre's polynomials

Application: Determine the Legendre polynomial $P_3(x)$

Since in $P_3(x)$, $n = k = 3$, then from the second part of equation (1), i.e. the odd powers of x:

$$y = a_1 \left\{ x - \dfrac{(k-1)(k+2)}{3!} x^3 + \dfrac{(k-1)(k-3)(k+2)(k+4)}{5!} x^5 - \cdots \right\}$$

i.e. $\quad y = a_1 \left\{ x - \dfrac{(2)(5)}{3!} x^3 + \dfrac{(2)(0)(5)(7)}{5!} x^5 \right\} = a_1 \left\{ x - \dfrac{5}{3} x^3 + 0 \right\}$

a_1 is chosen to make $y = 1$ when $x = 1$.

i.e. $\quad 1 = a_1 \left\{ 1 - \dfrac{5}{3} \right\} = a_1 \left(-\dfrac{2}{3} \right)$ from which, $a_1 = -\dfrac{3}{2}$

Hence, $\quad P_3(x) = -\dfrac{3}{2} \left(x - \dfrac{5}{3} x^3 \right)$ or $P_3(x) = \dfrac{1}{2} (5x^3 - 3x)$

Chapter 147 Power series methods of solving ordinary differential equations (6) – Rodrigue's formula

Rodrigue's formula

$$P_n(x) = \dfrac{1}{2^n n!} \dfrac{d^n (x^2 - 1)^n}{dx^n}$$

Application: Determine the Legendre polynomial $P_3(x)$ using Rodrigue's formula

In Rodrigue's formula, $P_n(x) = \dfrac{1}{2^n n!} \dfrac{d^n (x^2 - 1)^n}{dx^n}$ and when $n = 3$,

$$P_3(x) = \dfrac{1}{2^3 3!} \dfrac{d^3 (x^2 - 1)^3}{dx^3} = \dfrac{1}{2^3 (6)} \dfrac{d^3 (x^2 - 1)(x^4 - 2x^2 + 1)}{dx^3}$$

$$= \frac{1}{(8)(6)} \frac{d^3(x^6 - 3x^4 + 3x^2 - 1)}{dx^3}$$

$$\frac{d(x^6 - 3x^4 + 3x^2 - 1)}{dx} = 6x^5 - 12x^3 + 6x$$

$$\frac{d(6x^5 - 12x^3 + 6x)}{dx} = 30x^4 - 36x^2 + 6$$

and $\dfrac{d(30x^4 - 36x^2 + 6)}{dx} = 120x^3 - 72x$

Hence, $P_3(x) = \dfrac{1}{(8)(6)} \dfrac{d^3(x^6 - 3x^4 + 3x^2 - 1)}{dx^3}$

$$= \frac{1}{(8)(6)} (120x^3 - 72x) = \frac{1}{8}(20x^3 - 12x)$$

i.e. $P_3(x) = \dfrac{1}{2}(5x^3 - 3x)$ the same as in the previous chapter.

Chapter 148 Solution of partial differential equations (1) – by direct partial integration

Application: Solve the differential equation $\dfrac{\partial^2 u}{\partial x^2} = 6x^2 (2y - 1)$ given the boundary conditions that at $x = 0$, $\dfrac{\partial u}{\partial x} = \sin 2y$ and $u = \cos y$

Since $\dfrac{\partial^2 u}{\partial x^2} = 6x^2 (2y - 1)$ then integrating partially with respect to x gives:

$$\frac{\partial u}{\partial x} = \int 6x^2 (2y - 1)\, dx = (2y - 1)\int 6x^2\, dx = (2y - 1)\frac{6x^3}{3} + f(y)$$

$$= 2x^3 (2y - 1) + f(y)$$

where f(y) is an arbitrary function.

From the boundary conditions, when $x = 0$, $\dfrac{\partial u}{\partial x} = \sin 2y$

Hence, $\sin 2y = 2(0)^3 (2y - 1) + f(y)$ from which, $f(y) = \sin 2y$

Now $\dfrac{\partial u}{\partial x} = 2x^3 (2y - 1) + \sin 2y$

Integrating partially with respect to x gives:

$$u = \int [2x^3(2y - 1) + \sin 2y]\, dx = \frac{2x^4}{4}(2y - 1) + x(\sin 2y) + F(y)$$

From the boundary conditions, when x = 0, u = cos y, hence

$$\cos y = \frac{(0)^4}{2}(2y - 1) + (0)\sin 2y + F(y)$$

from which, F(y) = cos y

Hence, the solution of $\dfrac{\partial^2 u}{\partial x^2} = 6x^2(2y - 1)$ for the given boundary conditions is:

$$u = \frac{x^4}{2}(2y - 1) + x\sin 2y + \cos y$$

Chapter 149   Solution of partial differential equations (2) – the wave equation

The **wave equation** is given by: $\dfrac{\partial^2 u}{\partial x^2} = \dfrac{1}{c^2}\dfrac{\partial^2 u}{\partial t^2}$

where $c^2 = \dfrac{T}{\rho}$, with T being the tension in a string and ρ being the mass /unit length of the string.

Summary of solution of the wave equation

1. Identify clearly the initial and boundary conditions.
2. Assume a solution of the form u = XT and express the equations in terms of X and T and their derivatives.
3. Separate the variables by transposing the equation and equate each side to a constant, say, μ; two separate equations are obtained, one in x and the other in t.
4. Let $\mu = -p^2$ to give an oscillatory solution.
5. The two solutions are of the form: X = A cos px + B sin px
   and T = C cos cpt + D sin cpt
   Then u(x, t) = {A cos px + B sin px}{C cos cpt + D sin cpt}
6. Apply the boundary conditions to determine constants A and B.
7. Determine the general solution as an infinite sum.
8. Apply the remaining initial and boundary conditions and determine the coefficients A_n and B_n from equations (1) and (2) below:

$$A_n = \frac{2}{L}\int_0^L f(x)\sin\frac{n\pi x}{L}\,dx \quad \text{for } n = 1, 2, 3, \ldots \tag{1}$$

$$B_n = \frac{2}{cn\pi}\int_0^L g(x)\sin\frac{n\pi x}{L}\,dx \tag{2}$$

Application: Figure 149.1 shows a stretched string of length 50 cm which is set oscillating by displacing its mid-point a distance of 2 cm from its rest position and releasing it with zero velocity. Solve the wave equation:
$$\frac{\partial^2 u}{\partial x^2} = \frac{1}{c^2}\frac{\partial^2 u}{\partial t^2} \quad \text{where } c^2 = 1, \text{ to determine the resulting motion } u(x, t).$$

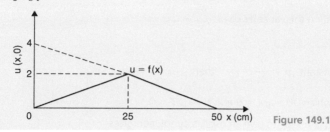

Figure 149.1

Following the above procedure:

1. The boundary and initial conditions given are:

$$\left.\begin{array}{l} u(0, t) = 0 \\ u(50, t) = 0 \end{array}\right\} \quad \text{i.e. fixed end points}$$

$$u(x, 0) = f(x) = \frac{2}{25}x \qquad\qquad 0 \le x \le 50$$

$$= -\frac{2}{25}x + 4 = \frac{100 - 2x}{25} \qquad 25 \le x \le 50$$

(Note: $y = mx + c$ is a straight line graph, so the gradient, m, between 0 and 25 is 2/25 and the y-axis intercept is zero, thus $y = f(x) = \frac{2}{25}x + 0$; between 25 and 50, the gradient $= -2/25$ and the y-axis intercept is at 4, thus $f(x) = -\frac{2}{25}x + 4$).

$$\left[\frac{\partial u}{\partial t}\right]_{t=0} = 0 \text{ i.e. zero initial velocity}$$

2. Assuming a solution $u = XT$, where X is a function of x only, and T is a function of t only,

then $\dfrac{\partial u}{\partial x} = X'T$ and $\dfrac{\partial^2 u}{\partial x^2} = X''T$ and $\dfrac{\partial u}{\partial y} = XT'$ and $\dfrac{\partial^2 u}{\partial y^2} = XT''$

Substituting into the partial differential equation,

$$\frac{\partial^2 u}{\partial x^2} = \frac{1}{c^2} \frac{\partial^2 u}{\partial t^2}$$

gives: $X''T = \dfrac{1}{c^2} XT''$ i.e. $X''T = XT''$ since $c^2 = 1$

3. Separating the variables gives: $\dfrac{X''}{X} = \dfrac{T''}{T}$

Let constant, $\mu = \dfrac{X''}{X} = \dfrac{T''}{T}$ then $\mu = \dfrac{X''}{X}$ and $\mu = \dfrac{T''}{T}$

from which, $X'' - \mu X = 0$ and $T'' - \mu T = 0$

4. Letting $\mu = -p^2$ to give an oscillatory solution gives

$X'' + p^2 X = 0$ and $T'' + p^2 T = 0$

The auxiliary equation for each is: $m^2 + p^2 = 0$ from which,

$$m = \sqrt{-p^2} = \pm jp$$

5. Solving each equation gives: $X = A\cos px + B\sin px$ and $T = C\cos pt + D\sin pt$

Thus, $u(x, t) = \{A\cos px + B\sin px\}\{C\cos pt + D\sin pt\}$

6. Applying the boundary conditions to determine constants A and B gives:

(i) $u(0, t) = 0$, hence $0 = A\{C\cos pt + D\sin pt\}$ from which we conclude that $A = 0$

Therefore, $u(x, t) = B\sin px\{C\cos pt + D\sin pt\}$ (a)

(ii) $u(50, t) = 0$, hence $0 = B\sin 50p\{C\cos pt + D\sin pt\}$

$B \neq 0$ hence $\sin 50p = 0$ from which, $50p = n\pi$ and $p = \dfrac{n\pi}{50}$

7. Substituting in equation (a) gives:

$$u(x, t) = B\sin\frac{n\pi x}{50}\left\{C\cos\frac{n\pi t}{50} + D\sin\frac{n\pi t}{50}\right\}$$

or, more generally,

$$u_n(x, t) = \sum_{n=1}^{\infty} \sin\frac{n\pi x}{50}\left\{A_n\cos\frac{n\pi t}{50} + B_n\sin\frac{n\pi t}{50}\right\}$$ (b)

where $A_n = BC$ and $B_n = BD$

8. From equation (1),

$$A_n = \frac{2}{L}\int_0^L f(x)\sin\frac{n\pi x}{L}\,dx$$

$$= \frac{2}{50}\left[\int_0^{25}\left(\frac{2}{25}x\right)\sin\frac{n\pi x}{50}\,dx + \int_{25}^{50}\left(\frac{100-2x}{25}\right)\sin\frac{n\pi x}{50}\,dx\right]$$

Each integral is determined using integration by parts (see chapter 121, page 334) with the result:

$$A_n = \frac{16}{n^2\pi^2}\sin\frac{n\pi}{2}$$

From equation (2), $B_n = \dfrac{2}{cn\pi}\displaystyle\int_0^L g(x)\sin\frac{n\pi x}{L}\,dx$

$$\left[\frac{\partial u}{\partial t}\right]_{t=0} = 0 = g(x) \text{ thus, } B_n = 0$$

Substituting into equation (b) gives:

$$u_n(x,t) = \sum_{n=1}^{\infty}\sin\frac{n\pi x}{50}\left\{A_n\cos\frac{n\pi t}{50} + B_n\sin\frac{n\pi t}{50}\right\}$$

$$= \sum_{n=1}^{\infty}\sin\frac{n\pi x}{50}\left\{\frac{16}{n^2\pi^2}\sin\frac{n\pi}{2}\cos\frac{n\pi t}{50} + (0)\sin\frac{n\pi t}{50}\right\}$$

Hence,

$$u(x,t) = \frac{16}{\pi^2}\sum_{n=1}^{\infty}\frac{1}{n^2}\sin\frac{n\pi x}{50}\sin\frac{n\pi}{2}\cos\frac{n\pi t}{50}$$

For stretched string problems as above, the main parts of the procedure are:

1. Determine A_n from equation (1).

   Note that $\dfrac{2}{L}\displaystyle\int_0^L f(x)\sin\frac{n\pi x}{L}\,dx$

   is **always** equal to $\dfrac{8d}{n^2\pi^2}\sin\dfrac{n\pi}{2}$ (see Figure 149.2)

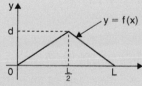

Figure 149.2

2. Determine B_n from equation (2)

3. Substitute in equation (b) to determine $u(x, t)$

Chapter 150   Solution of partial differential equations (3) – the heat conduction equation

The **heat conduction equation** is of the form: $\dfrac{\partial^2 u}{\partial x^2} = \dfrac{1}{c^2}\dfrac{\partial u}{\partial t}$ where $c^2 = \dfrac{h}{\sigma\rho}$, with h being the thermal conductivity of the material, σ the specific heat of the material, and ρ the mass/unit length of material.

Application: A metal bar, insulated along its sides, is 1 m long. It is initially at room temperature of 15°C and at time t = 0, the ends are placed into ice at 0°C. Find an expression for the temperature at a point P at a distance x m from one end at any time t seconds after t = 0

The temperature u along the length of bar is shown in Figure 150.1

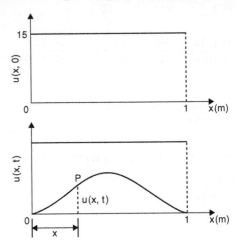

Figure 150.1

The heat conduction equation is $\dfrac{\partial^2 u}{\partial x^2} = \dfrac{1}{c^2}\dfrac{\partial u}{\partial t}$ and the given boundary conditions are:

$$u(0, t) = 0, \ u(1, t) = 0 \text{ and } u(x, 0) = 15$$

Assuming a solution of the form u = XT, then it may be shown that

$$X = A\cos px + B\sin px$$

and $T = ke^{-p^2 c^2 t}$

Thus, the general solution is given by:

$$u(x, t) = \{P\cos px + Q\sin px\}\, e^{-p^2 c^2 t}$$

$u(0, t) = 0$ thus $0 = P \, e^{-p^2c^2t}$ from which, $P = 0$
and $\quad u(x, t) = \{Q \sin px\} e^{-p^2c^2t}$

Also, $u(1, t) = 0$ thus $0 = \{Q \sin p\} \, e^{-p^2c^2t}$

Since $Q \neq 0$, $\sin p = 0$ from which, $p = n\pi$ where $n = 1, 2, 3, \ldots$

Hence, $u(x, t) = \displaystyle\sum_{n=1}^{\infty} \left\{ Q_n e^{-p^2c^2t} \sin n\pi x \right\}$

The final initial condition given was that at $t = 0$, $u = 15$,
i.e. $u(x, 0) = f(x) = 15$

Hence, $15 = \displaystyle\sum_{n=1}^{\infty} \left\{ Q_n \sin n\pi x \right\}$

where, from Fourier coefficients, $Q_n = 2 \times$ mean value of $15 \sin n\pi x$ from $x = 0$ to $x = 1$,

i.e. $\quad Q_n = \dfrac{2}{1} \displaystyle\int_0^1 15 \sin n\pi x \, dx = 30 \left[-\dfrac{\cos n\pi x}{n\pi} \right]_0^1$

$\qquad = -\dfrac{30}{n\pi} \left[\cos n\pi - \cos 0 \right] = \dfrac{30}{n\pi} (1 - \cos n\pi)$

$\qquad = 0$ (when n is even) and $\dfrac{60}{n\pi}$ (when n is odd)

Hence, the required solution is:

$$u(x, t) = \sum_{n=1}^{\infty} \left\{ Q_n e^{-p^2c^2t} \sin n\pi x \right\}$$

$$= \frac{60}{\pi} \sum_{n(\text{odd})=1}^{\infty} \frac{1}{n} (\sin n\pi x) e^{-n^2\pi^2c^2t}$$

Chapter 151   Solution of partial differential equations (4) – Laplace's equation

Laplace's equation, used extensively with electrostatic fields, is of the form:

$$\frac{\partial^2 u}{\partial x^2} + \frac{\partial^2 u}{\partial y^2} + \frac{\partial^2 u}{\partial z^2} = 0$$

Application: A square plate is bounded by the lines $x = 0$, $y = 0$, $x = 1$ and $y = 1$. Apply the Laplace equation $\dfrac{\partial^2 u}{\partial x^2} + \dfrac{\partial^2 u}{\partial y^2} = 0$ to determine the potential distribution $u(x, y)$ over the plate, subject to the following boundary conditions:

$u = 0$ when $x = 0$ $0 \leq y \leq 1$,

$u = 0$ when $x = 1$ $0 \leq y \leq 1$,

$u = 0$ when $y = 0$ $0 \leq x \leq 1$,

$u = 4$ when $y = 1$ $0 \leq x \leq 1$

Initially a solution of the form $u(x, y) = X(x)Y(y)$ is assumed, where X is a function of x only, and Y is a function of y only. Simplifying to $u = XY$, determining partial derivatives, and substituting into

$$\frac{\partial^2 u}{\partial x^2} + \frac{\partial^2 u}{\partial y^2} = 0 \text{ gives:} \qquad X''Y + XY'' = 0$$

Separating the variables gives: $\qquad \dfrac{X''}{X} = -\dfrac{Y''}{Y}$

Letting each side equal a constant, $-p^2$, gives the two equations:

$$X'' + p^2 X = 0 \quad \text{and} \quad Y'' - p^2 Y = 0$$

from which, $X = A \cos px + B \sin px$

and $Y = C e^{py} + D e^{-py}$ or $Y = C \cosh py + D \sinh py$ or $Y = E \sinh p(y + \phi)$

Hence $u(x, y) = XY = \{A \cos px + B \sin px\}\{E \sinh p(y + \phi)\}$

or $u(x, y) = \{P \cos px + Q \sin px\}\{\sinh p(y + \phi)\}$ where $P = AE$ and $Q = BE$

The first boundary condition is: $u(0, y) = 0$,

hence $0 = P \sinh p(y + \phi)$

from which, $P = 0$

Hence, $u(x, y) = Q \sin px \sinh p(y + \phi)$

The second boundary condition is: $u(1, y) = 0$,

hence $0 = Q \sin p(1) \sinh p(y + \phi)$

from which, $\sin p = 0$,

hence, $p = n\pi$ for $n = 1, 2, 3, \ldots$

The third boundary condition is: $u(x, 0) = 0$,

hence, $0 = Q \sin px \sinh p(\phi)$

from which, $\sinh p(\phi) = 0$ and $\phi = 0$

Hence, $u(x, y) = Q \sin px \sinh py$

Since there are many solutions for integer values of n,

$$u(x, y) = \sum_{n=1}^{\infty} Q_n \sin px \sinh py = \sum_{n=1}^{\infty} Q_n \sin n\pi x \sinh n\pi y \qquad \text{(a)}$$

The fourth boundary condition is: $u(x, 1) = 4 = f(x)$,

hence, $f(x) = \sum\limits_{n=1}^{\infty} Q_n \sin n\pi x \sinh n\pi(1)$

From Fourier series coefficients,

$Q_n \sinh n\pi = 2 \times$ the mean value of $f(x) \sin n\pi x$ from $x = 0$ to $x = 1$

$$= \frac{2}{1} \int_0^1 4 \sin n\pi x \, dx = 8 \left[-\frac{\cos n\pi x}{n\pi} \right]_0^1$$

$$= -\frac{8}{n\pi} (\cos n\pi - \cos 0) = \frac{8}{n\pi} (1 - \cos n\pi)$$

$$= 0 \text{ (for even values of n)}, \quad = \frac{16}{n\pi} \text{ (for odd values of n)}$$

Hence, $Q_n = \dfrac{16}{n\pi (\sinh n\pi)} = \dfrac{16}{n\pi} \text{cosech } n\pi$

Hence, from equation (a),

$$u(x, y) = \sum\limits_{n=1}^{\infty} Q_n \sin n\pi x \sinh n\pi y$$

$$= \frac{16}{\pi} \sum\limits_{n(\text{odd})=1}^{\infty} \frac{1}{n} (\text{cosech } n\pi \, \sin n\pi x \sinh n\pi y)$$

Laplace transforms

Why are Laplace transforms important?

The Laplace transform is a very powerful mathematical tool applied in various areas of engineering and science. With the increasing complexity of engineering problems, Laplace transforms help in solving complex problems with a very simple approach; the transform is an integral transform method which is particularly useful in solving linear ordinary differential equations. It has very wide applications in various areas of physics, electrical engineering, control engineering, optics, mathematics and signal processing, and in feedback control systems, such as in stability and control of aircraft systems.

The Heaviside unit step function is used in the mathematics of control theory and signal processing to represent a signal that switches on at a specified time and stays switched on indefinitely. It is also used in structural mechanics to describe different types of structural loads. The Heaviside function has applications in engineering where periodic functions are represented. In many physical situations things change suddenly; brakes are applied, a switch is thrown, collisions occur - and the Heaviside unit function is very useful for representing sudden change.

Laplace transforms and their inverses are mathematical techniques which allow us to solve differential equations and simultaneous differential equations, by primarily using algebraic methods. This simplification in the solving of equations, coupled with the ability to directly implement electrical components in their transformed form, makes the use of Laplace transforms widespread in both electrical engineering and control systems engineering.

Chapter 152   Standard Laplace transforms

Table 152.1

Time function f(t)	Laplace transform $\mathscr{L}\{f(t)\} = \int_0^\infty e^{-st} f(t)\, dt$
1. δ (unit impulse)	1
2. 1 (unit step function)	$\dfrac{1}{s}$
3. k (step function)	$\dfrac{k}{s}$
4. e^{at} (exponential function)	$\dfrac{1}{s-a}$
5. unit step delayed by T	$\dfrac{e^{-sT}}{s}$
6. $\sin \omega t$ (sine wave)	$\dfrac{\omega}{s^2 + \omega^2}$
7. $\cos \omega t$ (cosine wave)	$\dfrac{s}{s^2 + \omega^2}$
8. t (unit ramp function)	$\dfrac{1}{s^2}$
9. t^2	$\dfrac{2!}{s^3}$
10. t^n (n = positive integer)	$\dfrac{n!}{s^{n+1}}$
11. $\cosh \omega t$	$\dfrac{s}{s^2 - \omega^2}$
12. $\sinh \omega t$	$\dfrac{\omega}{s^2 - \omega^2}$
13. $e^{at} t^n$	$\dfrac{n!}{(s-a)^{n+1}}$
14. $e^{-at} \sin \omega t$ (damped sine wave)	$\dfrac{\omega}{(s+a)^2 + \omega^2}$
15. $e^{-at} \cos \omega t$ (damped cosine wave)	$\dfrac{s+a}{(s+a)^2 + \omega^2}$

Table 152.1 (cont.)

Time function f(t)	Laplace transform $\mathcal{L}\{f(t)\} = \int_0^\infty e^{-st}f(t)\,dt$
16. $e^{-at}\sinh \omega t$	$\dfrac{\omega}{(s+a)^2 - \omega^2}$
17. $e^{-at}\cosh \omega t$	$\dfrac{s+a}{(s+a)^2 - \omega^2}$

Common notations used for the Laplace transform

There are various commonly used notations for the Laplace transform of f(t) and these include:

(i) $\mathcal{L}\{f(t)\}$ or L{f(t)}

(ii) $\mathcal{L}(f)$ or Lf

(iii) $\overline{f}\,(s)$ or f(s)

Also, the letter p is sometimes used instead of s as the parameter.

Application: Determine $\mathcal{L}\left\{1 + 2t - \dfrac{1}{3}t^4\right\}$

$$\mathcal{L}\left\{1 + 2t - \frac{1}{3}t^4\right\} = \mathcal{L}\{1\} + 2\mathcal{L}\{t\} - \frac{1}{3}\mathcal{L}\{t^4\}$$

$$= \frac{1}{s} + 2\left(\frac{1}{s^2}\right) - \frac{1}{3}\left(\frac{4!}{s^{4+1}}\right) \quad \text{from 2, 8 and 10 of Table 152.1}$$

$$= \frac{1}{s} + \frac{2}{s^2} - \frac{1}{3}\left(\frac{4.3.2.1}{s^5}\right) = \frac{1}{s} + \frac{2}{s^2} - \frac{8}{s^5}$$

Application: Determine $\mathcal{L}\{5e^{2t} - 3e^{-t}\}$

$$\mathcal{L}\{5e^{2t} - 3e^{-t}\} = 5\mathcal{L}(e^{2t}) - 3\mathcal{L}\{e^{-t}\}$$

$$= 5\left(\frac{1}{s-2}\right) - 3\left(\frac{1}{s--1}\right) \quad \text{from 4 of Table 152.1}$$

$$= \frac{5}{s-2} - \frac{3}{s+1} = \frac{5(s+1) - 3(s-2)}{(s-2)(s+1)} = \frac{2s+11}{s^2 - s - 2}$$

Application: Determine $\mathcal{L}\{6\sin 3t - 4\cos 5t\}$

$\mathcal{L}\{6 \sin 3t - 4 \cos 5t\} = 6\mathcal{L}\{\sin 3t\} - 4\mathcal{L}\{\cos 5t\}$

$$= 6\left(\frac{3}{s^2 + 3^2}\right) - 4\left(\frac{s}{s^2 + 5^2}\right) \text{ from 6 and 7 of Table 152.1}$$

$$= \frac{18}{s^2 + 9} - \frac{4s}{s^2 + 25}$$

Application: Determine $\mathcal{L}\{2 \cosh 2\theta - \sinh 3\theta\}$

$\mathcal{L}\{2 \cosh 2\theta - \sinh 3\theta\} = 2\mathcal{L}\{\cosh 2\theta\} - \mathcal{L}\{\sinh 3\theta\}$

$$= 2\left(\frac{s}{s^2 - 2^2}\right) - \left(\frac{3}{s^2 - 3^2}\right) \text{ from 11 and 12 of Table 152.1}$$

$$= \frac{2s}{s^2 - 4} - \frac{3}{s^2 - 9}$$

Application: Determine $\mathcal{L}\{\sin^2 t\}$

$\mathcal{L}\{\sin^2 t\} = \mathcal{L}\left\{\frac{1}{2}(1 - \cos 2t)\right\}$ since $\cos 2t = 1 - 2 \sin^2 t$ and $\sin^2 t = \frac{1}{2}(1 - \cos 2t)$

$$= \frac{1}{2}\mathcal{L}\{1\} - \frac{1}{2}\mathcal{L}\{\cos 2t\}$$

$$= \frac{1}{2}\left(\frac{1}{s}\right) - \frac{1}{2}\left(\frac{s}{s^2 + 2^2}\right) \text{ from 2 and 7 of Table 152.1}$$

$$= \frac{(s^2 + 4) - s^2}{2s(s^2 + 4)} = \frac{4}{2s(s^2 + 4)} = \frac{2}{s(s^2 + 4)}$$

Application: Determine $\mathcal{L}\{2t^4 e^{3t}\}$

$\mathcal{L}\{2t^4 e^{3t}\} = 2\mathcal{L}\{t^4 e^{3t}\}$

$$= 2\left(\frac{4!}{(s - 3)^{4+1}}\right) \text{ from 13 of Table 152.1}$$

$$= \frac{2(4)(3)(2)}{(s - 3)^5} = \frac{48}{(s - 3)^5}$$

Application: Determine $\mathcal{L}\{4e^{3t} \cos 5t\}$

$\mathcal{L}\{4e^{3t} \cos 5t\} = 4\mathcal{L}\{e^{3t} \cos 5t\}$

$$= 4\left(\frac{s - 3}{(s - 3)^2 + 5^2}\right) \text{ from 15 of Table 152.1}$$

$$= \frac{4(s - 3)}{s^2 - 6s + 9 + 25} = \frac{4(s - 3)}{s^2 - 6s + 34}$$

Application: Determine $\mathscr{L}\{5e^{-3t} \sinh 2t\}$

$\mathscr{L}\{5e^{-3t} \sinh 2t\} = 5\mathscr{L}\{e^{-3t} \sinh 2t\}$

$$= 5\left(\frac{2}{(s - -3)^2 - 2^2}\right) \text{ from 16 of Table 152.1}$$

$$= \frac{10}{(s + 3)^2 - 2^2} = \frac{10}{s^2 + 6s + 9 - 4} = \frac{10}{s^2 + 6s + 5}$$

Application: Determine the Laplace transform of a step function of 10 volts which is delayed by $t = 5$ s, and sketch the function

The Laplace transform of a step function of 10 volts, shown in Figure 152.1(a), is given by:

$$\mathscr{L}\{10\} = \frac{10}{s} \qquad \text{from 3 of Table 152.1}$$

The Laplace transform of a step function of 10 volts which is delayed by $t = 5$ s is given by:

$$10\left(\frac{e^{-sT}}{s}\right) = 10\left(\frac{e^{-5s}}{s}\right) = \frac{10}{s}e^{-5s} \qquad \text{from 5 of Table 152.1}$$

The function is shown sketched in Figure 152.1(b).

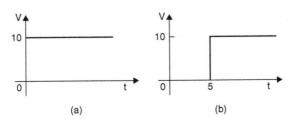

(a)   (b)   **Figure 152.1**

Application: Determine the Laplace transform of a ramp function which is delayed by 1 s and increases at 4 V/s. Sketch the function.

The Laplace transform of a ramp function which starts at zero and increases at 4 V/s, shown in Figure 152.2(a), is given by: $4\mathscr{L}\{t\} = \dfrac{4}{s^2}$ from 8 of Table 152.1

The Laplace transform of a ramp function which is delayed by 1 s and increases at 4 V/s is given by:

$$\left(\frac{4}{s^2}\right)e^{-s} \qquad \text{from 5 of Table 152.1}$$

A sketch of the ramp function is shown in Figure 152.2(b).

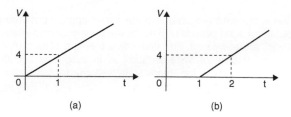

(a)   (b)   **Figure 152.2**

Application: Determine the Laplace transform of an impulse voltage of 8 volts which is delayed by 2 s. Sketch the function

The Laplace transform of an impulse voltage of 8V which starts at time $t = 0$, shown in Figure 152.3(a), is given by:

$$8\mathscr{L}\{\delta\} = 8 \quad \text{from 1 of Table 152.1}$$

The Laplace transform of an impulse voltage of 8 volts which is delayed by 2 s is given by:

$$8\,e^{-2s} \quad \text{from 5 of Table 152.1}$$

A sketch of the delayed impulse function is shown in Figure 152.3(b).

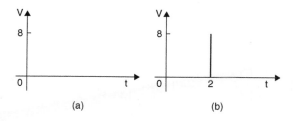

(a)   (b)   **Figure 152.3**

Chapter 153   The initial and final value theorems

The initial value theorem

$$\underset{t\to 0}{\text{lim it}} \; [f(t)] = \underset{s\to\infty}{\text{lim it}} \; [s\mathscr{L}\{f(t)\}]$$

The final value theorem

$$\underset{t\to\infty}{\text{limit}} \; [f(t)] = \underset{s\to 0}{\text{limit}} \; [s\mathscr{L}\{f(t)\}]$$

The initial and final value theorems are used in pulse circuit applications where the response of the circuit for small periods of time, or the behaviour immediately after the switch is closed, are of interest. The final value theorem is particularly useful in investigating the stability of systems (such as in automatic aircraft-landing systems) and is concerned with the steady state response for large values of time t, i.e. after all transient effects have died away.

Application: Verify the initial value theorem when $f(t) = 3e^{4t}$

If $f(t) = 3e^{4t}$ then $\mathscr{L}\{3e^{4t}\} = \dfrac{3}{s-4}$ from 4 of Table 152.1

By the initial value theorem, $\displaystyle\lim_{t\to 0} [3e^{4t}] = \lim_{s\to\infty}\left[s\left(\dfrac{3}{s-4}\right)\right]$

i.e. $3e^0 = \infty\left(\dfrac{3}{\infty - 4}\right)$

i.e.   **3 = 3**, which illustrates the theorem.

Application: Verify the initial value theorem for the voltage function $(5 + 2\cos 3t)$ volts

Let $f(t) = 5 + 2\cos 3t$

$\mathscr{L}\{f(t)\} = \mathscr{L}\{5 + 2\cos 3t\} = \dfrac{5}{s} + \dfrac{2s}{s^2 + 9}$ from 3 and 7 of Table 152.1

By the initial value theorem, $\displaystyle\lim_{t\to 0} [f(t)] = \lim_{s\to\infty} [s\mathscr{L}\{f(t)\}]$

i.e. $\displaystyle\lim_{t\to 0} [5 + 2\cos 3t] = \lim_{s\to\infty}\left[s\left(\dfrac{5}{s} + \dfrac{2s}{s^2 + 9}\right)\right] = \lim_{s\to\infty}\left[5 + \dfrac{2s^2}{s^2 + 9}\right]$

i.e. $5 + 2(1) = 5 + \dfrac{2\infty^2}{\infty^2 + 9} = 5 + 2$

i.e. **7 = 7**, which verifies the theorem in this case.

The initial value of the voltage is thus **7 V**

Application: Verify the final value theorem when $f(t) = 3e^{-4t}$

$$\lim_{t\to\infty} [3e^{-4t}] = \lim_{s\to 0}\left[s\left(\dfrac{3}{s+4}\right)\right]$$

$$3e^{-\infty} = (0)\left(\dfrac{3}{0+4}\right)$$

i.e.

i.e.   **0 = 0**, which illustrates the theorem.

Application: Verify the final value theorem for the function $(2 + 3e^{-2t} \sin 4t)$ cm, which represents the displacement of a particle

Let $f(t) = 2 + 3e^{-2t} \sin 4t$

$$\mathscr{L}\{f(t)\} = \mathscr{L}\{2 + 3e^{-2t} \sin 4t\} = \frac{2}{s} + 3\left(\frac{4}{(s--2)^2 + 4^2}\right)$$

$$= \frac{2}{s} + \frac{12}{(s+2)^2 + 16} \quad \text{from 3 and 14 of Table 152.1}$$

By the final value theorem, $\lim_{t\to\infty} [f(t)] = \lim_{s\to 0} [s\mathscr{L}\{f(t)\}]$

i.e. $\lim_{t\to\infty} [2 + 3e^{-2t} \sin 4t] = \lim_{s\to 0} \left[s\left(\frac{2}{s} + \frac{12}{(s+2)^2 + 16}\right)\right]$

$$= \lim_{s\to 0} \left[2 + \frac{12s}{(s+2)^2 + 16}\right]$$

i.e. $2 + 0 = 2 + 0$

i.e.   **2 = 2**, which verifies the theorem in this case.

The final value of the displacement is thus 2 cm.

Chapter 154   Inverse Laplace transforms

If the Laplace transform of a function $f(t)$ is $F(s)$, i.e. $\mathscr{L}\{f(t)\} = F(s)$, then $f(t)$ is called the **inverse Laplace transform** of $F(s)$ and is written as

$$f(t) = \mathscr{L}^{-1}\{F(s)\}$$

Table 152.1 on page 419 is used to determine inverse Laplace transforms.

Application: Determine $\mathscr{L}^{-1}\left\{\dfrac{1}{s^2 + 9}\right\}$

$$\mathscr{L}^{-1}\left\{\frac{1}{s^2 + 9}\right\} = \mathscr{L}^{-1}\left\{\frac{1}{s^2 + 3^2}\right\}$$

$$= \frac{1}{3}\mathscr{L}^{-1}\left\{\frac{3}{s^2 + 3^2}\right\} = \frac{1}{3}\sin 3t \quad \text{from 6 of Table 152.1}$$

Application: Determine $\mathscr{L}^{-1}\left\{\dfrac{5}{3s-1}\right\}$

$$\mathscr{L}^{-1}\left\{\frac{5}{3s-1}\right\} = \mathscr{L}^{-1}\left\{\frac{5}{3\left(s-\dfrac{1}{3}\right)}\right\}$$

$$= \frac{5}{3}\,\mathscr{L}^{-1}\left\{\frac{1}{\left(s-\dfrac{1}{3}\right)}\right\} = \frac{5}{3}\,e^{\frac{1}{3}t} \quad \text{from 4 of Table 152.1}$$

Application: Determine $\mathscr{L}^{-1}\left\{\dfrac{3}{s^4}\right\}$

$$\mathscr{L}^{-1}\left\{\frac{3}{s^4}\right\} = \frac{3}{3!}\,\mathscr{L}^{-1}\left\{\frac{3!}{s^{3+1}}\right\} = \frac{1}{2}\,t^3 \quad \text{from 10 of Table 152.1}$$

Application: Determine $\mathscr{L}^{-1}\left\{\dfrac{7s}{s^2+4}\right\}$

$$\mathscr{L}^{-1}\left\{\frac{7s}{s^2+4}\right\} = 7\mathscr{L}^{-1}\left\{\frac{s}{s^2+2^2}\right\} = 7\cos 2t \qquad \text{from 7 of Table 152.1}$$

Application: Determine $\mathscr{L}^{-1}\left\{\dfrac{3}{s^2-7}\right\}$

$$\mathscr{L}^{-1}\left\{\frac{3}{s^2-7}\right\} = 3\mathscr{L}^{-1}\left\{\frac{1}{s^2-(\sqrt{7})^2}\right\}$$

$$= \frac{3}{\sqrt{7}}\,\mathscr{L}^{-1}\left\{\frac{\sqrt{7}}{s^2-(\sqrt{7})^2}\right\}$$

$$= \frac{3}{\sqrt{7}}\,\sinh\sqrt{7}t \quad \text{from 12 of Table 152.1}$$

Application: Determine $\mathscr{L}^{-1}\left\{\dfrac{2}{(s-3)^5}\right\}$

$$\mathscr{L}^{-1}\left\{\frac{2}{(s-3)^5}\right\} = \frac{2}{4!}\,\mathscr{L}^{-1}\left\{\frac{4!}{(s-3)^{4+1}}\right\}$$

$$= \frac{1}{12}\,e^{3t}t^4 \qquad \text{from 13 of Table 152.1}$$

Application: Determine $\mathscr{L}^{-1}\left\{\dfrac{3}{s^2 - 4s + 13}\right\}$

$$\mathscr{L}^{-1}\left\{\frac{3}{s^2 - 4s + 13}\right\} = \mathscr{L}^{-1}\left\{\frac{3}{(s-2)^2 + 3^2}\right\}$$

$$= e^{2t}\sin 3t \qquad \text{from 14 of Table 152.1}$$

Application: Determine $\mathscr{L}^{-1}\left\{\dfrac{4s-3}{s^2 - 4s - 5}\right\}$

$$\mathscr{L}^{-1}\left\{\frac{4s-3}{s^2-4s-5}\right\} = \mathscr{L}^{-1}\left\{\frac{4s-3}{(s-2)^2-3^2}\right\} = \mathscr{L}^{-1}\left\{\frac{4(s-2)+5}{(s-2)^2-3^2}\right\}$$

$$= \mathscr{L}^{-1}\left\{\frac{4(s-2)}{(s-2)^2-3^2}\right\} + \mathscr{L}^{-1}\left\{\frac{5}{(s-2)^2-3^2}\right\}$$

$$= 4e^{2t}\cosh 3t + \mathscr{L}^{-1}\left\{\frac{\dfrac{5}{3}(3)}{(s-2)^2-3^2}\right\}$$

from 17 of Table 152.1

$$= 4e^{2t}\cosh 3t + \frac{5}{3}e^{2t}\sinh 3t \qquad \text{from 16 of Table 152.1}$$

Inverse Laplace transforms using partial fractions

Application: Determine $\mathscr{L}^{-1}\left\{\dfrac{4s-5}{s^2 - s - 2}\right\}$

Let $\dfrac{4s-5}{s^2-s-2} \equiv \dfrac{4s-5}{(s-2)(s+1)} \equiv \dfrac{A}{(s-2)} + \dfrac{B}{(s+1)} \equiv \dfrac{A(s+1)+B(s-2)}{(s-2)(s+1)}$

Hence, $4s - 5 \equiv A(s+1) + B(s-2)$

When $s = 2$, $3 = 3A$, from which, $A = 1$

When $s = -1$, $-9 = -3B$, from which, $B = 3$

Hence

$$\mathscr{L}^{-1}\left\{\frac{4s-5}{s^2-s-2}\right\} \equiv \mathscr{L}^{-1}\left\{\frac{1}{s-2}+\frac{3}{s+1}\right\} = \mathscr{L}^{-1}\left\{\frac{1}{s-2}\right\} + \mathscr{L}^{-1}\left\{\frac{3}{s+1}\right\}$$

$$= e^{2t} + 3e^{-t} \quad \text{from 4 of Table 152.1}$$

Application: Determine $\mathscr{L}^{-1}\left\{\dfrac{5s^2+8s-1}{(s+3)(s^2+1)}\right\}$

Let $\dfrac{5s^2+8s-1}{(s+3)(s^2+1)} \equiv \dfrac{A}{s+3} + \dfrac{Bs+C}{(s^2+1)} \equiv \dfrac{A(s^2+1)+(Bs+C)(s+3)}{(s+3)(s^2+1)}$

Hence, $5s^2+8s-1 \equiv A(s^2+1)+(Bs+C)(s+3)$

When $s = -3$, $20 = 10A$, from which, $A = 2$

Equating s^2 terms gives: $5 = A + B$, from which, $B = 3$, since $A = 2$

Equating s terms gives: $8 = 3B + C$, from which, $C = -1$, since $B = 3$

Hence

$$\mathscr{L}^{-1}\left\{\frac{5s^2+8s-1}{(s+3)(s^2+1)}\right\} \equiv \mathscr{L}^{-1}\left\{\frac{2}{s+3}+\frac{3s-1}{s^2+1}\right\}$$

$$\equiv \mathscr{L}^{-1}\left\{\frac{2}{s+3}\right\} + \mathscr{L}^{-1}\left\{\frac{3s}{s^2+1}\right\} - \mathscr{L}^{-1}\left\{\frac{1}{s^2+1}\right\}$$

$$= 2e^{-3t} + 3\cos t - \sin t \qquad \text{from 4, 7 and 6 of Table 152.1}$$

Chapter 155   Poles and zeros

It was seen in Chapter 154 that Laplace transforms, in general, have the form $f(s) = \dfrac{\phi(s)}{\theta(s)}$. This is the same form as most transfer functions for engineering systems, a **transfer function** being one that relates the response at a given pair of terminals to a source or stimulus at another pair of terminals.

Let a function in the s domain be given by: $f(s) = \dfrac{\phi(s)}{(s-a)(s-b)(s-c)}$ where $\phi(s)$ is of less degree than the denominator.

Poles: The values a, b, c, .. that makes the denominator zero, and hence f(s) infinite, are called the system poles of f(s).

If there are no repeated factors, the poles are **simple poles**.

If there are repeated factors, the poles are **multiple poles**.

Zeros: Values of s that make the numerator $\phi(s)$ zero, and hence f(s) zero, are called the system zeros of f(s).

For example: $\dfrac{s-4}{(s+1)(s-2)}$ has simple poles at s = −1 and s = +2, and a zero at s = 4

$\dfrac{s+3}{(s+1)^2(2s+5)}$ has a simple pole at $s = -\dfrac{5}{2}$ and double poles at s = −1, and a zero at s = −3

and $\dfrac{s+2}{s(s-1)(s+4)(2s+1)}$ has simple poles at s = 0, +1, −4, and$-\dfrac{1}{2}$ and a zero at s = −2

Pole-zero diagram

The poles and zeros of a function are values of complex frequency s and can therefore be plotted on the complex frequency or s-plane. The resulting plot is the **pole-zero diagram** or **pole-zero map**. On the rectangular axes, the real part is labelled the **σ-axis** and the imaginary part the **jω-axis**.

The location of a pole in the s-plane is denoted by a cross (×) and the location of a zero by a small circle (o). This is demonstrated in the following Applications.

From the pole-zero diagram it may be determined that the magnitude of the transfer function will be larger when it is closer to the poles and smaller when it is close to the zeros. This is important in understanding what the system does at various frequencies and is crucial in the study of **stability** and **control theory** in general.

Application: Determine for the transfer function:

$$R(s) = \frac{400\,(s+10)}{s\,(s+25)(s^2+10s+125)}$$

(a) the zero and (b) the poles. Show the poles and zero on a pole-zero diagram.

(a) For the numerator to be zero, (s + 10) = 0

Hence, **s = −10 is a zero** of R(s).

(b) For the denominator to be zero, s = 0 or s = −25 or (s² + 10s + 125) = 0

Using the quadratic formula, $s = \dfrac{-10 \pm \sqrt{10^2 - 4(1)(125)}}{2} = \dfrac{-10 \pm \sqrt{-400}}{2}$

$$= \frac{-10 \pm j20}{2}$$

$$= (-5 \pm j10)$$

Hence, **poles occur at s = 0, s = −25, (−5 + j10) and (−5 − j10)**

The pole-zero diagram is shown in Figure 155.1.

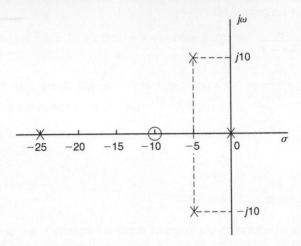

Figure 155.1

Application: Determine the poles and zeros for the function:

$F(s) = \dfrac{(s+3)(s-2)}{(s+4)(s^2+2s+2)}$ and plot them on a pole-zero map.

For the numerator to be zero, $(s+3) = 0$ and $(s-2) = 0$, hence **zeros occur at s = −3** and at **s = +2**

Poles occur when the denominator is zero, i.e. when $(s+4) = 0$, i.e. **s = −4,**

and when $s^2 + 2s + 2 = 0$, i.e. $s = \dfrac{-2 \pm \sqrt{2^2 - 4(1)(2)}}{2} = \dfrac{-2 \pm \sqrt{-4}}{2} = \dfrac{-2 \pm j2}{2}$

$$= (-1 + j) \text{ or } (-1 - j)$$

The poles and zeros are shown on the pole-zero map of F(s) in Figure 155.2.

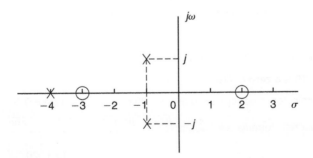

Figure 155.2

It is seen from these Applications that poles and zeros are always real or complex conjugate.

Chapter 156 The Laplace transform of the Heaviside function

Heaviside unit step function

In engineering applications, functions are frequently encountered whose values change abruptly at specified values of time t. One common example is when a voltage is switched on or off in an electrical circuit at a specified value of time t.

The switching process can be described mathematically by the function called the Unit Step Function – otherwise known as the Heaviside unit step function.

Figure 156.1 shows a function that maintains a zero value for all values of t up to t = c and a value of 1 for all values of t ≥ c. This is the Heaviside unit step function and is denoted by:

$$f(t) = H(t - c) \quad \text{or} \quad u(t - c)$$

where the c indicates the value of t at which the function changes from a value of zero to a value of unity (i.e. 1).

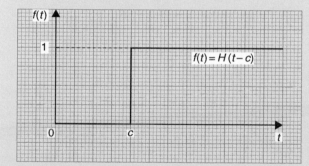

Figure 156.1

It follows that **f(t) = H(t − 5)** is as shown in Figure 156.2 and **f(t) = 3H(t − 4)** is a shown in Figure 156.3.

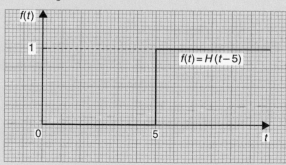

Figure 156.2

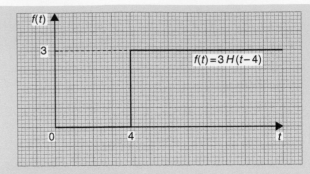

Figure 156.3

If the unit step occurs at the origin, then c = 0 and f(t) = H(t − 0), i.e. **H(t)** as shown in Figure 156.4.

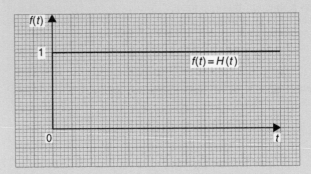

Figure 156.4

Figure 156.5(a) shows a graph of f(t) = t^2; the graph shown in Figure 156.5(b) is

$$f(t) = H(t − 2).t^2$$

where for t < 2, H(t − 2). t^2 = 0 and when t ≥ 2, H(t − 2). t^2 = t^2. The function H(t − 2) suppresses the function t^2 for all values of t up to t = 2 and then 'switches on' the function t^2 at t = 2.

A common situation in an electrical circuit is for a voltage V to be applied at a particular time, say, t = a, and removed later, say at t = b. Such a situation is written using step functions as:

$$V(t) = H(t − a) − H(t − b)$$

For example, Figure 156.6 shows the function **f(t) = H(t − 2) − H(t − 5)**

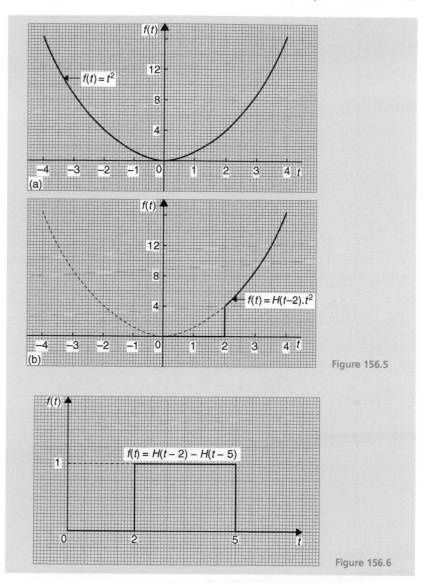

Figure 156.5

Figure 156.6

Application: A 12 V source is switched on at time t = 3 s. Sketch the waveform and write the function in terms of the Heaviside step function.

The function is shown sketched in Figure 156.7.

The Heaviside step function is: $V(t) = 12\,H(t - 3)$

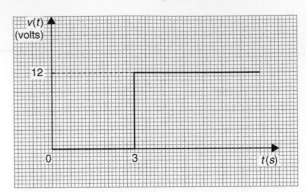

Figure 156.7

Application: Write the function $V(t) = \begin{cases} 1 & \text{for } 0 < t < a \\ 0 & \text{for } t > a \end{cases}$

in terms of the Heaviside step function and sketch the waveform.

The voltage has a value of 1 up until time $t = a$; then it is turned off.

The function is shown sketched in Figure 156.8.

The Heaviside step function is: **V(t) = H(t) − H(t − a)**

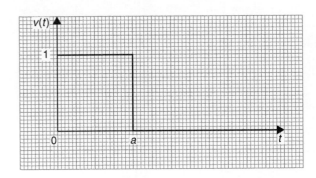

Figure 156.8

Application: Sketch the graph of $f(t) = H(t - \pi/3).\sin t$

Figure 156.9(a) shows a graph of $f(t) = \sin t$; the graph shown in Figure 156.9(b) is **$f(t) = H(t - \pi/3).\sin t$** where the graph of $\sin t$ does not 'switch on' until $t = \pi/3$

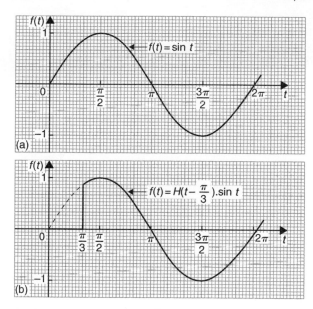

Figure 156.9

Application: Sketch the graph of $f(t) = 2\,H(t - 2\pi/3).\sin(t - \pi/6)$

Figure 156.10(a) shows a graph of $f(t) = 2\sin(t - \pi/6)$; the graph shown in Figure 156.10(b) is **$f(t) = 2\,H(t - 2\pi/3).\sin(t - \pi/6)$** where the graph of $2\sin(t - \pi/6)$ does not 'switch on' until $t = 2\pi/3$

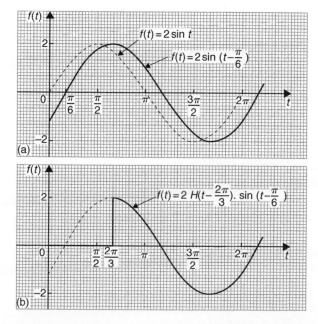

Figure 156.10

Laplace transform of H(t − c)

From the definition of a Laplace transform,

$$\mathscr{L}\{H(t - c)\} = \int_0^\infty e^{-st}H(t - c)\, dt$$

However, $e^{-st}H(t - c) = \begin{cases} 0 & \text{for } 0 \langle\, t\, \langle\, c \\ e^{-st} & \text{for } t \geq c \end{cases}$

Hence, $\mathscr{L}\{H(t - c)\} =$

$$\int_0^\infty e^{-st}H(t - c)\, dt = \int_c^\infty e^{-st}\, dt = \left[\frac{e^{-st}}{-s}\right]_c^\infty = \left[\frac{e^{-s(\infty)}}{-s} - \frac{e^{-sc}}{-s}\right]$$

$$= \left[0 - \frac{e^{-sc}}{-s}\right] = \frac{e^{-sc}}{s}$$

When c = 0 (i.e. a unit step at the origin), $\mathscr{L}\{H(t)\} = \dfrac{e^{-s(0)}}{s} = \dfrac{1}{s}$

Summarising, $\mathscr{L}\{\mathbf{H(t)}\} = \dfrac{1}{s}$ and $\mathscr{L}\{\mathbf{H(t - c)}\} = \dfrac{e^{-cs}}{s}$

From the definition of H(t): $\mathscr{L}\{\mathbf{1}\} = \{\mathbf{1.H(t)}\}$ $\mathscr{L}\{\mathbf{t}\} = \{\mathbf{t.H(t)}\}$

and $\mathscr{L}\{\mathbf{f(t)}\} = \{\mathbf{f(t).H(t)}\}$

Laplace transform of H(t − c).f(t − c)

It may be shown that:

$$\mathscr{L}\{\mathbf{H(t - c).f(t - c)}\} = \mathbf{e^{-cs}}\,\mathscr{L}\{\mathbf{f(t)}\} = \mathbf{e^{-cs}\,F(s)} \quad \text{where } F(s) = \mathscr{L}\{f(t)\}$$

Application: Determine $\mathscr{L}\{4\,H(t - 5)\}$

From above, $\mathscr{L}\{H(t - c).f(t - c)\} = e^{-cs}\,F(s)$ where in this case, $F(s) = \mathscr{L}\{4\}$ and c = 5

Hence, $\mathscr{L}\{4\,H(t - 5)\} = e^{-5s}\left(\dfrac{4}{s}\right) = \dfrac{4\,e^{-5s}}{s}$ from (ii) of Table 152.1, page 419

Application: Determine $\mathscr{L}\{H(t - 3).(t - 3)^2\}$

From above, $\mathcal{L}\{H(t - c).f(t - c)\} = e^{-cs}\,F(s)$ where in this case, $F(s) = \mathcal{L}\{t^2\}$ and $c = 3$

Note that $F(s)$ is the transform of t^2 and not of $(t - 3)^2$

Hence, $\mathcal{L}\{H(t - 3).f(t - 3)^2\} = e^{-3s}\left(\dfrac{2!}{s^3}\right) = \dfrac{2\,e^{-3s}}{s^3}$ from (vii) of Table 152.1, page 419

Application: Determine $\mathcal{L}\{H(t - 2).\sin(t - 2)\}$

From above, $\mathcal{L}\{H(t - c).f(t - c)\} = e^{-cs}F(s)$ where in this case, $F(s) = \mathcal{L}\{t^2\}$ and $c = 2$

Hence, $\mathcal{L}\{H(t - 2).\sin(t - 2)\} = e^{-2s}\left(\dfrac{1}{s^2 + 1^2}\right) = \dfrac{e^{-2s}}{s^2 + 1}$ from (iv) of Table 152.1, page 419

Application: Determine $\mathcal{L}\{H(t - 3).e^{t-3}\}$

From above, $\mathcal{L}\{H(t - c).f(t - c)\} = e^{-cs}F(s)$ where in this case, $F(s) = \mathcal{L}\{e^t\}$ and $c = 3$

Hence, $\mathcal{L}\{H(t - 3).e^{t-3}\} = e^{-3s}\left(\dfrac{1}{s - 1}\right) = \dfrac{e^{-3s}}{s - 1}$ from (iii) of Table 152.1, page 419

Inverse Laplace transforms of Heaviside functions

In the previous section it was stated that: $\mathcal{L}\{H(t - c).f(t - c)\} = e^{-cs}F(s)$ where $F(s) = \mathcal{L}\{f(t)\}$

Written in reverse, this becomes:

$$\text{if } F(s) = \mathcal{L}\{f(t)\}, \text{ then } e^{-cs}F(s) = \mathcal{L}\{H(t - c) . f(t - c)\}$$

This is known as the second shift theorem and is used when finding **inverse Laplace transforms**.

Application: Determine $\mathcal{L}^{-1}\left\{\dfrac{3\,e^{-2s}}{s}\right\}$

Part of the numerator corresponds to e^{-cs} where $c = 2$. This indicates $H(t - 2)$

Then $\dfrac{3}{s} = F(s) = \mathcal{L}\{3\}$ from (ii) of Table 152.1, page 419

Hence, $\mathcal{L}^{-1}\left\{\dfrac{3\,e^{-2s}}{s}\right\} = \mathbf{3H(t - 2)}$

Application: Determine the inverse of $\dfrac{e^{-3s}}{s^2}$

The numerator corresponds to e^{-cs} where $c = 3$. This indicates $H(t - 3)$

$\dfrac{1}{s^2} = F(s) = \mathscr{L}\{t\}$ from (vi) of Table 152.1, page 419

Then $\mathscr{L}^{-1}\left\{\dfrac{e^{-3s}}{s^2}\right\} = H(t - 3).(t - 3)$

Application: Determine $\mathscr{L}^{-1}\left\{\dfrac{8\,e^{-4s}}{s^2 + 4}\right\}$

Part of the numerator corresponds to e^{-cs} where $c = 4$. This indicates $H(t - 4)$

$\dfrac{8}{s^2 + 4}$ may be written as: $4\left(\dfrac{2}{s^2 + 2}\right)$

Then $4\left(\dfrac{2}{s^2 + 2}\right) = F(s) = \mathscr{L}\{4 \sin 2t\}$ from (iv) of Table 152.1, page 419

Hence, $\mathscr{L}^{-1}\left\{\dfrac{8\,e^{-4s}}{s^2 + 4}\right\} = H(t - 4).4 \sin 2(t - 4) = 4H(t - 4).\sin 2(t - 4)$

Application: Determine $\mathscr{L}^{-1}\left\{\dfrac{7\,e^{-3s}}{s^2 - 1}\right\}$

Part of the numerator corresponds to e^{-cs} where $c = 3$. This indicates $H(t - 3)$

$\dfrac{7}{s^2 - 1}$ may be written as: $7\left(\dfrac{1}{s^2 - 1^2}\right)$

Then $7\left(\dfrac{1}{s^2 - 1^2}\right) = F(s) = \mathscr{L}\{7 \sinh t\}$ from (x) of Table 152.1, page 419

Hence, $\mathscr{L}^{-1}\left\{\dfrac{7\,e^{-3s}}{s^2 - 1}\right\} = H(t - 3).7 \sinh (t - 3) = 7H(t - 3).\sinh (t - 3)$

Chapter 157   Solving differential equations using Laplace transforms

The Laplace transforms of derivatives

First derivatives: $\mathscr{L}\{f'(t)\} = s\mathscr{L}\{f(t)\} - f(0)$

or
$$\mathscr{L}\left\{\frac{dy}{dx}\right\} = s\mathscr{L}\{y\} - y(0) \tag{1}$$

where $y(0)$ is the value of y at $x = 0$

Second derivative: $\mathscr{L}\{f''(t)\} = s^2\mathscr{L}\{f(t)\} - sf(0) - f'(0)$

or
$$\mathscr{L}\left\{\frac{d^2y}{dx^2}\right\} = s^2\,\mathscr{L}\{y\} - s\,y(0) - y'(0) \tag{2}$$

where $y'(0)$ is the value of $\dfrac{dy}{dx}$ at $x = 0$

Higher derivatives:

$$\mathscr{L}\{f^n(t)\} = s^n\mathscr{L}\{f(t)\} - s^{n-1}f(0) - s^{n-2}f'(0) \ldots - f^{n-1}(0)$$

or $\mathscr{L}\left\{\dfrac{d^ny}{dx^n}\right\} = s^n\mathscr{L}\{y\} - s^{n-1}y(0) - s^{n-2}y'(0) \ldots - y^{n-1}(0)$

Procedure to solve differential equations by using Laplace transforms

1. Take the Laplace transform of both sides of the differential equation by applying the formulae for the Laplace transforms of derivatives (i.e. equations (1) and (2)) and, where necessary, using a list of standard Laplace transforms, as in Tables 152.1 on page 419

2. Put in the given initial conditions, i.e. $y(0)$ and $y'(0)$

3. Rearrange the equation to make $\mathscr{L}\{y\}$ the subject.

4. Determine y by using, where necessary, partial fractions, and taking the inverse of each term.

Application: Solve the differential equation $2\dfrac{d^2y}{dx^2} + 5\dfrac{dy}{dx} - 3y = 0$, given that when $x = 0$, $y = 4$ and $\dfrac{dy}{dx} = 9$

Using the above procedure:

1. $2\mathscr{L}\left\{\dfrac{d^2y}{dx^2}\right\} + 5\mathscr{L}\left\{\dfrac{dy}{dx}\right\} - 3\mathscr{L}\{y\} = \mathscr{L}\{0\}$

   $2[s^2\mathscr{L}\{y\} - sy(0) - y'(0)] + 5[s\mathscr{L}\{y\} - y(0)] - 3\mathscr{L}\{y\} = 0$,   from   equations   (1)   and (2)

2. $y(0) = 4$ and $y'(0) = 9$

   Thus $2[s^2\mathcal{L}\{y\} - 4s - 9] + 5[s\mathcal{L}\{y\} - 4] - 3\mathcal{L}\{y\} = 0$

   i.e. $2s^2\mathcal{L}\{y\} - 8s - 18 + 5s\mathcal{L}\{y\} - 20 - 3\mathcal{L}\{y\} = 0$

3. Rearranging gives: $(2s^2 + 5s - 3)\mathcal{L}\{y\} = 8s + 38$

   i.e. $\mathcal{L}\{y\} = \dfrac{8s + 38}{2s^2 + 5s - 3}$

4. $y = \mathcal{L}^{-1}\left\{\dfrac{8s + 38}{2s^2 + 5s - 3}\right\}$

   Let $\dfrac{8s + 38}{2s^2 + 5s - 3} \equiv \dfrac{8s + 38}{(2s - 1)(s + 3)}$

   $\equiv \dfrac{A}{2s - 1} + \dfrac{B}{s + 3} \equiv \dfrac{A(s + 3) + B(2s - 1)}{(2s - 1)(s + 3)}$

   Hence, $8s + 38 = A(s + 3) + B(2s - 1)$

   When $s = 0.5$, $42 = 3.5A$, from which, $A = 12$

   When $s = -3$, $14 = -7B$, from which, $B = -2$

   Hence, $y = \mathcal{L}^{-1}\left\{\dfrac{8s + 38}{2s^2 + 5s - 3}\right\} = \mathcal{L}^{-1}\left\{\dfrac{12}{2s - 1} - \dfrac{2}{s + 3}\right\}$

   $= \mathcal{L}^{-1}\left\{\dfrac{12}{2\left(s - \dfrac{1}{2}\right)}\right\} - \mathcal{L}^{-1}\left\{\dfrac{2}{s + 3}\right\}$

   Hence, $y = 6e^{\frac{1}{2}x} - 2e^{-3x}$ from 4 of Table 152.1 on page 419

Application: Solve $\dfrac{d^2y}{dx^2} - 3\dfrac{dy}{dx} = 9$, given that when $x = 0$, $y = 0$ and

$\dfrac{dy}{dx} = 0$

1. $\mathcal{L}\left\{\dfrac{d^2y}{dx^2}\right\} - 3\mathcal{L}\left\{\dfrac{dy}{dx}\right\} = \mathcal{L}\{9\}$

   Hence, $[s^2\mathcal{L}\{y\} - sy(0) - y'(0)] - 3[s\mathcal{L}\{y\} - y(0)] = \dfrac{9}{s}$

2. $y(0) = 0$ and $y'(0) = 0$

   Hence, $s^2\mathcal{L}\{y\} - 3s\mathcal{L}\{y\} = \dfrac{9}{s}$

3. Rearranging gives: $(s^2 - 3s)\mathcal{L}\{y\} = \dfrac{9}{s}$

i.e. $\mathcal{L}\{y\} = \dfrac{9}{s(s^2 - 3s)} = \dfrac{9}{s^2(s - 3)}$

4. $y = \mathcal{L}^{-1}\left\{\dfrac{9}{s^2(s - 3)}\right\}$

Let $\dfrac{9}{s^2(s - 3)} \equiv \dfrac{A}{s} + \dfrac{B}{s^2} + \dfrac{C}{s - 3} \equiv \dfrac{A(s)(s - 3) + B(s - 3) + Cs^2}{s^2(s - 3)}$

Hence, $9 \equiv A(s)(s - 3) + B(s - 3) + Cs^2$

When $s = 0, 9 = -3B$, from which, $B = -3$

When $s = 3, 9 = 9C$, from which, $C = 1$

Equating s^2 terms gives: $0 = A + C$, from which, $A = -1$, since $C = 1$

Hence, $\mathcal{L}^{-1}\left\{\dfrac{9}{s^2(s - 3)}\right\} = \mathcal{L}^{-1}\left\{-\dfrac{1}{s} - \dfrac{3}{s^2} + \dfrac{1}{s - 3}\right\}$

$\qquad\qquad\qquad\qquad = -1 - 3x + e^{3x}$ from 2, 8 and 4 of Table 152.1

i.e. $\qquad\qquad\qquad\qquad \mathbf{y = e^{3x} - 3x - 1}$

Application: Solve $\dfrac{d^2y}{dx^2} - 7\dfrac{dy}{dx} + 10y = e^{2x} + 20$, given that when $x = 0$,

$y = 0$ and $\dfrac{dy}{dx} = -\dfrac{1}{3}$

---

1. $\mathcal{L}\left\{\dfrac{d^2y}{dx^2}\right\} - 7\mathcal{L}\left\{\dfrac{dy}{dx}\right\} + 10\mathcal{L}\{y\} = \mathcal{L}\{e^{2x} + 20\}$

Hence, $[s^2\mathcal{L}\{y\} - sy(0) - y'(0)] - 7[s\mathcal{L}\{y\} - y(0)] + 10\mathcal{L}\{y\} = \dfrac{1}{s - 2} + \dfrac{20}{s}$

2. $y(0) = 0$ and $y'(0) = -\dfrac{1}{3}$

Hence, $s^2\mathcal{L}\{y\} - 0 - \left(-\dfrac{1}{3}\right) - 7s\mathcal{L}\{y\} + 0 + 10\mathcal{L}\{y\}$

$\qquad\qquad\qquad\qquad\qquad = \dfrac{s + 20(s - 2)}{s(s - 2)} = \dfrac{21s - 40}{s(s - 2)}$

3. $(s^2 - 7s + 10)\mathcal{L}\{y\}$

$\qquad = \dfrac{21s - 40}{s(s - 2)} - \dfrac{1}{3} = \dfrac{3(21s - 40) - s(s - 2)}{3s(s - 2)} = \dfrac{-s^2 + 65s - 120}{3s(s - 2)}$

Hence, $\qquad \mathcal{L}\{y\} = \dfrac{-s^2 + 65s - 120}{3s(s - 2)(s^2 - 7s + 10)} = \dfrac{1}{3}\left[\dfrac{-s^2 + 65s - 120}{s(s - 2)(s - 2)(s - 5)}\right]$

$\qquad\qquad\qquad\qquad = \dfrac{1}{3}\left[\dfrac{-s^2 + 65s - 120}{s(s - 5)(s - 2)^2}\right]$

4. $y = \dfrac{1}{3}\mathscr{L}^{-1}\left\{\dfrac{-s^2 + 65s - 120}{s(s-5)(s-2)^2}\right\}$

Let $\dfrac{-s^2 + 65s - 120}{s(s-5)(s-2)^2} \equiv \dfrac{A}{s} + \dfrac{B}{s-5} + \dfrac{C}{s-2} + \dfrac{D}{(s-2)^2}$

$$\equiv \dfrac{A(s-5)(s-2)^2 + B(s)(s-2)^2 + C(s)(s-5)(s-2) + D(s)(s-5)}{s(s-5)(s-2)^2}$$

Hence, $-s^2 + 65s - 120 \equiv A(s-5)(s-2)^2 + B(s)(s-2)^2 +$
$$C(s)(s-5)(s-2) + D(s)(s-5)$$

When $s = 0$, $-120 = -20A$, from which, $A = 6$

When $s = 5$, $180 = 45B$, from which, $B = 4$

When $s = 2$, $6 = -6D$, from which, $D = -1$

Equating s^3 terms gives: $0 = A + B + C$, from which, $C = -10$

Hence,

$$\dfrac{1}{3}\mathscr{L}^{-1}\left\{\dfrac{-s^2 + 65s - 120}{s(s-5)(s-2)^2}\right\} = \dfrac{1}{3}\mathscr{L}^{-1}\left\{\dfrac{6}{s} + \dfrac{4}{s-5} - \dfrac{10}{s-2} - \dfrac{1}{(s-2)^2}\right\}$$

$$= \dfrac{1}{3}\,[6 + 4e^{5x} - 10e^{2x} - xe^{2x}]$$

Thus, $y = 2 + \dfrac{4}{3}e^{5x} - \dfrac{10}{3}e^{2x} - \dfrac{x}{3}e^{2x}$

Chapter 158   Solving simultaneous differential equations using Laplace transforms

Procedure to solve simultaneous differential equations using Laplace transforms

1. Take the Laplace transform of both sides of each simultaneous equation by applying the formulae for the Laplace transforms of derivatives (i.e. equations (1) and (2), page 439) and using a list of standard Laplace transforms, as in Table 152.1, page 419

2. Put in the initial conditions, i.e. x(0), y(0), x'(0), y'(0)

3. Solve the simultaneous equations for $\mathscr{L}\{y\}$ and $\mathscr{L}\{x\}$ by the normal algebraic method.

4. Determine y and x by using, where necessary, partial fractions, and taking the inverse of each term.

Application: Solve the following pair of simultaneous differential equations

$$\frac{dy}{dt} + x = 1$$

$$\frac{dx}{dt} - y + 4e^t = 0$$

given that at $t = 0$, $x = 0$ and $y = 0$

Using the above procedure:

1. $\mathcal{L}\left\{\dfrac{dy}{dt}\right\} + \mathcal{L}\{x\} = \mathcal{L}\{1\}$ (1)

$\mathcal{L}\left\{\dfrac{dx}{dt}\right\} - \mathcal{L}\{y\} + 4\mathcal{L}\{e^t\} = \mathcal{L}\{0\}$ (2)

Equation (1) becomes:

$$[s\mathcal{L}\{y\} - y(0)] + \mathcal{L}\{x\} = \frac{1}{s} \tag{1'}$$

from equation (1), page 439 and Table 152.1

Equation (2) becomes:

$$[s\mathcal{L}\{x\} - x(0)] - \mathcal{L}\{y\} = -\frac{4}{s-1} \tag{2'}$$

2. $x(0) = 0$ and $y(0) = 0$ hence

Equation (1′) becomes:

$$s\mathcal{L}\{y\} + \mathcal{L}\{x\} = \frac{1}{s} \tag{1''}$$

and equation (2′) becomes: $s\mathcal{L}\{x\} - \mathcal{L}\{y\} = -\dfrac{4}{s-1}$

or $-\mathcal{L}\{y\} + s\mathcal{L}\{x\} = -\dfrac{4}{s-1}$ (2'')

3. $1 \times$ equation (1″) and $s \times$ equation (2″) gives:

$$s\mathcal{L}\{y\} + \mathcal{L}\{x\} = \frac{1}{s} \tag{3}$$

$$-s\mathcal{L}\{y\} + s^2\mathcal{L}\{x\} = -\frac{4s}{s-1} \tag{4}$$

Adding equations (3) and (4) gives:

$$(s^2 + 1)\mathcal{L}\{x\} = \frac{1}{s} - \frac{4s}{s-1} = \frac{(s-1) - s(4s)}{s(s-1)} = \frac{-4s^2 + s - 1}{s(s-1)}$$

from which, $\mathcal{L}\{x\} = \dfrac{-4s^2 + s - 1}{s(s-1)(s^2 + 1)}$

Using partial fractions

$$\frac{-4s^2 + s - 1}{s(s-1)(s^2 + 1)} \equiv \frac{A}{s} + \frac{B}{(s-1)} + \frac{Cs + D}{(s^2 + 1)}$$

$$= \frac{A(s-1)(s^2 + 1) + Bs(s^2 + 1) + (Cs + D)s(s-1)}{s(s-1)(s^2 + 1)}$$

Hence,
$$-4s^2 + s - 1 = A(s-1)(s^2 + 1) + Bs(s^2 + 1) + (Cs + D)s(s-1)$$

When s = 0, $-1 = -A$ hence, **A = 1**

When s = 1, $-4 = 2B$ hence, **B = −2**

Equating s^3 coefficients:

$$0 = A + B + C \quad \text{hence, } C = 1 \text{ (since } A = 1 \text{ and } B = -2)$$

Equating s^2 coefficients:

$$-4 = -A + D - C \quad \text{hence } D = -2 \text{ (since } A = 1 \text{ and } C = 1)$$

Thus, $\mathcal{L}\{x\} = \dfrac{-4s^2 + s - 1}{s(s-1)(s^2 + 1)} = \dfrac{1}{s} - \dfrac{2}{(s-1)} + \dfrac{s-2}{(s^2 + 1)}$

4. Hence, $x = \mathcal{L}^{-1}\left\{\dfrac{1}{s} - \dfrac{2}{(s-1)} + \dfrac{s-2}{(s^2 + 1)}\right\}$

$$= \mathcal{L}^{-1}\left\{\frac{1}{s} - \frac{2}{(s-1)} + \frac{s}{(s^2 + 1)} - \frac{2}{(s^2 + 1)}\right\}$$

i.e. **x = 1 − 2e^t + cos t − 2 sin t**   from Table 152.1, page 419

The second equation given originally is $\dfrac{dx}{dt} - y + 4e^t = 0$

from which, $y = \dfrac{dx}{dt} + 4e^t = \dfrac{d}{dt}\,(1 - 2e^t + \cos t - 2\sin t) + 4e^t$

$$= -2e^t - \sin t - 2\cos t + 4e^t$$

i.e.   **y = 2e^t− sin t −2 cos t**

[Alternatively, to determine y, return to equations (1″) and (2″)]

Application: Solve the following pair of simultaneous differential equations

$$\frac{d^2x}{dt^2} - x = y$$

$$\frac{d^2y}{dt^2} + y = -x$$

given that at $t = 0$, $x = 2$, $y = -1$, $\frac{dx}{dt} = 0$ and $\frac{dy}{dt} = 0$

1. $[s^2\mathscr{L}\{x\} - s\,x(0) - x'(0)] - \mathscr{L}\{x\} = \mathscr{L}\{y\}$ (1)

   $[s^2\mathscr{L}\{y\} - s\,y(0) - y'(0)] + \mathscr{L}\{y\} = -\mathscr{L}\{x\}$ (2)

2. $x(0) = 2$, $y(0) = -1$, $x'(0) = 0$ and $y'(0) = 0$

   hence $s^2\mathscr{L}\{x\} - 2s - \mathscr{L}\{x\} = \mathscr{L}\{y\}$ (1′)

         $s^2\mathscr{L}\{y\} + s + \mathscr{L}\{y\} = -\mathscr{L}\{x\}$ (2′)

3. Rearranging gives:

$$(s^2 - 1)\mathscr{L}\{x\} - \mathscr{L}\{y\} = 2s \tag{3}$$

$$\mathscr{L}\{x\} + (s^2 + 1)\mathscr{L}\{y\} = -s \tag{4}$$

Equation (3) $\times$ ($s^2 + 1$) and equation (4) $\times$ 1 gives:

$$(s^2 + 1)(s^2 - 1)\mathscr{L}\{x\} - (s^2 + 1)\mathscr{L}\{y\} = (s^2 + 1)2s \tag{5}$$

$$\mathscr{L}\{x\} + (s^2 + 1)\mathscr{L}\{y\} = -s \tag{6}$$

Adding equations (5) and (6) gives:

$$[(s^2 + 1)(s^2 - 1) + 1]\mathscr{L}\{x\} = (s^2 + 1)2s - s$$

i.e. $s^4\mathscr{L}\{x\} = 2s^3 + s = s(2s^2 + 1)$

from which, $\mathscr{L}\{x\} = \dfrac{s\,(2s^2 + 1)}{s^4} = \dfrac{2s^2 + 1}{s^3} = \dfrac{2s^2}{s^3} + \dfrac{1}{s^3} = \dfrac{2}{s} + \dfrac{1}{s^3}$

4. Hence $x = \mathscr{L}^{-1}\left\{\dfrac{2}{s} + \dfrac{1}{s^3}\right\}$

i.e. $x = 2 + \dfrac{1}{2}t^2$

Returning to equations (3) and (4) to determine y:

1 $\times$ equation (3) and ($s^2 - 1$) $\times$ equation (4) gives:

$$(s^2 - 1)\mathscr{L}\{x\} - \mathscr{L}\{y\} = 2s \tag{7}$$

$$(s^2 - 1)\mathscr{L}\{x\} + (s^2 - 1)(s^2 + 1)\mathscr{L}\{y\} = -s(s^2 - 1) \tag{8}$$

Equation (7) − equation (8) gives:

$$[-1 - (s^2 - 1)(s^2 + 1)]\mathcal{L}\{y\} = 2s + s(s^2 - 1)$$

i.e. $-s^4\mathcal{L}\{y\} = s^3 + s$

and $\mathcal{L}\{y\} = \dfrac{s^3 + s}{-s^4} = -\dfrac{1}{s} - \dfrac{1}{s^3}$

from which, $y = \mathcal{L}^{-1}\left\{-\dfrac{1}{s} - \dfrac{1}{s^3}\right\}$

i.e. $y = -1 - \dfrac{1}{2}t^2$

Z-transforms

Why are z-transforms important?

In mathematics and signal processing, the z-transform converts a discrete-time signal, which is a sequence of real or complex numbers, into a complex frequency domain representation. It can be considered as a discrete-time equivalent of the Laplace transform.

Laplace transform methods are widely used for analysis in linear systems and are used when a system is described by a linear differential equation, with constant coefficients. However, there are numerous systems that are described by difference equations – not differential equations – and these systems are common and different from those described by differential equations.

Systems that satisfy difference equations include computer-controlled systems - systems that take measurements with digital input/output boards or GPIB instruments (digital 8-bit parallel communications interface with data transfer rates up to 1 Mbyte/s), calculate an output voltage and output that voltage digitally. Frequently these systems run a program loop that executes in a fixed interval of time. Other systems that satisfy difference equations are those systems with digital filters – which are found anywhere digital signal processing/digital filtering is undertaken – that includes digital signal transmission systems like the telephone system or systems that process audio signals. A CD contains digital signal information, and when it is read off the CD, it is initially a digital signal that can be processed with a digital filter. There are an incredible number of systems used every day that have digital components which satisfy difference equations. In continuous systems Laplace transforms play a unique role. They allow system and circuit designers to analyse systems and predict performance, and to think in different terms – like frequency responses - to help understand linear continuous systems. They are a very powerful tool that shapes how engineers think about those systems. Z-transforms play the role in sampled systems that Laplace transforms play in continuous systems. In sampled systems, inputs and outputs are related by difference equations and z-transform techniques are used to solve those difference equations. In continuous systems, Laplace transforms are used to represent systems with transfer functions, while in sampled systems, z-transforms are used to represent systems with transfer functions.

Chapter 159   Sequences

The sequence $\dots, 2^{-3}, 2^{-2}, 2^{-1}, 2^0, 2^1, 2^2, 2^3, \dots$ has a general term of the form 2^k and this series can be written as $\{2^k\}_{-\infty}^{\infty}$ indicating that the power or index of the number 2 has a range from $-\infty$ to ∞.

The sum $\displaystyle\sum_{k=-\infty}^{\infty}\left(\frac{2}{z}\right)^k = \dots + \left(\frac{2}{z}\right)^{-3} + \left(\frac{2}{z}\right)^{-2} + \left(\frac{2}{z}\right)^{-1} + \left(\frac{2}{z}\right)^0 + \left(\frac{2}{z}\right)^1 + \left(\frac{2}{z}\right)^2 + \dots$

is called the z-transform of the sequence $Z\{2^k\}_{-\infty}^{\infty}$ and is denoted by $F(z)$, where the complex number z is chosen to ensure that the sum is finite.

Thus, $\{2^k\}_{-\infty}^{\infty}$ and $Z\{2^k\}_{-\infty}^{\infty} = F(z) = \displaystyle\sum_{k=-\infty}^{\infty}\left(\frac{2}{z}\right)^k$ form what is referred to as a 'z-transform pair'.

For simplicity, the range will be restricted to $Z\{x_k\}_0^{\infty}$ where $x_k = 0$ for $k < 0$ and denote it by $\{x_k\}$

i.e. $$Z\{x_k\} = F(z) = \sum_{k=0}^{\infty}\frac{x_k}{z^k}$$

This is the definition of the z-transform of the sequence $\{x_k\}$ and is used in the following applications.

Application: Determine the z-transform for the unit impulse $\{\delta_k\} = \{1, 0, 0, 0, \dots\}$

The z-transform of $\{\delta_k\}$ is given by:

$$Z\{\delta_k\} = F(z) = \sum_{k=0}^{\infty}\frac{\delta_k}{z^k} = \frac{1}{z^0} + \frac{0}{z^1} + \frac{0}{z^2} + \dots = 1$$

i.e. $Z\{\delta_k\} = 1$ valid for all values of z

Application: Determine the z-transform for the unit step sequence $\{u_k\} = \{1, 1, 1, 1, \dots\} = \{1\}$

The z-transform of $\{u_k\}$ is given by:

$$Z\{u_k\} = F(z) = \sum_{k=0}^{\infty}\frac{u_k}{z^k} = \sum_{k=0}^{\infty}\frac{1}{z^k} = \frac{1}{z^0} + \frac{1}{z^1} + \frac{1}{z^2} + \frac{1}{z^3} + \frac{1}{z^4} + \dots$$

i.e. $Z\{u_k\} = 1 + \dfrac{1}{z} + \dfrac{1}{z^2} + \dfrac{1}{z^3} + \dfrac{1}{z^4} + \dots$ (1)

Using the binomial theorem for $(1+x)^n$, the series expansion of $\dfrac{1}{1-x}$ may be determined:

$$\frac{1}{1-x} = (1-x)^{-1} = 1 + (-1)(-x) + \frac{(-1)(-2)}{2!}(-x)^2 + \frac{(-1)(-2)(-3)}{3!}(-x)^3 + \dots$$

$$= 1 + x + x^2 + x^3 + \dots \quad \text{valid for } |x| < 1 \quad (2)$$

Comparing equations (1) and (2) gives: $F(z) = \dfrac{1}{1 - \dfrac{1}{z}}$ provided $\left| \dfrac{1}{z} \right| < 1$

$\dfrac{1}{1 - \dfrac{1}{z}} = \dfrac{1}{\dfrac{z-1}{z}} = \dfrac{z}{z-1}$ hence, $\mathbf{z\{u_k\}} = \dfrac{z}{z-1}$ provided $|z| > 1$

Application: Show that the z-transform for the unit step sequence $\{x_k\} = \{1, a, a^2, a^3, a^4, ...\} = \{a^k\}$ is given by $\dfrac{z}{z-a}$

The z-transform of $\{x_k\}$ is given by:

$$Z\{x_k\} = Z\{a^k\} = \sum_{k=0}^{\infty} \frac{a^k}{z^k} = \sum_{k=0}^{\infty} \left(\frac{a}{z}\right)^k = \left(\frac{a}{z}\right)^0 + \left(\frac{a}{z}\right)^1 + \left(\frac{a}{z}\right)^2 + \left(\frac{a}{z}\right)^3 + \left(\frac{a}{z}\right)^4 + ...$$

$$= 1 + \frac{a}{z} + \left(\frac{a}{z}\right)^2 + \left(\frac{a}{z}\right)^3 + \left(\frac{a}{z}\right)^4 + ...$$

Comparing this with the series expansion of $\dfrac{1}{1-x} = 1 + x + x^2 + x^3 + ...$ which is valid for $|x| < 1$

i.e. equation (2) above, shows that:

$$F(z) = 1 + \frac{a}{z} + \left(\frac{a}{z}\right)^2 + \left(\frac{a}{z}\right)^3 + \left(\frac{a}{z}\right)^4 + ... = \frac{1}{1 - \dfrac{a}{z}} \text{ provided } \left| \frac{a}{z} \right| < 1$$

Hence, $\dfrac{1}{1 - \dfrac{a}{z}} = \dfrac{1}{\dfrac{z-a}{z}} = \dfrac{z}{z-a}$ and $\mathbf{F(z)} = \dfrac{\mathbf{z}}{\mathbf{z-a}}$ provided $|z| > |a|$

From the results obtained in the above applications, together with some additional results, a summary of some z-transforms is shown in Table 159.1, which may now be accepted and used.

Table 159.1

Sequence	Transform F(z)	
1. $\{\delta_k\} = \{1, 0, 0, ...\}$	1	for all values of z
2. $\{u_k\} = \{1, 1, 1, ...\}$	$\dfrac{z}{z-1}$	for $|z| > 1$
3. $\{k\} = \{0, 1, 2, 3, ...\}$	$\dfrac{z}{(z-1)^2}$	for $|z| > 1$
4. $\{k^2\} = \{0, 1, 4, 9, ...\}$	$\dfrac{z(z+1)}{(z-1)^3}$	for $|z| > 1$
5. $\{k^3\} = \{0, 1, 8, 27, ...\}$	$\dfrac{z(z^2 + 4z + 1)}{(z-1)^4}$	for $|z| > 1$

Table 159.1 Continued

Sequence	Transform F(z)	
6. $\{a^k\} = \{1, a, a^2, a^3, \ldots\}$	$\dfrac{z}{z-a}$	for $|z| > |a|$
7. $\{ka^k\} = \{0, a, 2a^2, 3a^3, \ldots\}$	$\dfrac{az}{(z-a)^2}$	for $|z| > |a|$
8. $\{k^2 a^k\} = \{0, a, 4a^2, 9a^3, \ldots\}$	$\dfrac{az(z+a)}{(z-a)^3}$	for $|z| > |a|$
9. $\{e^{-ak}\} = \{e^{-a}, e^{-2a}, e^{-3a}, \ldots\}$	$\dfrac{z}{z-e^{-a}}$	
10. $\sin ak = \{\sin a, \sin 2a, \ldots\}$	$\dfrac{z\sin a}{z^2 - 2z\cos a + 1}$	
11. $\cos ak = \{\cos a, \cos 2a, \ldots\}$	$\dfrac{z(z-\cos a)}{z^2 - 2z\cos a + 1}$	
12. $e^{-ak}\sin bk = \{e^{-a}\sin b, e^{-2a}\sin 2b, \ldots\}$	$\dfrac{ze^{-a}\sin b}{z^2 - 2ze^{-a}\cos b + e^{-2a}}$	
13. $e^{-ak}\cos bk = \{e^{-a}\cos b, e^{-2a}\cos 2b, \ldots\}$	$\dfrac{z^2 - ze^{-a}\cos b}{z^2 - 2ze^{-a}\cos b + e^{-2a}}$	

Application: Determine the z-transform of $5k^2$

From 4 in Table 159.1, $Z\left\{k^2\right\} = \dfrac{z(z+1)}{(z-1)^3}$

Hence, $\mathbf{Z}\left\{5k^2\right\} = 5Z\left\{k^2\right\} = \dfrac{5z(z+1)}{(z-1)^3}$

Application: Determine the z-transform of (a) 3^k (b) $(-3)^k$

(a) From 6 in Table 159.1, $Z\left\{a^k\right\} = \dfrac{z}{z-a}$

If $a = 3$, then $Z\left\{3^k\right\} = \dfrac{z}{z-3}$

(b) From 6 in Table 159.1, $Z\left\{a^k\right\} = \dfrac{z}{z-a}$

If $a = -3$, then $Z\left\{(-3)^k\right\} = \dfrac{z}{z--3} = \dfrac{z}{z+3}$

Application: Determine the z-transform of $2e^{-3k}$

From 9 in Table 159.1, $Z\{e^{-ak}\} = \dfrac{z}{z - e^{-a}}$

Hence, $Z\{2e^{-3k}\} = 2Z\{e^{-3k}\} = \dfrac{2z}{z - e^{-3}}$

Application: Determine $Z\{\cos 3k\}$

From 11 in Table 159.1, $Z\{\cos ak\} = \dfrac{z(z - \cos a)}{z^2 - 2z\cos a + 1}$

Hence, since a = 3, $Z\{\cos 3k\} = \dfrac{z(z - \cos 3)}{z^2 - 2z\cos 3 + 1}$

Application: Determine $Z\{e^{-2k}\cos 4k\}$

From 13 in Table 159.1, $Z\{e^{\,ak}\cos bk\} = \dfrac{z^2 - ze^{-a}\cos b}{z^2 - 2ze^{-a}\cos b + e^{-2a}}$

Hence, since a = 2 and b = 4, $Z\{e^{-2k}\cos 4k\} = \dfrac{z^2 - ze^{-2}\cos 4}{z^2 - 2ze^{-2}\cos 4 + e^{-4}}$

Chapter 160   Some properties of z-transforms

(a) Linearity property

The z-transform is a **linear transform**,

i.e. $Z(a\{x_k\} + b\{y_k\}) = aZ\{x_k\} + bZ\{y_k\}$ (1)

where a and b are constants

Application: Determine the z-transform of $2\{k\} - 3\{e^{-2k}\}$

Now, $2Z\{k\} = 2\left\{\dfrac{z}{(z-1)^2}\right\}$ from 4 in Table 159.1

and since $Z\{e^{-ak}\} = \dfrac{z}{z - e^{-a}}$ from 9 in Table 159.1,

then $Z\{e^{-2k}\} = \dfrac{z}{z - e^{-2}}$

Hence, $Z(2\{k\} - 3\{e^{-2k}\}) = 2Z\{k\} - 3Z\{e^{-2k}\}$

$$= 2\left\{\dfrac{z}{(z-1)^2}\right\} - 3\left\{\dfrac{z}{z - e^{-2}}\right\} \text{ from equation (1)}$$

$$= \left\{\dfrac{2z}{(z-1)^2}\right\} - \left\{\dfrac{3z}{z - e^{-2}}\right\}$$

$$= \frac{2z\left(z - e^{-2}\right) - 3z\left(z - 1\right)^{2}}{\left(z - 1\right)^{2}\left(z - e^{-2}\right)}$$

$$= \frac{2z^{2} - 2ze^{-2} - 3z\left(z^{2} - 2z + 1\right)}{\left(z - 1\right)^{2}\left(z - e^{-2}\right)}$$

$$= \frac{2z^{2} - 2ze^{-2} - 3z^{3} + 6z^{2} - 3z}{\left(z - 1\right)^{2}\left(z - e^{-2}\right)}$$

i.e. $$2Z\{k\} - 3Z\{e^{-2k}\} = \frac{-3z^{3} + 8z^{2} - z\left(2e^{-2} + 3\right)}{\left(z - 1\right)^{2}\left(z - e^{-2}\right)}$$

(b) First shift theorem (shifting to the left)

It may be shown by the **first shift theorem (shifting to the left)**, that

if $Z\{x_k\} = F(z)$

then $$Z\{x_{k+m}\} = z^{m}F(z) - [z^{m}x_0 + z^{m-1}x_1 + \ .... + zx_{m-1}] \tag{2}$$

is the z-transform of the sequence that has been shifted by m places to the left.

This theorem is often needed when solving difference equations (see Chapter 162).

Application: Determine $Z\{3^{k+2}\}$

Since from equation (2), $Z\{x_{k+m}\} = z^{m}F(z) - [z^{m}x_0 + z^{m-1}x_1 + \ .... + zx_{m-1}]$

then $$Z\{3^{k+2}\} = z^{2}Z\{3^{k}\} - [z^{2}3^{0} + z3^{1}] \tag{3}$$

From 6 of Table 159.1, $Z\{a^{k}\} = \dfrac{z}{z - a}$ thus $Z\{3^{k}\} = \dfrac{z}{z - 3}$

Hence, substituting in equation (3),

$$Z\{3^{k+2}\} = z^{2}\left(\frac{z}{z - 3}\right) - \left[z^{2} + 3z\right]$$

$$= \frac{z^{3}}{z - 3} - \left[z^{2} + 3z\right]$$

$$= \frac{z^{3} - \left(z - 3\right)\left[z^{2} + 3z\right]}{z - 3}$$

$$= \frac{z^{3} - \left[z^{3} + 3z^{2} - 3z^{2} - 9z\right]}{z - 3}$$

$$= \frac{z^{3} - \left[z^{3} - 9z\right]}{z - 3}$$

i.e.
$$Z\{3^{k+2}\} = \frac{9z}{z-3}$$

This is the z-transform of the sequence $\{9, 27, 81,...\}$ by shifting the sequence $\{1, 3, 9, 27,...\}$ two places to the left and losing the first two terms.

Application: Determine $Z\{k + 1\}$

From 3 of Table 159.1,
$$Z\{k\} = \frac{z}{(z-1)^2}$$

Since from equation (2), $Z\{x_{k+m}\} = z^m F(z) - [z^m x_0 + z^{m-1} x_1 + ... + z x_{m-1}]$

then
$$Z\{k + 1\} = z^1 Z(k) - [z^1 x_0]$$

$$= z^1 \left| \frac{z}{(z-1)^2} \right| - \left[z^1 \times 0 \right] \text{ from 3 of Table 159.1}$$

i.e.
$$Z\{k + 1\} = \frac{z^2}{(z-1)^2}$$

(c) Second shift theorem (shifting to the right)

It may be shown by the **second shift theorem (shifting to the right)**, that

if
$$Z\{x_k\} = F(z)$$

then
$$Z\{x_{k-m}\} = z^{-m} F(z) \tag{4}$$

is the z-transform of the sequence that has been shifted by m places to the right.

Application: Determine $Z\{x_{k-2}\}$

Since from equation (4), $Z\{x_{k-m}\} = z^{-m} F(z)$

then
$$Z\{x_{k-2}\} = z^{-2} F(z)$$

$$= z^{-2} \left(\frac{z}{z-1} \right) \text{ since } Z\{x_k\} = \frac{z}{z-1} \text{ from 2 of Table 159.1}$$

i.e.
$$Z\{x_{k-2}\} = \left(\frac{z^{-1}}{z-1} \right) = \frac{1}{z(z-1)}$$

This is the z-transform of the sequence $\{0, 0, 1, 1, 1, ...\}$ by shifting the sequence $\{1, 1, 1, 1, ...\}$ two places to the right and defining the first two terms as zeros.

Application: Determine $Z\{a^{k-1}\}$

Since from equation (4), $Z\{x_{k-m}\} = z^{-m}F(z)$

then $\qquad\qquad Z\{a^{k-1}\} = z^{-1}F(z)$

$$= z^{-1}\left(\frac{z}{z-a}\right) \text{ since } Z\{a^k\} = \frac{z}{z-a} \text{ from 6 of Table 159.1}$$

i.e. $\qquad\qquad \mathbf{Z\{a^{k-1}\}} = \left(\frac{z^{-1}\times z}{z-a}\right) = \frac{1}{(z-a)}$

which is the z-transform of $\{a^k\}$ shifted one place to the right.

(d) Translation

If the sequence $\{x_k\}$ has the z-transform $Z\{a^k x_k\} = F(z)$,

then the sequence $\{a^k x_k\}$ has the z-transform $\mathbf{Z\{a^k x_k\} = F(a^{-1}z)}$

Application: Determine $Z\{3^k k\}$

Since $\qquad Z(k) = \dfrac{z}{(z-1)^2}$ from 3 of Table 159.1

then by the translation property,

$$Z\{3^k k\} = F(3^{-1}z) = \frac{3^{-1}z}{(3^{-1}z-1)^2} = \frac{3^{-1}z}{\left[3^{-1}\left(z - \dfrac{1}{3^{-1}}\right)\right]^2}$$

$$= \frac{3^{-1}z}{3^{-2}\left(z-3\right)^2} = \frac{3^2 z}{3^1\left(z-3\right)^2}$$

i.e. $\qquad\qquad \mathbf{Z\{3^k k\}} = \dfrac{3z}{\left(z-3\right)^2}$

(e) Final value theorem

For the sequence $\{x_k\}$ with the z-transform $F(z)$,

$$\underset{k\to\infty}{Lim}\ x_k = \underset{z\to 1}{Lim}\left\{\left(\frac{z-1}{z}\right)F(z)\right\} \text{ provided that } \underset{k\to\infty}{Lim}\ x_k \text{ exists}$$

Application: Determine $\displaystyle\lim_{k\to\infty}\left\{\left(\frac{1}{3}\right)^k\right\}$

$F\{x^k\} = \dfrac{z}{z-1}$ and $\left\{\left(\dfrac{1}{3}\right)^k\right\} = \dfrac{z}{z-\dfrac{1}{3}} = \dfrac{3z}{3z-1}$

Now $\displaystyle\lim_{k\to\infty}\left\{\left(\frac{1}{3}\right)^k\right\} = \lim_{z\to 1}\left\{\left(\frac{z-1}{z}\right)F(z)\right\} = \lim_{z\to 1}\left\{\left(\frac{z-1}{z}\right)\frac{3z}{3z-1}\right\}$

$\qquad\qquad\qquad\qquad\qquad = \displaystyle\lim_{z\to 1}\left\{\frac{3(z-1)}{3z-1}\right\} = 0$

i.e. $\qquad\qquad\qquad \displaystyle\lim_{k\to\infty}\left\{\left(\frac{1}{3}\right)^k\right\} = 0$

Application: Determine $\displaystyle\lim_{k\to\infty}\left\{\frac{33z^2-25z}{(z-1)(3z-1)^2}\right\}$

By the final value theorem, $\displaystyle\lim_{k\to\infty}\left\{\frac{33z^2-25z}{(z-1)(3z-1)^2}\right\} = \lim_{z\to 1}\left\{\left(\frac{z-1}{z}\right)F(z)\right\}$

$\qquad\qquad\qquad\qquad\qquad = \displaystyle\lim_{z\to 1}\left\{\left(\frac{z-1}{z}\right)\frac{33z^2-25z}{(z-1)(3z-1)^2}\right\}$

$\qquad\qquad\qquad\qquad\qquad = \displaystyle\lim_{z\to 1}\left\{\frac{33z-25}{(3z-1)^2}\right\}$

$\qquad\qquad\qquad\qquad\qquad = \dfrac{33-25}{2^2} = \dfrac{8}{4} = 2$

(f) The initial value theorem

For the sequence $\{x_k\}$ with the z-transform $F(z)$,

$$\text{initial value, } x_0 = \lim_{z\to\infty}\{F(z)\}$$

Application: Determine $\displaystyle\lim_{z\to\infty}\left\{a^k\right\}$

$$F(z) = F\{a^k\} = \frac{z}{z-a} \text{ from 6 of Table 159.1}$$

and

$$\lim_{z \to \infty} \{F(z)\} = \lim_{z \to \infty} \left\{ \frac{z}{z-a} \right\}$$

$$= \lim_{z \to \infty} \left\{ \frac{\dfrac{d}{dz}(z)\,dz}{\dfrac{d}{dz}(z-a)\,dz} \right\} = \lim_{z \to \infty} \left\{ \frac{1}{1} \right\} = \begin{array}{l} \text{by L'Hopital's rule} \\ \text{(see Chapter 28)} \end{array}$$

i.e.

$$\lim_{z \to \infty} \{a^k\} = 1$$

(g) The derivative of the transform

If

$$Z\{x_k\} = F(z)$$

then

$$-zF'(z) = Z\{kx_k\}$$

Application: Determine the derivative of $Z\{ka^k\}$

$$F(z) = F\{a^k\} = \frac{z}{z-a} \text{ from 6 of Table 159.1}$$

Now $Z\{kx_k\} = -zF'(z)$ from above

i.e.

$$Z\{kx_k\} = -zF'\left(\frac{z}{z-a} \right) = -z \left[\frac{(z-a)(1)-z(1)}{(z-a)^2} \right] \text{ using the quotient rule}$$

$$= -z \left[\frac{z-a-z}{(z-a)^2} \right] = -z \left[\frac{-a}{(z-a)^2} \right] = \frac{az}{(z-a)^2}$$

i.e. **the derivative of $Z\{ka^k\}$** $= -zF'(z) = \dfrac{az}{(z-a)^2}$ which confirms result 7 in Table 159.1

Chapter 161   Inverse z-transforms

If the sequence $\{x_k\}$ has a Z transform $Z\{x_k\} = F(z)$,

then the inverse z-transform is defined as: $Z^{-1}F(z) = \{x_k\}$

In the following Applications, some inverses may be determined directly from Table 159.1, page 449–450, albeit with a little manipulation; others sometimes require the use of partial fractions – just as with inverse Laplace transforms.

Application: Determine the inverse z-transform of $F(z) = \dfrac{z}{z+5}$

From 6 in Table 159.1, $Z\{a^k\} = \dfrac{z}{z-a}$ hence $Z^{-1}\left\{\dfrac{z}{z-a}\right\} = a^k$

Comparing $\dfrac{z}{z+5}$ with $\dfrac{z}{z-a}$ shows that $a = -5$

Thus, $\qquad Z^{-1}\left\{\dfrac{z}{z+5}\right\} = (-5)^k$

Application: Determine the inverse z-transform of $F(z) = \dfrac{2z}{2z+1}$

$$\frac{2z}{2z+1} = \frac{2z}{2\left(z+\dfrac{1}{2}\right)} = \frac{z}{z+\dfrac{1}{2}}$$

From 6 in Table 159.1, $Z\{a^k\} = \dfrac{z}{z-a}$ hence $Z^{-1}\left\{\dfrac{z}{z-a}\right\} = a^k$

Comparing $\dfrac{z}{z+\dfrac{1}{2}}$ with $\dfrac{z}{z-a}$ shows that $a = -\dfrac{1}{2}$

Thus, $Z^{-1}\left\{\dfrac{2z}{2z+1}\right\} = Z^{-1}\left\{\dfrac{z}{z+\dfrac{1}{2}}\right\} = \left(-\dfrac{1}{2}\right)^k$

Application: If $F(z) = \dfrac{3z}{3z-1}$ determine $Z^{-1}F(z)$

$$\frac{3z}{3z-1} = \frac{3z}{3\left(z-\dfrac{1}{3}\right)} = \frac{z}{z-\dfrac{1}{3}}$$

From 6 in Table 159.1, $Z\{a^k\} = \dfrac{z}{z-a}$ hence $Z^{-1}\left\{\dfrac{z}{z-a}\right\} = a^k$

Comparing $\dfrac{z}{z-\dfrac{1}{3}}$ with $\dfrac{z}{z-a}$ shows that $a = \dfrac{1}{3}$

Thus, $Z^{-1}\left\{\dfrac{3z}{3z-1}\right\} = Z^{-1}\left\{\dfrac{z}{z-\dfrac{1}{3}}\right\} = \left(\dfrac{1}{3}\right)^k$

Application: Determine the inverse z-transform of $F(z) = \dfrac{z}{z-e^2}$

From 9 in Table 159.1, $Z\left\{e^{-ak}\right\} = \dfrac{z}{z - e^{-a}}$ hence $Z^{-1}\left\{\dfrac{z}{z - e^{-a}}\right\} = e^{-ak}$

Comparing $\dfrac{z}{z - e^2}$ with $\dfrac{z}{z - e^{-a}}$ shows that a = − 2

Thus, $\qquad Z^{-1}\left\{\dfrac{z}{z - e^2}\right\} = e^{--2k} = e^{2k}$

Application: If $F(z) = \dfrac{z}{z^2 + 1}$ determine $Z^{-1}\,F(z)$

In Table 159.1, results 10 and 11 have z^2 in their denominators.

If the numerator is to be z in result 10, then sin a has to equal 1, i.e. a = $\dfrac{\pi}{2}$

From 10 in Table 159.1, $Z\left\{\sin ak\right\} = \dfrac{z \sin a}{z^2 - 2z \cos a + 1}$

hence $Z^{-1}\left\{\dfrac{z \sin a}{z^2 - 2z \cos a + 1}\right\} = \sin ak$

When a = $\dfrac{\pi}{2}$, $\dfrac{z \sin a}{z^2 - 2z \cos a + 1} = \dfrac{z \sin \dfrac{\pi}{2}}{z^2 - 2z \cos \dfrac{\pi}{2} + 1} = \dfrac{z}{z^2 + 1}$

Thus, $Z^{-1}\left\{\dfrac{z}{z^2 + 1}\right\} = \sin \dfrac{\pi}{2} k$

Application: Determine the inverse z-transform of $F(z) = \dfrac{z}{z^2 - 7z + 12}$

Using partial fractions, let $\dfrac{z}{z^2 - 7z + 12} = \dfrac{z}{(z - 4)(z - 3)} = \dfrac{A}{(z - 4)} + \dfrac{B}{(z - 3)}$

$$= \dfrac{A(z - 3) + B(z - 4)}{(z - 4)(z - 3)}$$

from which, $\qquad z = A(z - 3) + B(z - 4)$

Letting z = 4 gives: $\qquad 4 = A$

Letting z = 3 gives: $\qquad 3 = -B$ i.e. B = − 3

Hence, $\qquad F(z) = \dfrac{z}{z^2 - 7z + 12} = \dfrac{4}{(z - 4)} - \dfrac{3}{(z - 3)}$

The nearest transform in Table 159.1 to either of these partial fractions is

$Z\{a^k\} = \dfrac{z}{z - a}$

Rearranging gives: $F(z) = \dfrac{4}{(z - 4)} - \dfrac{3}{(z - 3)} = \dfrac{4}{z} \times \dfrac{z}{(z - 4)} - \dfrac{3}{z} \times \dfrac{z}{(z - 3)}$

$$= 4 \times z^{-1}Z\{4^k\} - 3 \times z^{-1}Z\{3^k\}$$

Hence, $Z^{-1}F(z) = 4 \times \{4^{k-1}\} - 3 \times \{3^{k-1}\}$ by the second shift theorem

$$= \{4^k\} - \{3^k\} = \{4^k - 3^k\}$$

i.e.   **the sequence is $x_k = 4^k - 3^k$**

With the denominator of $F(z) = \dfrac{z}{z^2 - 7z + 12}$ being z, there is an alternative, and more straight-forward method of determining the inverse transform,

i.e. by initially rearranging as: $\dfrac{F(z)}{z} = \dfrac{1}{z^2 - 7z + 12}$

Using partial fractions, $\dfrac{1}{z^2 - 7z + 12} = \dfrac{1}{(z-4)(z-3)} = \dfrac{A}{(z-4)} + \dfrac{B}{(z-3)}$

$$= \dfrac{A(z-3) + B(z-4)}{(z-4)(z-3)}$$

from which, $1 = A(z-3) + B(z-4)$

Letting z = 4 gives: $1 = A$

Letting z = 3 gives: $1 = -B$ i.e. $B = -1$

Hence, $\dfrac{F(z)}{z} = \dfrac{1}{z^2 - 7z + 12} = \dfrac{1}{(z-4)} - \dfrac{1}{(z-3)}$

and $F(z) = \dfrac{z}{(z-4)} - \dfrac{z}{(z-3)}$

and $Z^{-1}F(z) = \{4^k\} - \{3^k\} = \{4^k - 3^k\}$ from 6 in Table 159.1

Application: Determine the inverse z-transform of $F(z) = \dfrac{z}{z^2 - 3z + 2}$

Since $F(z) = \dfrac{z}{z^2 - 3z + 2}$ then $\dfrac{F(z)}{z} = \dfrac{1}{z^2 - 3z + 2}$

Using partial fractions, let $\dfrac{1}{z^2 - 3z + 2} = \dfrac{1}{(z-1)(z-2)} = \dfrac{A}{(z-1)} + \dfrac{B}{(z-2)}$

$$= \dfrac{A(z-2) + B(z-1)}{(z-1)(z-2)}$$

from which, $1 = A(z-2) + B(z-1)$

Letting z = 1 gives: $1 = -A$ i.e. $A = -1$

Letting z = 2 gives: $1 = B$

Hence, $\dfrac{F(z)}{z} = \dfrac{-1}{(z-1)} + \dfrac{1}{(z-2)} = \dfrac{1}{(z-2)} - \dfrac{1}{(z-1)}$

and $F(z) = \dfrac{z}{(z-2)} - \dfrac{z}{(z-1)}$

Thus,
$$Z^{-1}F(z) = Z^{-1}\left\{\frac{z}{(z-2)} - \frac{z}{(z-1)}\right\}$$
$$= Z^{-1}\left\{\frac{z}{(z-2)}\right\} - Z^{-1}\left\{\frac{z}{(z-1)}\right\}$$

From 6 in Table 159.1, $Z^{-1}F(z) = (2)^k - (1)^k = \mathbf{(2)^k - 1}$

Chapter 162   Using z-transforms to solve difference equations

In Chapter 157, Laplace transforms were used to solve differential equations; in this section, the solution of difference equations using z-transforms is demonstrated.

Difference equations arise in several different ways – sometimes from the direct modelling of systems in discrete time, or as an approximation to a differential equation describing the behaviour of a system modelled as a continuous-time system. The z-transform method is based on the first shift theorem, (see earlier, page 452), and the method of solution is explained through the following applications.

Application: Solve the difference equation $x_{k+1} - 2x_k = 0$ given the initial condition that $x_0 = 3$

Taking the z-transform of each term gives:
$$Z\{x_{k+1}\} - 2Z\{x_k\} = Z\{0\}$$

Since from equation (2), Chapter 160, page 452
$$Z\{x_{k+m}\} = z^m F(z) - [z^m x_0 + z^{m-1}x_1 + \ldots + zx_{m-1}]$$

then $\left(z^1 Z\{k\} - \left[z^1(3)\right]\right) - 2Z\{x_k\} = 0$

i.e. $zZ\{x_k\} - 3z - 2Z\{x_k\} = 0$

i.e. $(z-2)Z\{x_k\} = 3z$

and $Z\{x_k\} = \dfrac{3z}{z-2}$

Taking the inverse z-transform gives: $\{x_k\} = Z^{-1}\left\{\dfrac{3z}{z-2}\right\} = 3Z^{-1}\left\{\dfrac{z}{z-2}\right\}$

i.e. $\{x_k\} = 3(2^k)$ from 6 of Table 159.1

Application: Solve the difference equation: $x_{k+2} - 3x_{k+1} + 2x_k = 1$ given that $x_0 = 0$ and $x_1 = 2$

Taking the z-transform of each term gives:

$$Z\{x_{k+2}\} - 3Z\{x_{k+1}\} + 2Z\{x_k\} = Z\{1\}$$

Since from equation (2), Chapter 160, page 452

$$Z\{x_{k+m}\} = z^m F(z) - [z^m x_0 + z^{m-1} x_1 + \ldots + z x_{m-1}]$$

$$(z^2 Z\{x_k\} - [z^2(0) + z^1(2)]) - 3(z^1 Z\{x_k\} - [z^1(0)]) + 2Z\{x_k\} = \frac{z}{z-1}$$

i.e.
$$z^2 Z\{x_k\} - 2z - 3zZ\left\{x_k\right\} + 2Z\left\{x_k\right\} = \frac{z}{z-1}$$

and
$$(z^2 - 3z + 2)Z\{x_k\} = \frac{z}{z-1} + 2z = \frac{z + 2z(z-1)}{z-1} = \frac{2z^2 - z}{z-1} = \frac{z(2z-1)}{z-1}$$

from which,
$$Z\{x_k\} = \frac{z(2z-1)}{(z-1)(z^2 - 3z + 2)} = \frac{z(2z-1)}{(z-1)(z-2)(z-1)}$$

or
$$\frac{Z\{x_k\}}{z} = \frac{(2z-1)}{(z-1)(z-2)(z-1)} = \frac{(2z-1)}{(z-1)^2(z-2)}$$

Using partial fractions, let $\dfrac{(2z-1)}{(z-1)^2(z-2)} = \dfrac{A}{(z-1)} + \dfrac{B}{(z-1)^2} + \dfrac{C}{(z-2)}$

$$= \frac{A(z-1)(z-2) + B(z-2) + C(z-1)^2}{(z-1)^2(z-2)}$$

and
$$2z - 1 = A(z-1)(z-2) + B(z-2) + C(z-1)^2$$

Letting z = 1 gives: $\qquad 1 = -B \quad$ i.e. $\quad B = -1$

Letting z = 2 gives: $\qquad 3 = C$

Equating z^2 coefficients gives: $\quad 0 = A + C \quad$ i.e. $\quad A = -3$

Hence,
$$\frac{Z\{x_k\}}{z} = \frac{(2z-1)}{(z-1)^2(z-2)} = \frac{-3}{(z-1)} + \frac{-1}{(z-1)^2} + \frac{3}{(z-2)}$$

Therefore,
$$Z\{x_k\} = 3\left(\frac{z}{(z-2)}\right) - 3\left(\frac{z}{(z-1)}\right) - \frac{z}{(z-1)^2}$$

Taking the inverse z-transform gives:

$$\{x_k\} = 3Z^{-1}\left(\frac{z}{(z-2)}\right) - 3Z^{-1}\left(\frac{z}{(z-1)}\right) - Z^{-1}\left(\frac{z}{(z-1)^2}\right)$$

$$= 3(2)^k - 3(1)^k - k \qquad \text{from 6 and 3 of Table 159.1}$$

i.e. $\qquad \mathbf{\{x_k\} = 3(2^k) - 3 - k}$

Application: Solve the difference equation: $x_{k+2} - x_k = 1$ given that $x_0 = 0$ and $x_1 = -1$

Taking the z-transform of each term gives:

$$Z\{x_{k+2}\} - Z\{x_k\} = Z\{1\}$$

Since from equation (2), Chapter 160, page 452

$$Z\{x_{k+m}\} = z^m F(z) - [z^m x_0 + z^{m-1} x_1 + \ldots + z x_{m-1}]$$

$$z^2 Z\{x_k\} - [z^2(0) + z^1(-1)] - Z\{x_k\} = \frac{z}{z-1}$$

i.e. $z^2 Z\{x_k\} + z - Z\{x_k\} = \dfrac{z}{z-1}$

and $(z^2 - 1)Z\{x_k\} = \dfrac{z}{z-1} - z = \dfrac{z - z(z-1)}{z-1} = \dfrac{2z - z^2}{z-1}$

from which, $Z\{x_k\} = \dfrac{2z - z^2}{(z-1)\left(z^2 - 1\right)} = \dfrac{2z - z^2}{(z-1)(z-1)(z+1)} = \dfrac{2z - z^2}{(z-1)^2(z+1)}$

and $\dfrac{Z\{x_k\}}{z} = \dfrac{2 - z}{(z-1)^2(z+1)}$

Using partial fractions, let $\dfrac{2-z}{(z-1)^2(z+1)} = \dfrac{A}{(z-1)} + \dfrac{B}{(z-1)^2} + \dfrac{C}{(z+1)}$

$$= \frac{A(z-1)(z+1) + B(z+1) + C(z-1)^2}{(z-1)^2(z+1)}$$

and $2 - z = A(z-1)(z+1) + B(z+1) + C(z-1)^2$

Letting $z = 1$ gives: $1 = 2B$ i.e. $B = 1/2$

Letting $z = -1$ gives: $3 = 4C$ i.e. $C = 3/4$

Equating z^2 coefficients gives: $0 = A + C$ i.e. $A = -3/4$

Hence, $\dfrac{Z\{x_k\}}{z} = \dfrac{2-z}{(z-1)^2(z+1)} = \dfrac{-3/4}{(z-1)} + \dfrac{1/2}{(z-1)^2} + \dfrac{3/4}{(z+1)}$

Therefore, $Z\{x_k\} = -\dfrac{3}{4}\left(\dfrac{z}{z-1}\right) + \dfrac{1}{2}\left(\dfrac{z}{(z-1)^2}\right) - \dfrac{3}{4}\left(\dfrac{z}{z+1}\right)$

Taking the inverse z-transform gives:

$$\{x_k\} = -\frac{3}{4} Z^{-1}\left(\frac{z}{(z-2)}\right) + \frac{1}{2} Z^{-1}\left(\frac{z}{(z-1)^2}\right) - \frac{3}{4} Z^{-1}\left(\frac{z}{z+1}\right)$$

i.e. $\{x_k\} = -\dfrac{3}{4}\left(2^k\right) + \dfrac{1}{2}k - \dfrac{3}{4}(-1)^k$ from 6 and 3 of Table 159.1

Application: Solve the difference equation: $x_{k+2} - 3x_{k+1} + 2x_k = 1$ given that $x_0 = 0$ and $x_1 = 1$

Taking the z-transform of both sides of the equation gives:

$$Z\{x_{k+2} - 3x_{K+1} + 2x_k\} = Z\{1\}$$

i.e. $Z\{x_{k+2}\} - 3Z\{x_{K+1}\} + 2Z\{x_k\} = Z\{1\}$

Using the first shift theorem and $Z\{x_k\} = F(z)$ gives:

$$\left(z^2F(z) - z^2x_0 - zx_1\right) - 3\left(zF(z) - zx_0\right) + 2F(z) = \frac{z}{z-1}$$

$x_0 = 0$ and $x_1 = 1$, hence $\left(z^2F(z) - z^2(0) - z(1)\right) - 3\left(zF(z) - z(0)\right) + 2F(z) = \frac{z}{z-1}$

i.e. $\qquad z^2F(z) - z - 3zF(z) + 2F(z) = \frac{z}{z-1}$

and $\qquad \left(z^2 - 3z + 2\right)F(z) = \frac{z}{z-1} + z$

$$= \frac{z}{z-1} + \frac{z}{1} = \frac{z + z(z-1)}{z-1} = \frac{z + z^2 - z}{z-1} = \frac{z^2}{z-1}$$

Hence, $\qquad F(z) = \frac{z^2}{\left(z^2 - 3z + 2\right)(z-1)} = \frac{z^2}{(z-2)(z-1)(z-1)} = \frac{z^2}{(z-2)(z-1)^2}$

and $\qquad \dfrac{F(z)}{z} = \dfrac{z}{(z-2)(z-1)^2}$

Using partial fractions, let $\dfrac{z}{(z-2)(z-1)^2} = \dfrac{A}{(z-2)} + \dfrac{B}{(z-1)} + \dfrac{C}{(z-1)^2}$

$$= \frac{A(z-1)^2 + B(z-2)(z-1) + C(z-2)}{(z-2)(z-1)^2}$$

from which, $\qquad z = A(z-1)^2 + B(z-2)(z-1) + C(z-2)$

Letting $z = 2$ gives: $\qquad 2 = A(1)^2 \qquad$ i.e. $A = 2$

Letting $z = 1$ gives: $\qquad 1 = C(-1) \qquad$ i.e. $C = -1$

Equating z^2 coefficients gives: $\quad 0 = A + B \qquad$ i.e. $B = -2$

Therefore, $\qquad \dfrac{F(z)}{z} = \dfrac{2}{(z-2)} - \dfrac{2}{(z-1)} - \dfrac{1}{(z-1)^2}$

or $\qquad F(z) = \dfrac{2z}{(z-2)} - \dfrac{2z}{(z-1)} - \dfrac{z}{(z-1)^2}$

Taking the inverse z-transform of $F(z)$ gives:

$$Z^{-1}F(z) = 2Z^{-1}\left[\frac{z}{(z-2)}\right] - 2Z^{-1}\left[\frac{z}{(z-1)}\right] - Z^{-1}\left[\frac{z}{(z-1)^2}\right]$$

$$= 2(2^k) - 2(1) - k \text{ from 2, 6 and 7 of Table 159.1}$$

i.e. $\qquad \{x_k\} = 2^{k+1} - 2 - k$

Section 16

Fourier series

Why are Fourier series important?

A Fourier series changes a periodic function into an infinite expansion of a function in terms of sines and cosines. In engineering and physics, expanding functions in terms of sines and cosines is useful because it makes it possible to more easily manipulate functions that are just too difficult to represent analytically. The fields of electronics, quantum mechanics and electrodynamics all make great use of Fourier series. The Fourier series has become one of the most widely used and useful mathematical tools available to any scientist.

There are many practical uses of Fourier series in science and engineering. The technique has practical applications in the resolution of sound waves into their different frequencies, for example, in an MP3 player, in telecommunications and Wi-Fi, in computer graphics and image processing, in climate variation, in water waves, and much more. Any field of physical science that uses sinusoidal signals, such as engineering, applied mathematics, and chemistry will make use of Fourier series. Applications are found in electrical engineering, such as in determining the harmonic components in ac waveforms, in vibration analysis, acoustics, optics, signal processing, image processing and in quantum mechanics. If it can be found 'on sight' that a function is even or odd, then determining the Fourier series becomes an easier exercise.

In communications, Fourier series are essential to understanding how a signal behaves when it passes through filters, amplifiers and communications channels. In astronomy, radar and digital signal processing, Fourier analysis is used to map the planet. In geology, seismic research uses Fourier analysis, and in optics, Fourier analysis is used in light diffraction.

In music, if a note has frequency f, integer multiples of that frequency, 2f, 3f, 4f, and so on, are known as harmonics. As a result, the mathematical study of overlapping waves is called harmonic analysis; this analysis is a diverse field and may be used to produce a Fourier series. Signal processing, medical imaging, astronomy, optics, and quantum mechanics are some of the fields that use harmonic analysis extensively.

A Fourier series may be represented not only as a sum of sines and cosines, but as a sum of complex exponentials. The complex exponentials provide a more convenient and compact way of expressing the Fourier series than the trigonometric form. It also allows the magnitude and phase spectra to be easily calculated. This form is widely used by engineers, for example, in circuit theory and in control theory.

Chapter 163   Fourier series for periodic functions of period 2π

The basis of a **Fourier series** is that all functions of practical significance which are defined in the interval $-\pi \leq x \leq \pi$ can be expressed in terms of a convergent trigonometric series of the form:

$$f(x) = a_0 + a_1 \cos x + a_2 \cos 2x + a_3 \cos 3x + \cdots + b_1 \sin x + b_2 \sin 2x$$
$$+ b_3 \sin 3x + \cdots$$

when $a_0, a_1, a_2, \ldots b_1, b_2, \ldots$ are real constants, i.e.

$$f(x) = a_0 + \sum_{n=1}^{\infty} (a_n \cos nx + b_n \sin nx) \qquad (1)$$

where for the range $-\pi$ to π:

$$a_0 = \frac{1}{2\pi} \int_{-\pi}^{\pi} f(x)\, dx$$

$$a_n = \frac{1}{\pi} \int_{-\pi}^{\pi} f(x) \cos nx \; dx \quad (n = 1, 2, 3, \ldots)$$

and $$b_n = \frac{1}{\pi} \int_{-\pi}^{\pi} f(x) \sin nx \; dx \quad (n = 1, 2, 3, \ldots)$$

Fourier series provides a method of analysing periodic functions into their constituent components. Alternating currents and voltages, displacement, velocity and acceleration of slider-crank mechanisms and acoustic waves are typical practical examples in engineering and science where periodic functions are involved and often require analysis.

For an exact representation of a complex wave, an infinite number of terms are, in general, required. In many practical cases, however, it is sufficient to take the first few terms only.

Application: Obtain a Fourier series for the periodic function $f(x)$ defined as:

$$f(x) = \begin{cases} -k, & \text{when} -\pi \langle x \langle 0 \\ +k, & \text{when } 0 \langle x \langle \pi \end{cases}$$

(The function is periodic outside of this range with period 2π)

The square wave function defined is shown in Figure 163.1. Since $f(x)$ is given by two different expressions in the two halves of the range the integration is performed in two parts, one from $-\pi$ to 0 and the other from 0 to π.

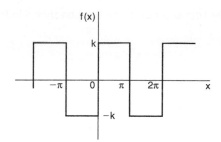

Figure 163.1

From above: $a_0 = \dfrac{1}{2\pi} \displaystyle\int_{-\pi}^{\pi} f(x)\,dx = \dfrac{1}{2\pi}\left[\int_{-\pi}^{0} -k\,dx + \int_{0}^{\pi} k\,dx\right]$

$\qquad\qquad\quad = \dfrac{1}{2\pi}\left\{[-kx]_{-\pi}^{0} + [kx]_{0}^{\pi}\right\}$

$\qquad\qquad\quad = 0$

[a_0 is in fact the **mean value** of the waveform over a complete period of 2π and this could have been deduced on sight from Figure 163.1]

$a_n = \dfrac{1}{\pi} \displaystyle\int_{-\pi}^{\pi} f(x)\cos nx\,\,dx = \dfrac{1}{\pi}\left\{\int_{-\pi}^{0} -k\cos nx\,\,dx + \int_{0}^{\pi} k\cos nx\,\,dx\right\}$

$\qquad = \dfrac{1}{\pi}\left\{\left[\dfrac{-k\sin nx}{n}\right]_{-\pi}^{0} + \left[\dfrac{k\sin nx}{n}\right]_{0}^{\pi}\right\}$

$\qquad = 0$

Hence $a_1, a_2, a_3, \ldots$ are all zero (since $\sin 0 = \sin(-n\pi) = \sin n\pi = 0$), and therefore no cosine terms will appear in the Fourier series.

$b_n = \dfrac{1}{\pi} \displaystyle\int_{-\pi}^{\pi} f(x)\sin nx\,\,dx = \dfrac{1}{\pi}\left\{\int_{-\pi}^{0} -k\sin nx\,\,dx + \int_{0}^{\pi} k\sin nx\,\,dx\right\}$

$\qquad = \dfrac{1}{\pi}\left\{\left[\dfrac{k\cos nx}{n}\right]_{-\pi}^{0} + \left[\dfrac{-k\cos nx}{n}\right]_{0}^{\pi}\right\}$

When n is odd:

$b_n = \dfrac{k}{\pi}\left\{\left[\left(\dfrac{1}{n}\right) - \left(-\dfrac{1}{n}\right)\right] + \left[-\left(-\dfrac{1}{n}\right) - \left(-\dfrac{1}{n}\right)\right]\right\} = \dfrac{k}{\pi}\left\{\dfrac{2}{n} + \dfrac{2}{n}\right\} = \dfrac{4k}{n\pi}$

Hence, $b_1 = \dfrac{4k}{\pi}$, $b_3 = \dfrac{4k}{3\pi}$, $b_5 = \dfrac{4k}{5\pi}$, and so on

When n is even: $b_n = \dfrac{k}{\pi}\left\{\left[\dfrac{1}{n} - \dfrac{1}{n}\right] + \left[-\dfrac{1}{n} - \left(-\dfrac{1}{n}\right)\right]\right\} = 0$

Hence, from equation (1), the Fourier series for the function shown in Figure 163.2 is given by:

$$f(x) = a_0 + \sum_{n=1}^{\infty}(a_n \cos nx + b_n \sin nx) = 0 + \sum_{n=1}^{\infty}(0 + b_n \sin nx)$$

i.e. $f(x) = \dfrac{4k}{\pi}\sin x + \dfrac{4k}{3\pi}\sin 3x + \dfrac{4k}{5\pi}\sin 5x + \cdots$

i.e. $f(x) = \dfrac{4k}{\pi}\left(\sin x + \dfrac{1}{3}\sin 3x + \dfrac{1}{5}\sin 5x + \cdots\right)$

If $k = \pi$ in the above Fourier series then:

$$f(x) = 4\left(\sin x + \dfrac{1}{3}\sin 3x + \dfrac{1}{5}\sin 5x + \cdots\right)$$

$4\sin x$ is termed the first partial sum of the Fourier series of $f(x)$,

$\left(4\sin x + \dfrac{4}{3}\sin 3x\right)$ is termed the second partial sum of the Fourier series, and

$\left(4\sin x + \dfrac{4}{3}\sin 3x + \dfrac{4}{5}\sin 5x\right)$ is termed the third partial sum, and so on.

Let $P_1 = 4\sin x$, $P_2 = \left(4\sin x + \dfrac{4}{3}\sin 3x\right)$ and

$P_3 = \left(4\sin x + \dfrac{4}{3}\sin 3x + \dfrac{4}{5}\sin 5x\right)$.

Graphs of P_1, P_2 and P_3, obtained by drawing up tables of values, and adding waveforms, are shown in Figures 163.2(a) to (c) and they show that the series is convergent, i.e. continually approximating towards a definite limit as more and more partial sums are taken, and in the limit will have the sum $f(x) = \pi$.

Even with just three partial sums, the waveform is starting to approach the **rectangular wave** the Fourier series is representing. Thus, a rectangular wave is comprised of a fundamental and an infinite number of odd harmonics.

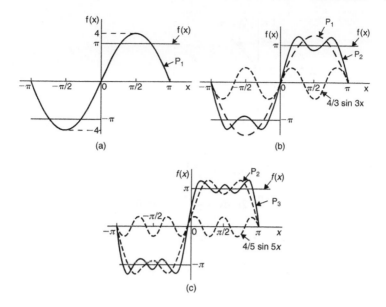

Figure 163.2

Chapter 164 Fourier series for a non-periodic function over range 2π

If a function f(x) is not periodic then it cannot be expanded in a Fourier series for **all** values of x.

However, it is possible to determine a Fourier series to represent the function over any range of width 2π.

For determining a Fourier series of a non-periodic function over a range 2π, exactly the same formulae for the Fourier coefficients are used as in equation (1), page 466.

Application: Determine the Fourier series to represent the function $f(x) = 2x$ in the range $-\pi$ to $+\pi$

The function $f(x) = 2x$ is not periodic. The function is shown in the range $-\pi$ to π in Figure 164.1 and is then constructed outside of that range so that it is periodic of period 2π (see broken lines) with the resulting saw-tooth waveform.

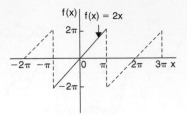

Figure 164.1

For a Fourier series: $f(x) = a_0 + \sum_{n=1}^{\infty} (a_n \cos nx + b_n \sin nx)$

$a_0 = \dfrac{1}{2\pi} \int_{-\pi}^{\pi} f(x)\,dx = \dfrac{1}{2\pi} \int_{-\pi}^{\pi} 2x\,dx = \dfrac{1}{2\pi} \left[x^2\right]_{-\pi}^{\pi} = 0$

$a_n = \dfrac{1}{\pi} \int_{-\pi}^{\pi} f(x) \cos nx\,dx = \dfrac{1}{\pi} \int_{-\pi}^{\pi} 2x \cos nx\,dx$

$\quad = \dfrac{2}{\pi} \left[\dfrac{x \sin nx}{n} - \int \dfrac{\sin nx}{n}\,dx \right]_{-\pi}^{\pi}$ by parts (see Chapter 121)

$\quad = \dfrac{2}{\pi} \left[\dfrac{x \sin nx}{n} + \dfrac{\cos nx}{n^2} \right]_{-\pi}^{\pi} = \dfrac{2}{\pi} \left[\left(0 + \dfrac{\cos n\pi}{n^2}\right) - \left(0 + \dfrac{\cos n(-\pi)}{n^2}\right) \right] = 0$

$b_n = \dfrac{1}{\pi} \int_{-\pi}^{\pi} f(x) \sin nx\,dx = \dfrac{1}{\pi} \int_{-\pi}^{\pi} 2x \sin nx\,dx$

$\quad = \dfrac{2}{\pi} \left[\dfrac{-x \cos nx}{n} - \int \left(\dfrac{-\cos nx}{n} \right)\,dx \right]_{-\pi}^{\pi}$ by parts

$\quad = \dfrac{2}{\pi} \left[\dfrac{-x \cos nx}{n} + \dfrac{\sin nx}{n^2} \right]_{-\pi}^{\pi}$

$\quad = \dfrac{2}{\pi} \left[\left(\dfrac{-\pi \cos n\pi}{n} + \dfrac{\sin n\pi}{n^2} \right) - \left(\dfrac{-(-\pi) \cos n(-\pi)}{n} + \dfrac{\sin n(-\pi)}{n^2} \right) \right]$

$\quad = \dfrac{2}{\pi} \left[\dfrac{-\pi \cos n\pi}{n} - \dfrac{\pi \cos(-n\pi)}{n} \right] = \dfrac{-4}{n} \cos n\pi$ since $\cos n\pi = \cos(-n\pi)$

When n is odd, $b_n = \dfrac{4}{n}$. Thus $b_1 = 4$, $b_3 = \dfrac{4}{3}$, $b_5 = \dfrac{4}{5}$, and so on.

When n is even, $b_n = -\dfrac{4}{n}$. Thus $b_2 = -\dfrac{4}{2}$, $b_4 = -\dfrac{4}{4}$, $b_6 = -\dfrac{4}{6}$, and so on.

Thus, $f(x) = 2x = 4 \sin x - \dfrac{4}{2} \sin 2x + \dfrac{4}{3} \sin 3x - \dfrac{4}{4} \sin 4x + \dfrac{4}{5} \sin 5x$
$\qquad\qquad\qquad\qquad\qquad\qquad - \dfrac{4}{6} \sin 6x + \cdots$

i.e. $\quad 2x = 4(\sin x - \dfrac{1}{2} \sin 2x + \dfrac{1}{3} \sin 3x - \dfrac{1}{4} \sin 4x$

$$+ \dfrac{1}{5} \sin 5x - \dfrac{1}{6} \sin 6x + \cdots)$$

for values of $f(x)$ between $-\pi$ and π.

Chapter 165 Even and odd functions

A function $y = f(x)$ is said to be **even** if $f(-x) = f(x)$ for all values of x. Graphs of even functions are always **symmetrical about the y-axis** (i.e. a mirror image). Two examples of even functions are $y = x^2$ and $y = \cos x$ as shown in Figure 63.2, page 198.

A function $y = f(x)$ is said to be **odd** if $f(-x) = -f(x)$ for all values of x. Graphs of odd functions are always **symmetrical about the origin.** Two examples of odd functions are $y = x^3$ and $y = \sin x$ as shown in Figure 63.3, page 198.

Many functions are neither even nor odd, two such examples being $y = \ln x$ and $y = e^x$.

Fourier cosine series

The Fourier series of an **even** periodic function $f(x)$ having period 2π contains **cosine terms only** (i.e. contains no sine terms) and may contain a constant term.

Hence $\quad \mathbf{f(x) = a_0 + \displaystyle\sum_{n=1}^{\infty} a_n \cos nx}$ $\hfill$ (1)

where $\quad a_0 = \dfrac{1}{2\pi} \displaystyle\int_{-\pi}^{\pi} f(x)\, dx = \dfrac{1}{\pi} \int_{0}^{\pi} \mathbf{f(x)\, dx}$ (due to symmetry)

and $\quad a_n = \dfrac{1}{\pi} \displaystyle\int_{-\pi}^{\pi} f(x) \cos nx\, dx = \dfrac{2}{\pi} \int_{0}^{\pi} \mathbf{f(x) \cos nx\, dx}$

Fourier sine series

The Fourier series of an **odd** periodic function $f(x)$ having period 2π contains **sine terms only** (i.e. contains no constant term and no cosine terms).

Hence $\quad \mathbf{f(x) = \displaystyle\sum_{n=1}^{\infty} b_n \sin nx}$ $\hfill$ (2)

where $\quad b_n = \dfrac{1}{\pi} \displaystyle\int_{-\pi}^{\pi} f(x) \sin nx\, dx = \dfrac{2}{\pi} \int_{0}^{\pi} \mathbf{f(x) \sin nx\, dx}$

Application: Determine the Fourier series for the periodic function defined by:

$$f(x) = \begin{cases} -2, \text{ when } -\pi < x < -\dfrac{\pi}{2} \\[2mm] 2, \text{ when } -\dfrac{\pi}{2} < x < \dfrac{\pi}{2} \\[2mm] -2, \text{ when } \dfrac{\pi}{2} < x < \pi \end{cases} \quad \text{and has a period of } 2\pi$$

The square wave shown in Figure 165.1 is an **even function** since it is symmetrical about the f(x) axis.

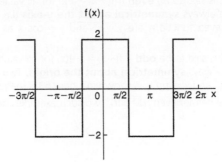

Figure 165.1

Hence from equation (1), the Fourier series is given by:

$$f(x) = a_0 + \sum_{n=1}^{\infty} a_n \cos nx \qquad \text{(i.e. the series contains no sine terms).}$$

$$a_0 = \frac{1}{\pi} \int_0^\pi f(x)\, dx = \frac{1}{\pi}\left\{ \int_0^{\pi/2} 2\, dx + \int_{\pi/2}^\pi -2\, dx \right\}$$

$$= \frac{1}{\pi}\left\{ \left[2x\right]_0^{\pi/2} + \left[-2x\right]_{\pi/2}^\pi \right\}$$

$$= \frac{1}{\pi}\left[(\pi) + \left[(-2\pi) - (-\pi)\right] \right]$$

$$= 0$$

$$a_n = \frac{2}{\pi} \int_0^\pi f(x) \cos nx\, dx = \frac{2}{\pi}\left\{ \int_0^{\pi/2} 2 \cos nx\, dx + \int_{\pi/2}^\pi -2 \cos nx\, dx \right\}$$

$$= \frac{4}{\pi}\left\{ \left[\frac{\sin nx}{n}\right]_0^{\pi/2} + \left[\frac{-\sin nx}{n}\right]_{\pi/2}^\pi \right\} = \frac{4}{\pi}\left\{ \left(\frac{\sin(\pi/2)n}{n} - 0\right) \right.$$

$$\left. + \left(0 - \frac{-\sin(\pi/2)n}{n}\right) \right\}$$

$$= \frac{4}{\pi}\left(\frac{2\sin(\pi/2)n}{n} \right) = \frac{8}{\pi n}\left(\sin\frac{n\pi}{2} \right)$$

When n is even, $a_n = 0$

When n is odd, $a_n = \dfrac{8}{\pi n}$ for n = 1, 5, 9,...

and $\qquad\qquad a_n = \dfrac{-8}{\pi n}$ for n = 3, 7, 11,...

Hence, $a_1 = \dfrac{8}{\pi}, a_3 = \dfrac{-8}{3\pi}, a_5 = \dfrac{8}{5\pi}$, and so on

Hence the Fourier series for the waveform of Figure 165.1 is given by:

$$f(x) = \frac{8}{\pi}\left\{\cos x - \frac{1}{3}\cos 3x + \frac{1}{5}\cos 5x - \frac{1}{7}\cos 7x + \cdots\right\}$$

Application: Obtain the Fourier series for the square wave shown in Figure 165.2.

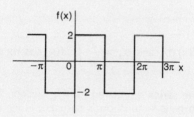

Figure 165.2

The square wave is an **odd function** since it is symmetrical about the origin.

Hence, from equation (2), the Fourier series is given by:

$$f(x) = \sum_{n=1}^{\infty} b_n \sin nx$$

The function is defined by: $f(x) = \begin{cases} -2, \text{ when } -\pi < x < 0 \\ 2, \text{ when } \quad 0 < x < \pi \end{cases}$

$$b_n = \frac{2}{\pi}\int_0^\pi f(x)\sin nx\, dx = \frac{2}{\pi}\int_0^\pi 2\sin nx\, dx = \frac{4}{\pi}\left[\frac{-\cos nx}{n}\right]_0^\pi$$

$$= \frac{4}{\pi}\left[\left(\frac{-\cos n\pi}{n}\right) - \left(-\frac{1}{n}\right)\right] = \frac{4}{\pi n}(1 - \cos n\pi)$$

When n is even, $b_n = 0$. When n is odd, $b_n = \dfrac{4}{\pi n}[1 - (-1)] = \dfrac{8}{\pi n}$

Hence, $b_1 = \dfrac{8}{\pi}, b_3 = \dfrac{8}{3\pi}, b_5 = \dfrac{8}{5\pi}$, and so on

Hence the Fourier series is:

$$f(x) = \frac{8}{\pi} \left(\sin x + \frac{1}{3} \sin 3x + \frac{1}{5} \sin 5x + \frac{1}{7} \sin 7x + \cdots \right)$$

Chapter 166   Half range Fourier series

When a function is defined over the range say 0 to π instead of from 0 to 2π it may be expanded in a series of sine terms only or of cosine terms only. The series produced is called a **half-range Fourier series**.

When a **half range cosine series** is required then:

$$f(x) = a_0 + \sum_{n=1}^{\infty} a_n \cos nx$$

where $a_0 = \dfrac{1}{\pi} \displaystyle\int_0^\pi f(x)\, dx$ and $a_n = \dfrac{2}{\pi} \displaystyle\int_0^\pi f(x) \cos nx\, dx$

If a **half-range cosine series** is required for the function $f(x) = x$ in the range 0 to π then an **even** periodic function is required. In Figure 166.1, $f(x) = x$ is shown plotted from $x = 0$ to $x = \pi$. Since an even function is symmetrical about the $f(x)$ axis the line AB is constructed as shown. If the triangular waveform produced is assumed to be periodic of period 2π outside of this range then the waveform is as shown in Figure 166.1.

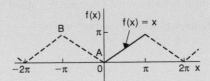

Figure 166.1

When a **half-range sine series** is required then the Fourier coefficient b_n is calculated as earlier, i.e.

$$f(x) = \sum_{n=1}^{\infty} b_n \sin nx$$

where $\qquad b_n = \dfrac{2}{\pi} \displaystyle\int_0^\pi f(x) \sin nx\, dx$

If a **half-range sine series** is required for the function $f(x) = x$ in the range 0 to π then an odd periodic function is required. In Figure 166.2, $f(x) = x$ is shown

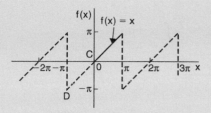

Figure 166.2

plotted from $x = 0$ to $x = \pi$. Since an odd function is symmetrical about the origin the line CD is constructed as shown. If the sawtooth waveform produced is assumed to be periodic of period 2π outside of this range, then the waveform is as shown in Figure 166.2

Application: Determine the half-range Fourier cosine series to represent the function $f(x) = x$ in the range $0 \le x \le \pi$

The function is shown in Figure 166.1.

When $f(x) = x$, $a_0 = \dfrac{1}{\pi}\displaystyle\int_0^\pi f(x)\,dx = \dfrac{1}{\pi}\displaystyle\int_0^\pi x\;dx = \dfrac{1}{\pi}\left[\dfrac{x^2}{2}\right]_0^\pi = \dfrac{\pi}{2}$

$$a_n = \dfrac{2}{\pi}\int_0^\pi f(x)\cos nx\,dx = \dfrac{2}{\pi}\int_0^\pi x \cos nx\,dx$$

$$= \dfrac{2}{\pi}\left[\dfrac{x \sin nx}{n} + \dfrac{\cos nx}{n^2}\right]_0^\pi \qquad \text{by parts}$$

$$= \dfrac{2}{\pi}\left[\left(\dfrac{\pi \sin n\pi}{n} + \dfrac{\cos n\pi}{n^2}\right) - \left(0 + \dfrac{\cos 0}{n^2}\right)\right]$$

$$= \dfrac{2}{\pi}\left(0 + \dfrac{\cos n\pi}{n^2} - \dfrac{\cos 0}{n^2}\right) = \dfrac{2}{\pi n^2}(\cos n\pi - 1)$$

When n is even, $a_n = 0$

When n is odd, $a_n = \dfrac{2}{\pi n^2}(-1-1) = \dfrac{-4}{\pi n^2}$

Hence, $a_1 = \dfrac{-4}{\pi}$, $a_3 = \dfrac{-4}{\pi 3^2}$, $a_5 = \dfrac{-4}{\pi 5^2}$, and so on

Hence, the half-range Fourier cosine series is given by:

$$f(x) = x = \dfrac{\pi}{2} - \dfrac{4}{\pi}\left(\cos x + \dfrac{1}{3^2}\cos 3x + \dfrac{1}{5^2}\cos 5x + \cdots\right)$$

Application: Determine the half-range Fourier sine series to represent the function $f(x) = x$ in the range $0 \le x \le \pi$

The function is shown in Figure 166.2.

When f(x) = x,

$$b_n = \frac{2}{\pi} \int_0^\pi f(x) \sin nx \, dx = \frac{2}{\pi} \int_0^\pi x \sin nx \, dx$$

$$= \frac{2}{\pi} \left[\frac{-x \cos nx}{n} + \frac{\sin nx}{n^2} \right]_0^\pi \qquad \text{by parts}$$

$$= \frac{2}{\pi} \left[\left(\frac{-\pi \cos n\pi}{n} + \frac{\sin n\pi}{n^2} \right) - (0 + 0) \right] = -\frac{2}{n} \cos n\pi$$

When n is odd, $b_n = \frac{2}{n}$. Hence, $b_1 = \frac{2}{1}$, $b_3 = \frac{2}{3}$, $b_5 = \frac{2}{5}$ and so on.

When n is even, $b_n = -\frac{2}{n}$. Hence $b_2 = -\frac{2}{2}$, $b_4 = -\frac{2}{4}$, $b_6 = -\frac{2}{6}$ and so on

Hence the half-range Fourier sine series is given by:

$$f(x) = x = 2 \left(\sin x - \frac{1}{2} \sin 2x + \frac{1}{3} \sin 3x - \frac{1}{4} \sin 4x + \frac{1}{5} \sin 5x - \ldots \right)$$

Chapter 167 Expansion of a periodic function of period L

If f(x) is a function of period L, then its Fourier series is given by:

$$f(x) = a_0 + \sum_{n=1}^{\infty} \left[a_n \cos\left(\frac{2\pi nx}{L} \right) + b_n \sin\left(\frac{2\pi nx}{L} \right) \right] \qquad (1)$$

where, in the range $-\frac{L}{2}$ to $+\frac{L}{2}$:

$$a_0 = \frac{1}{L} \int_{-L/2}^{L/2} f(x) dx, \quad a_n = \frac{2}{L} \int_{-L/2}^{L/2} f(x) \cos\left(\frac{2\pi nx}{L} \right) dx$$

and $$b_n = \frac{2}{L} \int_{-L/2}^{L/2} f(x) \sin\left(\frac{2\pi nx}{L} \right) dx$$

(The limits of integration may be replaced by any interval of length L, such as from 0 to L)

Application: The voltage from a square wave generator is of the form:

$$v(t) = \begin{cases} 0, & -4 < t < 0 \\ 10, & 0 < t < 4 \end{cases} \quad \text{and has a period of 8 ms.}$$

Find the Fourier series for this periodic function

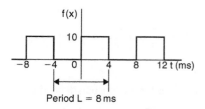

Period L = 8 ms

Figure 167.1

The square wave is shown in Figure 167.1. From above, the Fourier series is of the form:

$$v(t) = a_0 + \sum_{n=1}^{\infty} \left[a_n \cos\left(\frac{2\pi nt}{L}\right) + b_n \sin\left(\frac{2\pi nt}{L}\right) \right]$$

$$a_0 = \frac{1}{L} \int_{-L/2}^{L/2} v(t)dt = \frac{1}{8} \int_{-4}^{4} v(t)dt = \frac{1}{8}\left\{ \int_{-4}^{0} 0\,dt + \int_{0}^{4} 10\,dt \right\}$$

$$= \frac{1}{8}\left[10\,t\right]_{0}^{4} = 5$$

$$a_n = \frac{2}{L} \int_{-L/2}^{L/2} v(t) \cos\left(\frac{2\pi nt}{L}\right)dt = \frac{2}{8} \int_{-4}^{4} v(t) \cos\left(\frac{2\pi nt}{8}\right)dt$$

$$= \frac{1}{4}\left[\int_{-4}^{0} 0 \cos\left(\frac{\pi nt}{4}\right)dt + \int_{0}^{4} 10 \cos\left(\frac{\pi nt}{4}\right)dt \right]$$

$$= \frac{1}{4}\left[\frac{10 \sin\left(\dfrac{\pi nt}{4}\right)}{\left(\dfrac{\pi n}{4}\right)} \right]_{0}^{4} = \frac{10}{\pi n}[\sin \pi n - \sin 0] = 0$$

$$\text{for } n\ 1, 2, 3, \ldots$$

$$b_n = \frac{2}{L} \int_{-L/2}^{L/2} v(t) \sin\left(\frac{2\pi nt}{L}\right)dt$$

$$= \frac{2}{8} \int_{-4}^{4} v(t) \sin\left(\frac{2\pi nt}{8}\right)dt$$

$$= \frac{1}{4}\left\{ \int_{-4}^{0} 0 \sin\left(\frac{\pi nt}{4}\right)dt + \int_{0}^{4} 10 \sin\left(\frac{\pi nt}{4}\right)dt \right\}$$

$$= \frac{1}{4}\left[\frac{-10 \cos\left(\dfrac{\pi nt}{4}\right)}{\left(\dfrac{\pi n}{4}\right)} \right]_{0}^{4} = \frac{-10}{\pi n}[\cos \pi n - \cos 0]$$

When n is even, $b_n = 0$

When n is odd, $b_1 = \dfrac{-10}{\pi}(-1-1) = \dfrac{20}{\pi}$,

$b_3 = \dfrac{-10}{3\pi}(-1-1) = \dfrac{20}{3\pi}$, $b_5 = \dfrac{20}{5\pi}$, and so on

Thus the Fourier series for the function v(t) is given by:

$$v(t) = 5 + \frac{20}{\pi}\left[\sin\left(\frac{\pi t}{4}\right) + \frac{1}{3}\sin\left(\frac{3\pi t}{4}\right) + \frac{1}{5}\sin\left(\frac{5\pi t}{4}\right) + \cdots\right]$$

Application: Obtain the Fourier series for the function defined by:

$$f(x) = \begin{cases} 0, & \text{when } -2 \langle x \langle -1 \\ 5, & \text{when } -1 \langle x \langle 1 \\ 0, & \text{when } 1 \langle x \langle 2 \end{cases}$$

The function is periodic outside of this range of period 4

The function f(x) is shown in Figure 167.2 where period, $L = 4$. Since the function is symmetrical about the f(x) axis it is an even function and the Fourier series contains no sine terms (i.e. $b_n = 0$)

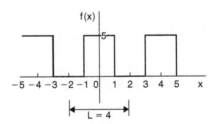

Figure 167.2

Thus, from equation (1), $f(x) = a_0 + \displaystyle\sum_{n=1}^{\infty} a_n \cos\left(\frac{2\pi n x}{L}\right)$

$$a_0 = \frac{1}{L}\int_{-L/2}^{L/2} f(x)\,dx = \frac{1}{4}\int_{-2}^{2} f(x)\,dx$$

$$= \frac{1}{4}\left\{\int_{-2}^{-1} 0\,dx + \int_{-1}^{1} 5\,dx + \int_{1}^{2} 0\,dx\right\}$$

$$= \frac{1}{4}[5x]_{-1}^{1} = \frac{1}{4}[(5) - (-5)] = \frac{10}{4} = \frac{5}{2}$$

$$a_n = \frac{2}{L}\int_{-L/2}^{L/2} f(x)\cos\left(\frac{2\pi n x}{L}\right)dx$$

$$= \frac{2}{4}\int_{-2}^{2} f(x)\cos\left(\frac{2\pi n x}{4}\right)dx$$

$$= \frac{1}{2} \left\{ \int_{-2}^{-1} 0 \cos\left(\frac{\pi n x}{2}\right) dx + \int_{-1}^{1} 5 \cos\left(\frac{\pi n x}{2}\right) dx + \int_{1}^{2} 0 \cos\left(\frac{\pi n x}{2}\right) dx \right\}$$

$$= \frac{5}{2} \left[\frac{\sin\dfrac{\pi n x}{2}}{\dfrac{\pi n}{2}} \right]_{-1}^{1} = \frac{5}{\pi n}\left[\sin\left(\frac{\pi n}{2}\right) - \sin\left(\frac{-\pi n}{2}\right) \right]$$

When n is even, $a_n = 0$

When n is odd, $a_1 = \dfrac{5}{\pi}(1 - -1) = \dfrac{10}{\pi}$, $a_3 = \dfrac{5}{3\pi}(-1 - 1) = \dfrac{-10}{3\pi}$,

$a_5 = \dfrac{5}{5\pi}(1 - -1) = \dfrac{10}{5\pi}$, and so on

Hence the Fourier series for the function $f(x)$ is given by:

$$f(x) = \frac{5}{2} + \frac{10}{\pi}\left[\cos\left(\frac{\pi x}{2}\right) - \frac{1}{3}\cos\left(\frac{3\pi x}{2}\right) + \frac{1}{5}\cos\left(\frac{5\pi x}{2}\right) - \frac{1}{7}\cos\left(\frac{7\pi x}{2}\right) + \dots \right]$$

Chapter 168 Half-range Fourier series for functions defined over range L

A **half-range cosine series** in the range 0 to L can be expanded as:

$$f(x) = a_0 + \sum_{n=1}^{\infty} a_n \cos\left(\frac{n\pi x}{L}\right) \tag{1}$$

where

$$a_0 = \frac{1}{L}\int_0^L f(x)\, dx \quad \text{and} \quad a_n = \frac{2}{L}\int_0^L f(x)\cos\left(\frac{n\pi x}{L}\right) dx$$

A **half-range sine series** in the range 0 to L can be expanded as:

$$f(x) = \sum_{n=1}^{\infty} b_n \sin\left(\frac{n\pi x}{L}\right) \tag{2}$$

where

$$b_n = \frac{2}{L}\int_0^L f(x)\sin\left(\frac{n\pi x}{L}\right) dx$$

Application: Determine the half-range Fourier cosine series for the function $f(x) = x$ in the range $0 \le x \le 2$

A half-range Fourier cosine series indicates an even function. Thus the graph of $f(x) = x$ in the range 0 to 2 is shown in Figure 168.1 and is extended outside of this range so as to be symmetrical about the $f(x)$ axis as shown by the broken lines.

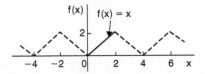

Figure 168.1

For a half-range cosine series: $f(x) = a_0 + \sum\limits_{n=1}^{\infty} a_n \cos\left(\dfrac{n\pi x}{L}\right)$ from equation (1)

$$a_0 = \frac{1}{L}\int_0^L f(x)\,dx = \frac{1}{2}\int_0^2 x\,dx = \frac{1}{2}\left[\frac{x^2}{2}\right]_0^2 = 1$$

$$a_n = \frac{2}{L}\int_0^L f(x)\cos\left(\frac{n\pi x}{L}\right)dx$$

$$= \frac{2}{2}\int_0^2 x\cos\left(\frac{n\pi x}{2}\right)dx = \left[\frac{x\sin\left(\dfrac{n\pi x}{2}\right)}{\left(\dfrac{n\pi}{2}\right)} + \frac{\cos\left(\dfrac{n\pi x}{2}\right)}{\left(\dfrac{n\pi}{2}\right)^2}\right]_0^2$$

$$= \left[\left(\frac{2\sin n\pi}{\left(\dfrac{n\pi}{2}\right)} + \frac{\cos n\pi}{\left(\dfrac{n\pi}{2}\right)^2}\right) - \left(0 + \frac{\cos 0}{\left(\dfrac{n\pi}{2}\right)^2}\right)\right] = \left[\frac{\cos n\pi}{\left(\dfrac{n\pi}{2}\right)^2} - \frac{1}{\left(\dfrac{n\pi}{2}\right)^2}\right]$$

$$= \left(\frac{2}{\pi n}\right)^2 (\cos n\pi - 1)$$

When n is even, $a_n = 0$, $a_1 = \dfrac{-8}{\pi^2}$, $a_3 = \dfrac{-8}{\pi^2 3^2}$, $a_5 = \dfrac{-8}{\pi^2 5^2}$, and so on.

Hence the half-range Fourier cosine series for $f(x)$ in the range 0 to 2 is given by:

$$f(x) = 1 - \frac{8}{\pi^2}\left[\cos\left(\frac{\pi x}{2}\right) + \frac{1}{3^2}\cos\left(\frac{3\pi x}{2}\right) + \frac{1}{5^2}\cos\left(\frac{5\pi x}{2}\right) + \cdots\right]$$

Application: Determine the half-range Fourier sine series for the function $f(x) = x$ in the range $0 \leq x \leq 2$

A half-range Fourier sine series indicates an odd function. Thus the graph of $f(x) = x$ in the range 0 to 2 is shown in Figure 168.2 and is extended outside of this range so as to be symmetrical about the origin, as shown by the broken lines.

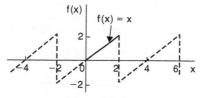

Figure 168.2

For a half-range sine series: $f(x) = \sum_{n=1}^{\infty} b_n \sin\left(\frac{n\pi x}{L}\right)$ from equation (2)

$$b_n = \frac{2}{L} \int_0^L f(x) \sin\left(\frac{n\pi x}{L}\right) dx = \frac{2}{2} \int_0^2 x \sin\left(\frac{n\pi x}{L}\right) dx$$

$$= \left[\frac{-x \cos\left(\frac{n\pi x}{2}\right)}{\left(\frac{n\pi}{2}\right)} + \frac{\sin\left(\frac{n\pi x}{2}\right)}{\left(\frac{n\pi}{2}\right)^2} \right]_0^2$$

$$= \left[\left(\frac{-2\cos n\pi}{\left(\frac{n\pi}{2}\right)} + \frac{\sin n\pi}{\left(\frac{n\pi}{2}\right)^2} \right) - \left(0 + \frac{\sin 0}{\left(\frac{n\pi}{2}\right)^2} \right) \right] = \frac{-2\cos n\pi}{\frac{n\pi}{2}} = \frac{-4}{n\pi} \cos n\pi$$

Hence, $b_1 = \frac{-4}{\pi}(-1) = \frac{4}{\pi}$, $b_2 = \frac{-4}{2\pi}(1) = \frac{-4}{2\pi}$, $b_3 = \frac{-4}{3\pi}(-1) = \frac{4}{3\pi}$, and so on

Thus the half-range Fourier sine series in the range 0 to 2 is given by:

$$f(x) = \frac{4}{\pi}\left[\sin\left(\frac{\pi x}{2}\right) - \frac{1}{2}\sin\left(\frac{2\pi x}{2}\right) + \frac{1}{3}\sin\left(\frac{3\pi x}{2}\right) - \frac{1}{4}\sin\left(\frac{4\pi x}{2}\right) + \cdots \right]$$

Chapter 169   The complex or exponential form of a Fourier series

The form used for the Fourier series considered previously consisted of cosine and sine terms. However, there is another form that is commonly used – one that directly gives the amplitude terms in the frequency spectrum and relates to phasor notation. This form involves the use of complex numbers (see Chapter 69). It is called the **exponential** or **complex form** of a Fourier series.

$$e^{j\theta} = \cos\theta + j\sin\theta \quad \text{and} \quad e^{-j\theta} = \cos\theta - j\sin\theta$$

$$e^{j\theta} + e^{-j\theta} = 2\cos\theta \quad \text{from which,} \quad \cos\theta = \frac{e^{j\theta} + e^{-j\theta}}{2} \tag{1}$$

$$e^{j\theta} - e^{-j\theta} = 2j\sin\theta \quad \text{from which,} \quad \sin\theta = \frac{e^{j\theta} - e^{-j\theta}}{2j} \tag{2}$$

The **complex** or **exponential form** of the Fourier series.

$$f(x) = \sum_{n=-\infty}^{\infty} c_n e^{j\frac{2\pi nx}{L}} \tag{3}$$

where

$$c_n = \frac{1}{L} \int_{-\frac{L}{2}}^{\frac{L}{2}} f(x)\, e^{-j\frac{2\pi nx}{L}}\, dx \tag{4}$$

Care needs to be taken when determining c_0. If n appears in the denominator of an expression the expansion can be invalid when $n = 0$. In such circumstances it is usually simpler to evaluate c_0 by using the relationship:

$$c_0 = a_0 = \frac{1}{L} \int_{-\frac{L}{2}}^{\frac{L}{2}} f(x)\, dx \tag{5}$$

Application: Determine the complex Fourier series for the function defined by:

$$f(x) = \begin{cases} 0, & \text{when } -2 \le x \le -1 \\ 5, & \text{when } -1 \le x \le 1 \\ 0, & \text{when } 1 \le x \le 2 \end{cases}$$

The function is periodic outside this range of period 4.

This is the same Application Problem as on page 478 and we can use this to demonstrate that the two forms of Fourier series are equivalent.

The function $f(x)$ was shown in Figure 167.2, where the period, $L = 4$.

From equation (3), the complex Fourier series is given by:

$$f(x) = \sum_{n=-\infty}^{\infty} c_n e^{j\frac{2\pi nx}{L}}$$

where c_n is given by: $c_n = \dfrac{1}{L}\displaystyle\int_{-\frac{L}{2}}^{\frac{L}{2}} f(x)\, e^{-j\frac{2\pi nx}{L}}\, dx$ (from equation (4))

With reference to Figure 167.2, when $L = 4$,

$$c_n = \frac{1}{4}\left\{\int_{-2}^{-1} 0\, dx + \int_{-1}^{1} 5e^{-j\frac{2\pi nx}{4}}\, dx + \int_{1}^{2} 0\, dx\right\}$$

$$= \frac{1}{4}\int_{-1}^{1} 5e^{-j\frac{\pi nx}{2}}\, dx = \frac{5}{4}\left[\frac{e^{-j\frac{\pi nx}{2}}}{-\dfrac{j\pi n}{2}}\right]_{-1}^{1} = \frac{-5}{j2\pi n}\left[e^{-j\frac{\pi nx}{2}}\right]_{-1}^{1}$$

$$= \frac{-5}{j2\pi n}\left(e^{-j\frac{\pi n}{2}} - e^{j\frac{\pi n}{2}}\right) = \frac{5}{\pi n}\left(\frac{e^{j\frac{\pi n}{2}} - e^{-j\frac{\pi n}{2}}}{2j}\right)$$

$$= \frac{5}{\pi n}\sin\frac{\pi n}{2} \qquad \text{(from equation (2))}$$

Hence, from equation (3), **the complex form of the Fourier series** is given by:

$$f(x) = \sum_{n=-\infty}^{\infty} c_n\, e^{j\frac{2\pi nx}{L}} = \sum_{n=-\infty}^{\infty} \frac{5}{\pi n}\sin\frac{\pi n}{2}\, e^{j\frac{\pi nx}{2}} \tag{6}$$

Let us show how this result is equivalent to the result involving sine and cosine terms determined on page 479.

From equation (5),

$$c_0 = a_0 = \frac{1}{L}\int_{-\frac{L}{2}}^{\frac{L}{2}} f(x)\, dx = \frac{1}{4}\int_{-1}^{1} 5\, dx = \frac{5}{4}[x]_{-1}^{1} = \frac{5}{4}[1 - -1] = \frac{5}{2}$$

Since $c_n = \dfrac{5}{\pi n}\sin\dfrac{\pi n}{2}$, then $c_1 = \dfrac{5}{\pi}\sin\dfrac{\pi}{2} = \dfrac{5}{\pi}$

$$c_2 = \frac{5}{2\pi}\sin\pi = 0 \qquad \text{(in fact, all even terms will be zero since } \sin n\pi = 0\text{)}$$

$$c_3 = \frac{5}{\pi n}\sin\frac{\pi n}{2} = \frac{5}{3\pi}\sin\frac{3\pi}{2} = -\frac{5}{3\pi}$$

By similar substitution, $c_5 = \dfrac{5}{5\pi}$, $c_7 = -\dfrac{5}{7\pi}$, and so on.

Similarly, $c_{-1} = \dfrac{5}{-\pi}\sin\dfrac{-\pi}{2} = \dfrac{5}{\pi}$

$$c_{-2} = -\frac{5}{2\pi}\sin\frac{-2\pi}{2} = 0 = c_{-4} = c_{-6}, \text{ and so on}$$

$$c_{-3} = -\frac{5}{3\pi}\sin\frac{-3\pi}{2} = -\frac{5}{3\pi}$$

$$c_{-5} = -\frac{5}{5\pi}\sin\frac{-5\pi}{2} = \frac{5}{5\pi}, \text{ and so on.}$$

Hence, the extended complex form of the Fourier series shown in equation (6) becomes:

$$f(x) = \frac{5}{2} + \frac{5}{\pi}e^{j\frac{\pi x}{2}} - \frac{5}{3\pi}e^{j\frac{3\pi x}{2}} + \frac{5}{5\pi}e^{j\frac{5\pi x}{2}} - \frac{5}{7\pi}e^{j\frac{7\pi x}{2}} + \cdots$$

$$+ \frac{5}{\pi}e^{-j\frac{\pi x}{2}} - \frac{5}{3\pi}e^{-j\frac{3\pi x}{2}} + \frac{5}{5\pi}e^{-j\frac{5\pi x}{2}} - \frac{5}{7\pi}e^{-j\frac{7\pi x}{2}} + \cdots$$

$$= \frac{5}{2} + \frac{5}{\pi}\left(e^{j\frac{\pi x}{2}} + e^{-j\frac{\pi x}{2}}\right) - \frac{5}{3\pi}\left(e^{j\frac{3\pi x}{2}} + e^{-j\frac{3\pi x}{2}}\right) + \frac{5}{5\pi}\left(e^{j\frac{5\pi x}{2}} + e^{-j\frac{5\pi x}{2}}\right) - \cdots$$

$$= \frac{5}{2} + \frac{5}{\pi}(2)\left(\frac{e^{j\frac{\pi x}{2}} + e^{-j\frac{\pi x}{2}}}{2}\right) - \frac{5}{3\pi}(2)\left(\frac{e^{j\frac{3\pi x}{2}} + e^{-j\frac{3\pi x}{2}}}{2}\right)$$

$$+ \frac{5}{5\pi}(2)\left(\frac{e^{j\frac{5\pi x}{2}} + e^{-j\frac{5\pi x}{2}}}{2}\right) - \cdots$$

$$= \frac{5}{2} + \frac{10}{\pi}\cos\left(\frac{\pi x}{2}\right) - \frac{10}{3\pi}\cos\left(\frac{3\pi x}{2}\right) + \frac{10}{5\pi}\cos\left(\frac{5\pi x}{2}\right) - \cdots$$

(from equation 1)

i.e. $$f(x) = \frac{5}{2} + \frac{10}{\pi}\left[\cos\left(\frac{\pi x}{2}\right) - \frac{1}{3}\cos\left(\frac{3\pi x}{2}\right) + \frac{1}{5}\cos\left(\frac{5\pi x}{2}\right) - \cdots\right]$$

which is the same as obtained on page 479.

Hence, $\displaystyle\sum_{n=-\infty}^{\infty}\frac{5}{\pi n}\sin\frac{n\pi}{2}e^{j\frac{\pi n x}{2}}$ is equivalent to:

$$\frac{5}{2} + \frac{10}{\pi}\left[\cos\left(\frac{\pi x}{2}\right) - \frac{1}{3}\cos\left(\frac{3\pi x}{2}\right) + \frac{1}{5}\cos\left(\frac{5\pi x}{2}\right) - \cdots\right]$$

Symmetry relationships

If even or odd symmetry is noted in a function, then time can be saved in determining coefficients.

The Fourier coefficients present in the complex Fourier series form are affected by symmetry.

For **even symmetry**:

$$c_n = \frac{a_n}{2} = \frac{2}{L}\int_0^{\frac{L}{2}} f(x)\cos\left(\frac{2\pi nx}{L}\right)dx \qquad (7)$$

For **odd symmetry**:

$$c_n = \frac{-jb_n}{2} = -j\frac{2}{L}\int_0^{\frac{L}{2}} f(x)\sin\left(\frac{2\pi nx}{L}\right)dx \qquad (8)$$

For example, in the Application Problem on page 482, the function f(x) is even, since the waveform is symmetrical about the f(x) axis. Thus equation (7) could have been used, giving:

$$c_n = \frac{2}{L}\int_0^{\frac{L}{2}} f(x)\cos\left(\frac{2\pi nx}{L}\right)dx$$

$$= \frac{2}{4}\int_0^2 f(x)\cos\left(\frac{2\pi nx}{4}\right)dx = \frac{1}{2}\left\{\int_0^1 5\cos\left(\frac{\pi nx}{2}\right)dx + \int_1^2 0\ dx\right\}$$

$$= \frac{5}{2}\left[\frac{\sin\left(\frac{\pi nx}{2}\right)}{\frac{\pi n}{2}}\right]_0^1 = \frac{5}{2}\left(\frac{2}{\pi n}\right)\left(\sin\frac{n\pi}{2} - 0\right) = \frac{5}{\pi n}\sin\frac{n\pi}{2}$$

which is the same answer as on page 483; however, a knowledge of even functions has produced the coefficient more quickly.

Application: Obtain the Fourier series, in complex form, for the square wave shown in Figure 169.1

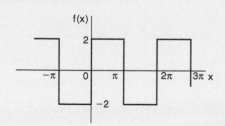

Figure 169.1

The square wave shown in Figure 169.1 is an **odd function** since it is symmetrical about the origin.

The period of the waveform, L = 2π.

Thus, using equation (8): $c_n = -j\frac{2}{L}\int_0^{\frac{L}{2}} f(x)\sin\left(\frac{2\pi nx}{L}\right)dx$

$$= -j\frac{2}{2\pi}\int_0^\pi 2\sin\left(\frac{2\pi nx}{2\pi}\right)dx$$

$$= -j\frac{2}{\pi}\int_0^\pi \sin nx\ dx$$

$$= -j\frac{2}{\pi}\left[\frac{-\cos nx}{n}\right]_0^\pi$$

$$= -j\frac{2}{\pi n}\left((-\cos \pi n)-(-\cos 0)\right)$$

i.e.
$$c_n = -j\frac{2}{\pi n}\left[1-\cos \pi n\right]\qquad(9)$$

From equation (3), the complex Fourier series is given by:

$$f(x) = \sum_{n=-\infty}^\infty c_n e^{j\frac{2\pi nx}{L}} = \sum_{n=-\infty}^\infty -j\frac{2}{n\pi}(1-\cos n\pi)e^{jnx}\qquad(10)$$

This is the same as that obtained on page 474, i.e.

$$f(x) = \frac{8}{\pi}\left(\sin x + \frac{1}{3}\sin 3x + \frac{1}{5}\sin 5x + \frac{1}{7}\sin 7x + \cdots\right)$$

which is demonstrated below.

From equation (9), $c_n = -j\frac{2}{n\pi}\left(1-\cos n\pi\right)$

When $n = 1$, $c_1 = -j\frac{2}{(1)\pi}\left(1-\cos \pi\right) = -j\frac{2}{\pi}\left(1--1\right) = -\frac{j4}{\pi}$

When $n = 2$, $c_2 = -j\frac{2}{2\pi}\left(1-\cos 2\pi\right) = 0$; in fact, all even values of c_n will be zero.

When $n = 3$, $c_3 = -j\frac{2}{3\pi}\left(1-\cos 3\pi\right) = -j\frac{2}{3\pi}\left(1--1\right) = -\frac{j4}{3\pi}$

By similar reasoning, $c_5 = -\frac{j4}{5\pi}, c_7 = -\frac{j4}{7\pi}$, and so on.

When $n = -1$, $c_{-1} = -j\frac{2}{(-1)\pi}\left(1-\cos(-\pi)\right) = +j\frac{2}{\pi}\left(1--1\right) = +\frac{j4}{\pi}$

When $n = -3$, $c_{-3} = -j\frac{2}{(-3)\pi}\left(1-\cos(-3\pi)\right) = +j\frac{2}{3\pi}\left(1--1\right) = +\frac{j4}{3\pi}$

By similar reasoning, $c_{-5} = +\frac{j4}{5\pi}, c_{-7} = +\frac{j4}{7\pi}$, and so on.

Since the waveform is odd, $c_0 = a_0 = 0$

From equation (10), $f(x) = \sum_{n=-\infty}^\infty -j\frac{2}{n\pi}\left(1-\cos n\pi\right)e^{jnx}$

Hence,

$$f(x) = -\frac{j4}{\pi} e^{jx} - \frac{j4}{3\pi} e^{j3x} - \frac{j4}{5\pi} e^{j5x} - \frac{j4}{7\pi} e^{j7x} - \dots$$
$$+ \frac{j4}{\pi} e^{-jx} + \frac{j4}{3\pi} e^{-j3x} + \frac{j4}{5\pi} e^{-j5x} + \frac{j4}{7\pi} e^{-j7x} + \dots$$

$$= \left(-\frac{j4}{\pi} e^{jx} + \frac{j4}{\pi} e^{-jx}\right) + \left(-\frac{j4}{3\pi} e^{3x} + \frac{j4}{3\pi} e^{-3x}\right)$$
$$+ \left(-\frac{j4}{5\pi} e^{5x} + \frac{j4}{5\pi} e^{-5x}\right) + \dots$$

$$= -\frac{j4}{\pi}\left(e^{jx} - e^{-jx}\right) - \frac{j4}{3\pi}\left(e^{3x} - e^{-3x}\right) - \frac{j4}{5\pi}\left(e^{5x} - e^{-5x}\right) + \dots$$

$$= \frac{4}{j\pi}(e^{jx} - e^{-jx}) + \frac{4}{j3\pi}(e^{3x} - e^{-3x}) + \frac{4}{j5\pi}(e^{5x} - e^{-5x}) + \dots$$

by multiplying top and bottom by j

$$= \frac{8}{\pi}\left(\frac{e^{jx} - e^{-jx}}{2j}\right) + \frac{8}{3\pi}\left(\frac{e^{j3x} - e^{-j3}}{2j}\right) + \frac{8}{5\pi}\left(\frac{e^{j5x} - e^{-j5x}}{2j}\right) + \dots \quad \text{by rearranging}$$

$$= \frac{8}{\pi}\sin x + \frac{8}{3\pi}\sin 3x + \frac{8}{3x}\sin 5x + \dots \qquad \text{from equation 2, page 482}$$

i.e. $\qquad f(x) = \frac{8}{\pi}\left(\sin x + \frac{1}{3}\sin 3x + \frac{1}{5}\sin 5x + \frac{1}{7}\sin 7x + \cdots\right)$

Hence, $\; f(x) = \sum_{n=-\infty}^{\infty} -j\frac{2}{n\pi}\left(1 - \cos n\pi\right) e^{jnx}$
$$\equiv \frac{8}{\pi}\left(\sin x + \frac{1}{3}\sin 3x + \frac{1}{5}\sin 5x + \frac{1}{7}\sin 7x + \cdots\right)$$

Chapter 170 A numerical method of harmonic analysis

Many practical waveforms can be represented by simple mathematical expressions, and, by using Fourier series, the magnitude of their harmonic components determined, as above. For waveforms not in this category, analysis may be achieved by numerical methods.

Harmonic analysis is the process of resolving a periodic, non-sinusoidal quantity into a series of sinusoidal components of ascending order of frequency.

The **trapezoidal rule** can be used to evaluate the Fourier coefficients, which are given by:

$$a_0 \approx \frac{1}{p}\sum_{k=1}^{p} y_k \qquad (1)$$

$$a_n \approx \frac{2}{P} \sum_{k=1}^{p} y_k \cos nx_k \qquad (2)$$

$$b_n \approx \frac{2}{P} \sum_{k=1}^{p} y_k \sin nx_k \qquad (3)$$

Application: A graph of voltage V against angle θ is shown in Figure 170.1. Determine a Fourier series to represent the graph.

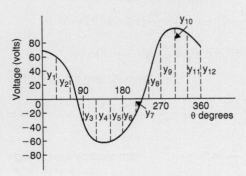

Figure 170.1

The values of the ordinates y_1, y_2, y_3, ... are 62, 35, −38, −64, −63, −52, −28, 24, 80, 96, 90 and 70, the 12 equal intervals each being of width 30°. (If a larger number of intervals are used, results having a greater accuracy are achieved).

The voltage may be analysed into its first three constituent components as follows:

The data is tabulated in the proforma shown in Table 170.1.

From equation (1), $a_0 \approx \dfrac{1}{p} \sum_{k=1}^{p} y_k = \dfrac{1}{12}(212) = 17.67$ (since $p = 12$)

From equation (2), $a_n \approx \dfrac{2}{p} \sum_{k=1}^{p} y_k \cos nx_k$

hence $a_1 \approx \dfrac{2}{12}(417.94) = 69.66$

$$a_2 \approx \frac{2}{12}(-39) = -6.50 \quad \text{and} \quad a_3 \approx \frac{2}{12}(-49) = -8.17$$

From equation (3), $b_n \approx \dfrac{2}{p} \sum_{k=1}^{p} y_k \sin nx_k$

hence $b_1 \approx \dfrac{2}{12}(-278.53) = -46.42$

$$b_2 \approx \frac{2}{12}(29.43) = 4.91 \quad \text{and} \quad b_3 \approx \frac{2}{12}(55) = 9.17$$

Table 170.1

Ordinates	θ	V	cos θ	V cos θ	sin θ	V sin θ	cos 2θ	V cos 2θ	sin 2θ	V sin 2θ	cos 3θ	V cos 3θ	sin 3θ	V sin 3θ
y_1	30	62	0.866	53.69	0.5	31	0.5	31	0.866	53.69	0	0	1	62
y_2	60	35	0.5	17.5	0.866	30.31	−0.5	−17.5	0.866	30.31	−1	−35	0	0
y_3	90	−38	0	0	1	−38	−1	38	0	0	0	0	−1	38
y_4	120	−64	−0.5	32	0.866	−55.42	−0.5	32	−0.866	55.42	1	−64	0	0
y_5	150	−63	−0.866	54.56	0.5	−31.5	0.5	−31.5	−0.866	54.56	0	0	1	−63
y_6	180	−52	−1	52	0	0	1	−52	0	0	−1	52	0	0
y_7	210	−28	−0.866	24.25	−0.5	14	0.5	−14	0.866	−24.25	0	0	−1	28
y_8	240	24	−0.5	−12	−0.866	−20.78	−0.5	−12	0.866	20.78	1	24	0	0
y_9	270	80	0	0	−1	−80	−1	−80	0	0	0	0	1	80
y_{10}	300	96	0.5	48	−0.866	−83.14	−0.5	−48	−0.866	−83.14	−1	−96	0	0
y_{11}	330	90	0.866	77.94	−0.5	−45	0.5	45	−0.866	−77.94	0	0	−1	−90
y_{12}	360	70	1	70	0	0	1	70	0	0	1	70	0	0
	$\sum_{k=1}^{12} y_k = 212$		$\sum_{k=1}^{12} y_k \cos\theta_k$ $= 417.94$		$\sum_{k=1}^{12} y_k \sin\theta_k$ $= -278.53$		$\sum_{k=1}^{12} y_k \cos 2\theta_k$ $= -39$		$\sum_{k=1}^{12} y_k \sin 2\theta_k$ $= 29.43$		$\sum_{k=1}^{12} y_k \cos 3\theta_k$ $= -49$		$\sum_{k=1}^{12} y_k \sin 3\theta_k$ $= 55$	

Substituting these values into the Fourier series:

$$f(x) = a_0 + \sum_{n=1}^{\infty} (a_n \cos nx + b_n \sin nx)$$

gives: **v = 17.67 + 69.66 cos θ − 6.50 cos 2θ − 8.17 cos 3θ + ...**

$$-46.42 \sin \theta + 4.91 \sin 2\theta + 9.17 \sin 3\theta + \cdots \tag{4}$$

Note that in equation (4), $(-46.42 \sin\theta + 69.66 \cos\theta)$ comprises the fundamental, $(4.91 \sin 2\theta - 6.50 \cos 2\theta)$ comprises the second harmonic and $(9.17 \sin 3\theta - 8.17 \cos 3\theta)$ comprises the third harmonic.

It is shown in Chapter 54 that: $a \sin \omega t + b \cos \omega t = R \sin(\omega t + \alpha)$

where $a = R \cos \alpha$, $b = R \sin \alpha$, $R = \sqrt{a^2 + b^2}$ and $\alpha = \tan^{-1} \dfrac{b}{a}$

For the fundamental, $R = \sqrt{(-46.42)^2 + (69.66)^2} = 83.71$

If $a = R \cos \alpha$, then $\cos \alpha = \dfrac{a}{R} = \dfrac{-46.42}{83.71}$ which is negative,

and if $b = R \sin \alpha$, then $\sin \alpha = \dfrac{b}{R} = \dfrac{69.66}{83.71}$ which is positive.

The only quadrant where $\cos \alpha$ is negative and $\sin \alpha$ is positive is the second quadrant.

Hence, $\alpha = \tan^{-1} \dfrac{b}{a} = \tan^{-1} \dfrac{69.66}{-46.42} = 2.16$ rad

Thus, $(-46.42 \sin \theta + 69.66 \cos \theta) = 83.71 \sin(\theta + 2.16)$

By a similar method it may be shown that the second harmonic

$(4.91 \sin 2\theta - 6.50 \cos 2\theta) = 8.15 \sin(2\theta - 0.92)$ and the third harmonic

$(9.17 \sin 3\theta - 8.17 \cos 3\theta) = 12.28 \sin(3\theta - 0.73)$

Hence equation (4) may be re-written as:

$v = 17.67 + 83.71 \sin(\theta + 2.16) + 8.15 \sin(2\theta - 0.92) + 12.28 \sin(3\theta - 0.73)$ volts

which is the form normally used with complex waveforms.

Chapter 171   Complex waveform considerations

It is sometimes possible to predict the harmonic content of a waveform on inspection of particular waveform characteristics.

1. If a periodic waveform is such that the area above the horizontal axis is equal to the area below then the mean value is zero. Hence $a_0 = 0$ (see Figure 171.1(a)).

2. An **even function** is symmetrical about the vertical axis and contains **no sine terms** (see Figure 171.1(b)).

3. An **odd function** is symmetrical about the origin and contains **no cosine terms** (see Figure 171.1(c)).

4. $f(x) = f(x + \pi)$ represents a waveform which repeats after half a cycle and **only even harmonics** are present (see Figure 171.1(d)).

5. $f(x) = -f(x + \pi)$ represents a waveform for which the positive and negative cycles are identical in shape and **only odd harmonics** are present (see Figure 171.1(e)).

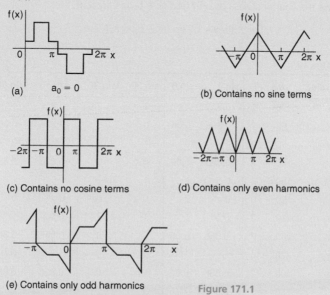

Figure 171.1

Application: An alternating current i amperes is shown in Figure 171.2. Analyse the waveform into its constituent harmonics as far as and including the fifth harmonic, taking 30° intervals.

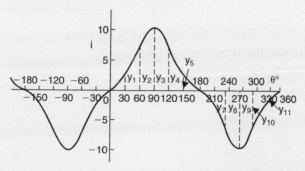

Figure 171.2

With reference to Figure 171.2, the following characteristics are noted:

(i) The mean value is zero since the area above the θ axis is equal to the area below it. Thus the constant term, or d.c. component, $a_0 = 0$

(ii) Since the waveform is symmetrical about the origin the function i is odd, which means that there are no cosine terms present in the Fourier series.

(iii) The waveform is of the form $f(\theta) = -f(\theta + \pi)$ which means that only odd harmonics are present.

Investigating waveform characteristics has thus saved unnecessary calculations and in this case the Fourier series has only odd sine terms present, i.e.

$$i = b_1 \sin \theta + b_3 \sin 3\theta + b_5 \sin 5\theta + \cdots$$

Table 171.1

Ordinate	θ	i	$\sin \theta$	$i \sin \theta$	$\sin 3\theta$	$i \sin 3\theta$	$\sin 5\theta$	$i \sin 5\theta$
Y_1	30	2	0.5	1	1	2	0.5	1
Y_2	60	7	0.866	6.06	0	0	−0.866	−6.06
Y_3	90	10	1	10	−1	−10	1	10
Y_4	120	7	0.866	6.06	0	0	−0.866	−6.06
Y_5	150	2	0.5	1	1	2	0.5	1
Y_6	180	0	0	0	0	0	0	0
Y_7	210	−2	−0.5	1	−1	2	−0.5	1
Y_8	240	−7	−0.866	6.06	0	0	0.866	−6.06
Y_9	270	−10	−1	10	1	−10	−1	10
Y_{10}	300	−7	−0.866	6.06	0	0	0.866	−6.06
Y_{11}	330	−2	−0.5	1	−1	2	−0.5	1
Y_{12}	360	0	0	0	0	0	0	0
			$\sum\limits_{k=1}^{12} y_k \sin \theta_k$		$\sum\limits_{k=1}^{12} y_k \sin 3\theta_k$		$\sum\limits_{k=1}^{12} y_k \sin 5\theta_k$	
			= 48.24		= −12		= −0.24	

A proforma, similar to Table 170.1, page 489, but without the 'cosine terms' columns and without the 'even sine terms' columns is shown in Table 171.1 up to, and including, the fifth harmonic, from which the Fourier coefficients b_1, b_3 and b_5 can be determined. Twelve co-ordinates are chosen and labelled y_1, y_2, y_3, .. y_{12} as shown in Figure 171.2.

From equation (3), $b_n = \dfrac{2}{p} \sum\limits_{k=1}^{p} i_k \sin n\theta_k$ where $p = 12$

Hence, $b_1 \approx \dfrac{2}{12}(48.24) = 8.04$, $b_3 \approx \dfrac{2}{12}(-12) = -2.00$ and

$b_5 \approx \dfrac{2}{12}(-0.24) = -0.04$

Thus the Fourier series for current i is given by:

$$i = 8.04 \sin \theta - 2.00 \sin 3\theta - 0.04 \sin 5\theta$$

Statistics and probability

Why are statistics and probability important?

Statistics is the study of the collection, organisation, analysis, and interpretation of data. It deals with all aspects of this, including the planning of data collection in terms of the design of surveys and experiments. Statistics is applicable to a wide variety of academic disciplines, including natural and social sciences, engineering, government, and business. Statistical methods can be used for summarising or describing a collection of data. Engineering statistics combines engineering and statistics. Design of experiments is a methodology for formulating scientific and engineering problems using statistical models. Quality control and process control use statistics as a tool to manage conformance to specifications of manufacturing processes and their products. Time and methods engineering use statistics to study repetitive operations in manufacturing in order to set standards and find optimum manufacturing procedures. Reliability engineering measures the ability of a system to perform for its intended function (and time) and has tools for improving performance. Probabilistic design involves the use of probability in product and system design. System identification uses statistical methods to build mathematical models of dynamical systems from measured data. System identification also includes the optimal design of experiments for efficiently generating informative data for fitting such models.

In many real-life situations, it is helpful to describe data by a single number that is most representative of the entire collection of numbers. Such a number is called a *measure of central tendency*; the most commonly used measures are mean, median, mode and standard deviation, the latter being the average distance between the actual data and the mean. Statistics is important in the field of engineering since it provides tools to analyse collected data. For example, a chemical engineer may wish to analyse temperature measurements from a mixing tank. Statistical methods can be used to determine how reliable and reproducible the temperature measurements are, how much the temperature varies within the data set, what future temperatures of the tank may be, and how confident the engineer can be in the temperature measurements made. When performing statistical analysis on a set of data, the mean, median, mode, and standard deviation are all helpful values to calculate.

Engineers deal with uncertainty in their work, often with precision and analysis, and probability theory is widely used to model systems in engineering and scientific applications. There are several examples where probability is used in engineering. For example, with electronic circuits, scaling down the power and energy of such circuits reduces the reliability and predictability of many individual elements, but the circuits must nevertheless be engineered so that the overall circuit is reliable. Centres for disease control need to decide whether to institute massive vaccination or other preventative measures in the face of globally threatening, possibly mutating diseases in humans and animals. System designers must weigh the costs and benefits of measures for reliability and security, such as levels of backups and firewalls, in the face of uncertainty about threats from equipment failures or malicious attackers. Models incorporating probability theory have been developed and are continuously being improved for understanding the brain, gene pools within populations, weather and climate forecasts, microelectronic devices, and imaging systems such as computer aided tomography (CAT) scan and radar. The electric power grid, including power generating stations, transmission lines, and consumers, is a complex system with many redundancies; however, breakdowns occur, and guidance for investment comes from modelling the most likely sequences of events that could cause outage. Similar planning and analysis are done for communication networks, transportation networks, water, and other infrastructure. Probabilities, permutations and combinations are used daily in many different fields that range from gambling and games, to mechanical or structural failure rates, to rates of detection in medical screening. Uncertainty is clearly all around us, in our daily lives and in many professions. Use of standard deviation is widely used when results of opinion polls are described. The language of probability theory lets people break down complex problems, and to argue about pieces of them with each other, and then aggregate information about subsystems to analyse a whole system.

The binomial distribution is used only when both of two conditions are met – the test has only two possible outcomes, and the sample must be random. If both conditions are met, then this distribution may be used to predict the probability of a desired result. For example, a binomial distribution may be used in determining whether a new drug being tested has or has not contributed to alleviating symptoms of a disease. Common applications of this distribution range from scientific and engineering applications to military and medical ones, in quality assurance, genetics and in experimental design.

A Poisson distribution has several applications and is essentially a derived limiting case of the binomial distribution. It is most applicably relevant to a situation in which the total number of successes is known, but the number of trials is not. An example of such a situation would be if the mean expected number of cancer cells present per sample is known and it was required to determine the probability of finding 1.5 times that number of cells in any given sample; this is an example when the Poisson distribution would be used. The Poisson distribution has widespread applications in analysing traffic flow, in fault prediction on electric cables, in the prediction of randomly occurring accidents, and in reliability engineering.

A normal distribution is a very important statistical data distribution pattern occurring in many natural phenomena, such as height, blood pressure, lengths of objects produced by machines, marks in a test, errors in measurements, and so on. In general, when data is gathered, we expect to see a particular pattern to the data, called a *normal distribution*. This is a distribution where the data is evenly distributed around the mean in a very regular way, which when plotted as a histogram will result in a *bell curve*. The normal distribution is the most important of all probability distributions; it is applied directly to many practical problems in every engineering discipline. There are two principal applications of the normal distribution to engineering and reliability. One application deals with the analysis of items which exhibit failure due to wear, such as mechanical devices - frequently the wear-out failure distribution is sufficiently close to normal that the use of this distribution for predicting or assessing reliability is valid. Another application is in the analysis of manufactured items and their ability to meet specifications. No two parts made to the same specification are exactly alike; the variability of parts leads to a variability in systems composed of those parts. The design must take this variability into account, otherwise the system may not meet the specification requirement due to the combined effect of part variability.

Correlation coefficients measure the strength of association between two variables. The most common correlation coefficient, called the product-moment correlation coefficient, measures the strength of the *linear association* between variables. A positive value indicates a positive correlation and the higher the value, the stronger the correlation. Similarly, a negative value indicates a negative correlation and the lower the value the stronger the correlation.

The general process of fitting data to a linear combination of basic functions is termed linear regression. Linear least squares regression is by far the most widely used modelling method; it is what most people mean when they say they have used 'regression', 'linear regression' or 'least squares' to fit a model to their data. Not only is linear least squares regression the most widely used modelling method, but it has been adapted to a broad range of situations that are outside its direct scope. It plays a strong underlying role in many other modelling methods.

Estimation theory is a branch of statistics and signal processing that deals with estimating the values of parameters based on measured/empirical data that has a random component. Estimation theory can be found at the heart of many electronic signal processing systems designed to extract information; these systems include radar, sonar, speech, image, communications, control and seismology. In statistical testing, a result is called statistically significant if it is unlikely to have occurred by chance, and hence provides enough evidence to reject the hypothesis of 'no effect'. The tests involve comparing the observed values with theoretical values. The tests establish whether there is a relationship between the variables, or whether pure chance could produce the observed results. For most scientific research, a statistical significance test eliminates the possibility that the results arose by chance, allowing a rejection of the null hypothesis.

Chi-square and distribution-free tests are used in science and engineering. Chi-square is a statistical test commonly used to compare observed data with data we would expect to obtain according to a specific hypothesis. Distribution-free methods do not rely on assumptions that the data are drawn from a given probability distribution. Non-parametric methods are widely used for studying populations that take on a ranked order.

Chapter 172 Presentation of ungrouped data

Ungrouped data can be presented diagrammatically by:

(a) **pictograms**, in which pictorial symbols are used to represent quantities,

(b) **horizontal bar charts**, having data represented by equally spaced horizontal rectangles,

(c) **vertical bar charts**, in which data are represented by equally spaced vertical rectangles,

(d) **percentage component bar chart**, where rectangles are subdivided into values corresponding to the percentage relative frequencies of the members, and

(e) **pie diagrams**, where the area of a circle represents the whole, and the areas of the sectors of the circle are made proportional to the parts that make up the whole.

Application: The number of television sets repaired in a workshop by a technician in six, one-month periods is as shown below.

Month	January	February	March	April	May	June
Number repaired	11	6	15	9	13	8

Present the data in a pictogram

This data is represented as a pictogram as shown in Figure 172.1 where each symbol represents two television sets repaired. Thus, in January, $5\frac{1}{2}$ symbols are used to represent the 11 sets repaired, in February, 3 symbols are used to represent the 6 sets repaired, and so on.

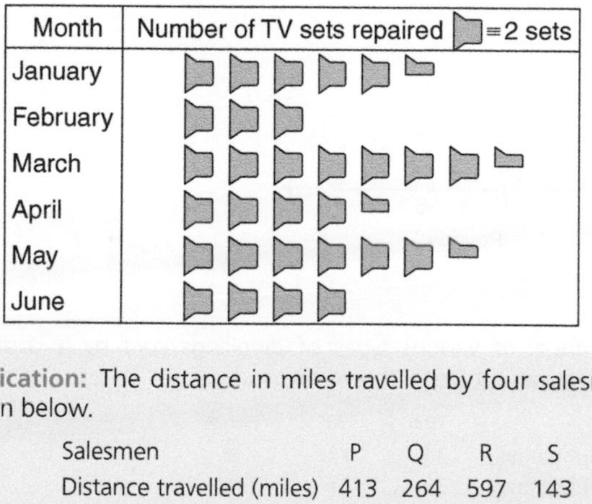

Figure 172.1

Application: The distance in miles travelled by four salesmen in a week are as shown below.

Salesmen	P	Q	R	S
Distance travelled (miles)	413	264	597	143

Represent the data by a horizontal bar chart

To represent these data diagrammatically by a horizontal bar chart, equally spaced horizontal rectangles of any width, but whose length is proportional to the distance travelled, are used. Thus, the length of the rectangle for salesman P is proportional to 413 miles, and so on. The horizontal bar chart depicting these data is shown in Figure 172.2.

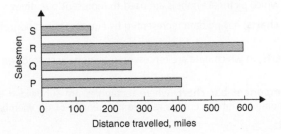

Figure 172.2

Application: The number of issues of tools or materials from a store in a factory is observed for seven, one-hour periods in a day, and the results of the survey are as follows:

Period	1	2	3	4	5	6	7
Number of issues	34	17	9	5	27	13	6

Represent the data by a vertical bar chart

In a vertical bar chart, equally spaced vertical rectangles of any width, but whose height is proportional to the quantity being represented, are used. Thus the height of the rectangle for period 1 is proportional to 34 units, and so on. The vertical bar chart depicting these data is shown in Figure 172.3.

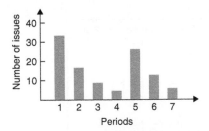

Figure 172.3

Application: The numbers of various types of dwellings sold by a company annually over a three-year period are as shown below.

	Year 1	Year 2	Year 3
4-roomed bungalows	24	17	7
5-roomed bungalows	38	71	118
4-roomed houses	44	50	53

| 5-roomed houses | 64 | 82 | 147 |
| 6-roomed houses | 30 | 30 | 25 |

Draw a percentage component bar chart to represent the above data

To draw a percentage component bar chart to present these data, a table of percentage relative frequency values, correct to the nearest 1%, is the first requirement. Since, percentage relative frequency $= \dfrac{\text{frequency of member} \times 100}{\text{total frequency}}$ then for 4-roomed bungalows in year 1:

$$\text{percentage relative frequency} = \frac{24 \times 100}{24 + 38 + 44 + 64 + 30} = 12\%$$

The percentage relative frequencies of the other types of dwellings for each of the three years are similarly calculated and the results are as shown in the table below

	Year 1	Year 2	Year 3
4-roomed bungalows	12%	7%	2%
5-roomed bungalows	19%	28%	34%
4-roomed houses	22%	20%	15%
5-roomed houses	32%	33%	42%
6-roomed houses	15%	12%	7%

The percentage component bar chart is produced by constructing three equally spaced rectangles of any width, corresponding to the three years. The heights of the rectangles correspond to 100% relative frequency, and are subdivided into the values in the table of percentages shown above. A key is used (different types of shading or different colour schemes) to indicate corresponding percentage values in the rows of the table of percentages. The percentage component bar chart is shown in Figure 172.4.

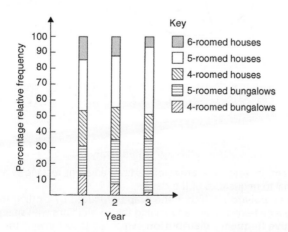

Figure 172.4

Application: The retail price of a product costing £2 is made up as follows: materials 10p, labour 20p, research and development 40p, overheads 70p, profit 60p.

Present this data on a pie diagram

To present these data on a pie diagram, a circle of any radius is drawn, and the area of the circle represents the whole, which in this case is £2. The circle is subdivided into sectors so that the areas of the sectors are proportional to the parts, i.e. the parts that make up the total retail price. For the area of a sector to be proportional to a part, the angle at the centre of the circle must be proportional to that part. The whole, £2 or 200p, corresponds to 360°. Therefore,

$$10p \text{ corresponds to } 360 \times \frac{10}{200} \text{ degrees, i.e. } 18°$$

$$20p \text{ corresponds to } 360 \times \frac{20}{200} \text{ degrees, i.e. } 36°$$

and so on, giving the angles at the centre of the circle for the parts of the retail price as: 18°, 36°, 72°, 126° and 108°, respectively.

The pie diagram is shown in Figure 172.5.

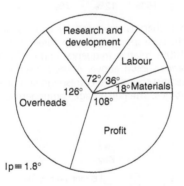

Figure 172.5

Chapter 173   Presentation of grouped data

Grouped data can be presented diagrammatically by:

(a) a **histogram**, in which the **areas** of vertical, adjacent rectangles are made proportional to frequencies of the classes,

(b) a **frequency polygon**, which is the graph produced by plotting frequency against class mid-point values and joining the co-ordinates with straight lines,

(c) a **cumulative frequency distribution**, which is a table showing the cumulative frequency for each value of upper class boundary, and

(d) an **ogive** or a **cumulative frequency distribution curve**, which is a curve obtained by joining the co-ordinates of cumulative frequency (vertically) against upper class boundary (horizontally).

Application: The masses of 50 ingots, in kilograms, are measured correct to the nearest 0.1 kg and the results are as shown below.

8.0	8.6	8.2	7.5	8.0	9.1	8.1	7.6	8.2	7.8
8.3	7.1	8.1	8.3	8.7	7.8	8.7	8.5	8.4	8.5
7.7	8.4	7.9	8.8	7.2	8.1	7.8	8.2	7.7	7.5
8.1	7.4	8.8	8.0	8.4	8.5	8.1	7.3	9.0	8.6
7.4	8.2	8.4	7.7	8.3	8.2	7.9	8.5	7.9	8.0

Produce for this data (a) a frequency distribution for 7 classes, (b) a frequency polygon, (c) a histogram, (d) a cumulative frequency distribution, and (e) an ogive.

(a) The **range** of the data is the member having the largest value minus the member having the smallest value. Inspection of the set of data shows that: range = $9.1 - 7.1 = 2.0$

The size of each class is given approximately by

$$\frac{\text{range}}{\text{number of classes}}$$

If about seven classes are required, the size of each class is 2.0/7, that is approximately 0.3, and thus the **class limits** are selected as 7.1 to 7.3, 7.4 to 7.6, 7.7 to 7.9, and so on.

The **class mid-point** for the 7.1 to 7.3 class is $\frac{7.35 + 7.05}{2}$, i.e. 7.2, for the 7.4 to 7.6 class is $\frac{7.65 + 7.35}{2}$ i.e. 7.5, and so on.

To assist with accurately determining the number in each class, a **tally diagram** is produced as shown in Table 173.1. This is obtained by listing the classes in the left-hand column and then inspecting each of the 50 members of the set of data in turn and allocating it to the appropriate class by putting a '1' in the appropriate row. Each fifth '1' allocated to a particular row is marked as an oblique line to help with final counting.

A **frequency distribution** for the data is shown in Table 173.2 and lists classes and their corresponding frequencies. Class mid-points are also shown in this table, since they are used when constructing the frequency polygon and histogram.

(b) A **frequency polygon** is shown in Figure 173.1, the co-ordinates corresponding to the class mid-point/frequency values, given in Table 173.2. The co-ordinates are joined by straight lines and the polygon is 'anchored-down' at each end by joining to the next class mid-point value and zero frequency.

(c) A **histogram** is shown in Figure 173.2, the width of a rectangle corresponding to (upper class boundary value – lower class boundary value) and height corresponding to the class frequency. The easiest way to draw a histogram is to mark class mid-point values on the horizontal scale and to draw the rectangles symmetrically about the appropriate class mid-point values and touching one another. A histogram for the data given in Table 173.2 is shown in Figure 173.2.

Table 173.1

Class	Tally
7.1 to 7.3	111
7.4 to 7.6	1̶1̶1̶1̶
7.7 to 7.9	1̶1̶1̶1̶ 1111
8.0 to 8.2	1̶1̶1̶1̶ 1̶1̶1̶1̶ 1111
8.3 to 8.5	1̶1̶1̶1̶ 1̶1̶1̶1̶ 1
8.6 to 8.8	1̶1̶1̶1̶ 1
8.9 to 9.1	11

Table 173.2

Class	Class mid-point	Frequency
7.1 to 7.3	7.2	3
7.4 to 7.6	7.5	5
7.7 to 7.9	7.8	9
8.0 to 8.2	8.1	14
8.3 to 8.5	8.4	11
8.6 to 8.8	8.7	6
8.9 to 9.1	9.0	2

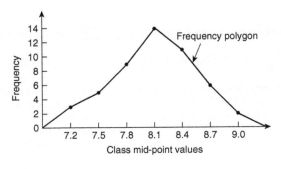

Figure 173.1

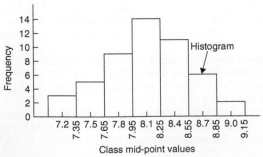

Figure 173.2

(d) A **cumulative frequency distribution** is a table giving values of cumulative frequency for the values of upper class boundaries, and is shown in Table 173.3. Columns 1 and 2 show the classes and their frequencies. Column 3 lists the upper class boundary values for the classes given in column 1. Column 4 gives the cumulative frequency values for all frequencies less than the upper class boundary values given in column 3. Thus, for example, for the 7.7 to 7.9 class shown in row 3, the cumulative frequency value is the sum of all frequencies having values of less than 7.95, i.e. 3 + 5 + 9 = 17, and so on.

(e) The **ogive** for the cumulative frequency distribution given in Table 173.3 is shown in Figure 173.3. The co-ordinates corresponding to each upper class boundary/cumulative frequency value are plotted and the co-ordinates are joined by straight lines (– not the best curve drawn through the co-ordinates as in experimental work). The ogive is 'anchored' at its start by adding the co-ordinate (7.05, 0).

Table 173.3

1 Class	2 Frequency	3 Upper class boundary	4 Cumulative frequency
		Less than	
7.1–7.3	3	7.35	3
7.4–7.6	5	7.65	8
7.7–7.9	9	7.95	17
8.0–8.2	14	8.25	31
8.3–8.5	11	8.55	42
8.6–8.8	6	8.85	48
8.9–9.1	2	9.15	50

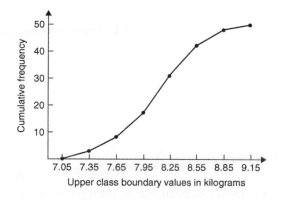

Upper class boundary values in kilograms

Figure 173.3

Chapter 174   Measures of central tendency

(a) Discrete data

mean value, $\bar{x} = \dfrac{\sum x}{n}$

the **median** is the middle term of a ranked set of data,

the **mode** is the most commonly occurring value in a set of data, and

standard deviation, $\sigma = \sqrt{\left[\dfrac{\sum (x - \bar{x})^2}{n}\right]}$

Application: Find the median of the set {7, 5, 74, 10}

The set: {7, 5, 74, 10} is ranked as {5, 7, 10, 74}, and since it contains an even number of members (four in this case), the mean of 7 and 10 is taken, giving a median value of **8.5**

Application: Find the median of the set {3, 81, 15, 7, 14}

The set: {3, 81, 15, 7, 14} is ranked as {3, 7, 14, 15, 81} and the median value is the value of the middle member, i.e. **14**

Application: Find the modal value of the set {5, 6, 8, 2, 5, 4, 6, 5, 3}

The set: {5, 6, 8, 2, 5, 4, 6, 5, 3} has a modal value of **5**, since the member having a value of 5 occurs three times.

Application: Find the mean, median and modal values for the set {2, 3, 7, 5, 5, 13, 1, 7, 4, 8, 3, 4, 3}

For the set {2, 3, 7, 5, 5, 13, 1, 7, 4, 8, 3, 4, 3}

mean value,

$$\bar{x} = \frac{2 + 3 + 7 + 5 + 5 + 13 + 1 + 7 + 4 + 8 + 3 + 4 + 3}{13} = \frac{65}{13} = 5$$

To obtain the median value the set is ranked, that is, placed in ascending order of magnitude, and since the set contains an odd number of members the value of the middle member is the median value. Ranking the set gives:

{1, 2, 3, 3, 3, 4, 4, 5, 5, 7, 7, 8, 13}

The middle term is the seventh member, i.e. 4, thus the **median value is 4**.

The **modal value** is the value of the most commonly occurring member and is **3**, which occurs three times, all other members only occurring once or twice.

Application: Determine the standard deviation from the mean of the set of numbers:

$$\{5, 6, 8, 4, 10, 3\}, \text{ correct to 4 significant figures}$$

The arithmetic mean, $\bar{x} = \dfrac{\sum x}{n} = \dfrac{5+6+8+4+10+3}{6} = 6$

Standard deviation, $\sigma = \sqrt{\left[\dfrac{\sum(x - \bar{x})^2}{n}\right]}$

The $(x - \bar{x})^2$ values are: $(5 - 6)^2$, $(6 - 6)^2$, $(8 - 6)^2$, $(4 - 6)^2$, $(10 - 6)^2$ and $(3 - 6)^2$

The sum of the $(x - \bar{x})^2$ values,

i.e. $\sum(x - \bar{x})^2 = 1 + 0 + 4 + 4 + 16 + 9 = 34$

and $\dfrac{\sum(x - \bar{x})^2}{n} = \dfrac{34}{6} = 5.\dot{6}$ since there are 6 members in the set.

Hence, **standard deviation**,

$$\sigma = \sqrt{\left[\dfrac{\sum(x - \bar{x})^2}{n}\right]} = \sqrt{5.\dot{6}}$$

$$= \mathbf{2.380}, \text{ correct to 4 significant figures.}$$

(b) Grouped data

$$\text{mean value, } \bar{x} = \dfrac{\sum(fx)}{\sum f}$$

$$\text{standard deviation, } \sigma = \sqrt{\left[\dfrac{\sum\{f(x - \bar{x})^2\}}{\sum f}\right]}$$

Application: Find (a) the mean value, and (b) the standard deviation for the following values of resistance, in ohms, of 48 resistors:

20.5–20.9 3,	21.0–21.4 10,	21.5–21.9 11,
22.0–22.4 13,	22.5–22.9 9,	23.0–23.4 2

(a) The class mid-point/frequency values are:

20.7 3,   21.2 10,   21.7 11,   22.2 13,   22.7 9 and 23.2 2

For grouped data, the mean value is given by: $\bar{x} = \dfrac{\sum(f\,x)}{\sum f}$

where f is the class frequency and x is the class mid-point value. Hence

$$\text{mean value, } \bar{x} = \dfrac{\begin{array}{c}(3 \times 20.7) + (10 \times 21.2) + (11 \times 21.7) \\ + (13 \times 22.2) + (9 \times 22.7) + (2 \times 23.2)\end{array}}{48}$$

$$= \dfrac{1052.1}{48} = 21.919..$$

i.e. **the mean value is 21.9 ohms**, correct to 3 significant figures.

(b) From part (a), mean value, $\bar{x} = 21.92$, correct to 4 significant figures.

The 'x-values' are the class mid-point values, i.e. 20.7, 21.2, 21.7, ...

Thus the $(x - \bar{x})^2$ values are $(20.7 - 21.92)^2$, $(21.2 - 21.92)^2$, $(21.7 - 21.92)^2$, ...,

and the $f(x - \bar{x})^2$ values are $3(20.7 - 21.92)^2$, $10(21.2 - 21.92)^2$, $11(21.7 - 21.92)^2$, ...

The $\sum f(x - \bar{x})^2$ values are $4.4652 + 5.1840 + 0.5324 + 1.0192 + 5.4756 + 3.2768 = 19.9532$

$$\dfrac{\sum\{f(x - \bar{x})^2\}}{\sum f} = \dfrac{19.9532}{48} = 0.41569$$

and **standard deviation,**

$$\sigma = \sqrt{\left[\dfrac{\sum\left\{f(x - \bar{x})^2\right\}}{\sum f}\right]} = \sqrt{0.41569}$$

$$= 0.645, \text{ correct to 3 significant figures}$$

Application: The time taken in minutes to assemble a device is measured 50 times and the results are as shown below:

14.5–15.5	5,	16.5–17.5	8,	18.5–19.5	16,
20.5–21.5	12,	22.5–23.5	6,	24.5–25.5	3

Determine the mean, median and modal values of the distribution by depicting the data on a histogram

The histogram is shown in Figure 174.1. The mean value lies at the centroid of the histogram. With reference to any arbitrary axis, say YY shown at a time of 14 minutes, the position of the horizontal value of the centroid can be obtained from the relationship AM = $\sum$(am), where A is the area of the histogram, M is the horizontal distance of the centroid from the axis YY, a is the area of a rectangle of the histogram and m is the distance of the centroid of the rectangle from YY. The areas of the individual

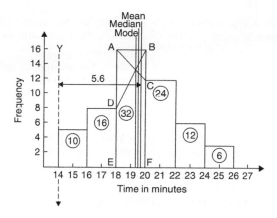

Figure 174.1

rectangles are shown circled on the histogram giving a total area of 100 square units. The positions, m, of the centroids of the individual rectangles are 1, 3, 5, ... units from YY. Thus

$$100\,M = (10 \times 1) + (16 \times 3) + (32 \times 5) + (24 \times 7)$$
$$+ (12 \times 9) + (6 \times 11)$$

i.e.

$$M = \frac{560}{100} = 5.6 \text{ units from YY}$$

Thus the position of the **mean** with reference to the time scale is 14 + 5.6, i.e. **19.6 minutes**.

The median is the value of time corresponding to a vertical line dividing the total area of the histogram into two equal parts. The total area is 100 square units; hence the vertical line must be drawn to give 50 units of area on each side. To achieve this with reference to Figure 174.1, rectangle ABFE must be split so that 50 − (10 + 16) units of area lie on one side and 50 − (24 + 12 + 6) units of area lie on the other. This shows that the area of ABFE is split so that 24 units of area lie to the left of the line and 8 units of area lie to the right, i.e. the vertical line must pass through 19.5 minutes. Thus the **median value** of the distribution is **19.5 minutes**.

The mode is obtained by dividing the line AB, which is the height of the highest rectangle, proportionally to the heights of the adjacent rectangles. With reference to Figure 174.1, this is done by joining AC and BD and drawing a vertical line through the point of intersection of these two lines. This gives the **mode** of the distribution and is **19.3 minutes**.

Chapter 175 Quartiles, deciles and percentiles

The **quartile values** of a set of discrete data are obtained by selecting the values of members which divide the set into four equal parts.

When a set contains a large number of members, the set can be split into ten parts, each containing an equal number of members; these ten parts are then called **deciles**.

For sets containing a very large number of members, the set may be split into one hundred parts, each containing an equal number of members; one of these parts is called a **percentile**.

Application: The frequency distribution given below refers to the overtime worked by a group of craftsmen during each of 48 working weeks in a year.

    25–29   5,   30–34   4,   35–39   7,   40–44   11,
    45–49   12,   50–54   8,   55–59   1

Draw an ogive for this data and hence determine the quartiles values

---

The cumulative frequency distribution (i.e. upper class boundary/cumulative frequency values) is:

    29.5   5,   34.5   9,   39.5   16,   44.5   27,
    49.5   39,   54.5   47,   59.5   48

The ogive is formed by plotting these values on a graph, as shown in Figure 175.1.

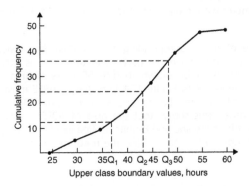

Figure 175.1

The total frequency is divided into four equal parts, each having a range of 48/4, i.e. 12. This gives cumulative frequency values of 0 to 12 corresponding to the first quartile, 12 to 24 corresponding to the second quartile, 24 to 36 corresponding to the third quartile and 36 to 48 corresponding to the fourth quartile of the distribution, i.e. the distribution is divided into four equal parts. The quartile values are those of the variable corresponding to cumulative frequency values of 12, 24 and 36, marked Q_1, Q_2 and Q_3 in Figure 175.1. These values, correct to the nearest hour, are **37 hours, 43 hours and 48 hours**, respectively. The Q_2 value is also equal to the median value of the distribution. One measure of the dispersion of a distribution is called the **semi-interquartile range** and is given by $(Q_3 - Q_1)/2$, and is $(48 - 37)/2$ in this case, i.e. **5½ hours**.

Application: Determine the numbers contained in the (a) 41st to 50th percentile group, and (b) 8th decile group of the set of numbers shown below:

$$14 \quad 22 \quad 17 \quad 21 \quad 30 \quad 28 \quad 37 \quad 7 \quad 23 \quad 32$$
$$24 \quad 17 \quad 20 \quad 22 \quad 27 \quad 19 \quad 26 \quad 21 \quad 15 \quad 29$$

The set is ranked, giving:

$$7 \quad 14 \quad 15 \quad 17 \quad 17 \quad 19 \quad 20 \quad 21 \quad 21 \quad 22$$
$$22 \quad 23 \quad 24 \quad 26 \quad 27 \quad 28 \quad 29 \quad 30 \quad 32 \quad 37$$

(a) There are 20 numbers in the set, hence the first 10% will be the two numbers 7 and 14, the second 10% will be 15 and 17, and so on

Thus the 41st to 50th percentile group will be the numbers **21 and 22**

(b) The first decile group is obtained by splitting the ranked set into 10 equal groups and selecting the first group, i.e. the numbers 7 and 14. The second decile group are the numbers 15 and 17, and so on.

Thus the 8th decile group contains the numbers **27 and 28**

Chapter 176 Probability

The probability of events **A or B or C or ..... N** happening is given by

$$p_A + p_B + p_C + \cdots + p_N$$

The probability of events **A and B and C and ... N** happening is given by

$$p_A \times p_B \times p_C \times \cdots \times p_N$$

Application: Determine the probability of selecting at random the winning horse in a race in which 10 horses are running

Since only one of the ten horses can win, the probability of selecting at random the winning horse is $\dfrac{\text{number of winners}}{\text{number of horses}}$, i.e. $\dfrac{1}{10}$ or **0.10**

Application: Determine the probability of selecting at random the winning horses in both the first and second races if there are 10 horses in each race

The probability of selecting the winning horse in the first race is $\dfrac{1}{10}$

The probability of selecting the winning horse in the second race is $\dfrac{1}{10}$

The probability of selecting the winning horses in the first **and** second race is given by the multiplication law of probability, i.e.

$$\text{probability} = \frac{1}{10} \times \frac{1}{10} = \frac{1}{100} \quad \text{or} \quad \textbf{0.01}$$

Application: The probability of a component failing in one year due to excessive temperature is $\frac{1}{20}$, due to excessive vibration is $\frac{1}{25}$ and due to excessive humidity is $\frac{1}{50}$. Determine the probabilities that during a one year period a component: (a) fails due to excessive temperature and excessive vibration, (b) fails due to excessive vibration or excessive humidity, and (c) will not fail due to excessive temperature and excessive humidity

Let p_A be the probability of failure due to excessive temperature, then

$$p_A = \frac{1}{20} \text{ and } \bar{p}_A = \frac{19}{20} \quad \text{(where } \bar{p}_A \text{ is the probability of not failing)}$$

Let p_B be the probability of failure due to excessive vibration, then

$$p_B = \frac{1}{25} \quad \text{and} \quad \bar{p}_B = \frac{24}{25}$$

Let p_C be the probability of failure due to excessive humidity, then

$$p_C = \frac{1}{50} \quad \text{and} \quad \bar{p}_C = \frac{49}{50}$$

(a) The probability of a component failing due to excessive temperature **and** excessive vibration is given by:

$$p_A \times p_B = \frac{1}{20} \times \frac{1}{25} = \frac{1}{500} \quad \text{or} \quad \textbf{0.002}$$

(b) The probability of a component failing due to excessive vibration **or** excessive humidity is:

$$p_B + p_C = \frac{1}{25} + \frac{1}{50} = \frac{3}{50} \quad \text{or} \quad \textbf{0.06}$$

(c) The probability that a component will not fail due to excessive temperature **and** will not fail due to excess humidity is:

$$\bar{p}_A \times \bar{p}_C = \frac{19}{20} \times \frac{49}{50} = \frac{931}{1000} \quad \text{or} \quad \textbf{0.931}$$

Application: A batch of 40 components contains 5 which are defective. If a component is drawn at random from the batch and tested and then a second component is drawn at random, calculate the probability of having one defective component, both with and without replacement.

The probability of having one defective component can be achieved in two ways. If p is the probability of drawing a defective component and q is the probability of drawing a satisfactory component, then the probability of having one defective component is given by drawing a satisfactory component and then a defective component **or** by drawing a defective component and then a satisfactory one, i.e. by $q \times p + p \times q$

With replacement:

$$p = \frac{5}{40} = \frac{1}{8} \text{ and } q = \frac{35}{40} = \frac{7}{8}$$

Hence, probbility of having one defective component is:

$$\frac{1}{8} \times \frac{7}{8} + \frac{7}{8} \times \frac{1}{8} \text{, i.e. } \frac{7}{64} + \frac{7}{64} = \frac{7}{32} \text{ or } \textbf{0.2188}$$

Without replacement:

$p_1 = \dfrac{1}{8}$ and $q_1 = \dfrac{7}{8}$ on the first of the two draws. The batch number is now 39 for

the second draw, thus, $p_2 = \dfrac{5}{39}$ and $q_2 = \dfrac{35}{39}$

$$p_1 q_2 + q_1 p_2 = \frac{1}{8} \times \frac{35}{39} + \frac{7}{8} \times \frac{5}{39} = \frac{35 + 35}{312} = \frac{70}{312} \text{ or } \textbf{0.2244}$$

Chapter 177 Permutations and combinations

Permutations

If n different objects are available, they can be arranged in different orders of selection. Each different ordered arrangement is called a *permutation*.

For example, permutations of the three letters X, Y and Z taken together are:

$$\text{XYZ, XZY, YXZ, YZX, ZXY and ZYX}$$

This can be expressed as $^3P_3 = 6$, the upper 3 denoting the number of items from which the arrangements are made, and the lower 3 indicating the number of items used in each arrangement.

If we take the same three letters XYZ two at a time the permutations XY, YZ, XZ, ZX, YZ, ZY can be found, and denoted by $^3P_2 = 6$

(Note that the order of the letters matters in permutations, i.e. YX is a different permutation from XY).

In general, $^nP_r = n(n-1)(n-2)(n-r+1)$ or $^nP_r = \dfrac{n!}{(n-r)!}$

For example, $^5P_4 = 5(4)(3)(2) = 120$ or $^5P_4 = \dfrac{5!}{(5-4)!} = \dfrac{5!}{1!} = (5)(4)(3)(2) = 120$

Also, $^3P_3 = 6$ from above; using $^nP_r = \dfrac{n!}{(n-r)!}$ gives $^3P_3 = \dfrac{3!}{(3-3)!} = \dfrac{6}{0!}$.

Since this must equal 6, then $0! = 1$ (check this with your calculator).

Combinations

If selections of the three letters X, Y, Z are made without regard to the order of the letters in each group, i.e. XY is now the same as YX for example, then each group is called a **combination**. The number of possible combinations is denoted by ^nC_r where n is the total number of items and r is the number in each selection.

In general, $^nC_r = \dfrac{n!}{r!(n-r)!}$

For example, $^5C_4 = \dfrac{5!}{4!(5-4)!} = \dfrac{5!}{4!} = \dfrac{5 \times 4 \times 3 \times 2 \times 1}{4 \times 3 \times 2 \times 1} = 5$

Application: Calculate the number of permutations there are of: (a) 5 distinct objects taken 2 at a time, (b) 4 distinct objects taken 2 at a time.

(a) $^5P_2 = \dfrac{5!}{(5-2)!} = \dfrac{5!}{3!} = \dfrac{5 \times 4 \times 3 \times 2}{3 \times 2} = \mathbf{20}$

(b) $^4P_2 = \dfrac{4!}{(4-2)!} = \dfrac{4!}{2!} = \mathbf{12}$

Application: Calculate the number of combinations there are of: (a) 5 distinct objects taken 2 at a time, (b) 4 distinct objects taken 2 at a time.

(a) $^5C_2 = \dfrac{5!}{2!(5-2)!} = \dfrac{5!}{2!3!} = \dfrac{5 \times 4 \times 3 \times 2 \times 1}{(2 \times 1)(3 \times 2 \times 1)} = \mathbf{10}$

(b) $^4C_2 = \dfrac{4!}{2!(4-2)!} = \dfrac{4!}{2!2!} = \mathbf{6}$

Application: A class has 24 students. 4 can represent the class at an exam board. How many combinations are possible when choosing this group?

Number of combinations possible, $^nC_r = \dfrac{n!}{r!(n-r)!}$

i.e. $^{24}C_4 = \dfrac{24!}{4!(24-4)!} = \dfrac{24!}{4!20!} = \mathbf{10626}$

Application: In how many ways can a team of eleven be picked from sixteen possible players?

$$\text{Number of ways} = {}^{n}C_r = {}^{16}C_{11} = \frac{16!}{11!(16-11)!} = \frac{16!}{11!5!} = \mathbf{4368}$$

Chapter 178   Bayes' theorem

Bayes' theorem is one of probability theory (originally stated by the Reverend Thomas Bayes), and may be seen as a way of understanding how the probability that a theory is true is affected by a new piece of evidence. The theorem has been used in a wide variety of contexts, ranging from marine biology to the development of 'Bayesian' spam blockers for email systems; in science, it has been used to try to clarify the relationship between theory and evidence. Insights in the philosophy of science involving confirmation, falsification and other topics can be made more precise, and sometimes extended or corrected, by using Bayes' theorem.

Bayes' theorem may be stated mathematically as:

$$P\left(A_1|B\right) = \frac{P\left(B|A_1\right)P\left(A_1\right)}{P\left(B|A_1\right)P\left(A_1\right) + P\left(B|A_2\right)P\left(A_2\right) + \dots}$$

or

$$P\left(A_i|B\right) = \frac{P\left(B|A_i\right)P\left(A_i\right)}{\sum_{j=1}^{n}P\left(B|A_j\right)P\left(A_j\right)} \quad (i = 1, 2, \dots, n)$$

where   P(A|B) is the probability of A given B, i.e. *after* B is observed

P(A) and P(B) are the probabilities of A and B without regard to each other

and   P(B|A) is the probability of observing event B given that A is true

In the Bayes theorem formula, 'A' represents a theory or hypothesis that is to be tested, and 'B' represents a new piece of evidence that seems to confirm or disprove the theory.

Application: An outdoor degree ceremony is taking place tomorrow, 5 July, in the hot climate of Dubai. In recent years it has rained only 2 days in the four-month period June to September. However, the weather forecaster has predicted rain for tomorrow. When it actually rains, the weatherman correctly forecasts rain 85% of the time. When it doesn't rain, he incorrectly forecasts rain 15% of the time. Determine the probability that it will rain tomorrow.

There are two possible mutually-exclusive events occurring here – it either rains or it does not rain.

Also, a third event occurs when the weatherman predicts rain.

Let the notation for these events be: Event A_1 It rains at the ceremony

Event A_2 It does not rain at the ceremony

Event B   The weatherman predicts rain

The probability values are:

$$P\left(A_1\right) = \frac{2}{30+31+31+30} = \frac{1}{61} \quad \text{(i.e. it rains 2 days in the months June to}$$

September)

$$P\left(A_2\right) = \frac{120}{30+31+31+30} = \frac{60}{61} \quad \text{(i.e. it does not rain for 120 of the 122 days}$$

in the months June to September)

$P(B|A_1) = 0.85$ (i.e. when it rains, the weatherman predicts rain 85% of the time)

$P(B|A_2) = 0.15$ (i.e. when it does not rain, the weatherman predicts rain 15% of the time)

Using Bayes' theorem to determine the probability that it will rain tomorrow, given the forecast of rain by the weatherman:

$$P\left(A_1|B\right) = \frac{P\left(B|A_1\right)P\left(A_1\right)}{P\left(B|A_1\right)P\left(A_1\right) + P\left(B|A_2\right)P\left(A_2\right)}$$

$$= \frac{(0.85)\left(\dfrac{1}{61}\right)}{0.85 \times \dfrac{1}{61} + 0.15 \times \dfrac{60}{61}} = \frac{0.0139344}{0.1614754} = \textbf{0.0863 or 8.63\%}$$

Even when the weatherman predicts rain, it rains only between 8% and 9% of the time. **Hence, there is a good chance it will not rain tomorrow in Dubai for the degree ceremony**.

Chapter 179 The binomial distribution

If p is the probability that an event will happen and q is the probability that the event will not happen, then the probabilities that the event will happen 0, 1, 2, 3,..., n times in n trials are given by the successive terms of the expansion of $(q + p)^n$, taken from left to right, i.e.

$$q^n, \quad nq^{n-1}p, \quad \frac{n(n-1)}{2!}q^{n-2}p^2, \quad \frac{n(n-1)(n-2)}{3!}q^{n-3}p^3, \quad \dots$$

Industrial inspection

The probabilities that 0, 1, 2, 3, ... , n components are defective in a sample of n components, drawn at random from a large batch of components, are given by the successive terms of the expansion of $(q + p)^n$, taken from left to right.

Application: A die is rolled 9 times. Find the probabilities of having a 4 upwards (a) 3 times and (b) less than 4 times

Let p be the probability of having a 4 upwards. Then $p = 1/6$, since dice have six sides.

Let q be the probability of not having a 4 upwards. Then $q = 5/6$. The probabilities of having a 4 upwards 0, 1, 2.. n times are given by the successive terms of the expansion of $(q + p)^n$, taken from left to right.

From the binomial expansion (see Chapter 26):

$$(q + q)^9 = q^9 + 9q^8p + 36q^7p^2 + 84q^6p^3 + ...$$

The probability of having a 4 upwards no times is
$$q^9 = (5/6)^9 = \mathbf{0.1938}$$

The probability of having a 4 upwards once is $9q^8p = 9(5/6)^8 (1/6)$
$$= \mathbf{0.3489}$$

The probability of having a 4 upwards twice is
$$36q^7p^2 = 36(5/6)^7(1/6)^2 = \mathbf{0.2791}$$

(a) The probability of having a 4 upwards 3 times is
$$84q^6p^3 = 84(5/6)^6 (1/6)^3 = \mathbf{0.1302}$$

(b) The probability of having a 4 upwards less than 4 times is the sum of the probabilities of having a 4 upwards 0, 1, 2, and 3 times, i.e.

$$0.1938 + 0.3489 + 0.2791 + 0.1302 = \mathbf{0.9520}$$

Application: A package contains 50 similar components and inspection shows that four have been damaged during transit. If six components are drawn at random from the contents of the package, determine the probabilities that in this sample (a) one and (b) less than three are damaged

The probability of a component being damaged, p, is 4 in 50, i.e. 0.08 per unit. Thus, the probability of a component not being damaged, q, is $1 - 0.08$, i.e. 0.92

The probability of there being 0, 1, 2,..., 6 damaged components is given by the successive terms of $(q + p)^6$, taken from left to right.

$$(q + p)^6 = q^6 + 6q^5p + 15q^4p^2 + 20q^3p^3 + \cdots$$

(a) The probability of one damaged component is
$$6q^5p = 6 \times 0.92^5 \times 0.08 = \mathbf{0.3164}$$

(b) The probability of less than three damaged components is given by the sum of the probabilities of 0, 1 and 2 damaged components, i.e.

$$q^6 + 6q^5p + 15q^4p^2 = 0.92^6 + 6 \times 0.92^5 \times 0.08 + 15 \times 0.92^4 \times 0.08^2$$
$$= 0.6064 + 0.3164 + 0.0688 = \mathbf{0.9916}$$

Chapter 180   The Poisson distribution

If λ is the expectation of the occurrence of an event then the probability of 0, 1, 2, 3, ... occurrences is given by:

$$e^{-\lambda},\ \lambda e^{-\lambda},\ \lambda^2 \frac{e^{-\lambda}}{2!},\ \lambda^3 \frac{e^{-\lambda}}{3!},\ ...$$

Application: If 3% of the gearwheels produced by a company are defective, determine the probabilities that in a sample of 80 gearwheels (a) two and (b) more than two will be defective

The sample number, n, is large, the probability of a defective gearwheel, p, is small and the product np is 80×0.03, i.e. 2.4, which is less than 5. Hence a Poisson approximation to a binomial distribution may be used. The expectation of a defective gearwheel, $\lambda = np = 2.4$

The probabilities of 0, 1, 2,... defective gearwheels are given by the successive terms of the expression $e^{-\lambda}\left(1 + \lambda + \frac{\lambda^2}{2!} + \frac{\lambda^3}{3!} + ...\right)$ taken from left to right, i.e. by $e^{-\lambda},\ \lambda e^{-\lambda},\ \frac{\lambda^2 e^{-\lambda}}{2!},\ ..$

The probability of no defective gearwheels is $e^{-\lambda} = e^{-2.4} = \mathbf{0.0907}$

The probability of 1 defective gearwheel is $\lambda e^{-\lambda} = 2.4e^{-2.4}$

$$= \mathbf{0.2177}$$

(a) the probability of 2 defective gearwheels is $\dfrac{\lambda^2 e^{-\lambda}}{2!} = \dfrac{2.4^2 e^{-2.4}}{2 \times 1}$

$$= \mathbf{0.2613}$$

(b) The probability of having more than 2 defective gearwheels is 1 − (the sum of the probabilities of having 0, 1, and 2 defective gearwheels), i.e.

$$1 - (0.0907 + 0.2177 + 0.2613), \text{ that is, } \mathbf{0.4303}$$

Application: A production department has 35 similar milling machines. The number of breakdowns on each machine averages 0.06 per week. Determine the probabilities of having (a) one, and (b) less than three machines breaking down in any week

Since the average occurrence of a breakdown is known but the number of times when a machine did not break down is unknown, a Poisson distribution must be used.

The expectation of a breakdown for 35 machines is 35×0.06, i.e. 2.1 breakdowns per week. The probabilities of a breakdown occurring 0, 1, 2, ... times are given by the successive terms of the expression $e^{-\lambda}\left(1 + \lambda + \dfrac{\lambda^2}{2!} + \dfrac{\lambda^3}{3!} + \cdots\right)$, taken from left to right.

Hence the probability of no breakdowns $e^{-\lambda} = e^{-2.1} = \mathbf{0.1225}$

(a) The probability of 1 breakdown is $\lambda e^{-\lambda} = 2.1e^{-2.1} = \mathbf{0.2572}$

(b) The probability of 2 breakdowns is $\dfrac{\lambda^2 e^{-\lambda}}{2!} = \dfrac{2.1^2 e^{-2.1}}{2 \times 1} = 0.2700$

The probability of less than 3 breakdowns per week is the sum of the probabilities of 0, 1 and 2 breakdowns per week,

i.e. $0.1225 + 0.2572 + 0.2700 = \mathbf{0.6497}$

Chapter 181   The normal distribution

A table of partial areas under the standardised normal curve is shown in Table 181.1.

Application: The mean height of 500 people is 170 cm and the standard deviation is 9 cm. Assuming the heights are normally distributed, determine (a) the number of people likely to have heights between 150 cm and 195 cm, (b) the number of people likely to have heights of less than 165 cm, and (c) the number of people likely to have heights of more than 194 cm

(a) The mean value, $\bar{x}$, is 170 cm and corresponds to a normal standard variate value, z, of zero on the standardised normal curve. A height of 150 cm has a z-value given by $z = \dfrac{x - \bar{x}}{\sigma}$ standard deviations, i.e. $\dfrac{150 - 170}{9}$ or -2.22 standard deviations.

Using a table of partial areas beneath the standardised normal curve (see Table 181.1), a z-value of -2.22 corresponds to an area of 0.4868 between the mean value and the ordinate $z = -2.22$.

The negative z-value shows that it lies to the left of the $z = 0$ ordinate.

This area is shown shaded in Figure 181.1(a). Similarly, 195 cm has a z-value of $\dfrac{195 - 170}{9}$ that is 2.78 standard deviations. From Table 181.1, this value of z

Table 181.1   Partial areas under the standardised normal curve

$z = \dfrac{x - \bar{x}}{\sigma}$	0	1	2	3	4	5	6	7	8	9
0.0	0.0000	0.0040	0.0080	0.0120	0.0159	0.0199	0.0239	0.0279	0.0319	0.0359
0.1	0.0398	0.0438	0.0478	0.0517	0.0557	0.0596	0.0636	0.0678	0.0714	0.0753
0.2	0.0793	0.0832	0.0871	0.0910	0.0948	0.0987	0.1026	0.1064	0.1103	0.1141
0.3	0.1179	0.1217	0.1255	0.1293	0.1331	0.1388	0.1406	0.1443	0.1480	0.1517
0.4	0.1554	0.1591	0.1628	0.1664	0.1700	0.1736	0.1772	0.1808	0.1844	0.1879
0.5	0.1915	0.1950	0.1985	0.2019	0.2054	0.2086	0.2123	0.2157	0.2190	0.2224
0.6	0.2257	0.2291	0.2324	0.2357	0.2389	0.2422	0.2454	0.2486	0.2517	0.2549
0.7	0.2580	0.2611	0.2642	0.2673	0.2704	0.2734	0.2760	0.2794	0.2823	0.2852
0.8	0.2881	0.2910	0.2939	0.2967	0.2995	0.3023	0.3051	0.3078	0.3106	0.3133
0.9	0.3159	0.3186	0.3212	0.3238	0.3264	0.3289	0.3315	0.3340	0.3365	0.3389
1.0	0.3413	0.3438	0.3451	0.3485	0.3508	0.3531	0.3554	0.3577	0.3599	0.3621
1.1	0.3643	0.3665	0.3686	0.3708	0.3729	0.3749	0.3770	0.3790	0.3810	0.3830
1.2	0.3849	0.3869	0.3888	0.3907	0.3925	0.3944	0.3962	0.3980	0.3997	0.4015
1.3	0.4032	0.4049	0.4066	0.4082	0.4099	0.4115	0.4131	0.4147	0.4162	0.4177

z										
1.4	0.4192	0.4207	0.4222	0.4236	0.4251	0.4265	0.4279	0.4292	0.4306	0.4319
1.5	0.4332	0.4345	0.4357	0.4370	0.4382	0.4394	0.4406	0.4418	0.4430	0.4441
1.6	0.4452	0.4463	0.4474	0.4484	0.4495	0.4505	0.4515	0.4525	0.4535	0.4545
1.7	0.4554	0.4564	0.4573	0.4582	0.4591	0.4599	0.4608	0.4616	0.4625	0.4633
1.8	0.4641	0.4649	0.4656	0.4664	0.4671	0.4678	0.4686	0.4693	0.4699	0.4706
1.9	0.4713	0.4719	0.4726	0.4732	0.4738	0.4744	0.4750	0.4756	0.4762	0.4767
2.0	0.4772	0.4778	0.4783	0.4785	0.4793	0.4798	0.4803	0.4808	0.4812	0.4817
2.1	0.4821	0.4826	0.4830	0.4834	0.4838	0.4842	0.4846	0.4850	0.4854	0.4857
2.2	0.4861	0.4864	0.4868	0.4871	0.4875	0.4878	0.4881	0.4884	0.4887	0.4890
2.3	0.4893	0.4896	0.4898	0.4901	0.4904	0.4906	0.4909	0.4911	0.4913	0.4916
2.4	0.4918	0.4920	0.4922	0.4925	0.4927	0.4929	0.4931	0.4932	0.4934	0.4936
2.5	0.4938	0.4940	0.4941	0.4943	0.4945	0.4946	0.4948	0.4949	0.4951	0.4952
2.6	0.4953	0.4955	0.4956	0.4957	0.4959	0.4960	0.4961	0.4962	0.4963	0.4964
2.7	0.4965	0.4966	0.4967	0.4968	0.4969	0.4970	0.4971	0.4972	0.4973	0.4974
2.8	0.4974	0.4975	0.4976	0.4977	0.4977	0.4978	0.4979	0.4980	0.4980	0.4981
2.9	0.4981	0.4982	0.4982	0.4983	0.4984	0.4984	0.4985	0.4985	0.4986	0.4986
3.0	0.4987	0.4987	0.4987	0.4988	0.4988	0.4989	0.4989	0.4989	0.4990	0.4990
3.1	0.4990	0.4991	0.4991	0.4991	0.4992	0.4992	0.4992	0.4992	0.4993	0.4993
3.2	0.4993	0.4993	0.4994	0.4994	0.4994	0.4994	0.4994	0.4995	0.4995	0.4995
3.3	0.4995	0.4995	0.4995	0.4996	0.4996	0.4996	0.4996	0.4996	0.4996	0.4997
3.4	0.4997	0.4997	0.4997	0.4997	0.4997	0.4997	0.4997	0.4997	0.4997	0.4998
3.5	0.4998	0.4998	0.4998	0.4998	0.4998	0.4998	0.4998	0.4998	0.4998	0.4998
3.6	0.4998	0.4998	0.4999	0.4999	0.4999	0.4999	0.4999	0.4999	0.4999	0.4999
3.7	0.4999	0.4999	0.4999	0.4999	0.4999	0.4999	0.4999	0.4999	0.4999	0.4999
3.8	0.4999	0.4999	0.4999	0.4999	0.4999	0.4999	0.4999	0.4999	0.4999	0.4999
3.9	0.5000	0.5000	0.5000	0.5000	0.5000	0.5000	0.5000	0.5000	0.5000	0.5000

corresponds to an area of 0.4973, the positive value of z showing that it lies to the right of the z = 0 ordinate. This area is shown shaded in Figure 181.1(b). The total area shaded in Figures 181.1(a) and (b) is shown in Figure 181.1(c) and is 0.4868 + 0.4973, i.e. 0.9841 of the total area beneath the curve.

However, the area is directly proportional to probability. Thus, the probability that a person will have a height of between 150 and 195 cm is 0.9841. For a group of 500 people, 500 × 0.9841, i.e. **492 people are likely to have heights in this range**.

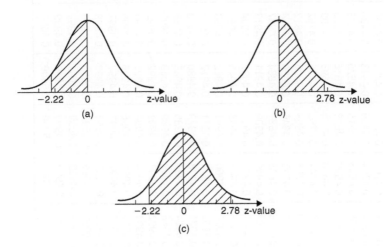

Figure 181.1

(b) A height of 165 cm corresponds to $\dfrac{165 - 170}{9}$, i.e. −0.56 standard deviations.

The area between z = 0 and z = −0.56 (from Table 181.1) is 0.2123, shown shaded in Figure 181.2(a). The total area under the standardised normal curve is unity and since the curve is symmetrical, it follows that the total area to the left of the z = 0 ordinate is 0.5000. Thus the area to the left of the z = −0.56 ordinate ('left' means 'less than', 'right' means 'more than') is 0.5000 − 0.2123, i.e. 0.2877 of the total area, which is shown shaded in Figure 181.2(b). The area is directly proportional to probability and since the total area beneath the standardised normal curve is unity, the probability of a person's height being less than 165 cm is 0.2877. For a group of 500 people, 500 × 0.2877, i.e. **144 people are likely to have heights of less than 165 cm.**

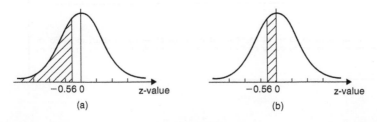

Figure 181.2

(c) 194 cm corresponds to a z-value of $\dfrac{194 - 170}{9}$ that is, 2.67 standard deviations.

From Table 181.1, the area between z = 0, z = 2.67 and the standardised normal curve is 0.4962, shown shaded in Figure 181.3(a). Since the standardised normal curve is symmetrical, the total area to the right of the z = 0 ordinate is 0.5000, hence the shaded area shown in Figure 181.3(b) is 0.5000 − 0.4962, i.e. 0.0038. This area represents the probability of a person having a height of more than 194 cm, and for 500 people, the number of people likely to have a height of more than 194 cm is 0.0038 × 500, i.e. **2 people**.

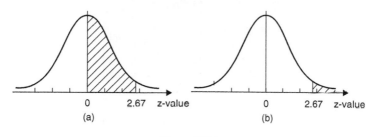

(a)   (b)

Figure 181.3

Testing for a normal distribution

Application: Use normal probability paper to determine whether the data given below, which refers to the masses of 50 copper ingots, is approximately normally distributed. If the data is normally distributed, determine the mean and standard deviation of the data from the graph drawn.

Class mid-point value (kg)	29.5	30.5	31.5	32.5	33.5	34.5	35.5	36.5	37.5	38.5
Frequency	2	4	6	8	9	8	6	4	2	1

To test the normality of a distribution, the upper class boundary/percentage cumulative frequency values are plotted on normal probability paper. The upper class boundary values are: 30, 31, 32,..., 38, 39. The corresponding cumulative frequency values (for 'less than' the upper class boundary values) are: 2, (4 + 2) = 6, (6 + 4+2) = 12, 20, 29, 37, 43, 47, 49 and 50. The corresponding percentage cumulative frequency values are $\dfrac{2}{50} \times 100 = 4$, $\dfrac{6}{50} \times 100 = 12$, 24, 40, 58, 74, 86, 94, 98 and 100%

The co-ordinates of upper class boundary/percentage cumulative frequency values are plotted as shown in Figure 181.4. When plotting these values, it will always be found that the co-ordinate for the 100% cumulative frequency value cannot be plotted, since the maximum value on the probability scale is 99.99. **Since the points plotted in Figure 181.4 lie very nearly in a straight line, the data is approximately normally distributed.**

The mean value and standard deviation can be determined from Figure 181.4. Since a normal curve is symmetrical, the mean value is the value of the variable corre-

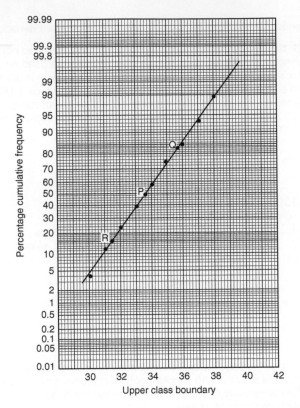

Figure 181.4

sponding to a 50% cumulative frequency value, shown as point P on the graph. This shows that **the mean value is 33.6 kg**. The standard deviation is determined using the 84% and 16% cumulative frequency values, shown as Q and R in Figure 181.4. The variable values for Q and R are 35.7 and 31.4 respectively; thus two standard deviations correspond to 35.7 − 31.4, i.e. 4.3, showing that the standard deviation of the distribution is approximately $\frac{4.3}{2}$ i.e. **2.15 standard deviations**.

Chapter 182 Linear correlation

The Pearson product-moment formula for determining the linear correlation coefficient states,

coefficient of correlation, $r = \dfrac{\sum xy}{\sqrt{\left\{\left(\sum x^2\right)\left(\sum y^2\right)\right\}}}$

where $x = (X - \bar{X})$ and $y = (Y - \bar{Y})$

Application: In an experiment to determine the relationship between force on a wire and the resulting extension, the following data is obtained:

Force (N)	10	20	30	40	50	60	70
Extension (mm)	0.22	0.40	0.61	0.85	1.20	1.45	1.70

Determine the linear coefficient of correlation for this data

Let X be the variable force values and Y be the dependent variable extension values, respectively. Using a tabular method to determine the quantities of this formula gives:

X	Y	$x = (X - \bar{X})$	$y = (Y - \bar{Y})$	xy	x^2	y^2
10	0.22	−30	−0.699	20.97	900	0.489
20	0.40	−20	−0.519	10.38	400	0.269
30	0.61	−10	−0.309	3.09	100	0.095
40	0.85	0	−0.069	0	0	0.005
50	1.20	10	0.281	2.81	100	0.079
60	1.45	20	0.531	10.62	400	0.282
70	1.70	30	0.781	23.43	900	0.610
$\Sigma X = 280$ $\bar{X} = \dfrac{280}{7}$ $= 40$	$\Sigma Y = 6.43$ $\bar{Y} = \dfrac{6.43}{7}$ $= 0.919$			$\Sigma xy = 71.30$	$\Sigma x^2 = 2800$	$\Sigma y^2 = 1.829$

Thus, **coefficient of correlation,**

$$r = \frac{\sum xy}{\sqrt{\left\{\left(\sum x^2\right)\left(\sum y^2\right)\right\}}} = \frac{71.3}{\sqrt{[2800 \times 1.829]}} = 0.996$$

This shows that a **very good direct correlation exists** between the values of force and extension.

Chapter 183 Linear regression

The least-squares regression lines

If the equation of the least-squares regression line is of the form: $Y = a_0 + a_1X$ the values of regression coefficient a_0 and a_1 are obtained from the equations:

$$\sum Y = a_0 N + a_1 \sum X \qquad (1)$$

$$\sum (XY) = a_0 \sum X + a_1 \sum X^2 \qquad (2)$$

If the equation of the regression line is of the form: $X = b_0 + b_1Y$ the values of regression coefficient b_0 and b_1 are obtained from the equations:

$$\sum X = b_0 N + b_1 \sum Y \qquad (3)$$

$$\sum (XY) = b_0 \sum Y + b_1 \sum Y^2 \qquad (4)$$

Application: The experimental values relating centripetal force and radius, for a mass travelling at constant velocity in a circle, are as shown:

Force (N)	5	10	15	20	25	30	35	40
Radius (cm)	55	30	16	12	11	9	7	5

Determine the equations of (a) the regression line of force on radius and (b) the regression line of radius on force. Hence, calculate the force at a radius of 40 cm and the radius corresponding to a force of 32 N

(a) Let the radius be the independent variable X, and the force be the dependent variable Y.

The equation of the regression line of force on radius is of the form $Y = a_0 + a_1X$

Using a tabular approach to determine the values of the summations gives:

Radius, X	Force, Y	X^2	XY	Y^2
55	5	3025	275	25
30	10	900	300	100
16	15	256	240	225
12	20	144	240	400
11	25	121	275	625
9	30	81	270	900
7	35	49	245	1225
5	40	25	200	1600
$\sum X = 145$	$\sum Y = 180$	$\sum X^2 = 4601$	$\sum XY = 2045$	$\sum Y^2 = 5100$

Thus, from equations (1) and (2), $180 = 8a_0 + 145a_1$
and $2045 = 145a_0 + 4601a_1$

Solving these simultaneous equations gives $a_0 = 33.7$ and $a_1 = -0.617$, correct to 3 significant figures. Thus the equation of the regression line of force on radius is:

$$Y = 33.7 - 0.617\,X$$

Thus the force, Y, at a radius of 40 cm, is:

$$Y = 33.7 - 0.617(40) = 9.02$$

i.e.   **the force at a radius of 40 cm is 9.02 N**

(b) The equation of the regression line of radius on force is of the form $X = b_0 + b_1 Y$

From equations (3) and (4), $145 = 8b_0 + 180b_1$
and $2045 = 180b_0 + 5100b_1$

Solving these simultaneous equations gives $b_0 = 44.2$ and $b_1 = -1.16$, correct to 3 significant figures. Thus the equation of the regression line of radius on force is:

$$X = 44.2 - 1.16Y$$

Thus, the radius, X, when the force is 32 N is:

$$X = 44.2 - 1.16(32) = 7.08,$$

i.e. **the radius when the force is 32 N is 7.08 cm**

Chapter 184   Sampling and estimation theories

Theorem 1

If all possible samples of size N are drawn from a finite population, N_p, without replacement, and the standard deviation of the mean values of the sampling distribution of means is determined, then:

standard error of the means, $\sigma_{\bar{x}} = \dfrac{\sigma}{\sqrt{N}} \sqrt{\left(\dfrac{N_p - N}{N_p - 1}\right)}$ (1)

where $\sigma_{\bar{x}}$ is the standard deviation of the sampling distribution of means and σ is the standard deviation of the population

For an infinite population and/or for sampling with replacement:

$$\sigma_{\bar{x}} = \dfrac{\sigma}{\sqrt{N}}$$ (2)

Theorem 2

If all possible samples of size N are drawn from a population of size N_p and the mean value of the sampling distribution of means $\mu_{\bar{x}}$ is determined then

$$\mu_{\bar{x}} = \mu \qquad (3)$$

where μ is the mean value of the population

Application: The heights of 3000 people are normally distributed with a mean of 175 cm and a standard deviation of 8 cm. If random samples are taken of 40 people, predict the standard deviation and the mean of the sampling distribution of means if sampling is done (a) with replacement, and (b) without replacement

For the population: number of members, $N_p = 3000$;

standard deviation, $\sigma = 8$ cm; mean, $\mu = 175$ cm

For the samples: number in each sample, $N = 40$

(a) When sampling is done **with replacement**, the total number of possible samples (two or more can be the same) is infinite. Hence, from equation (2) the **standard error of the mean (i.e. the standard deviation of the sampling distribution of means)**

$$\sigma_{\bar{x}} = \frac{\sigma}{\sqrt{N}} = \frac{8}{\sqrt{40}} = 1.265 \text{ cm}$$

From equation (3), **the mean of the sampling distribution**

$$\mu_{\bar{x}} = \mu = 175 \text{ cm}$$

(b) When sampling is done **without replacement**, the total number of possible samples is finite and hence equation (1) applies. Thus **the standard error of the means,**

$$\sigma_{\bar{x}} = \frac{\sigma}{\sqrt{N}} \sqrt{\left(\frac{N_p - N}{N_p - 1}\right)} = \frac{8}{\sqrt{40}} \sqrt{\left(\frac{3000 - 40}{3000 - 1}\right)}$$

$$= (1.265)(0.9935) = 1.257 \text{ cm}$$

Provided the sample size is large, the mean of the sampling distribution of means is the same for both finite and infinite populations. Hence, from equation (3), $\mu_{\bar{x}} = 175$ cm

The estimation of population parameters based on a large sample size

Table 184.1 **Confidence levels**

Confidence level, %	99	98	96	95	90	80	50
Confidence coefficient, z_C	2.58	2.33	2.05	1.96	1.645	1.28	0.6745

Application: Determine the confidence coefficient corresponding to a confidence level of 98.5%

98.5% is equivalent to a per unit value of 0.9850. This indicates that the area under the standardised normal curve between $-z_C$ and $+z_C$, i.e. corresponding to $2z_C$, is 0.9850 of the total area. Hence the area between the mean value and z_C is 0.9850/2 i.e. 0.4925 of the total area. The z-value corresponding to a partial area of 0.4925 is 2.43 standard deviations from Table 181.1 on page 518. Thus, **the confidence coefficient corresponding to a confidence level of 98.5% is 2.43**

Estimating the mean of a population when the standard deviation of the population is known

The confidence limits of the mean of a population are:

$$\bar{x} \pm \frac{z_C \, \sigma}{\sqrt{N}} \sqrt{\left(\frac{N_p - N}{N_p - 1} \right)} \tag{4}$$

for a finite population of size N_p

The **confidence limits for the mean of the population are:**

$$\bar{x} \pm \frac{z_C \, \sigma}{\sqrt{N}} \tag{5}$$

for an infinite population.

Application: It is found that the standard deviation of the diameters of rivets produces by a certain machine over a long period of time is 0.018 cm. The diameters of a random sample of 100 rivets produced by this machine in a day have a mean value of 0.476 cm. If the machine produces 2500 rivets a day, determine (a) the 90% confidence limits, and (b) the 97% confidence limits for an estimate of the mean diameter of all the rivets produced by the machine in a day

For the population: standard deviation, $\sigma = 0.018$ cm

                         number in the population, $N_p = 2500$

For the sample: number in the sample, $N = 100$

                         mean, $\bar{x} = 0.476$ cm

There is a finite population and the standard deviation of the population is known, hence expression (4) is used.

(a) For a 90% confidence level, the value of z_C, the confidence coefficient, is 1.645 from Table 184.1. Hence, the estimate of the confidence limits of the population mean, μ, is:

$$0.476 \pm \left(\frac{(1.645)(0.018)}{\sqrt{100}} \right) \sqrt{\left(\frac{2500 - 100}{2500 - 1} \right)}$$

i.e. $0.476 \pm (0.00296)(0.9800) = 0.476 \pm 0.0029 \, \text{cm}$

Thus, **the 90% confidence limits are 0.473 cm and 0.479 cm**

This indicates that if the mean diameter of a sample of 100 rivets is 0.476 cm, then it is predicted that the mean diameter of all the rivets will be between 0.473 cm and 0.479 cm and this prediction is made with confidence that it will be correct nine times out of ten.

(b) For a 97% confidence level, the value of z_C has to be determined from a table of partial areas under the standardised normal curve given in Table 181.1, as it is not one of the values given in Table 184.1. The total area between ordinates drawn at $-z_C$ and $+ z_C$ has to be 0.9700. Because the standardised normal curve is symmetrical, the area between $z_C = 0$ and z_C is 0.9700/2, i.e. 0.4850. From Table 181.1 an area of 0.4850 corresponds to a z_C value of 2.17.

Hence, the estimated value of the confidence limits of the population mean is between

$$\bar{x} \pm \frac{z_C \, \sigma}{\sqrt{N}} \sqrt{\left(\frac{N_p - N}{N_p - 1} \right)} = 0.476 \pm \left(\frac{(2.17)(0.018)}{\sqrt{100}} \right) \sqrt{\left(\frac{2500 - 100}{2500 - 1} \right)}$$

$$= 0.476 \pm (0.0039)(0.9800)$$

$$= 0.476 \pm 0.0038$$

Thus, **the 97% confidence limits are 0.472 cm and 0.480 cm**

It can be seen that the higher value of confidence level required in part (b) results in a larger confidence interval.

Estimating the mean and standard deviation of a population from sample data

The confidence limits of the mean value of the population, μ, are given by:

$$\mu_{\bar{x}} \pm z_C \, \sigma_{\bar{x}} \tag{6}$$

If s is the standard deviation of a sample, then the confidence limits of the standard deviation of the population are given by:

$$s \pm z_C \, \sigma_s \tag{7}$$

Application: Several samples of 50 fuses selected at random from a large batch are tested when operating at a 10% overload current and the mean time of the sampling distribution before the fuses failed is 16.50 minutes. The standard error of the means is 1.4 minutes. Determine the estimated mean time to failure of the batch of fuses for a confidence level of 90%

For the sampling distribution: the mean, $\mu_{\bar{x}} = 16.50$,

the standard error of the means, $\sigma_{\bar{x}} = 1.4$

The estimated mean of the population is based on sampling distribution data only and so expression (6) is used.

For an 90% confidence level, $z_C = 1.645$ (from Table 184.1),

thus $\mu_{\bar{x}} \pm z_C\, \sigma_{\bar{x}} = 16.50 \pm (1.645)(1.4) = 16.50 \pm 2.30$ minutes.

Thus, **the 90% confidence level of the mean time to failure is from 14.20 minutes to 18.80 minutes.**

Estimating the mean of a population based on a small sample size

Table 184.2 **Percentile values (t_p) for Student's t distribution with ν degrees of freedom (shaded area = p)**

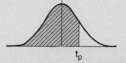

ν	$t_{0.995}$	$t_{0.99}$	$t_{0.975}$	$t_{0.95}$	$t_{0.90}$	$t_{0.80}$	$t_{0.75}$	$t_{0.70}$	$t_{0.60}$	$t_{0.55}$
1	63.66	31.82	12.71	6.31	3.08	1.376	1.000	0.727	0.325	0.158
2	9.92	6.96	4.30	2.92	1.89	1.061	0.816	0.617	0.289	0.142
3	5.84	4.54	3.18	2.35	1.64	0.978	0.765	0.584	0.277	0.137
4	4.60	3.75	2.78	2.13	1.53	0.941	0.741	0.569	0.271	0.134
5	4.03	3.36	2.57	2.02	1.48	0.920	0.727	0.559	0.267	0.132
6	3.71	3.14	2.45	1.94	1.44	0.906	0.718	0.553	0.265	0.131
7	3.50	3.00	2.36	1.90	1.42	0.896	0.711	0.549	0.263	0.130
8	3.36	2.90	2.31	1.86	1.40	0.889	0.706	0.546	0.262	0.130
9	3.25	2.82	2.26	1.83	1.38	0.883	0.703	0.543	0.261	0.129
10	3.17	2.76	2.23	1.81	1.37	0.879	0.700	0.542	0.260	0.129
11	3.11	2.72	2.20	1.80	1.36	0.876	0.697	0.540	0.260	0.129
12	3.06	2.68	2.18	1.78	1.36	0.873	0.695	0.539	0.259	0.128
13	3.01	2.65	2.16	1.77	1.35	0.870	0.694	0.538	0.259	0.128
14	2.98	2.62	2.14	1.76	1.34	0.868	0.692	0.537	0.258	0.128
15	2.95	2.60	2.13	1.75	1.34	0.866	0.691	0.536	0.258	0.128
16	2.92	2.58	2.12	1.75	1.34	0.865	0.690	0.535	0.258	0.128
17	2.90	2.57	2.11	1.74	1.33	0.863	0.689	0.534	0.257	0.128

Table 184.2   **Continued**

ν	$t_{0.995}$	$t_{0.99}$	$t_{0.975}$	$t_{0.95}$	$t_{0.90}$	$t_{0.80}$	$t_{0.75}$	$t_{0.70}$	$t_{0.60}$	$t_{0.55}$
18	2.88	2.55	2.10	1.73	1.33	0.862	0.688	0.534	0.257	0.127
19	2.86	2.54	2.09	1.73	1.33	0.861	0.688	0.533	0.257	0.127
20	2.84	2.53	2.09	1.72	1.32	0.860	0.687	0.533	0.257	0.127
21	2.83	2.52	2.08	1.72	1.32	0.859	0.686	0.532	0.257	0.127
22	2.82	2.51	2.07	1.72	1.32	0.858	0.686	0.532	0.256	0.127
23	2.81	2.50	2.07	1.71	1.32	0.858	0.685	0.532	0.256	0.127
24	2.80	2.49	2.06	1.71	1.32	0.857	0.685	0.531	0.256	0.127
25	2.79	2.48	2.06	1.71	1.32	0.856	0.684	0.531	0.256	0.127
26	2.78	2.48	2.06	1.71	1.32	0.856	0.684	0.531	0.256	0.127
27	2.77	2.47	2.05	1.70	1.31	0.855	0.684	0.531	0.256	0.127
28	2.76	2.47	2.05	1.70	1.31	0.855	0.683	0.530	0.256	0.127
29	2.76	2.46	2.04	1.70	1.31	0.854	0.683	0.530	0.256	0.127
30	2.75	2.46	2.04	1.70	1.31	0.854	0.683	0.530	0.256	0.127
40	2.70	2.42	2.02	1.68	1.30	0.851	0.681	0.529	0.255	0.126
60	2.66	2.39	2.00	1.67	1.30	0.848	0.679	0.527	0.254	0.126
120	2.62	2.36	1.98	1.66	1.29	0.845	0.677	0.526	0.254	0.126
∞	2.58	2.33	1.96	1.645	1.28	0.842	0.674	0.524	0.253	0.126

The confidence limits of the mean value of a population based on a small sample drawn at random from the population are given by

$$\bar{x} \pm \frac{t_c s}{\sqrt{(N-1)}} \tag{8}$$

Application: A sample of 12 measurements of the diameter of a bar are made and the mean of the sample is 1.850 cm. The standard deviation of the samples is 0.16 mm. Determine (a) the 90% confidence limits and (b) the 70% confidence limits for an estimate of the actual diameter of the bar

For the sample: the sample size, $N = 12$; mean, $\bar{x} = 1.850$ cm; standard deviation, $s = 0.16$ mm $= 0.016$ cm

Since the sample number is less than 30, the small sample estimate as given in expression (8) must be used. The number of degrees of freedom, i.e. sample size minus the number of estimations of population parameters to be made, is $12 - 1$, i.e. 11

(a) The confidence coefficient value corresponding to a percentile value of $t_{0.90}$ and a degree of freedom value of $\nu = 11$ can be found by using Table 184.2, and is 1.36, i.e. $t_c = 1.36$. The estimated value of the mean of the population is given by:

$$\bar{x} \pm \frac{t_c s}{\sqrt{(N-1)}} = 1.850 \pm \frac{(1.36)(0.016)}{\sqrt{11}}$$

$$= 1.850 \pm 0.0066 \text{ cm}$$

Thus, **the 90% confidence limits are 1.843 cm and 1.857 cm**

This indicates that the actual diameter is likely to lie between 1.843 cm and 1.857 cm and that this prediction stands a 90% chance of being correct.

(b) The confidence coefficient value corresponding to $t_{0.70}$ and to $\nu = 11$ is obtained from Table 184.2, and is 0.540, i.e. $t_C = 0.540$.

The estimated value of the 70% confidence limits is given by:

$$\bar{x} \pm \frac{t_C\, s}{\sqrt{(N-1)}} = 1.850 \pm \frac{(0.540)(0.016)}{\sqrt{11}}$$

$$= 1.850 \pm 0.0026 \text{ cm}$$

Thus, **the 70% confidence limits are 1.847 cm and 1.853 cm**, i.e. the actual diameter of the bar is between 1.847 cm and 1.853 cm and this result has a 70% probability of being correct.

Table 185.1 **Chi-square distribution**

Percentile values (χ_p^2) for the Chi-square distribution with ν degrees of freedom

ν	$\chi_{0.995}^2$	$\chi_{0.99}^2$	$\chi_{0.975}^2$	$\chi_{0.95}^2$	$\chi_{0.90}^2$	$\chi_{0.75}^2$	$\chi_{0.50}^2$	$\chi_{0.25}^2$	$\chi_{0.10}^2$	$\chi_{0.05}^2$	$\chi_{0.025}^2$	$\chi_{0.001}^2$	$\chi_{0.005}^2$
1	7.88	6.63	5.02	3.84	2.71	1.32	0.455	0.102	0.0158	0.0039	0.0010	0.0002	0.0000
2	10.6	9.21	7.38	5.99	4.61	2.77	1.39	0.575	0.211	0.103	0.0506	0.0201	0.0100
3	12.8	11.3	9.35	7.81	6.25	4.11	2.37	1.21	0.584	0.352	0.216	0.115	0.072
4	14.9	13.3	11.1	9.49	7.78	5.39	3.36	1.92	1.06	0.711	0.484	0.297	0.207
5	16.7	15.1	12.8	11.1	9.24	6.63	4.35	2.67	1.61	1.15	0.831	0.554	0.412
6	18.5	16.8	14.4	12.6	10.6	7.84	5.35	3.45	2.20	1.64	1.24	0.872	0.676
7	20.3	18.5	16.0	14.1	12.0	9.04	6.35	4.25	2.83	2.17	1.69	1.24	0.989
8	22.0	20.1	17.5	15.5	13.4	10.2	7.34	5.07	3.49	2.73	2.18	1.65	1.34
9	23.6	21.7	19.0	16.9	14.7	11.4	8.34	5.90	4.17	3.33	2.70	2.09	1.73
10	25.2	23.2	20.5	18.3	16.0	12.5	9.34	6.74	4.87	3.94	3.25	2.56	2.16
11	26.8	24.7	21.9	19.7	17.3	13.7	10.3	7.58	5.58	4.57	3.82	3.05	2.60
12	28.3	26.2	23.3	21.0	18.5	14.8	11.3	8.44	6.30	5.23	4.40	3.57	3.07

13	29.8	27.7	24.7	22.4	19.8	16.0	12.3	9.30	7.04	5.89	5.01	4.11	3.57
14	31.3	29.1	26.1	23.7	21.1	17.1	13.3	10.2	7.79	6.57	5.63	4.66	4.07
15	32.8	30.6	27.5	25.0	22.3	18.2	14.3	11.0	8.55	7.26	6.26	5.23	4.60
16	34.3	32.0	28.8	26.3	23.5	19.4	15.3	11.9	9.31	7.96	6.91	5.81	5.14
17	35.7	33.4	30.2	27.6	24.8	20.5	16.3	12.8	10.1	8.67	7.56	6.41	5.70
18	37.2	34.8	31.5	28.9	26.0	21.6	17.3	13.7	10.9	9.39	8.23	7.01	6.26
19	38.6	36.2	32.9	30.1	27.2	22.7	18.3	14.6	11.7	10.1	8.91	7.63	6.84
20	40.0	37.6	34.4	31.4	28.4	23.8	19.3	15.5	12.4	10.9	9.59	8.26	7.43
21	41.4	38.9	35.5	32.7	29.6	24.9	20.3	16.3	13.2	11.6	10.3	8.90	8.03
22	42.8	40.3	36.8	33.9	30.8	26.0	21.3	17.2	14.0	12.3	11.0	9.54	8.64
23	44.2	41.6	38.1	35.2	32.0	27.1	22.3	18.1	14.8	13.1	11.7	10.2	9.26
24	45.6	43.0	39.4	36.4	33.2	28.2	23.3	19.0	15.7	13.8	12.4	10.9	9.89
25	46.9	44.3	40.6	37.7	34.4	29.3	24.3	19.9	16.5	14.6	13.1	11.5	10.5
26	48.3	45.9	41.9	38.9	35.6	30.4	25.3	20.8	17.3	15.4	13.8	12.2	11.2
27	49.6	47.0	43.2	40.1	36.7	31.5	26.3	21.7	18.1	16.2	14.6	12.9	11.8
28	51.0	48.3	44.5	41.3	37.9	32.6	27.3	22.7	18.9	16.9	15.3	13.6	12.5
29	52.3	49.6	45.7	42.6	39.1	33.7	28.3	23.6	19.8	17.7	16.0	14.3	13.1
30	53.7	50.9	47.7	43.8	40.3	34.8	29.3	24.5	20.6	18.5	16.8	15.0	13.8
40	66.8	63.7	59.3	55.8	51.8	45.6	39.3	33.7	29.1	26.5	24.4	22.2	20.7
50	79.5	76.2	71.4	67.5	63.2	56.3	49.3	42.9	37.7	34.8	32.4	29.7	28.0
60	92.0	88.4	83.3	79.1	74.4	67.0	59.3	52.3	46.5	43.2	40.5	37.5	35.5
70	104.2	100.4	95.0	90.5	85.5	77.6	69.3	61.7	55.3	51.7	48.8	45.4	43.3
80	116.3	112.3	106.6	101.9	96.6	88.1	79.3	71.1	64.3	60.4	57.2	53.5	51.2
90	128.3	124.1	118.1	113.1	107.6	98.6	89.3	80.6	73.3	69.1	65.6	61.8	59.2
100	140.2	135.8	129.6	124.3	118.5	109.1	99.3	90.1	82.4	77.9	74.2	70.1	67.3

Application: As a result of a survey carried out of 200 families, each with five children, the distribution shown below was produced. Test the null hypothesis that the observed frequencies are consistent with male and female births being equally probable, assuming a binomial distribution, a level of significance of 0.05 and a 'too good to be true' fit at a confidence level of 95%

Number of boys (B) and girls (G)	5B,0G	4B,1G	3B,2G	2B,3G	1B,4G,	0B,5G
Number of families	11	35	69	55	25	5

To determine the expected frequencies

Using the usual binomial distribution symbols, let p be the probability of a male birth and $q = 1 - p$ be the probability of a female birth. The probabilities of having 5 boys, 4 boys,.., 0 boys are given by the successive terms of the expansion of $(q + p)^n$. Since there are 5 children in each family, $n = 5$, and $(q + p)^5 = q^5 + 5q^4 p + 10q^3 p^2 + 10q^2 p^3 + 5qp^4 + p^5$

When $q = p = 0.5$, the probabilities of 5 boys, 4 boys,..., 0 boys are 0.03125, 0.15625, 0.3125, 0.3125, 0.15625 and 0.03125

For 200 families, the expected frequencies, rounded off to the nearest whole number are: 6, 31, 63, 63, 31 and 6 respectively.

To determine the χ^2-value

Using a tabular approach, the χ^2-value is calculated using

$$\chi^2 = \sum \left\{ \frac{(o - e)^2}{e} \right\}$$

Number of boys(B) and girls(G)	Observed frequency, o	Expected frequency, e	o − e	$(o − e)^2$	$\dfrac{(o − e)^2}{e}$
5B, 0G	11	6	5	25	4.167
4B, 1G	35	31	4	16	0.516
3B, 2G	69	63	6	36	0.571
2B, 3G	55	63	−8	64	1.016
1B, 4G	25	31	−6	36	1.161
0B, 5G	5	6	−1	1	0.167

$$\chi^2 = \sum \left\{ \frac{(o - e)^2}{e} \right\} = 7.598$$

To test the significance of the χ^2-value

The number of degrees of freedom is given by $v = N - 1$ where N is the number of rows in the table above, thus $v = 6 - 1 = 5$. For a level of significance of 0.05, the confidence level is 95%, i.e. 0.95 per unit. From Table 185.1, for the $\chi^2_{0.95}$, $v = 5$ value, the percentile value χ^2_p is 11.1. Since the calculated value of χ^2 is less than

χ_p^2 the null hypothesis that the observed frequencies are consistent with male and female births being equally probable is accepted.

For a confidence level of 95%, the $\chi_{0.05}^2$, $\nu = 5$ value from Table 185.1 is 1.15 and because the calculated value of χ^2 (i.e. 7.598) is greater than this value, **the fit is not so good as to be unbelievable**.

Chapter 186   The sign test

Table 186.1   **Critical values for the sign test**

n	$\alpha_1 = 5\%$ $\alpha_2 = 10\%$	$2\frac{1}{2}\%$ 5%	1% 2%	$\frac{1}{2}\%$ 1%
1	—	—	—	—
2	—	—	—	—
3	—	—	—	—
4	—	—	—	—
5	0	—	—	—
6	0	0	—	—
7	0	0	0	—
8	1	0	0	0
9	1	1	0	0
10	1	1	0	0
11	2	1	1	0
12	2	2	1	1
12	2	2	1	1
13	3	2	1	1
14	3	2	2	1
15	3	3	2	2
16	4	3	2	2
17	4	4	3	2
18	5	4	3	3
19	5	4	4	3
20	5	5	4	3
21	6	5	4	4
22	6	5	5	4
23	7	6	5	4
24	7	6	5	5
25	7	7	6	5
26	8	7	6	6
27	8	7	7	6
28	9	8	7	6
29	9	8	7	7
30	10	9	8	7
31	10	9	8	7

Table 186.1 **Continued**

n	$\alpha_1 = 5\%$ $\alpha_2 = 10\%$	$2\frac{1}{2}\%$ 5%	1% 2%	$\frac{1}{2}\%$ 1%
32	10	9	8	8
33	11	10	9	8
34	11	10	9	9
35	12	11	10	9
36	12	11	10	9
37	13	12	10	10
38	13	12	11	10
39	13	12	11	11
40	14	13	12	11
41	14	13	12	11
42	15	14	13	12
43	15	14	13	12
44	16	15	13	13
45	16	15	14	13
46	16	15	14	13
47	17	16	15	14
48	17	16	15	14
49	18	17	15	15
50	18	17	16	15

Procedure for sign test

1. State for the data the null and alternative hypotheses, H_0 and H_1

2. Know whether the stated significance level, α, is for a one-tailed or a two-tailed test. Let, for example, H_0: $x = \phi$, then if H_1: $x \neq \phi$ then a two-tailed test is suggested because x could be less than or more than ϕ (thus use α_2 in Table 186.1), but if say H_1: $x < \phi$ or H_1: $x > \phi$ then a one-tailed test is suggested (thus use α_1 in Table 186.1)

3. Assign plus or minus signs to each piece of data – compared with ϕ or assign plus and minus signs to the difference for paired observations

4. Sum either the number of plus signs or the number of minus signs. For the two-tailed test, whichever is the smallest is taken; for a one-tailed test, the one which would be expected to have the smaller value when H_1 is true is used. The sum decided upon is denoted by S

5. Use Table 186.1 for given values of n, and α_1 or α_2 to read the critical region of S. For example, if, say, n = 16 and $\alpha_1 = 5\%$, then from Table 186.1, $S \leq 4$. Thus if S in part (iv) is greater than 4 we accept the null hypothesis H_0 and if S is less than or equal to 4 we accept the alternative hypothesis H_1

Application: A manager of a manufacturer is concerned about suspected slow progress in dealing with orders. He wants at least half of the orders received to be processed within a working day (i.e. 7 hours). A little later he decides to time 17 orders selected at random, to check if his request had been met.

The times spent by the 17 orders being processed were as follows:

$$4\frac{3}{4}h \quad 9\frac{3}{4}h \quad 15\frac{1}{2}h \quad 11h \quad 8\frac{1}{4}h \quad 6\frac{1}{2}h \quad 9h \quad 8\frac{3}{4}h \quad 10\frac{3}{4}h$$

$$3\frac{1}{2}h \quad 8\frac{1}{2}h \quad 9\frac{1}{2}h \quad 15\frac{1}{4}h \quad 13h \quad 8h \quad 7\frac{3}{4}h \quad 6\frac{3}{4}h$$

Use the sign test at a significance level of 5% to check if the managers request for quicker processing is being met

Using the above procedure:

1. The hypotheses are H_0: $t = 7\,h$ and H_1: $t > 7\,h$, where t is time.

2. Since H_1 is $t > 7\,h$, a one-tail test is assumed, i.e. $\alpha_1 = 5\%$

3. In the sign test each value of data is assigned a + or − sign. For the above data let us assign a + for times greater than 7 hours and a − for less than 7 hours. This gives the following pattern:

$$- \; + \; + \; + \; + \; - \; + \; + \; +$$
$$- \; + \; + \; + \; + \; + \; + \; -$$

4. The test statistic, S, in this case is the number of minus signs (−if H_0 were true there would be an equal number of + and − signs). Table 186.1 gives critical values for the sign test and is given in terms of small values; hence in this case S is the number of − signs, i.e. **S = 4**

5. From Table 186.1, with a sample size n = 17, for a significance level of $\alpha_1 = 5\%$, **S ≤ 4**. Since S = 4 in our data, the result **is significant** at $\alpha_1 = 5\%$, i.e. **the alternative hypothesis is accepted – it appears that the managers request for quicker processing of orders is not being met.**

Chapter 187   Wilcoxon signed-rank test

Table 187.1   Critical values for the Wilcoxon signed-rank test

n	$\alpha_1 = 5\%$ $\alpha_2 = 10\%$	$2\frac{1}{2}\%$ 5%	1% 2%	$\frac{1}{2}\%$ 1%
1	—	—	—	—
2	—	—	—	—
3	—	—	—	—

Table 187.1   **Continued**

n	$\alpha_1 = 5\%$ $\alpha_2 = 10\%$	$2\frac{1}{2}\%$ 5%	1% 2%	$\frac{1}{2}\%$ 1%
4	—	—	—	—
5	0	—	—	—
6	2	0	—	—
7	3	2	0	—
8	5	3	1	0
9	8	5	3	1
10	10	8	5	3
11	13	10	7	5
12	17	13	9	7
13	21	17	12	9
14	25	21	15	12
15	30	25	19	15
16	35	29	23	19
17	41	34	27	23
18	47	40	32	27
19	53	46	37	32
20	60	52	43	37
21	67	58	49	42
22	75	65	55	48
23	83	73	62	54
24	91	81	69	61
25	100	89	76	68
26	110	98	84	75
27	119	107	92	83
28	130	116	101	91
29	140	126	110	100
30	151	137	120	109
31	163	147	130	118
32	175	159	140	128
33	187	170	151	138
34	200	182	162	148
35	213	195	173	159
36	227	208	185	171
37	241	221	198	182
38	256	235	211	194
39	271	249	224	207
40	286	264	238	220
41	302	279	252	233
42	319	294	266	247
43	336	310	281	261
44	353	327	296	276
45	371	343	312	291

Table 187.1   **Continued**

n	$\alpha_1 = 5\%$ $\alpha_2 = 10\%$	$2\frac{1}{2}\%$ 5%	1% 2%	$\frac{1}{2}\%$ 1%
46	389	361	328	307
47	407	378	345	322
48	426	396	362	339
49	446	415	379	355
50	466	434	397	373

Procedure for the Wilcoxon signed-rank test

1. State for the data the null and alternative hypotheses, H_0 and H_1
2. Know whether the stated significance level, α, is for a one-tailed or a two-tailed test (see 2. in the procedure for the sign test on page 536)
3. Find the difference of each piece of data compared with the null hypothesis or assign plus and minus signs to the difference for paired observations
4. Rank the differences, ignoring whether they are positive or negative
5. The Wilcoxon signed-rank statistic T is calculated as the sum of the ranks of either the positive differences or the negative differences – whichever is the smaller for a two-tailed test, and the one which would be expected to have the smaller value when H_1 is true for a one-tailed test
6. Use Table 187.1 for given values of n, and α_1 or α_2 to read the critical region of T. For example, if, say, n = 16 and α_1 = 5%, then from Table 187.1, $t \le 35$. Thus if T in part 5 is greater than 35 we accept the null hypothesis H_0 and if T is less than or equal to 35 we accept the alternative hypothesis H_1

Application: The following data represents the number of hours that a portable car vacuum cleaner operates before recharging is required.

Operating time (h)   1.4 2.3 0.8 1.4 1.8 1.5 1.9 1.4 2.1 1.1 1.6

Use the Wilcoxon signed-rank test to test the hypothesis, at a 5% level of significance, that this particular vacuum cleaner operates, on average, 1.7 hours before needing a recharge

Using the above procedure:

1. H_0: $t = 1.7\,h$ and H_1: $t \ne 1.7\,h$

2. Significance level, $\alpha_2 = 5\%$ (since this is a two-tailed test)

3. Taking the difference between each operating time and 1.7h gives:

$-0.3\,h$ $+0.6\,h$ $-0.9\,h$ $-0.3\,h$ $+0.1\,h$ $-0.2\,h$
$+0.2\,h$ $-0.3\,h$ $+0.4\,h$ $-0.6\,h$ $-0.1\,h$

4. These differences may now be ranked from 1 to 11 (ignoring whether they are positive or negative).

Some of the differences are equal to each other. For example, there are two 0.1's (ignoring signs) that would occupy positions 1 and 2 when ordered. We average these as far as rankings are concerned i.e. each is assigned a ranking of $\frac{1+2}{2}$ i.e. 1.5. Similarly the two 0.2 values in positions 3 and 4 when ordered are each assigned rankings of $\frac{3+4}{2}$ i.e. 3.5, and the three 0.3 values in positions 5, 6, and 7 are each assigned a ranking of $\frac{5+6+7}{3}$ i.e. 6, and so on. The rankings are therefore:

Rank	1.5	1.5	3.5	3.5	6	6
Difference	+0.1	−0.1	−0.2	+0.2	−0.3	−0.3

Rank	6	8	9.5	9.5	11
Difference	−0.3	+0.4	+0.6	−0.6	−0.9

5. There are 4 positive terms and 7 negative terms. Taking the smaller number, the four positive terms have rankings of 1.5, 3.5, 8 and 9.5.
   Summing the positive ranks gives: $\mathbf{T} = 1.5 + 3.5 + 8 + 9.5 = \mathbf{22.5}$

6. From Table 187.1, when n = 11 and α_2 = 5%, $\mathbf{T \le 10}$

   Since T = 22.5 falls in the acceptance region (i.e. in this case is greater than 10), **the null hypothesis is accepted, i.e. the average operating time is not significantly different from 1.7 h**

   [Note that if, say, a piece of the given data was 1.7 h, such that the difference was zero, that data is ignored and n would be 10 instead of 11 in this case.]

Chapter 188   The Mann-Whitney test

Table 188.1   **Critical values for the Mann-Whitney test**

n_1	n_2	$\alpha_1 = 5\%$ $\alpha_2 = 10\%$	$2\frac{1}{2}\%$ 5%	1% 2%	$\frac{1}{2}\%$ 1%
2	2	—	—	—	—
2	3	—	—	—	—
2	4	—	—	—	—
2	5	0	—	—	—
2	6	0	—	—	—
2	7	0	—	—	—
2	8	1	0	—	—
2	9	1	0	—	—
2	10	1	0	—	—
2	11	1	0	—	—
2	12	2	1	—	—
2	13	2	1	0	—

Table 188.1 **Continued**

n_1	n_2	$\alpha_1 = 5\%$ $\alpha_2 = 10\%$	$2\frac{1}{2}\%$ 5%	1% 2%	$\frac{1}{2}\%$ 1%
2	14	3	1	0	—
2	15	3	1	0	—
2	16	3	1	0	—
2	17	3	2	0	—
2	18	4	2	0	—
2	19	4	2	1	0
2	20	4	2	1	0
3	3	0	—	—	—
3	4	0	—	—	—
3	5	1	0	—	—
3	6	2	1	—	—
3	7	2	1	0	—
3	8	3	2	0	—
3	9	4	2	1	0
3	10	4	3	1	0
3	11	5	3	1	0
3	12	5	4	2	1
3	13	6	4	2	1
3	14	7	5	2	1
3	15	7	5	3	2
3	16	8	6	3	2
3	17	9	6	4	2
3	18	9	7	4	2
3	19	10	7	4	3
3	20	11	8	5	3
4	4	1	0	—	—
4	5	2	1	0	—
4	6	3	2	1	0
4	7	4	3	1	0
4	8	5	4	2	1
4	9	6	4	3	1
4	10	7	5	3	2
4	11	8	6	4	2
4	12	9	7	5	3
4	13	10	8	5	3
4	14	11	9	6	4
4	15	12	10	7	5
4	16	14	11	7	5
4	17	15	11	8	6
4	18	16	12	9	6
4	19	17	13	9	7
4	20	18	14	10	8
5	5	4	2	1	0
5	6	5	3	2	1
5	7	6	5	3	1

Table 188.1 Continued

n_1	n_2	$\alpha_1 = 5\%$ $\alpha_2 = 10\%$	$2\frac{1}{2}\%$ 5%	1% 2%	$\frac{1}{2}\%$ 1%
5	8	8	6	4	2
5	9	9	7	5	3
5	10	11	8	6	4
5	11	12	9	7	5
5	12	13	11	8	6
5	13	15	12	9	7
5	14	16	13	10	7
5	15	18	14	11	8
5	16	19	15	12	9
5	17	20	17	13	10
5	18	22	18	14	11
5	19	23	19	15	12
5	20	25	20	16	13
6	6	7	5	3	2
6	7	8	6	4	3
6	8	10	8	6	4
6	9	12	10	7	5
6	10	14	11	8	6
6	11	16	13	9	7
6	12	17	14	11	9
6	13	19	16	12	10
6	14	21	17	13	11
6	15	23	19	15	12
6	16	25	21	16	13
6	17	26	22	18	15
6	18	28	24	19	16
6	19	30	25	20	17
6	20	32	27	22	18
7	7	11	8	6	4
7	8	13	10	7	6
7	9	15	12	9	7
7	10	17	14	11	9
7	11	19	16	12	10
7	12	21	18	14	12
7	13	24	20	16	13
7	14	26	22	17	15
7	15	28	24	19	16
7	16	30	26	21	18
7	17	33	28	23	19
7	18	35	30	24	21
7	19	37	32	26	22
7	20	39	34	28	24
8	8	15	13	9	7
8	9	18	15	11	9
8	10	20	17	13	11

Table 188.1 **Continued**

n_1	n_2	$\alpha_1 = 5\%$ $\alpha_2 = 10\%$	$2\frac{1}{2}\%$ 5%	1% 2%	$\frac{1}{2}\%$ 1%
8	11	23	19	15	13
8	12	26	22	17	15
8	13	28	24	20	17
8	14	31	26	22	18
8	15	33	29	24	20
8	16	36	31	26	22
8	17	39	34	28	24
8	18	41	36	30	26
8	19	44	38	32	28
8	20	47	41	34	30
9	9	21	17	14	11
9	10	24	20	16	13
9	11	27	23	18	16
9	12	30	26	21	18
9	13	33	28	23	20
9	14	36	31	26	22
9	15	39	34	28	24
9	16	42	37	31	27
9	17	45	39	33	29
9	18	48	42	36	31
9	19	51	45	38	33
9	20	54	48	40	36
10	10	27	23	19	16
10	11	31	26	22	18
10	12	34	29	24	21
10	13	37	33	27	24
10	14	41	36	30	26
10	15	44	39	33	29
10	16	48	42	36	31
10	17	51	45	38	34
10	18	55	48	41	37
10	19	58	52	44	39
10	20	62	55	47	42
11	11	34	30	25	21
11	12	38	33	28	24
11	13	42	37	31	27
11	14	46	40	34	30
11	15	50	44	37	33
11	16	54	47	41	36
11	17	57	51	44	39
11	18	61	55	47	42
11	19	65	58	50	45
11	20	69	62	53	48
12	12	42	37	31	27
12	13	47	41	35	31

Table 188.1   **Continued**

n_1	n_2	$\alpha_1 = 5\%$ $\alpha_2 = 10\%$	$2\frac{1}{2}\%$ 5%	1% 2%	$\frac{1}{2}\%$ 1%
12	14	51	45	38	34
12	15	55	49	42	37
12	16	60	53	46	41
12	17	64	57	49	44
12	18	68	61	53	47
12	19	72	65	56	51
12	20	77	69	60	54
13	13	51	45	39	34
13	14	56	50	43	38
13	15	61	54	47	42
13	16	65	59	51	45
13	17	70	63	55	49
13	18	75	67	59	53
13	19	80	72	63	57
13	20	84	76	67	60
14	14	61	55	47	42
14	15	66	59	51	46
14	16	71	64	56	50
14	17	77	69	60	54
14	18	82	74	65	58
14	19	87	78	69	63
14	20	92	83	73	67
15	15	72	64	56	51
15	16	77	70	61	55
15	17	83	75	66	60
15	18	88	80	70	64
15	19	94	85	75	69
15	20	100	90	80	73
16	16	83	75	66	60
16	17	89	81	71	65
16	18	95	86	76	70
16	19	101	92	82	74
16	20	107	98	87	79
17	17	96	87	77	70
17	18	102	92	82	75
17	19	109	99	88	81
17	20	115	105	93	86
18	18	109	99	88	81
18	19	116	106	94	87
18	20	123	112	100	92
19	19	123	112	101	93
19	20	130	119	107	99
20	20	138	127	114	105

Procedure for the Mann-Whitney test

1. State for the data the null and alternative hypotheses, H_0 and H_1
2. Know whether the stated significance level, α, is for a one-tailed or a two-tailed test (see 2. in the procedure for the sign test on page 536)
3. Arrange all the data in ascending order whilst retaining their separate identities
4. If the data is now a mixture of, say, A's and B's, write under each letter A the number of B's that precede it in the sequence (or vice-versa)
5. Add together the numbers obtained from 4 and denote total by U.
   U is defined as whichever type of count would be expected to be smallest when H_1 is true
6. Use Table 188.1 for given values of n_1 and n_2, and α_1 or α_2 to read the critical region of U. For example, if, say, $n_1 = 10$ and $n_2 = 16$ and $\alpha_2 = 5\%$, then from Table 188.1, $U \leq 42$. If U in part 5 is greater than 42 we accept the null hypothesis H_0, and if U is equal or less than 42, we accept the alternative hypothesis H_1

Application: 10 British cars and 8 non-British cars are compared for faults during their first 10000 miles of use. The percentage of cars of each type developing faults were as follows:

Non-British cars,	P	5	8	14	10	15	7	12	4		
British cars,	Q	18	9	25	6	21	20	28	11	16	34

Use the Mann-Whitney test, at a level of significance of 1%, to test whether non-British cars have better average reliability than British models

Using the above procedure:

1. The hypotheses are:
   H_0: Equal proportions of British and non-British cars have breakdowns
   H_1: A higher proportion of British cars have breakdowns
2. Level of significance $\alpha_1 = 1\%$
3. Let the sizes of the samples be n_P and n_Q, where $n_P = 8$ and $n_Q = 10$
   The Mann-Whitney test compares every item in sample P in turn with every item in sample Q, a record being kept of the number of times, say, that the item from P is greater than Q, or vice-versa. In this case there are $n_P n_Q$, i.e. $(8)(10) = 80$ comparisons to be made. All the data is arranged into ascending order whilst retaining their separate identities – an easy way is to arrange a linear scale as shown in Figure 188.1.

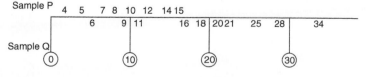

Figure 188.1

From Figure 188.1, a list of P's and Q's can be ranked giving:

P P Q P P Q P Q P P P Q Q Q Q Q Q Q

4. Write under each letter P the number of Q's that precede it in the sequence, giving:

P P Q P P Q P Q P P P Q Q Q Q Q Q Q
0 0   1 1   2   3 3 3

5. Add together these 8 numbers, denoting the sum by U, i.e.

$$U = 0 + 0 + 1 + 1 + 2 + 3 + 3 + 3 = 13$$

6. The critical regions are of the form U ≤ critical region
From Table 188.1, for a sample size 8 and 10 at significance level $\alpha_1 = 1\%$ the critical regions is **U ≤ 13**
The value of U in our case, from 5, is 13 which is significant at 1% significance level.

The Mann-Whitney test has therefore confirmed that **there is evidence that the non-British cars have better reliability than the British cars in the first 10,000 miles, i.e. the alternative hypothesis applies.**

Index

Printed and bound by CPI Group (UK) Ltd, Croydon, CR0 4YY

23/10/2024

01778263-0002